T0176596

The Probability Companion for Engineering and Computer Science

This friendly guide is the companion you need to convert pure mathematics into understanding and facility with a host of probabilistic tools. The book provides a high-level view of probability and its most powerful applications. It begins with the basic rules of probability and quickly progresses to some of the most sophisticated modern techniques in use, including Kalman filters, Monte Carlo techniques, machine learning methods, Bayesian inference and stochastic processes. It draws on 30 years of experience in applying probabilistic methods to problems in computational science and engineering, and numerous practical examples illustrate where these techniques are used in the real world. Topics of discussion range from carbon dating to Wasserstein GANs, one of the most recent developments in Deep Learning. The underlying mathematics is presented in full, but clarity takes priority over complete rigour, making this text a starting reference source for researchers and a readable overview for students.

ADAM PRÜGEL-BENNETT is Professor of Electronics and Computer Science at the University of Southampton. He received his PhD in Statistical Physics at the University of Edinburgh, where he became interested in disordered and complex systems. He currently researches in the area of mathematical modelling, optimisation and machine learning and has published many papers on these subjects.

The Probability Companion for Engineering and Computer Science

Adam Prügel-Bennett
University of Southampton

CAMBRIDGE
UNIVERSITY PRESS

CAMBRIDGE
UNIVERSITY PRESS

University Printing House, Cambridge CB2 8BS, United Kingdom

One Liberty Plaza, 20th Floor, New York, NY 10006, USA

477 Williamstown Road, Port Melbourne, VIC 3207, Australia

314–321, 3rd Floor, Plot 3, Splendor Forum, Jasola District Centre, New Delhi – 110025, India

79 Anson Road, #06–04/06, Singapore 079906

Cambridge University Press is part of the University of Cambridge.

It furthers the University's mission by disseminating knowledge in the pursuit of education, learning, and research at the highest international levels of excellence.

www.cambridge.org
Information on this title: www.cambridge.org/9781108480536
DOI: 10.1017/9781108635349

First published 2019

Printed in the United Kingdom by TJ International, Padstow, Cornwall

A catalogue record for this publication is available from the British Library.

Library of Congress Cataloging-in-Publication Data
Names: Prügel-Bennett, Adam, 1963– author.
Title: The probability companion for engineering and computer science / Adam Prügel-Bennett, University of Southampton.
Description: Cambridge, United Kingdom ; New York, NY, USA : University Printing House, 2019. | Includes bibliographical references and index.
Identifiers: LCCN 2019015914 | ISBN 9781108480536 (hardback : alk. paper) | ISBN 9781108727709 (paperback : alk. paper)
Subjects: LCSH: Engineering–Statistical methods. | Computer science–Statistical methods. | Probabilities.
Classification: LCC TA340 .P84 2019 | DDC 519.2–dc23
LC record available at https://lccn.loc.gov/2019015914

ISBN 978-1-108-48053-6 Hardback
ISBN 978-1-108-72770-9 Paperback

Contents

Preface

Probability provides by far the most powerful and successful calculus for dealing with uncertainty. The rules of probability are reasonably quick to master; much of the interest comes from the tools and techniques that have been developed to apply probability in different areas. This book provides a high-level guide to probability theory and the tool set that has developed around it. The text started life as notes for a course aimed at research students in engineering and science. I hope the book has retained some of that original spirit. Perhaps inevitably, the book has grown and many details added. I would, however, encourage the reader not to get bogged down in the details. I believe that you can pick up the technical details when you come to use the tools to solve your problem, but it is important to have some feel for what tools are out there. The book reflects my personal interests and knowledge. No doubt there are important areas I have missed. The one benefit of my ignorance is that it keeps the book to manageable proportions. There are likely to be areas which are over-represented due to the quirks of my personal interest. I hope the balance I've struck is not too idiosyncratic and gives a reasonable overview of the practical applications of probability.

I personally dislike books that demand of their readers that they do all the problems. Consequently, I had initially intended to avoid providing exercises. In the end, however, I reconsidered when a student explained that he learns through doing. I have therefore provided exercises at the end of each chapter. Because I dislike exercises where I don't know if I have the right solution, I have supplied complete solutions to all the problems. The reader is invited to treat the exercises in any way they wish. You may want to ignore the exercises altogether, just read the solutions, or carefully work through them yourself. For those who wish to do even more exercises you may like to consult Grimmett and Stirzaker (2001b) or Mosteller (1988).

This book intentionally focuses on giving an intuitive understanding of the techniques rather than providing a mathematically rigorous treatment. I found it difficult, however, to just present formula and I have mostly tried to give complete derivations of important results. To avoid expanding the text too much I have consigned some of the technical material to appendices. I have tried to correct the text as much as I can, but I possess in abundance the human disposition to err. If errors remain (and I am sure they will), I hope they are not too off-putting. One useful lesson (though one I would prefer not to teach) is never believe things just because they are in print. This means being able to check results for consistency and derive them from first principles. Of course, it is useful to have a relatively reliable source rather than check everything from scratch. The only reward I can offer is the knowledge that I will put any corrections I receive into any new editions.

I am deeply indebted to Robert Piché from Tampere University of Technology, who as a reviewer of the book did me the great honour of providing a very detailed list of improvements and corrections. Not only did he perform the Herculean task of correcting my English, but he provided a lot of technical guidance, introducing me, for example, to a cleaner proof of Jensen's inequality, among many other significant improvements. I would also like to thank Dr Jim Bennett for carefully reading the manuscript and pointing out additional errors and confusions.

Nomenclature

$x, y, \ldots$ Normal variables are written in italics

$\boldsymbol{x}, \boldsymbol{y}, \ldots$ Vectors are written in bold roman script

$X, Y, \ldots$ Random variables are written as capitals

$\boldsymbol{X}, \boldsymbol{Y}, \ldots$ Random vectors are written as bold capitals

$\mathsf{\mathbf{M}}, \mathsf{\mathbf{A}}, \ldots$ Matrices are written as bold sans-serif capitals

$F_X(x)$ Cumulative probability function of a random variable X, page 12

$f_X(x)$ Probability density of a continuous random variable X, page 12

$\mathcal{N}(\boldsymbol{x}|\boldsymbol{\mu}, \boldsymbol{\Sigma})$ Multivariate normal distribution with mean $\boldsymbol{\mu}$ and covariance $\boldsymbol{\Sigma}$, see Equation (2.12), page 37

$\mathcal{N}(x|\mu, \sigma^2)$ Normal (or Gaussian) distribution, see Equation (2.4), page 31

$\text{Bern}(X|\mu)$ Bernoulli distribution for binary variables, see Equation (4.1), page 60

$\text{Bet}(x|a, b)$ Beta distribution, see Equation (2.7), page 34

$\text{Bin}(m|n, p)$ Binomial distribution, see Equation (2.1), page 26

$\text{Cat}(\boldsymbol{X}|\boldsymbol{p})$ Categorical distribution, see Equation (4.2), page 68

$\text{Cau}(x)$ Cauchy distribution, see Equation (2.8), page 35

$\text{Dir}(\boldsymbol{x}|\boldsymbol{\alpha})$ Dirichlet distribution, see Equation (2.13), page 38

$\text{Exp}(x|b)$ Exponential distribution, see Equation (2.6), page 32

$\text{Gam}(x|a, b)$ Gamma distribution, see Equation (2.5), page 31

$\text{Hyp}(k|N, m, n)$ Hypergeometric distribution, see Equation (2.2), page 27

$\text{LogNorm}(x|\mu, \sigma)$ Log-normal distribution, see Equation (5.2), page 85

$\text{Mult}(\boldsymbol{n}|n, \boldsymbol{p})$ Multinomial distribution, see Equation (2.10), page 36

$\mathrm{Poi}(m|\mu)$ Poisson distribution, see Equation (2.3), page 28

$\mathrm{U}(x|a,b)$ Uniform distribution in the interval (a,b), page 48

$\mathrm{Wei}(x|\lambda,k)$ Weibull distribution, see Equation (2.6), page 33

$\emptyset$ The empty set

$\hat{x}$ Estimator of the quantity x, see Equation (4.1), page 61

Λ^k The $(k-1)$-dimensional unit simplex (i.e. the set of k-component vectors with non-negative elements that sum to 1), see Equation (2.9), page 36

Λ_n^k The k-dimensional discrete (integer) simplex that sums to n (i.e. the set of k non-negative integers that sum to n), see Equation (2.11), page 36

$\log(x)$ Denotes the natural logarithm of x

$\mathbb{N}$ The set of natural numbers (i.e. integers greater than 0)

$\mathbb{R}$ The set of real numbers

Ω The set of all possible elementary events

$[\![\mathit{predicate}]\!]$ indicator function returning 1 if *predicate* is true and zero otherwise, see Equation (1.8), page 16

$|\mathbf{A}|$ Determinant of matrix $\mathbf{A}$, see Equation (5.2), page 97

$X \sim f_X$ The random variable X is drawn from the distribution $f_X(x)$, see Equation (3.2), page 47

$\mathbb{E}_X\left[g(X)\right]$ Expectation with respect to random variable X of some function $g(X)$, see Equation (1.6), page 15

$\mathbb{E}\left[g(X)\right]$ Short for $\mathbb{E}_X\left[g(X)\right]$ when there is no ambiguity which variable is being marginalised (averaged) over, see Equation (1.6), page 15

$\mathbb{C}\mathrm{ov}\left[X,Y\right]$ The covariance of two random variables defined as $\mathbb{E}\left[X\,Y\right] - \mathbb{E}\left[X\right]\,\mathbb{E}\left[Y\right]$, see Equation (1.10), page 18

$\mathbb{C}\mathrm{ov}\left[\boldsymbol{X},\boldsymbol{Y}\right]$ The covariance matrix of two random vectors $\boldsymbol{X}$ and $\boldsymbol{Y}$ defined so that the matrix $\mathbf{C} = \mathbb{C}\mathrm{ov}\left[\boldsymbol{X},\boldsymbol{Y}\right]$ has components $C_{ij} = \mathbb{C}\mathrm{ov}\left[X_i,Y_j\right]$, see Equation (1.10), page 18

$\mathbb{C}\mathrm{ov}\left[\boldsymbol{X}\right]$ Short form for the covariance matrix $\mathbb{C}\mathrm{ov}\left[\boldsymbol{X},\boldsymbol{X}\right]$, see Equation (1.10), page 18

$\mathbb{V}\mathrm{ar}\left[X\right]$ The variance of variable X given by $\mathbb{E}\left[X^2\right] - \mathbb{E}\left[X\right]^2$, see Equation (1.8), page 17

$\neg A$ Not the event A (logical negation)

$A \vee B$ The event A or B (logical or)

$A \wedge B$ The event A and B (logical and)

$\mathbb{P}(A)$ Probability of event A happening, see Equation (1.0), page 3

$\mathbb{P}(A, B)$ Joint probability of event A and event B both happening, see Equation (1.0), page 5

$\mathbb{P}(A|B)$ Conditional probability of event A happening given event B happens, see Equation (1.1), page 6

1

Introduction

Contents

This book is a survey of the mathematical tools and techniques in probability and is aimed primarily at scientists and engineers. The intention is to give a broad overview of the subject, starting from the very basics but covering techniques used at the forefront of research in probabilistic modelling. Before we get to some of the more advanced tools it is necessary to understand the language of probability and some of the foundational concepts.

This chapter sets up the mathematical language we need in order to discuss probabilities and their properties. Most of this consists of definitions and simple mathematics, but it is a prerequisite for talking about the more interesting tools that we meet in later chapters.

1.1 Why Probabilities?

We live in a world full of uncertainties. The toss of a coin is sufficiently uncertain that it is regularly used to decide who goes first in many sporting competitions. Yet even with coin tosses we can make strong predictions. Thus if we toss a coin 1000 times with overwhelming probability, we are likely to get between 450 and 550 heads. The mathematical language that allows us to reason under uncertainty is probability theory. This book aims to provide a broad-brush overview of the mathematical and computational tools used by engineers and scientists to make sense of uncertainties.

In some situations what we observe is the consequence of so many unobserved and uncertain events that we can make extremely precise predictions that are taken as laws of physics even though they are just statements about what is overwhelmingly probable. The field of statistical physics (aka statistical mechanics) is founded on probability. However, uncertainty is ubiquitous and often does not involve a sufficient number of events to enable precise predictions. In these situations probability theory can be important in understanding experiments and making predictions. For example, if we wish to distinguish between natural fluctuations in the weather and the effects of climate change it is vital that we can reason accurately about uncertainty. However, probability can only answer these pressing scientific questions in combination with accurate models of the world. Such models are the subject matter of the scientific disciplines. Probability theory acts as a unifying glue, allowing us to make the best possible predictions or extract the most amount of information from observations. Although probability is not a prerequisite for doing good science, in almost any discipline in science, engineering, or social science it enhances a practitioner's armoury. I hope to give a spirit of the range of applications through examples sprinkled across the text.

Becoming a researcher in any field involves developing a toolkit of techniques that can be brought out as needed to tackle new problems. To be an accomplished user, the researcher has to acquire experience through practical application of the tools. This text cannot replace that step; rather, its intention is to make new researchers aware of what probabilistic tools exist and provide enough intuition to be able to judge the usefulness of the tool. In many ways this text is my personal compilation of tricks I've learned over many years of probabilistic modelling. The subject, and consequently this book, is mathematical, and in places I go into detail, but I recommend that you skip sections when you are getting bogged down or feel you have to push on even though you don't understand everything. This is a high-level tour; when there is a technique you really want to use you can come back and spend the time necessary to master that technique.

We start slowly by carefully, defining the key concepts we use and point out possible misunderstandings. Apologies to those who find this too elementary; however, we will quickly get into more advanced material. Without any more fuss let's get started.

1.2 Events and Probabilities

It is useful to know the mathematical language and formalism of probability. There are two main reasons for this: firstly, it allows you to read and understand the literature; secondly, when you write papers or your thesis, it is necessary to be able to talk the talk. For example, if you have a quantity you are treating as a random variable, you should call it a random variable, but this also requires you to know precisely what is meant by the term.

1.2.1 Events

The standard mathematical formulation of probability considers a set of *elementary events*, Ω, consisting of all possible outcomes in the world we are considering. For example, we might want to model the situation of tossing a coin, in which case the set of outcomes are $\Omega = \{\text{heads, tails}\}$, or if we roll a dice the elementary events would be $\Omega = \{1, 2, 3, 4, 5, 6\}$. We take an *event*, A, to be a subset of the space of elementary events $A \subset \Omega$ (note that there is a distinction made between the terms 'elementary event' and 'event', although an event could be an elementary event). In rolling a dice, the event, A, might be the dice landing on a six, $A = \{6\}$, or a number greater than 3, $A = \{4, 5, 6\}$. The probability of an event is denoted by $\mathbb{P}(A)$. Probabilities take values between zero and one

$$0 \leq \mathbb{P}(A) \leq 1,$$

with the interpretation that $\mathbb{P}(A) = 0$ means that the event never occurs and $\mathbb{P}(A) = 1$ meaning that the event always occurs. In this *set theory* view of probabilities the probability of no event occurring is 0, i.e. $\mathbb{P}(\emptyset) = 0$ where $\emptyset = \{\}$ is the empty set. In contrast, one elementary event must happen so that $\mathbb{P}(\Omega) = 1$. For a fair coin we expect $\mathbb{P}(\{\text{head}\}) = \mathbb{P}(\{\text{tail}\}) = 1/2$.

Talking about events gets us immediately into a linguistic dilemma. We can either consider an event to be a set of elementary events (the *set theory* point of view), or we can take it as a true–false proposition (the *propositional logic* point of view). Thus, when talking about a pair of events, A and B, we can view the event, C, of both A and B occurring as a set theoretic statement

We use the standard notation of $\cup$ and $\cap$ to denote union and intersection or sets and $\vee$ and $\wedge$ to denote 'logical or' and 'logical and'.

$$C = \{\omega | \omega \in A \text{ and } \omega \in B\} = A \cap B$$

or alternatively as a logical statement $C = A \wedge B$ about predicates (both A and B are true). The event, D, that either A or B (or possibly both) are true can be viewed as a set theoretic statement, $D = A \cup B$, or as a propositional statement, $D = A \vee B$. Similarly the event, E, that A does not occur can either be written in set language as $E = A^c = \Omega - A = \{\omega | \omega \notin A\}$ or as the logical statement $E = \neg A$. Both languages have advantages. The set theoretic language makes many simple results in probability transparent that are more obscure when using the language of propositional logic. However, often when modelling a system it is much easier to think in terms of propositions. In this chapter we will tend to migrate from a language of set theory to the language of propositions.

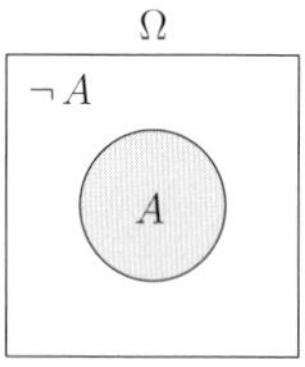

Returning to the axioms of probability, denoting the event 'A does not occur' by $\neg A$ then

$$\mathbb{P}\left(A\right) + \mathbb{P}\left(\neg A\right) = \mathbb{P}\left(\Omega\right) = 1$$

with the intuitively clear meaning that the event will either occur or not (a coin is either a head or not a head, or a dice is either a six or not a six). If we consider a set of *exhaustive* and *mutually exclusive* events, $\{A_i | i \in I\}$, where $I \in \mathbb{N}$ is an index set (that is a set of integers that label the events) then

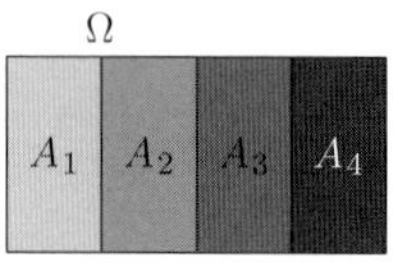

$$\sum_{i \in I} \mathbb{P}\left(A_i\right) = 1.$$

$\bigcup_{i \in I} A_i = A_1 \cup A_2 \cup \cdots \cup A_{|I|}$

Here *exhaustive* means that we have covered every possible outcome (i.e. $\bigcup_{i \in I} A_i = \Omega$) and *mutually exclusive* means that $A_i \cap A_j = \emptyset$ for all distinct pairs i and j. In the example of the dice, the events $\{1, 6\}$, $\{2, 5\}$, and $\{3, 4\}$ form an exhaustive and mutually exclusive set of events. When we roll a dice one of these events will occur. Note that real coins and real dice behave differently from mathematicians' coins and dice. A real coin might land on its edge, or it might roll away and get lost. Probability, like all applied mathematics, is a *model* of reality. It is the responsibility of the user of probability theory to ensure that their model captures the features that they care about.

Many mathematical texts formalise probabilities in terms of a probability space, consisting of a state space (or set of elementary events), Ω, a family of all possible events, $\mathcal{A}$, which will frequently be the set of all subsets of elementary events ($\mathcal{A} = 2^{\Omega}$, i.e. the power set of Ω) and a probability, $\mathbb{P}\left(A\right)$, associated with each event $A \in \mathcal{A}$. Thus a formal way of referring to probabilities is as a triple $(\Omega, \mathcal{A}, \mathbb{P})$. Don't be put off: this is just how mathematicians like to set things up.

What happens when the set of events are not denumerable? This would occur if the events took a continuous value, for example, *what will the temperature be tomorrow?*, or *where does a dart land?* This leads to a difficulty: the probability of any elementary event may well be zero! Worse, the family of all events, $\mathcal{A}$, can potentially become precarious, as it involves subsets of a non-denumerable sets. In over 30 years of working with probabilities I am yet to meet a case where anything precarious actually happened. To rigorously formulate probabilities in a way to avoid contradictions, even when working with the most complex of sets Andrei Kolmogorov borrowed ideas from mathematical analysis known as measure theory. If you pick up a mathematics text on probability you will get a good dose of measures (sigma), σ-fields (a generalisation of power sets), filtrations (a hierarchy of events), etc. However, don't panic. This formalism is massive overkill for nearly all situations. In most of engineering or science you will never face the pathological functions that keep mathematicians awake at night and require measure theory. I have never come across a practical application where the mechanics of measure theory was at all necessary. This is not to put down the importance of putting probability on a firm theoretical footing. However, in my experience, measure theory is neither necessary nor even helpful in applying probability to real-world problems. If you want to know more about measure theory and the type of problems which necessitate it refer to Appendix 1.A.

1.2.2 Assigning Probabilities

In the mathematical set-up above probabilities must have certain properties, but they are assumed to be given. One of the first tasks for engineers and scientist is to assign meaningful probabilities. How then do we do this and what do mean by probability? These questions have raised considerable debate among the philosophically minded. A seeming common-sense answer would be that it is the expected frequency of occurrence of an outcome in a large number of independent trials. Indeed, some have argued that is the only rational way of viewing probabilities. However, doubters have pointed out that there are many problems with uncertainty that are never repeated (who will win next year's Wimbledon final?). Thus, the argument goes that probabilities should be viewed as our degree of belief about something happening.

This philosophical debate, however, throws little light onto the practical question of how we should assign probabilities (however, for those interested we return to this debate in Chapter 8). Suppose we want to assign a probability to the outcome of a coin toss. If the coin has two distinct sides, then I, like most people, would happily assume that it has equal probability of being either a head or tail. Pushed on why, my first response would be that I believe that I would get heads as often as tails (a frequentist's explanation). Pushed further, I would argue that the outcome is likely to depend mainly on the speed of rotation and the time the coin has to rotate, which is beyond most people's ability to control precisely. I would find it very unlikely that the design on the faces of the coin would significantly bias the outcome. Pushed still further, I might resort to the conservation of angle momentum and small amount of air resistance or perhaps I might just shrug my shoulders and say 'that's my model of the world and I'm happy with it'. Of course, it would be possible to determine empirically the result of many coin tosses. Although I have never done this, the fact that it's not common knowledge whether heads is more likely than tails or the other way around suggests that the probability is indeed close to a half. In practice, probabilities are often assigned using a symmetry argument. That is, all outcomes at some level are considered equally likely.

Alas, one of the drawbacks of this is that the initial task of allocating probabilities often comes down to counting possible outcomes (combinatorics). Many people consequently view probability as hard and maybe even boring – personally I find combinatorics fascinating and beautiful, although I concede that it is an acquired taste. It is certainly true that combinatorics quickly becomes difficult and it is very easy to get wrong. However, it is only a very small part of probability. We are about to see that manipulating probabilities is actually relatively straightforward, and I hope this book will convince the reader that there is much, much more to probability than just counting combinations.

1.2.3 Joint and Conditional Probabilities

Probabilities of single events are somewhat boring. There is little to say about them. The interest comes when we have two or more events. To reason about this we need to set up a formalism for calculating with multiple events: a calculus of probability. It turns out that once you understand how to handle two events then the generalisation to more events is simple.

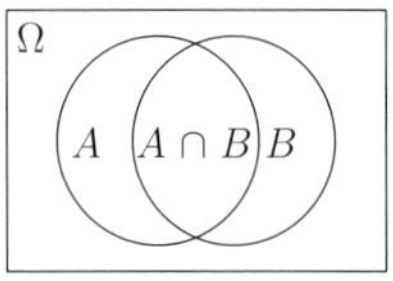

The *joint probability* of two events, A and B, both occurring is denoted $\mathbb{P}(A, B)$. If we think of events $A \subset \Omega$ and $B \subset \Omega$ being subsets of elementary events, then

$$\mathbb{P}(A, B) = \mathbb{P}(A \cap B).$$

That is, it is the probability of the intersection of the two subsets.

Example 1.1 Rolling an Honest Dice
If we consider the probability of the event, A, of an honest dice landing on an even number ($A = \{2, 4, 6\}$) and the event, B, of the number being greater than 3 ($B = \{4, 5, 6\}$), then $\mathbb{P}(A, B) = \mathbb{P}(A \cap B) = \mathbb{P}(\{4, 6\}) = 1/3$.

From elementary set theory $A = A \cap \Omega = A \cap (B \cup \neg B) = (A \cap B) \cup (A \cap \neg B)$. However, $A \cap B$ and $A \cap \neg B$ are non-overlapping sets (that is $(A \cap B) \cap (A \cap \neg B) = \emptyset$), so that

$$\mathbb{P}(A) = \mathbb{P}(A \cap B) + \mathbb{P}(A \cap \neg B) = \mathbb{P}(A, B) + \mathbb{P}(A, \neg B). \tag{1.1}$$

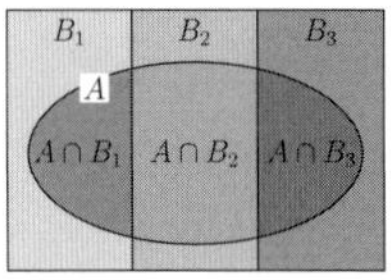

This is sometimes known as the 'additive law of probability'. It trivially generalises to many events. If $\{B_i | i \in I\}$ forms an *exhaustive* and *mutually exclusive* set of events, so that $A = \bigcup_{i \in \mathcal{I}} A \cap B_i$, then

$$\mathbb{P}(A) = \sum_{i \in \mathcal{I}} \mathbb{P}(A \cap B_i) = \sum_{i \in \mathcal{I}} \mathbb{P}(A, B_i).$$

This is an example of the *law of total probability*. Although it is possible to formalise probability in terms of sets (we could, for example, use $\mathbb{P}(A \cap B)$ as our standard notation for the joint probabilities), when we come to modelling the real world, it is more natural to think of the events as logical (true-false) statements (or predicates). The set notation then looks rather confusing. Thus, it is more usual to think of the additive law of probability as an axiom that we can exploit when necessary.

Note, while joint probabilities are symmetric in the sense

$$\mathbb{P}(A, B) = \mathbb{P}(B, A)$$

conditional probabilities aren't

$$\mathbb{P}(A|B) \neq \mathbb{P}(B|A)$$

unless

$$\mathbb{P}(A) = \mathbb{P}(B).$$

The second building block for reasoning about multiple events is the *conditional probability* of event A occurring *given* that event B occurs. It is denoted $\mathbb{P}(A|B)$. The conditional probability is equal to the joint probability of both events occurring divided by the probability that event B occurs

$$\mathbb{P}(A|B) = \frac{\mathbb{P}(A, B)}{\mathbb{P}(B)}. \tag{1.2}$$

It is a probability for A with all the usual properties of a probability, for example

$$\mathbb{P}(A|B) + \mathbb{P}(\neg A|B) = 1.$$

(Note that it is *not* a probability for B so that $\mathbb{P}(A|B) + \mathbb{P}(A|\neg B)$ will not generally be equal to 1.) The conditional probability is not defined if $\mathbb{P}(B) = 0$ (although this doesn't usually worry us as we tend not to care about events that will never happen). Given Equation (1.2) it might seem that conditional probabilities are secondary to joint probabilities, but when it comes to modelling real systems it is often the case that we can more easily specify the conditional probability. That is, if A depends on B in some way then $\mathbb{P}(A|B)$ is the probability of A when you know B has happened and this is often easy to model. However, it is wrong to think that conditional probabilities always express causality. If $\mathbb{P}(A) > 0$ and $\mathbb{P}(B) > 0$, then $\mathbb{P}(A|B)$ and $\mathbb{P}(B|A)$ are both meaningful whatever the causal relationship (e.g. $\mathbb{P}(A|B)$ is well defined even if A causes B). A consequence of Equation (1.2) is that

$$\mathbb{P}(A, B) = \mathbb{P}(A|B)\,\mathbb{P}(B) = \mathbb{P}(B|A)\,\mathbb{P}(A). \tag{1.3}$$

This is sometimes known as the 'multiplicative law of probabilities'. Equations (1.1) and (1.3) provide the cornerstone to developing a calculus for reasoning about probabilities.

Extending these laws to more events is simple. The trick is to split all the events into two groups. These groups of event can be considered as single compound events. We can then apply the laws of probability given above to the compound events. Thus, for example,

We interpret $\mathbb{P}(A|B,C)$ as $\mathbb{P}(A|(B,C))$, that is the probability of A given B and C (the comma has a higher precedence – binds stronger – than the bar).

$$\mathbb{P}(A|B,C) + \mathbb{P}(\neg A|B,C) = 1 \qquad \text{treating } B \wedge C \text{ as a single event}$$
$$\mathbb{P}(A,B,C) + \mathbb{P}(A,B,\neg C) = \mathbb{P}(A,B) \qquad \text{treating } A \wedge B \text{ as a single event.}$$

With three events there are a large number of identities between joint and conditional probabilities, e.g.

$$\begin{aligned}\mathbb{P}(A,B,C) &= \mathbb{P}(A,B|C)\,\mathbb{P}(C) = \mathbb{P}(A|B,C)\,\mathbb{P}(B|C)\,\mathbb{P}(C)\\ &= \mathbb{P}(A,C|B)\,\mathbb{P}(B) = \mathbb{P}(A|B,C)\,\mathbb{P}(C|B)\,\mathbb{P}(B)\\ &= \mathbb{P}(B,C|A)\,\mathbb{P}(A) = \mathbb{P}(C|A,B)\,\mathbb{P}(B|A)\,\mathbb{P}(A)\end{aligned}$$

etc. This is not difficult, but some care is required to make sure that what you do is valid.

An obvious consequence of Equation (1.3) is the identity

$$\mathbb{P}(A|B) = \frac{\mathbb{P}(B|A)\,\mathbb{P}(A)}{\mathbb{P}(B)}. \tag{1.4}$$

This formula provides a means of going from one conditional probability, $\mathbb{P}(B|A)$, to the reverse conditional probability, $\mathbb{P}(A|B)$. This seemingly innocuous equation, known as *Bayes' rule*, is the basis for one of the most powerful formalism in probabilistic inference known as the Bayesian approach. We return to this many times, particular in Chapter 8 which is devoted to Bayesian inference.

Example 1.2 Manipulating Probabilities
To understand the rules for manipulating probabilities consider two events, A and B, where the joint probabilities of these events and the negation of these events (i.e. the outcome when the event does not happen) are given by

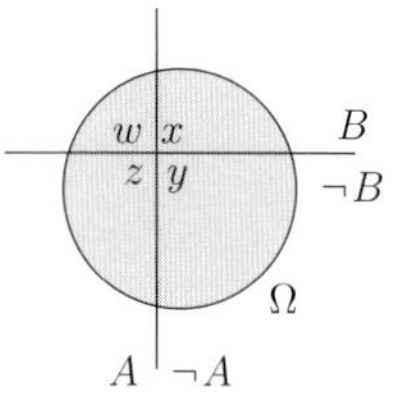

$$\mathbb{P}\left(A, B\right) = w \qquad \mathbb{P}\left(A, \neg B\right) = z$$
$$\mathbb{P}\left(\neg A, B\right) = x \qquad \mathbb{P}\left(\neg A, \neg B\right) = y$$

where $w + x + y + z = 1$. The probabilities of events A and B are given by

$$\mathbb{P}\left(A\right) = \mathbb{P}\left(A, B\right) + \mathbb{P}\left(A, \neg B\right) = w + z$$
$$\mathbb{P}\left(B\right) = \mathbb{P}\left(A, B\right) + \mathbb{P}\left(\neg A, B\right) = w + x.$$

Then (some of) the joint probabilities are given by

$$\mathbb{P}\left(A|B\right) = \frac{\mathbb{P}\left(A, B\right)}{\mathbb{P}\left(B\right)} = \frac{w}{w + x} \qquad \mathbb{P}\left(\neg A|B\right) = \frac{\mathbb{P}\left(\neg A, B\right)}{\mathbb{P}\left(B\right)} = \frac{x}{w + x}$$
$$\mathbb{P}\left(B|A\right) = \frac{\mathbb{P}\left(A, B\right)}{\mathbb{P}\left(A\right)} = \frac{w}{w + z} \qquad \mathbb{P}\left(A|\neg B\right) = \frac{\mathbb{P}\left(A, \neg B\right)}{\mathbb{P}\left(\neg B\right)} = \frac{z}{z + y}.$$

An example of the 'addition law of probability' is

$$\mathbb{P}\left(A|B\right) + \mathbb{P}\left(\neg A|B\right) = \frac{w}{w + x} + \frac{x}{w + x} = 1$$

and the 'multiplicative law of probability' is

$$\mathbb{P}\left(A, B\right) = \mathbb{P}\left(A|B\right) \mathbb{P}\left(B\right) = \mathbb{P}\left(B|A\right) \mathbb{P}\left(A\right)$$
$$w = \frac{w}{w + x}\left(w + x\right) = \frac{w}{w + z}\left(w + z\right).$$

Note, however, that

$$\mathbb{P}\left(A|B\right) + \mathbb{P}\left(A|\neg B\right) = \frac{w}{w + x} + \frac{z}{z + y} \neq 1 \quad \text{(in general).}$$

The laws of probability are very simple, but it is very easy to get confused about exactly what terms are what. Thus care is necessary.

1.2.4 Independence

Two events, A and B, are said to be *independent* if

$$\mathbb{P}\left(A, B\right) = \mathbb{P}\left(A\right) \mathbb{P}\left(B\right). \tag{1.5}$$

Note that independence does *not* imply $A \cap B = \emptyset$, (i.e. $\mathbb{P}\left(A, B\right) = 0$) – which says rather that event A and B cannot both happen, i.e. they are mutually exclusive. Independence is a rather more subtle, but nevertheless a strong statement about two events. Its utility is that it means that you can treat the events in isolation. From Equations (1.3) and (1.5) it follows that for independent events $\mathbb{P}(A|B) = \mathbb{P}(A)$ – i.e. the probability of event A happening is blind to whether event B happens. Equation (1.5) shows that independence is a symmetric relation and is well defined even when $\mathbb{P}(A) = 0$ or $\mathbb{P}(B) = 0$ (where the conditional probability is not defined). Using the formula $\mathbb{P}(A|B) = \mathbb{P}(A)$ as a definition of independence doesn't explicitly show the symmetry and might not be applicable. However, it is very often how we would use independence.

If two events are causally independent (e.g. the event of tossing heads and rolling a 6), then they will be probabilistically independent

$$\mathbb{P}(\text{Coin} = H, \text{Dice} = 6) = \mathbb{P}(\text{Coin} = H) \times \mathbb{P}(\text{Dice} = 6).$$

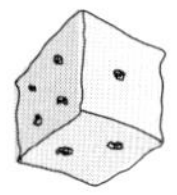

However, probabilistic independence is a mathematical relationship $\mathbb{P}(A, B) = \mathbb{P}(A)\,\mathbb{P}(B)$, which doesn't require A and B to be causally independent.

Example 1.3 Probabilistic Independence
Consider the (clearly manufactured) situation where we toss a coin with a probability p of getting heads. If we get heads, then we choose an honest dice which we throw. Otherwise we choose a biased dice $\mathbb{P}(D = 1) = \mathbb{P}(D = 2) = \mathbb{P}(D = 3) = 1/12$, $\mathbb{P}(D = 4) = \mathbb{P}(D = 5) = \mathbb{P}(D = 6) = 1/4$, where D denotes the number rolled.

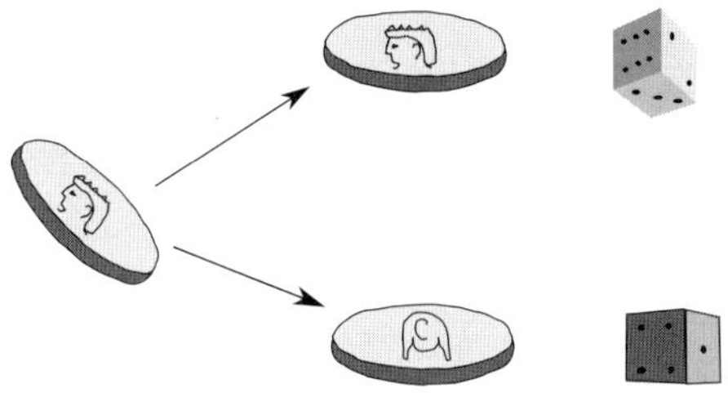

Let A be the event of getting tails, and B be the event of getting either a 1 or 6. The probability of getting tails is $\mathbb{P}(A) = 1 - p$ (depending on the bias of the coin). A simple calculation shows

$$\mathbb{P}(A, B) = (1 - p) \times \left(\frac{1}{12} + \frac{1}{4}\right) = \frac{1}{3}(1 - p)$$

$$\mathbb{P}(\neg A, B) = p \times \left(\frac{1}{6} + \frac{1}{6}\right) = \frac{1}{3}p$$

so that $\mathbb{P}(B) = \mathbb{P}(A, B) + \mathbb{P}(\neg A, B) = 1/3$. Thus, $\mathbb{P}(A, B) = (1 - p)/3 = \mathbb{P}(A)\,\mathbb{P}(B)$; so the events are independent, though they are clearly not causally independent (insofar as which dice we roll depends on the outcome of event A).

We can generalise the idea of independence to a family of events $\{A_i | i \in I\}$ (for some index set $I \subset \mathbb{N}$). The family of events are said to be independent if *for all* $I' \subseteq I$

$$\mathbb{P}\left(\bigwedge_{i\in I'} A_i\right) = \prod_{i\in I'} \mathbb{P}\left(A_i\right)$$

where the left-hand side denotes the joint probability of all events A_i for which $i \in I'$. This is a much stronger statement than *pairwise independent* (i.e. each pair of events are independent). It is possible for a family of events to be pairwise independent without itself being independent.

Example 1.4 Eight-Sided Dice
Consider an eight-sided honest dice so that $\Omega = \{1, 2, 3, 4, 5, 6, 7, 8\}$ and consider the family of events $A = \{1, 2, 3, 4\}$, $B = \{2, 4, 6, 8\}$, and $C = \{2, 4, 5, 7\}$. Since the dice is honest $\mathbb{P}\left(A\right) = \mathbb{P}\left(B\right) = \mathbb{P}\left(C\right) = 1/2$. Similarly,

$$\begin{aligned}\mathbb{P}\left(A, B\right) &= \mathbb{P}\left(\{2, 4\}\right) = 1/4 = \mathbb{P}\left(A\right)\mathbb{P}\left(B\right)\\ \mathbb{P}\left(A, C\right) &= \mathbb{P}\left(\{2, 4\}\right) = 1/4 = \mathbb{P}\left(A\right)\mathbb{P}\left(C\right)\\ \mathbb{P}\left(B, C\right) &= \mathbb{P}\left(\{2, 4\}\right) = 1/4 = \mathbb{P}\left(B\right)\mathbb{P}\left(C\right)\end{aligned}$$

so that the events are pairwise independent, but

$$\mathbb{P}\left(A, B, C\right) = \mathbb{P}\left(\{2, 4\}\right) = \tfrac{1}{4} \qquad \mathbb{P}\left(A\right)\mathbb{P}\left(B\right)\mathbb{P}\left(C\right) = \tfrac{1}{8}$$

Thus the family of events are not independent.

Often we meet events, A and B, which depend on each other through some intermediate event, C. We define events as being *conditionally independent* if

$$\mathbb{P}\left(A, B|C\right) = \mathbb{P}\left(A|C\right)\mathbb{P}\left(B|C\right).$$

Example 1.5 Plumbers and Cold Weather
Consider the case where, if it is very cold, a pipe might freeze and burst, and I am highly likely to call a plumber. Thus, the event of it being cold and calling a plumber are dependent on each other so that

$$\mathbb{P}\left(\text{cold, call plumber}\right) \neq \mathbb{P}\left(\text{cold}\right)\mathbb{P}\left(\text{call plumber}\right).$$

However, they are linked (at least in my simplified version of the world) through the burst pipe. If I know the pipe is burst, then it is irrelevant whether it is cold or not. In this example, the events of calling a plumber and it being cold are conditionally independent given the event that we have a burst pipe

$$\mathbb{P}\left(\text{cold, call plumber}|\text{burst pipe}\right) = \mathbb{P}\left(\text{cold}|\text{burst pipe}\right)\\ \mathbb{P}\left(\text{call plumber}|\text{burst pipe}\right).$$

Conditional independence is not as strong a condition as full independence. We cannot ignore the dependence of A and B unless we know C. Nevertheless, there are times when conditional independence can considerably simplify otherwise complicated relationships. The idea of conditional independence plays an important role in, for example, graphical models (see Section 8.5.2) and Markov chains (see Chapter 11).

1.3 Random Variables

A *random variable*, X, is a number associated with an outcome. That is, it maps each elementary event to a real number, $X : \Omega \to \mathbb{R}$. Sometimes there is a very natural mapping from the outcomes to the random variable. For example, in rolling a dice, X might denote the number shown on the dice. Or in a series of 100 coin tosses, X might denote the number of heads. In the first example each elementary event is mapped to a unique number, while in the second many elementary events (i.e. sequences of heads and tails) will be mapped to the same number. However, for any mapping from events to numbers we can define a random variables. As a consequence, if X is a random variable then any function $Y = g(X)$ is also a random variable.

Example 1.6 Random Variables
In any situation with uncertainty we can define random variables. For example, when throwing two dice we could assign the total number of dots to a random variable X that takes values from 2 to 12. However, we might want to assign a different value to our random variable. For example, in a simplified game of craps (a gambling game involving rolling a dice) we might win if we roll a 7 or 11 and lose otherwise. In this case, we might want to use a different random variable, Y, where we assign the event of rolling a 7 or 11 a value of 1 and all other events the value -1. Y is just a function of X. They are both random variables.

A convention that is often used with random variables (and used throughout this text) is to write them in upper case. This is to show that random variables are rather special mathematical objects. The values taken by the random variables are written in lower case. Although this is a common convention, many authors will have their own variants of it, such as denoting random variables by Greek letters. The probability that a random variable X takes a value x is written as $\mathbb{P}(X = x)$. Again this can seem confusing, especially since we are using the symbol x as an unknown variable. It can also be confusing when considering samples as it is sometimes unclear whether we have an observed number x or a random variable X. Some authors make a distinction between potential observations (pre-statistics) and the actual observations or realisations of the observations. We will stick with our dichotomy between random and non-random variables and leave it up to the intelligence of the reader to make sense of notation when

Note that we also use upper-case letters to denote quantities other than random variables. Thus, we have being denoting events (subsets of elementary events) with upper-case letters.

the situation is more complex. Nevertheless, there are good reasons for treating random variable differently, especially when taking expectations (averages over probability distributions), since, as we will see in Section 1.4, they don't always obey the same laws as numbers.

Random variables provide a partitioning of the elementary event space into sets labelled by the value of the random variables. When the values taken by the random variables are discrete (i.e. $X \in \mathcal{X} = \{x_1, x_2, \ldots\}$) we define the *probability mass function*, $f_X(x) = \mathbb{P}\left(X = x\right)$ which has the property

$$\sum_{x \in \mathcal{X}} f_X(x) = \sum_{x \in \mathcal{X}} \mathbb{P}\left(X = x\right) = 1.$$

In Section 2.1 we consider some probability mass distributions for discrete variables. We cover discrete distribution in more detail in the Chapter 4.

We can also consider situations in which more than one random variable has been defined. For example, imagine rolling two dice. Then we can define $X \in \{1, 2, 3, 4, 5, 6\}$ as a random variable whose value is equal to the number rolled by dice 1 and $Y \in \{1, 2, 3, 4, 5, 6\}$ as the number rolled by dice 2. In such situations we can consider the joint probability mass function to be $f_{X,Y}(x, y) = \mathbb{P}\left(X = x \text{ and } Y = y\right)$.

1.4 Probability Densities

A difficulty arises when we consider a continuous or non-denumerable random variable, X. For example, X might be a random number that can take a value anywhere in the range 0 to 1. The probability of the random variable taking any particular value, $X = x$, is (usually) zero (i.e. $\mathbb{P}\left(X = x\right) = 0$). We therefore have to proceed a bit more cautiously. Instead of considering the probability of X taking a particular value we can consider the *cumulative distribution function* (CDF)

$$F_X(x) = \mathbb{P}\left(X \leq x\right).$$

This is a function that starts from $F(-\infty) = 0$ and finishes with $F(\infty) = 1$ (e.g. see Figure 1.1).

It is a well-defined function for any situations (at least, if it is not well defined you have a pathological distribution – something I've yet to meet in any real-world situation). If there are two random variables we can define the two dimensional cumulative probability function

$$F_{X,Y}(x, y) = \mathbb{P}\left(X \leq x \text{ and } Y \leq y\right).$$

This generalises to distributions of more than two variables.

Figure 1.1 Example of a cumulative probability function.

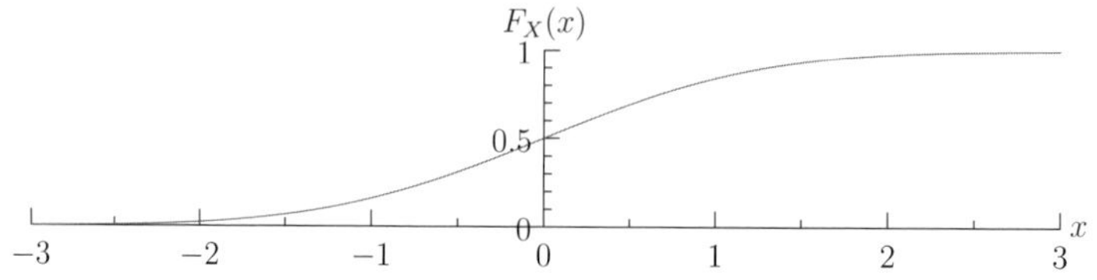

Rather than using cumulative probability functions it is often more natural and more useful to work with the *probability density function* (PDF) $f_X(x)$ defined such that

$$F_X(x) = \mathbb{P}\left(X \le x\right) = \int_{-\infty}^{x} f_X(y)\,\mathrm{d}y.$$

If $F_X(x)$ is differentiable we have

$$f_X(x) = \frac{\mathrm{d}F_X(x)}{\mathrm{d}x} = \lim_{\delta x \to 0} \frac{\mathbb{P}\left(x \le X < x + \delta x\right)}{\delta x}.$$

Probability densities play a similar role to probability masses for discrete distribution. However, it is important to realise that probability densities are *not* probabilities! In particular, they are not necessarily less than 1. The value of a probability density doesn't tell you that the point is likely to be visited. The quantity that acts likes a probability is $f_X(x)\,\mathrm{d}x$, which can be thought of as the probability of X being in an infinitesimal interval around x. The name density is apt. If we think of probability as a mass that sums to 1, then $F_X(x)$ corresponds to the mass up to the point x; and $f_X(x)$ is the density (probability mass per unit volume) at the point x. The probability mass between two points a and b is

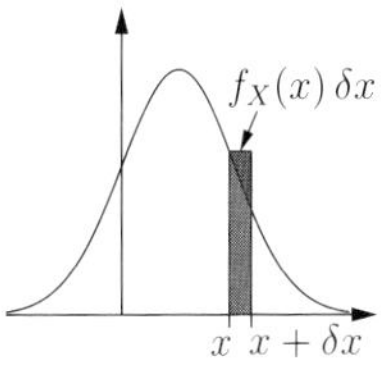

$$\text{probability mass} = \mathbb{P}\left(a \le X \le b\right) = \int_{a}^{b} f_X(x)\,\mathrm{d}x = F[b] - F[a].$$

Change of Variables

Probability densities behave rather peculiarly under changes of the variables because they are densities. Consider a transformation $T : \boldsymbol{X} \to \boldsymbol{Y}$ from one set of variables $\boldsymbol{X}$ to another set of variables $\boldsymbol{Y}(\boldsymbol{X})$. In general, changes of variables don't preserve local volumes. As a result, to preserve probability in every volume we must modify the probability density. Denote the probability density with respect to these variables by $f_{\boldsymbol{X}}(\boldsymbol{x})$ and $f_{\boldsymbol{Y}}(\boldsymbol{y})$ respectively. Consider a small volume δV around an arbitrarily chosen point $\boldsymbol{x}^*$ in $\boldsymbol{X}$ space. This volume gets transformed to a volume $T(\delta V)$ around $\boldsymbol{y}(\boldsymbol{x}^*)$ in $\boldsymbol{Y}$ space. Because it is the same event, we require that the probability of being in that volume should be the same in either coordinate system so that

The density per cm² will be different to the density per inch².

$$\int_{\boldsymbol{x}\in\delta V} f_{\boldsymbol{X}}(\boldsymbol{x})\,\mathrm{d}\boldsymbol{x} = \int_{\boldsymbol{y}\in T(\delta V)} f_{\boldsymbol{Y}}(\boldsymbol{y})\,\mathrm{d}\boldsymbol{y}. \tag{1.6}$$

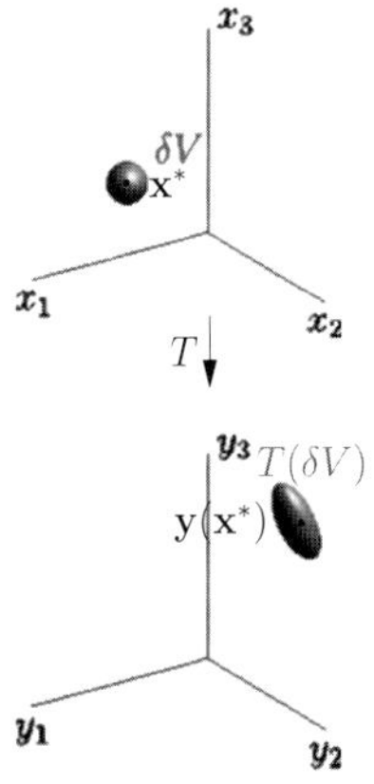

Assuming that the densities are sufficiently smooth and δV sufficiently small that the density is approximately constant in δV and $T(\delta V)$, then

$$\int_{\boldsymbol{x}\in\delta V} f_{\boldsymbol{X}}(\boldsymbol{x})\,\mathrm{d}\boldsymbol{x} \approx f_{\boldsymbol{X}}(\boldsymbol{x}^*)\,\delta V, \qquad \int_{\boldsymbol{y}\in T(\delta V)} f_{\boldsymbol{Y}}(\boldsymbol{y})\,\mathrm{d}\boldsymbol{y} \approx f_{\boldsymbol{Y}}(\boldsymbol{y}(\boldsymbol{x}^*))\,T(\delta V).$$

The approximations become exact in the limit $\delta V \to 0$. In other words,

$$f_{\boldsymbol{Y}}(\boldsymbol{y}(\boldsymbol{x}^*)) = f_{\boldsymbol{X}}(\boldsymbol{x}^*) \lim_{\delta V \to 0} \frac{\delta V}{T(\delta V)}.$$

The ratio of these two volumes is given by the absolute value of the determinant of the *Jacobian*. That is,

$$\lim_{\delta V \to 0} \frac{\delta V}{T(\delta V)} = |J_{\boldsymbol{y}\to\boldsymbol{x}}(\boldsymbol{y}(\boldsymbol{x}^*))|$$

where the *Jacobian determinant* for the mapping from $\boldsymbol{y}$ to $\boldsymbol{x}(\boldsymbol{y})$ evaluated at the point $\boldsymbol{y}$ is given by

$$J_{\boldsymbol{y}\to\boldsymbol{x}}(\boldsymbol{y}) = \left|\frac{\partial(x_1, x_2, \ldots, x_n)}{\partial(y_1, y_2, \ldots, y_n)}\right| = \begin{vmatrix} \frac{\partial x_1(\boldsymbol{y})}{\partial y_1} & \frac{\partial x_2(\boldsymbol{y})}{\partial y_1} & \cdots & \frac{\partial x_n(\boldsymbol{y})}{\partial y_1} \\ \frac{\partial x_1(\boldsymbol{y})}{\partial y_2} & \frac{\partial x_2(\boldsymbol{y})}{\partial y_2} & \cdots & \frac{\partial x_n(\boldsymbol{y})}{\partial y_2} \\ \vdots & \vdots & \ddots & \vdots \\ \frac{\partial x_1(\boldsymbol{y})}{\partial y_n} & \frac{\partial x_2(\boldsymbol{y})}{\partial y_n} & \cdots & \frac{\partial x_n(\boldsymbol{y})}{\partial y_n} \end{vmatrix}.$$

The determinant of a square matrix M, denoted by $|\mathsf{M}|$, measures (up to a sign) the change of any volume under the linear coordinate transformation defined by M. The Jacobian determinant measures the local change in volume of the infinitesimal volume with sides $\mathrm{d}x_i$ compared with the infinitesimal volume with sides $\mathrm{d}y_i$. This is the same term that appears in the standard rule of calculus for changing variables

$$\int_{\boldsymbol{x}\in\delta V} g(\boldsymbol{x})\,\mathrm{d}\boldsymbol{x} = \int_{\boldsymbol{y}\in T(\delta V)} g(\boldsymbol{x}(\boldsymbol{y}))\,|J_{\boldsymbol{y}\to\boldsymbol{x}}(\boldsymbol{y})|\,\mathrm{d}\boldsymbol{y}$$

(the absolute part of the Jacobian determinant appears because any sign change in the coordinate transform is absorbed into a change in the limits of the integral). Note that the Jacobian for the inverse transform is just given by the reciprocal of the Jacobian for the original transform, that is,

$$J_{\boldsymbol{x}\to\boldsymbol{y}}(\boldsymbol{x}) = \left|\frac{\partial(y_1, y_2, \ldots, y_n)}{\partial(x_1, x_2, \ldots, x_n)}\right| = \frac{1}{J_{\boldsymbol{y}\to\boldsymbol{x}}(\boldsymbol{y}(\boldsymbol{x}))}.$$

Returning to the change in the densities. The condition that the probability is preserved in any volume of space under a coordinate mapping requires

$$f_{\boldsymbol{Y}}(\boldsymbol{y}) = f_{\boldsymbol{X}}(\boldsymbol{x}(\boldsymbol{y}))\,|J_{\boldsymbol{y}\to\boldsymbol{x}}(\boldsymbol{y})|.$$

In one dimension

$$f_Y(y) = f_X(x(y))\left|\frac{\mathrm{d}x}{\mathrm{d}y}\right|.$$

A useful mnemonic for remembering which way around to put the Jacobian is to write

$$f_Y(y)\,|\mathrm{d}y| = f_X(x)\,|\mathrm{d}x|,$$

which is to be interpreted as the probability in a small element.

Example 1.7 Normal and Log-Normal Distributions

We assume that X is normally distributed according to

$$f_X(x) = \mathcal{N}(x|\mu, \sigma^2) = \frac{1}{\sqrt{2\pi}\,\sigma}\mathrm{e}^{-(x-\mu)^2/(2\sigma^2)}$$

If $Y = e^X$ or ($X = \log(Y)$) then

$$f_Y(y) = f_X(x(y)) \frac{\mathrm{d}x}{\mathrm{d}y} = f_x(\log(y)) \frac{1}{y} = \frac{1}{\sqrt{2\pi}\,\sigma\, y} e^{-(\log(y)-\mu)^2/(2\sigma^2)}.$$

This distribution is known as the *log-normal distribution*. We will see it many times.

1.5 Expectations

To provide some unity between discrete and continuous random variables, it is customary to have a unified notation for averaging over a probability distribution. This averaging is known as the *expectation* and is denoted by mathematicians by $\mathbb{E}_X\left[\cdots\right]$. For an arbitrary function $g(X)$

$$\mathbb{E}_X\left[g(X)\right] = \begin{cases} \sum_{x\in\mathcal{X}} g(x)\, f_X(x) & \text{if } X \text{ is a discrete variable} \\ \int_{-\infty}^{\infty} g(x)\, f_X(x)\,\mathrm{d}x & \text{if } X \text{ is a continuous variable.} \end{cases} \tag{1.7}$$

(Physicists often use angled brackets to denote expectations, i.e. $\langle g(X)\rangle_X = \mathbb{E}_X\left[g(X)\right]$. They are also more likely to call expectations averages.) When it is obvious what we are averaging over we often drop the subscript and write $\mathbb{E}\left[X\right]$ or $\langle X\rangle$.

Are the expressions in (1.7) obvious? There is an old joke about a professor being asked whether what he has written on the board is obvious. He thinks for a while, leaves the room and after half an hour returns and responds to the questioner 'yes', before carrying on where he left off. The equations in (1.7) are sometimes known as *the unconscious statistician theorem* as for many statisticians the results seems so obvious they require no proof. To see if this is justified let's do the proof, at least, for the discrete case. What do we mean by $\mathbb{E}\left[g(X)\right]$ if we don't mean Equation (1.7)? Recall that we said any function of a random variable is itself just a random variable. Let's denote $g(X)$ by Y. We can surely agree that

$$\mathbb{E}_X\left[g(X)\right] = \mathbb{E}_Y\left[Y\right] = \sum_{y\in\mathcal{Y}} y\, f_Y(y)$$

where $\mathcal{Y}$ is the set of all possible values $Y = g(X)$ can take. That is, we can take the defining property of the expectation operator to be that the expectation of a random variable is its mean. We should also be able to convince ourselves that

$$f_Y(y) = \sum_{x:g(x)=y} f_X(x),$$

where the sum is over all values of x for which $g(x) = y$. Thus,

$$\mathbb{E}_X\left[g(X)\right] = \mathbb{E}_Y\left[Y\right] = \sum_{y\in\mathcal{Y}} y \sum_{x:g(x)=y} f_X(x) = \sum_{y\in\mathcal{Y}} \sum_{x:g(x)=y} g(x)\, f_X(x)$$

where we have taken y inside the second sum and used that $y = g(x)$. Now for every value of x we have that $g(x) = y$ for some particular y, therefore

$$\sum_{y \in \mathcal{Y}} \sum_{x : g(x) = y} = \sum_{x \in \mathcal{X}}$$

where $\mathcal{X}$ is the set of all possible values taken by x. Using this we obtain Equation (1.7) for discrete variables. The generalisation to continuous variables follows a similar line of arguments (you can find a page-long proof showing this in full measure theoretic glory, although the main idea is the same as that given). So is the result obvious? I allow you to decide, but many seemingly obvious results are wrong, so asking the question is always worthwhile.

Expectation is one of the most useful operations when dealing with random variables. It is a linear operator, meaning that

$$\mathbb{E}\left[c\,X\right] = c\,\mathbb{E}\left[X\right]$$
$$\mathbb{E}\left[X + Y\right] = \mathbb{E}\left[X\right] + \mathbb{E}\left[Y\right].$$

In addition, $\mathbb{E}\left[c\right] = c$, where c is a scalar (i.e. a number rather than a random variable). These results follow immediately from the linearity of summations and integrals. Note that it is important in taking expectations to distinguish between *scalar* quantities, such as c (even though they may be arbitrary), and random variables, such as X and Y.

If $g(X) = a\,X + b$ is a linear function then $\mathbb{E}\left[g(X)\right] = a\,\mathbb{E}\left[X\right] + b$. That is, the expectation of a linear function only depends on the mean and not on other quantities such as the variance (measuring the variation in the random variable). Note, that in general $\mathbb{E}\left[g(X)\right] \neq g(\mathbb{E}\left[X\right])$.

1.5.1 Indicator Functions

A useful devices to reason about probabilities is the indicator function. In this book we denote indicator functions by

$$[\![predicate]\!] = \begin{cases} 1 & predicate \text{ is true} \\ 0 & \text{otherwise} \end{cases} \tag{1.8}$$

where *predicate* is a statement that is true or false. There are different conventions for writing indicator functions, the one used here is close to that popularised by Donald Knuth (Knuth, 1997a; Graham et al., 1989) who attributes the notation to Kenneth Iverson. Indicator functions allow you to write a large number of otherwise complex functions in a compact form.

We can relate probabilities and expectations through an indicator function. Let A be an event, then the probability of A occurring is

$$\mathbb{P}\left(A\right) = \mathbb{E}\left[[\![A \text{ occurs}]\!]\right] = \mathbb{E}\left[[\![\omega \in A]\!]\right].$$

For example, if we want to compute the probability that the random variable, X, is equal to 0, then we can equivalently compute the expectation $\mathbb{E}\left[[\![X = 0]\!]\right]$.

If we denote the joint predicate A and B by $A \wedge B$ (in predicate logic $\wedge$ denotes *logical and* while $\vee$ denotes *logical or*) then we can manipulate predicates using

$$[\![A \wedge B]\!] = [\![A]\!]\,[\![B]\!] \qquad\qquad [\![A \vee B]\!] = [\![A]\!] + [\![B]\!] - [\![A \wedge B]\!].$$

A	B	$A \wedge B$	$A \vee B$
F	F	F	F
T	F	F	T
F	T	F	T
T	T	T	T

These are easily checked by enumerating all possible values that A and B can take (note that A and B are predicates that just take values of true or false). If we take expectations of the last of these formula we obtain a useful identity for the probability of either event A or event B occurring

$$\mathbb{P}\left(A \vee B\right) = \mathbb{P}\left(A\right) + \mathbb{P}\left(B\right) - \mathbb{P}\left(A, B\right).$$

This example shows how we can use expectations over indicator functions to obtain useful results. We can, of course, obtain this result directly from set theory. The event $A \cap B$ occurs both in A and in B so we have to subtract $\mathbb{P}\left(A, B\right)$ to prevent over-counting.

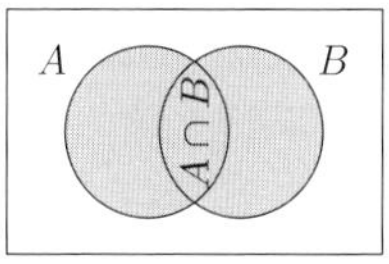

In Question 1.3 on page 22 we ask you to repeat the proof of the unconscious statistician theorem using indicator functions.

1.5.2 Statistics

Distributions of random variables are often rather complicated objects and we would like to get simple numerical results capturing their main features. That is, we want to know some *statistics* of the random variable. By far the most useful statistic is its *mean*. We will often denote means by μ (mu),

$$\mu = \mathbb{E}_X\left[X\right].$$

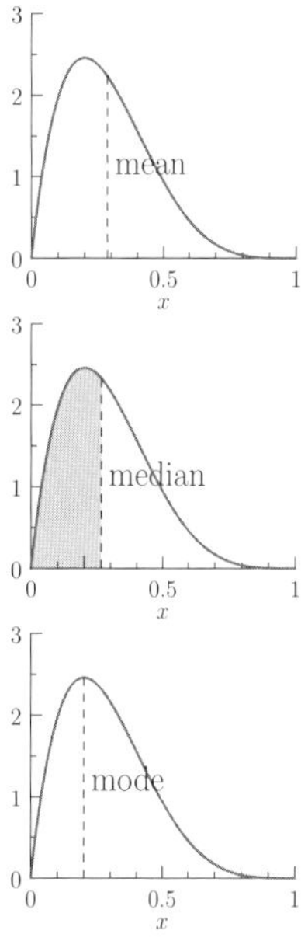

Although the mean is by far the most useful average, there are a couple of other averages that you might meet. The *median* is the centre value. That is, it is the value of x such that $\mathbb{P}\left(X < x\right) < 1/2$ and $\mathbb{P}\left(X > x\right) \geq 1/2$. The median is often useful when dealing with random variables whose distributions have *long tails*. This happens when there are rare events which have unusually large (or small) values. In this case, the mean can be quite misleading about what happens typically. The median tells you more about what you can expect typically. Because medians are not distorted by rare events or *outliers*, they are sometime used as a *robust statistic* in engineering frameworks that have to make decisions based on noisy data. The other average that is occasionally used is the *mode* which is the value x for which $f_X(x)$ is a maximum. That is, it is the most commonly occurring value of the random variable. The median and mode do not generally have very nice mathematical properties (they are not even necessarily uniquely defined) so are more rarely used than the means.

Variance. The next most important statistic after the mean in describing a random variable is the *variance*, which measures the variation or the size of the fluctuations. The variance is defined as

$$\mathbb{V}\text{ar}\left[X\right] = \mathbb{E}\left[X^2\right] - \mathbb{E}\left[X\right]^2. \tag{1.9}$$

This is an example where we want to distinguish between a random variable X and the value it takes x. For any scalar (normal number), c, $\mathbb{V}\mathrm{ar}\left[c\right] = 0$, while for any random variable, X, that has fluctuations, $\mathbb{V}\mathrm{ar}\left[X\right] \neq 0$. Thus, it is useful to distinguish between random variables and numbers – hence the convention of using capital letters to denote random variables and small letters to denote numbers. The dimensions of the variance is the square of the dimension of the mean and thus the two are not directly comparable. To get an idea of the size of the fluctuations it is usual to consider the *standard deviation* defined as the square root of the variance. Just as the mean is often denoted by μ, the standard deviation is commonly denoted by σ (sigma) and the variance is often written as σ^2. We will return to discussions of the mean and variance many times in this book.

As discussed, we call random variables *independent* if for any two random variables X and Y

$$f_{X,Y}(x,y) = f_X(x)\, f_Y(y).$$

An important property that follows from this is that for two independent random variables X and Y

$$\text{if } X \text{ and } Y \text{ are independent} \quad \Rightarrow \quad \mathbb{V}\mathrm{ar}\left[X+Y\right] = \mathbb{V}\mathrm{ar}\left[X\right] + \mathbb{V}\mathrm{ar}\left[Y\right]. \qquad (1.10)$$

We will use this property extensively throughout this text. This property is also true for a family of statistics known as cumulants. The mean is equal to the first cumulant, the variances are equal to the second cumulant. Higher order cumulants involve the expectation of X^n for $n > 2$. We will look cumulants in more detail in later chapters.

When we have more than one random variable we are often interested in how they vary with respect to each other. The crudest measure of the interdependence of two variables is the *covariance*. The covariance between X and Y is defined as

$$\mathbb{C}\mathrm{ov}\left[X,Y\right] = \mathbb{E}\left[X\,Y\right] - \mathbb{E}\left[X\right]\,\mathbb{E}\left[Y\right].$$

In general

$$\mathbb{V}\mathrm{ar}\left[X+Y\right] = \mathbb{V}\mathrm{ar}\left[X\right] + \mathbb{V}\mathrm{ar}\left[Y\right] + 2\,\mathbb{C}\mathrm{ov}\left[X,Y\right].$$

If X and Y are independent then their covariance will be zero (having zero covariance does not, however, guarantee that X and Y are independent). The variance and covariance are sometimes called *second-order statistics* as they involve terms that are quadratic in the random variables. We will consider even higher order statistics in Chapter 4. When $\boldsymbol{X}$ and $\boldsymbol{Y}$ are vectors of random variables then we can define the covariance matrix

$$\mathbb{C}\mathrm{ov}\left[\boldsymbol{X},\boldsymbol{Y}\right] = \mathbb{E}\left[\boldsymbol{X}\,\boldsymbol{Y}^{\mathsf{T}}\right] - \mathbb{E}\left[\boldsymbol{X}\right]\,\mathbb{E}\left[\boldsymbol{Y}^{\mathsf{T}}\right]$$

so that the matrix $\mathbf{C} = \mathbb{C}\mathrm{ov}\left[\boldsymbol{X},\boldsymbol{Y}\right]$ has components $C_{ij} = \mathbb{C}\mathrm{ov}\left[X_i, Y_j\right]$. In a slight abuse of notation we will sometimes use $\mathbb{C}\mathrm{ov}\left[\boldsymbol{X}\right]$ to denote the covariance matrix

$$\mathbb{C}\mathrm{ov}\left[\boldsymbol{X}\right] = \mathbb{C}\mathrm{ov}\left[\boldsymbol{X},\boldsymbol{X}\right] = \mathbb{E}\left[\boldsymbol{X}\,\boldsymbol{X}^{\mathsf{T}}\right] - \mathbb{E}\left[\boldsymbol{X}\right]\,\mathbb{E}\left[\boldsymbol{X}^{\mathsf{T}}\right]$$

so that the matrix $\mathbf{C} = \mathbb{C}\text{ov}\left[\boldsymbol{X}\right]$ has components $C_{ij} = \mathbb{C}\text{ov}\left[X_i, X_j\right]$ (the diagonal components are then the variances $C_{ii} = \mathbb{V}\text{ar}\left[X_i\right]$). This notation is useful when the random vector $\boldsymbol{X}$ takes a lot of space to express.

The *statistical correlation* (also called the Pearson's correlation) between two random variables X and Y is defined as

$$\rho(X,Y) = \frac{\mathbb{C}\text{ov}\left[X,Y\right]}{\sqrt{\mathbb{V}\text{ar}\left[x\right]\ \mathbb{V}\text{ar}\left[y\right]}}.$$

It is a number between -1 and 1. If $X = aY + b$ then the correlation between X and Y will either be equal to 1 if $a > 0$ or equal to -1 if $a < 0$. If X and Y are independent the correlation will be zero, although zero correlation does not guarantee statistical independence.

1.6 Probabilistic Inference

One of the major applications of probability in engineering and science is to learn from empirical (experimental) data in order to construct a model of the system. This is then often used to make future predictions. A common method to achieve this within a probabilistic framework is to consider an empirical observation, X, to be a random variable from some distribution function $f_X(x|\theta)$, where θ is a parameter to be inferred. Both x and θ may be multidimensional vectors (that is, we might have multiple quantities we observe, $\boldsymbol{X} = (X_1, X_2, \ldots, X_n)$ and/or a probability distribution that depends on multiple parameters $\boldsymbol{\theta} = (\theta_1, \theta_2, \ldots, \theta_m)$). The distribution function is part of the model that we have to supply. It should capture our understanding of the physical mechanism which generates the data. In Chapter 2 we discuss some of the classic distribution functions that are used when the data is generated by a simple mechanism.

The inference problem is to estimate the parameter(s) θ of the distribution $f_X(x|\theta)$ from random samples drawn from that distribution. Any estimate of these parameters is called an *estimator* – there are lots of different estimators as we will discover. We follow the convention of denoting an estimator using a hat, $\hat{\theta}$. An estimator will be a function of the empirical data X – the exact functional form will depend on the distribution $f(x|\theta)$. The distribution $f(x|\theta)$ is referred to as the *likelihood* of the data, given the parameters θ. One of the most commonly used estimators is the maximum likelihood estimator – that is the value of θ that maximises the likelihood of the data (see Section 6.4 on page 123 for examples of computing maximum likelihood estimators). A more sophisticated approach is to treat the unknown parameter as a random variable and seek the distribution of the parameters given the data $f(\theta|x)$. Using Bayes' rule, Equation (1.4), we can write this probability as

$$f(\theta|x) = \frac{f(x|\theta)\, f(\theta)}{f(x)}$$

which allows us to determine $f(\theta|x)$ in terms of the likelihood $f(x|\theta)$ and a 'prior' probability distribution $f(\theta)$ coding our belief about θ before making the measurement. (The denominator, $f(x)$, is just a normalisation constant which we can compute from the likelihood and prior.) This approach is known as Bayesian inference and is discussed at length in Chapter 8.

Frequently in collecting data we make a series of observations $\mathcal{D} = (X_1, X_2, X_3, \ldots . X_n)$. A common assumption is that these observations are independent of each other. Independence requires that the value of the ith observation does not influence the value of any other observation, X_j $(j \neq i)$. This is often a very natural assumption which usually holds with reasonable accuracy. In the case when the observations are independent, the likelihood of the observations factorise

$$f(\mathcal{D}|\theta) = \prod_{i=1}^{n} f(x_i|\theta).$$

(This assumption is so common that occasionally researchers forget that it relies on an assumption about the data, and regrettably it gets applied in situations where it is not justified.) When we are given a collection of independent data points the maximum likelihood estimator is that value $\hat{\theta}$ which maximises $f(\mathcal{D}|\theta)$.

For those people who like being awkward (or deep) it is possible to question whether any set of observation can be considered independent. After all, the observations are all about the same phenomena so are not independent of the test being carried out. At best we can hope they are conditionally independent. One attempt to make this idea of independent observations more rigorous was made by Bruno de Finetti in his representation theorem. He showed that if the probability density for an infinite series of observations was symmetric under any permutation then the observations would be conditionally independent of each other given some parameters describing the probability distribution. Thus, if the observations are *exchangeable* then they can be treated as conditionally independent given the parameter(s) describing their underlying probability distribution. Some people find it is easier to reason in terms of exchangeability than to decide if the observed data could affect each other.

Although it's tempting to dismiss the debate about independence of observations as paranoia, there are times where caution is required. If we built a machine to toss a coin that was so precise that it always provided exactly the same force for each toss so that the coin spun exactly the same number of times before landing, then the outcome would depend on the initial conditions. If we used the result of the previous toss as the initial condition of the next toss then the results of each experiment would be far from independent. That is, if the machine rotated the coin an even number of times we would get either the outcome $HHHHHHHH\ldots$ or $TTTTTTTT\ldots$ depending on the initial set-up. Even without such a machine we can question whether, in doing a series of coin tosses, we should consider each trial as independent. Most of us are trusting enough to believe that the outcome of a coin toss is barely influenced by which side the coin starts, but that may need testing. When carrying out a series of (often expensive) experiments, statisticians, rightly, go to considerable lengths to ensure the experiments are as independent as possible.

Marginalisation. In complex situations it is often useful to model the likelihood with a large number of parameters $\boldsymbol{\Theta} = (\Theta_1, \Theta_2, \ldots, \Theta_m)$ (here we take the Bayesian viewpoint of considering these parameters as random variables). We may be uninterested in some of parameters as they are never observed. These parameters arise because we build into our model some hidden variables that we believe capture reality, but whose value we are not able to observe. Such parameters are sometimes referred to as nuisance parameters or latent variables. If we have a probability distribution over all the parameters $f_{\boldsymbol{\Theta}}(\boldsymbol{\theta})$ then we will often want to average over all possible values of the nuisance parameters. For example, we might be interested in the distribution of Θ_1 alone, which is given by

$$f_{\Theta_1}(\theta_1) = \sum_{\theta_2, \theta_3, \ldots, \theta_n} f_{\boldsymbol{\Theta}}(\boldsymbol{\theta})$$

or we might be interested in the first two random parameters

$$f_{\Theta_1 \Theta_2}(\theta_1, \theta_2) = \sum_{\theta_3, \ldots, \theta_n} f_{\boldsymbol{\Theta}}(\boldsymbol{\theta}).$$

(If the parameters were continuous we would replace the sums with integrals.) This process of averaging over parameters we are not interested in is known as *marginalisation*.

The material covered here forms the foundation of probability as a mathematical subject. Conceptually there is nothing too hard, but there are some terms (e.g. random variables, independence, expectations, etc.) with precise mathematical meanings that any practical engineer or scientist should know. In formulating probabilistic models it is quite easy to get horribly confused. As with all mathematics the trick to avoid confusion is to identify and name the key components of the model (often random variables) and to be able to break down sophisticated manipulations to simple rules such as the additive and multiplicative laws of probabilities. A quick way to identify errors is to check for consistency (do all probabilities sum to one, are they all positive, etc.).

Additional Reading

There is a huge number of books devoted to probability and its many extensions. A classic treatment of the subject is given in the two-volume epic by Feller (1968a,b). A modern text that provides a background to probability and random processes is Grimmett and Stirzaker (2001a).

Exercise for Chapter 1

Exercises are provided for readers who learn through doing. If life is too short or you just want to get on with the plot, feel free to skip them altogether or to jump to the solutions provided. Much of what we covered in this chapter were definitions. In consequence this section is short.

Exercise 1.1 (answer on page 392)

Consider an honest dice and the two events $L = \{1, 2, 3\}$ and $M = \{3, 4\}$. Compute the joint probabilities $\mathbb{P}\left(L, M\right)$, $\mathbb{P}\left(\neg L, M\right)$, $\mathbb{P}\left(L, \neg M\right)$, and the conditional probabilities $\mathbb{P}\left(L|M\right)$, $\mathbb{P}\left(M|L\right)$, $\mathbb{P}\left(\neg M|L\right)$, and $\mathbb{P}\left(M|\neg L\right)$. Compute (i) $\mathbb{P}\left(L, M\right) + \mathbb{P}\left(L, \neg M\right)$, (ii) $\mathbb{P}\left(M|L\right) + \mathbb{P}\left(\neg M|L\right)$, and (iii) $\mathbb{P}\left(M|L\right) + \mathbb{P}\left(M|\neg L\right)$.

Exercise 1.2 (answer on page 392)

Let D_1 and D_2 denote the number rolled by two independent and honest dice. Enumerate all the values of $S = D_1 + D_2$ and compute their probabilities. Compute $\mathbb{E}\left[S\right]$, $\mathbb{E}\left[S^2\right]$, and $\mathbb{V}\text{ar}\left[S\right]$. Also compute $\mathbb{E}\left[D_1\right] = \mathbb{E}\left[D_2\right]$ and $\mathbb{V}\text{ar}\left[D_1\right] = \mathbb{V}\text{ar}\left[D_2\right]$ and verify that $\mathbb{E}\left[S\right] = \mathbb{E}\left[D_1\right] + \mathbb{E}\left[D_2\right]$ and $\mathbb{V}\text{ar}\left[S\right] = \mathbb{V}\text{ar}\left[D_1\right] + \mathbb{V}\text{ar}\left[D_2\right]$.

Exercise 1.3 (answer on page 393)

Repeat the proof given on page 15 of the unconscious statistician theorem

$$\mathbb{E}_X\left[g(X)\right] = \sum_{X \in \mathcal{X}} g(X)\, f_X(X)$$

starting from $\mathbb{E}_Y\left[Y\right]$, but this time using indicator functions.

Exercise 1.4 (answer on page 394)

A classic result in machine learning is known as the *bias-variance dilemma*, where it is shown that the expected generalisation error can be viewed as the sum of a bias term and a variance term. The derivation is an interesting exercise in simplifying expectations. We consider a regression problem where we have a feature vector $\boldsymbol{x}$ and we wish to predict some underlying function $g(\boldsymbol{x})$. The function, $g(\boldsymbol{x})$, is unknown, although we are given a finite training set $\mathcal{D} = \left((\boldsymbol{x}_i, y_i) | i = 1, 2, \ldots, n\right)$, where $y_i = g(\boldsymbol{x}_i)$. We use $\mathcal{D}$ to train a learning machine $g(\boldsymbol{x}|\mathbf{W})$, where $\mathbf{W}$ is a set of weights that depend on the training set. The generalisation error is given by

$$E(\mathbf{W}) = \mathbb{E}_{\boldsymbol{x}}\left[\left(g(\boldsymbol{x}) - g(\boldsymbol{x}|\mathbf{W})\right)^2\right].$$

That is, it is the mean squared difference between the true result, $g(\boldsymbol{x})$, and the prediction of learning machine, $g(\boldsymbol{x}|\mathbf{W})$. The expectation is with respect to some underlying probability of the data. The generalisation error depends on the set of weights, $\mathbf{W}$, which in turn depends on the training set, $\mathcal{D}$. We are interested in the expected error averaged over all data sets of size n,

$$E = \mathbb{E}_{\mathcal{D}}\left[E(\mathbf{W})\right] = \mathbb{E}_{\mathcal{D}}\left[\mathbb{E}_{\boldsymbol{x}}\left[\left(g(\boldsymbol{x}) - g(\boldsymbol{x}|\mathbf{W})\right)^2\right]\right].$$

In the bias-variance dilemma, we consider the response of the mean machine, $\mathbb{E}_{\mathcal{D}}\left[g(\boldsymbol{x}|\mathbf{W})\right]$. In particular, we show that the expected error, E, is the sum of a *bias* term,

$$\mathbb{E}_{\boldsymbol{x}}\left[\left(g(\boldsymbol{x}) - \mathbb{E}_{\mathcal{D}}\left[g(\boldsymbol{x}|\mathbf{W})\right]\right)^2\right],$$

that measures the different between the true function and the response of the mean machine (i.e. the response we get by averaging the responses of an ensemble of learning machines trained on every possible training set of size n), plus a *variance* term,

$$\mathbb{E}_{\boldsymbol{x}}\left[\mathbb{E}_{\mathcal{D}}\left[\left(g(\boldsymbol{x}|\mathbf{W}) - \mathbb{E}_{\mathcal{D}}\left[g(\boldsymbol{x}|\mathbf{W})\right]\right)^2\right]\right]$$

that measures the expected variance in the responses of the machines. Hint: add and subtract the response of the mean machine in the definition of the expected generalisation.

Appendix 1.A Measure for Measure

This appendix provides a brief account of the use of measure theory in probabilities and explains why complete ignorance of the subject is unlikely to set you back.

Part of the legacy of Andrei Kolmogorov's foundational work on probability is that a huge number of books and articles on probability start with the incantation 'Consider the triple $(\Omega, \mathcal{F}, \mu)$ …'. For non-mathematicians this is quite intimidating, but it is rarely important to follow this. In some ways it's not that difficult. Ω is the space of elementary events. This might be the set of outcomes, {heads, tails} or it might be some continuous interval. It can even be something more complex like the set of functions, as we will see in Chapter 12 on stochastic processes. The next part $\mathcal{F}$ has the very posh name of a σ-algebra or sometimes filtration. It is usually just the set of events that you are interested in. It is a complete set in the sense that it includes the intersection and union of all other events. The elements of $\mathcal{F}$ are just subsets of Ω, but the σ-algebra name says it should be a reasonably well-behaved set – we will come back to this later in this section. Finally, μ is the probability measure, but for almost all problems you are likely to meet (though not all problems that a mathematician can dream up) this really just means there is something like a probability density or probability mass assigned to the elementary events.

So what's the big deal? The problem is that if you try hard you can cook up examples where weird things happen. Because one of the duties of mathematicians is to lay solid foundations they go to some lengths to prevent weird things. One weird thing that can happen is that, if we are unlucky, taking expectations can give us apparently inconsistent answers. The cause of this is how we perform integration. What Kolmogorov did was borrow an idea from integration theory due to Henri Lebesgue (pronounced 'Luhbeg') known as measure theory. To understand why this is necessary we consider the classical interpretation of integration due to Riemann. According to Riemann, we can think of an integral as a limit process where we add up the area of small strips, where we take the area of the strips to be their width times the function value at some point in the interval of the strip. The value of the integral is the limit where the strip sizes go to zero. This will work for all the integrals you are every likely to meet in practice. But for mathematicians that's not enough.

The Riemann Integral

$$\int_a^b f(x)\,\mathrm{d}x = \lim_{|x_{i+1}-x_i|\to 0} \sum_{i=1}^{n-1} f_i \times (x_{i+1} - x_i)$$

$$f_i = f(x) \text{ for } x \in [x_i, x_{i+1}]$$

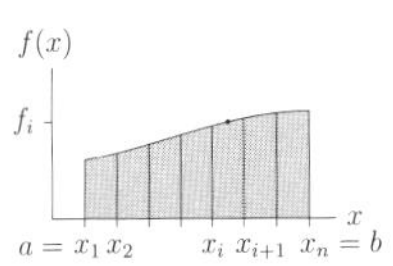

Suppose we had a function that was equal to 1 at every rational number and 0 at each irrational number (a function first proposed by Dirichlet). Consider integrating such a function from 0 to 1. Now mathematicians have long known that rational numbers are not dense. That is, if you choose a randomly selected real number between 0 and 1, with overwhelming probability it will be irrational. Thus, it is reasonable to think that this integral should be equal to 0. However, let us consider applying Riemann's method for computing an integral. We can split the interval into strips of size 2^{-n} and take the mid-point as the representative value of the function. However, this representative value will be a rational value so equal 1. This remains true in the limit $n \to \infty$ so this algorithm for computing integrals gives the answer 1. Of course, if we had chosen a random point in the interval or a point at distance $\sqrt{1/2}$ along the interval we would have got the answer 0. This is a problem for probability theory, as well as integration theory, because if I wanted to be difficult then I could assign different probabilities to points depending on whether they are rational or irrational. To overcome this problem for integrals, Lebesgue founded measure theory where he devised a different interpretation of the integral which is consistent with Riemann's integral whenever Riemann's integral is well defined, but is able to handle pathological functions such as the one proposed by Dirichlet. One important benefit of measure theory is that it provides consistent rules for deciding when it is valid to interchange the order of performing integrals or of exchanging limits. Kolmogorov borrowed the ideas of measure theory in providing an axiomatic basis for probabilities.

This is very worthy and fulfils the responsibility of mathematicians to make the foundations of their subject secure. But it is not really relevant if you want to solve real-world problems, at least, in my experience. The challenge from the real world does not come from pathological functions, but from the need to evaluate quite tricky formulae. What most scientists and engineers are interested in, and the topic of this book, are practical applications, very often computing numerical quantities. These tasks are both fascinating and challenging, but aren't helped by measure theory. Perhaps for highly sophisticated mathematicians the language of measure theory may help them think about their problem is a more general way – I am afraid I am not in a position to judge. For the rest of us measure theory tends to obscure rather than elucidate. In my experience most mathematical books on probability quietly drop the measure theory after the first chapter, leaving the reader wondering why they bothered introducing it in the first place. The fact that measure theory is not necessary for most real-world applications should in no way diminish Kolmogorov, who solved the important open problem of laying the axiomatic foundations of probability. Kolmogorov also took great interest in solving many practical problems and made a large number of important contributions to applied probability and science in general.

2

Survey of Distributions

Contents

In this chapter we give a survey of some of the most frequently encountered distributions. In Chapters 4 and 5 we will cover some of these in more detail. Many of the common distributions belong to the *exponential* family of distributions. We present this family and some of its properties in Section 2.4. We conclude the chapter by two appendices covering the gamma and beta functions; both functions that arise in some of the common distributions.

If probability is the language for discussing uncertainties, then Chapter 1 could be viewed as learning the basic grammar and simple verbs, while this chapter is more like learning the important nouns or objects. With these in place, we will be in a position in subsequent chapters to use our language to convey ideas.

2.1 Discrete Distributions

It is easy to be put off by the seemingly endless number of probability distributions, but there are only a handful of distributions that keep on cropping up. The sooner you make friends with them the easier your life is going to be. We start with discrete distributions, which are those that involve random variables which take values that lie in a countable set.

2.1.1 Binomial Distribution

One of the frequently met probability distributions that pops up in a huge number of applications is the *binomial distribution*. It arises when we sample n objects that belong to two classes, A and B say. We assume that the probability of choosing an object of class A is p. This does not change over time. We can think of randomly choosing red and blue balls from a bag where the ratio of red to blue balls is p. Each time we choose a ball we put it back and mix up the balls before drawing the next sample. The probability of choosing m objects of class A in n trials is given by

$$\mathbb{P}\left(X = m|n, p\right) = \text{Bin}(m|n, p) = \binom{n}{m} p^m \left(1 - p\right)^{n-m}, \tag{2.1}$$

where $\binom{n}{m} = \frac{n!}{m!(n-m)!}$ is the binomial coefficient (often referred to as 'n choose m'). Figure 2.1 shows examples of the binomial distribution.

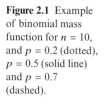

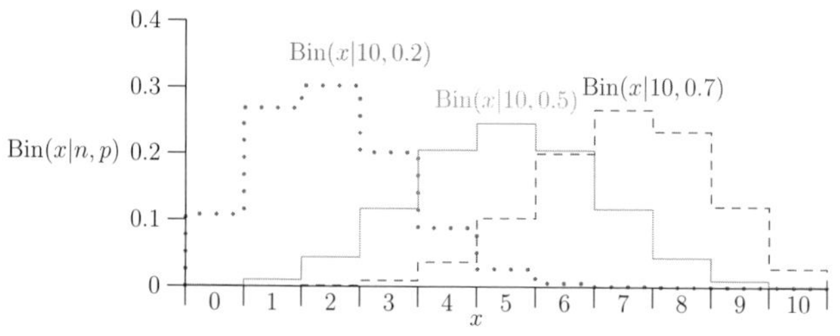

Figure 2.1 Example of binomial mass function for $n = 10$, and $p = 0.2$ (dotted), $p = 0.5$ (solid line) and $p = 0.7$ (dashed).

The mean of a binomial distribution is $n\,p$ and the variance is $n\,p\,(1 - p)$. We return to the binomial distribution in Chapter 4. A large number of distributions are in some way related to the binomial distribution: the hypergeometric distribution describes the situation of sampling without replacement; the Poisson distribution corresponds to a limit of the binomial as $p \to 0$ and $n \to \infty$, but with $p/n \to \mu$, a constant; the multinomial distribution is a generalisation of the binomial distribution to the case where there are more than two classes; finally the Gaussian distribution is a limit of the binomial distribution as $n \to \infty$.

Example 2.1 Rolling Dice

What is the probability of getting three sixes in 10 rolls of an honest dice?

This situation describes a case of repeating independent random binary trials, which gives rise to a binomial probability. In this case, we have a success probability of $p = 1/6$ and have $n = 10$ trials so the probability of three successes is

$$\text{Bin}(3|10, 1/6) = \binom{10}{3}\left(\frac{1}{6}\right)^3\left(1 - \frac{1}{6}\right)^7 = 0.155.$$

Thus we can expect this to happen around 15% of the times we attempt it.

2.1.2 Hypergeometric Distribution

Although binomial distributions are the most common type of discrete distribution when dealing with two classes, they are not the only one. The hypergeometric distribution describes the probability of choosing k samples of class A out of n attempts, given that there is a total of N objects, m of which are of class A. For example, if you have a bag of N balls of which m are red and the rest are blue, and you sample n balls from the bag without replacement, then the hypergeometric distribution

$$\mathbb{P}\left(K = k|N, m, n\right) = \text{Hyp}(k|N, m, n) = \frac{\binom{m}{k}\binom{N-m}{n-k}}{\binom{N}{n}} \tag{2.2}$$

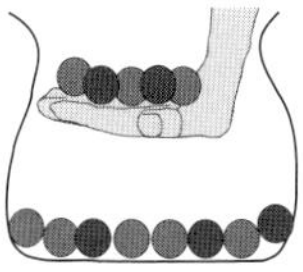

$N = 13$, $m = 8$
$n = 5$, $k = 3$

tells you the probability that k of the drawn balls are red. (N is the total number of balls and is not a random variable – we use a capital to follow a commonly used convention for describing this distribution.) This probability is just the number of different ways of choosing k red balls from the m red balls, times the number of ways of choosing $n - k$ blue balls from the $N - m$ blue balls, divided by the total number of ways of choosing n balls from N balls. There are a number of surprising symmetries, for example, $\text{Hyp}(k|N, m, n) = \text{Hyp}(k|N, n, m)$ (that is, we get the same probability when we exchange the number of red balls and the number of balls that we sample). These arise due to the many identities involving binomial coefficients.

The mean value of K is nm/N and its variance is $n(m/N)(1 - m/N)(N - n)/(N - 1)$. Typical probability masses are shown in Figure 2.2. We observe that these figures are not too dissimilar to those for the binomial distribution. Indeed, in the limit $N, m \to \infty$ such that $m/N \to p$ the hypergeometric distribution converges to the binomial distribution (see Exercise 2.4 on page 41).

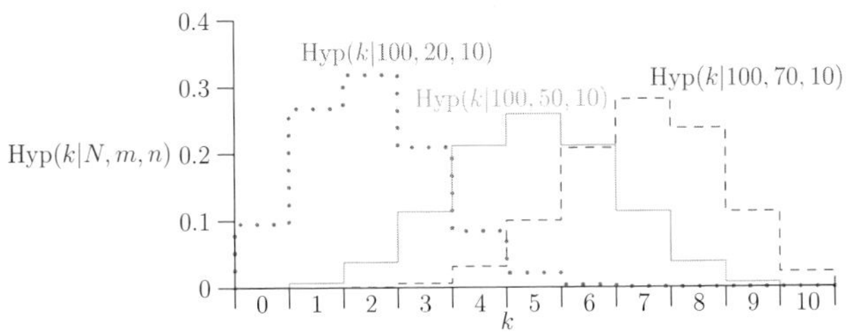

Figure 2.2 Examples of hypergeometric distributions for $N = 100$, $n = 10$, and $m = 30$ (dotted), $m = 50$ (solid line) and $m = 70$ (dashed).

An application of this distribution is when there is a shipment of N objects, m of which are defective. If we sample n of the objects, then the hypergeometric distribution tells us the probability that k of the samples are defective. This is also the distribution you need to use if you want to calculate the probability of winning a prize in the UK National Lottery (see Exercise 2.5 on page 41). The hypergeometric distribution is not so well known as its ubiquity deserves. Possibly its low profile is a consequence of the fact that it is not always that easy to deal with analytically.

Example 2.2 Bridge

In the game of bridge, each player is dealt 13 cards. What is the probability that player 1 has three aces?

Treating a bridge hand as a random sample of 13 cards from a pack of 52 cards where we don't replace the cards, then we see that this is a job for the hypergeometric distribution. The total number of cards is $N = 52$, the number of aces is $m = 4$. A hand is a sample of $n = 13$ cards so that the probability we seek is

$$\text{Hyp}(3|52, 4, 13) = \frac{\binom{4}{3}\binom{48}{10}}{\binom{52}{13}} = \frac{858}{20\,825} = 0.0412.$$

That is, a player can expect such a bridge hand around 4% of the time.

2.1.3 Poisson Distribution

The Poisson distribution can be regarded as a limiting case of the binomial distribution when $p \to 0$ and $n \to \infty$, but with $p\,n = \mu$. In this limit, with $m \ll n$, the binomial coefficient simplifies

$$\binom{n}{m} = \frac{n\,(n-1)\,\ldots\,(n-m)}{m!} \approx \frac{n^m}{m!},$$

and

$$(1-p)^{n-m} = \mathrm{e}^{(n-m)\log(1-p)} \approx \mathrm{e}^{-p(n-m)} \approx \mathrm{e}^{-p\,n} = \mathrm{e}^{-\mu}$$

where we have used the Taylor expansion $\log(1-p) = -p + O(p^2)$. Thus in this limit

$$\lim_{\substack{p\to 0\\ p\times n=\mu}} \binom{n}{m} p^m (1-p)^{n-m} = \frac{(n\,p)^m \mathrm{e}^{-\mu}}{m!} = \frac{\mu^m \mathrm{e}^{-\mu}}{m!} = \text{Poi}(m|\mu). \qquad (2.3)$$

This is the definition of the Poisson distribution. Its somewhat simpler form than the binomial distribution makes it easier to use. It is also important because it describes the distribution of independent point events that occur in space or time (we return to this in Section 12.3). The Poisson distribution arises, for

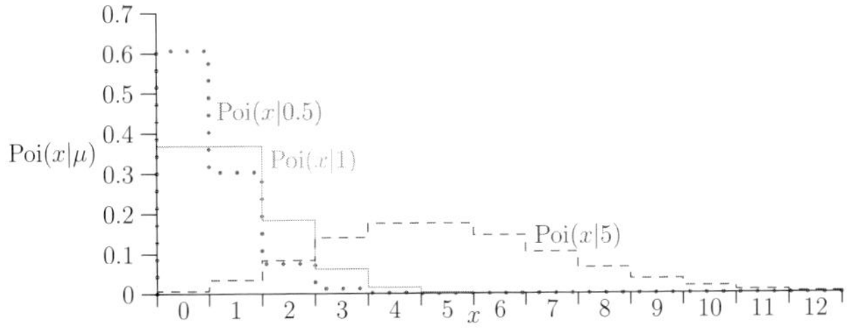

Figure 2.3 Examples of the Poisson distribution for $\mu = 0.5$ (dotted), $\mu = 1$ (solid line), and $\mu = 2$ (dashed).

example, if you want to know the probability of a Geiger counter having 10 counts in a minute given that the background radiation level is six counts per minute (answer 0.0413). Both the mean and variance of the Poisson distribution are equal to μ. Typical examples of the distribution are shown in Figure 2.3.

Example 2.3 Carbon Dating

Carbon dating is traditionally based on counting the number of beta particle emissions associated with the radioactive decay of carbon 14. Carbon 14 is an radioactive isotope of carbon with a relatively short half-life of 5730 years. All carbon 14 that initially existed in the early earth would have decayed a long time ago. However, it is constantly being replenished through neutron capture caused by cosmic rays creating neutrons that react with nitrogen in the atmosphere.

$$^{14}\text{N} + {}^{1}n \rightarrow {}^{14}\text{C} + {}^{1}p.$$

where ^{1}n is a neutron and ^{1}p a proton. As a consequence, carbon in the atmosphere (CO_2) has around one part per trillion of carbon 14. This is equivalent to 60 billion atoms per mole of carbon. This carbon is then taken up by plants through photosynthesis. By measuring the ratio of carbon 14 we are then able to deduce its age.

The probability of an atom of carbon decaying in one year is $\lambda = 1.245 \times 10^{-4}$. The number of carbon 14 atoms in a 1 mole sample (i.e. approximately 12 g of carbon) is

$$N = N_0\, \mathrm{e}^{-\lambda t}$$

where $N_0 = 6 \times 10^{10}$ is the estimated number of atoms absorbed from the atmosphere through photosynthesis and t is the age of the sample. The expected number of decays in a time Δt is $\mu = \lambda N \Delta t$. Now suppose we observe $n = 100$ decays of carbon 14 in one hour. As radioactive decays are well approximated by a Poisson distribution, the probability of the decay is

$$\mathbb{P}\left(n = 100\right) = \frac{\mu^{100}}{100!}\mathrm{e}^{-\mu}$$

where $\mu = \lambda\, \Delta t\, N_0\, \mathrm{e}^{-\lambda t} = 852.7\, \mathrm{e}^{-\lambda t}$ (recall that we know λ, Δt, and N_0, but we don't know the age of the sample t). In Figure 2.4 we show

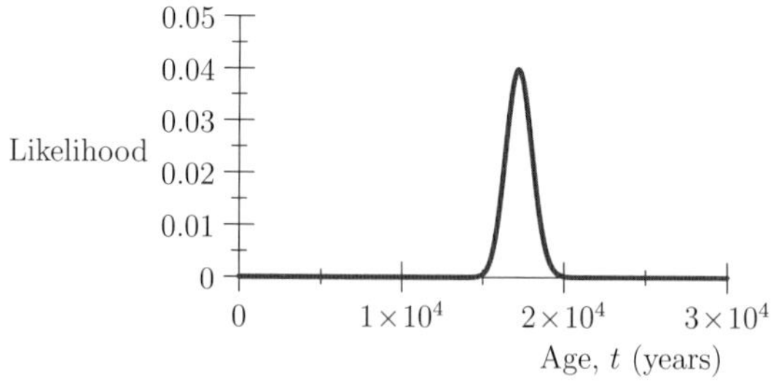

Figure 2.4 Probability of observing 100 decays in a sample with 1 mole of carbon atoms an hour versus the age of the sample.

the probability of observing 100 decays in the sample versus the age t in years. We see that with high likelihood the age of the sample would be between 15,000 and 20,000 years.

We assumed that the proportion of carbon 14 in the atmosphere is constant over time. This is not true (it is not even true over location), thus to obtain a more accurate estimate, the concentration of carbon 14 is calibrated against tree samples that can be aged by counting rings. However, as our probabilistic model shows, there is also a natural uncertainty caused by the underlying Poisson nature of radioactive decay. To obtain precise dates for a small sample requires that measurements over a very long time interval be made. Modern carbon dating tends to measure the proportion of carbon 14 directly using a mass spectrometer to reduce the uncertainty caused by that randomness of beta decays.

2.2 Continuous Distributions

These distributions describe random variables that take on continuous values. By far the most important distribution in this class is the Gaussian or normal distribution. However, there are a number of other continuous distributions that are common enough that they are worth getting to know.

2.2.1 Normal Distribution

The *normal distribution* – also called the *Gaussian distribution* – is by far the most frequently encountered continuous distribution. There are a number of reasons for this. The central limit theorem (see Section 5.3 on page 81) tells us that the distribution of the sum of many random variables (under mild conditions) will converge to a normal distribution as the number of elements in the sum increase. Many of the other distributions converge to the Gaussian distribution as their parameters increase. This means that in practical situations many quantities will be approximately normally distributed. If all you know about a random variable is its mean and variance then there is a line of reasoning (the so-called maximum entropy argument, see Section 9.2.2 on page 267) that says that of all possible distributions with the observed mean and variance the normal distribution is

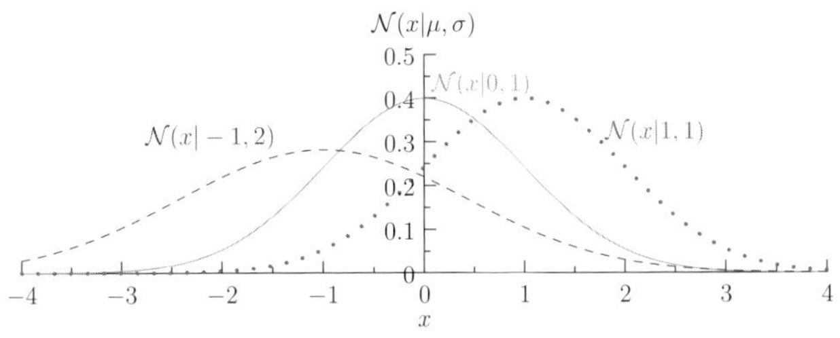

Figure 2.5 Examples of the normal distribution for $(\mu, \sigma) = (0, 1)$ (solid line), $(\mu, \sigma) = (1, 1)$ (dotted), and $(\mu, \sigma) = (-1, 2)$ (dashed).

overwhelmingly the most likely. Thus, assuming it is normally distributed is, in some sense, the optimal decision. However, before you use this argument you need to understand the small print (i.e. you've made a strong assumption about your variables that *ain't necessarily so*). A further reason why normal distributions are so commonly used is simply because they are easy to manipulate mathematically – this is a less contemptible motivation than it may at first appear. All models are abstractions from reality, and an approximate, but an easily solvable model is often much more useful than a more accurate but complex or intractable model.

The probability density for the normal distribution is defined as

$$\mathcal{N}(x|\mu.\sigma^2) = \frac{1}{\sqrt{2\pi}\,\sigma}\, e^{-\frac{(x-\mu)^2}{2\sigma^2}}, \tag{2.4}$$

which has mean μ and variance σ^2. Examples of the normal probability density functions are shown in Figure 2.5. We will have much more to say about the normal distribution in Chapter 5.

2.2.2 Gamma Distribution

When considering problems where a continuous random variable only takes positive values, the normal distribution can provide a poor model. Often a more appropriate model is the *gamma distribution* defined for $X > 0$ through the probability density

$$\text{Gam}(x|a,b) = \frac{b^a x^{a-1} e^{-b\,x}}{\Gamma(a)}, \tag{2.5}$$

where $\Gamma(a)$ is the *gamma function* defined (for real $a > 0$) by

$$\Gamma(a) = \int_0^{\infty} x^{a-1} e^{-x} dx$$

The gamma function is actually defined throughout the complex plane except where a is equal to 0 or a negative integer.

(see Appendix 2.A). It can easily be verified using integration by parts that $\Gamma(a+1) = a\,\Gamma(a)$. For positive integers, $n > 0$, the gamma function is given by $\Gamma(n) = (n-1)!$ (factorial). In some texts the gamma distribution is defined with parameters $\alpha = a$ and $\beta = 1/b$. The mean of the gamma distribution is given by a/b (or $\alpha\,\beta$) while the variance is given by a/b^2 (or $\alpha\,\beta^2$). Examples of the distribution are shown in Figure 2.6.

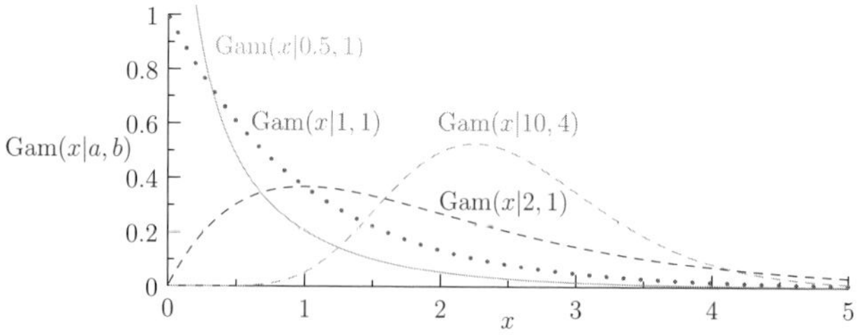

Figure 2.6 Examples of the gamma distribution for $b = 1$ and $a = 0.5$ (solid line), $a = 1$ (dotted) and $a = 2$ (dashed).

The gamma distribution is often used to empirically fit data that are known to always take positive values. For example, if you wish to model the intensity of light from different stars or the sizes of different countries then the gamma distribution is often a reasonable fit. Given an empirically measured mean $\hat{\mu}$ and variance $\hat{\sigma}^2$ a simple fit is to choose $a = \hat{\mu}^2/\hat{\sigma}^2$ and $b = \hat{\mu}/\hat{\sigma}^2$. (Although this gives a reasonably good fit, it is not the maximum likelihood estimator for a and b.) Gamma distributions also arise naturally in many problems. We discuss a few examples here.

The *chi-squared* (or χ^2) distribution is a particular form of the gamma distribution. The distribution arises in sums such as

$$S_k = \sum_{i=1}^{k} X_i^2,$$

where X_i are normally distributed variables with mean 0 and variance 1. Then S_k is distributed according to

$$f_{S_k}(s) = \chi_k(s) = \mathrm{Gam}\left(s \middle| \tfrac{k}{2}, \tfrac{1}{2}\right).$$

The χ^2-distribution arises when evaluating the expected errors in curve fitting.

In the special case of $a = 1$, the gamma distribution reduces to the exponential distribution

$$\mathrm{Exp}(x|b) = \mathrm{Gam}(x|1, b) = b\,\mathrm{e}^{-b\,x}. \tag{2.6}$$

The exponential distribution describes the waiting times between events in a Poisson process (see Section 12.3.1 on page 380).

The velocity of particles in an ideal gas (a model for a real gas which is often very accurate) are normally distributed, such that the components of the velocity V_x, V_y, and V_z have distribution $\mathcal{N}(V_i|0, m/(2\,k\,T))$, where k is the Boltzmann constant and T the temperature. The speed $V = \|\boldsymbol{V}\|$ is consequently distributed according to the Maxwell–Boltzmann distribution, which is related to the gamma distribution

$$\begin{aligned}\mathbb{P}\left(v \le V < v + \mathrm{d}v\right) = f_V(v)\,\mathrm{d}v &= 4\,\pi \left(\frac{m}{2\,\pi\,k\,T}\right)^{3/2} v^2 \mathrm{e}^{-\dfrac{m\,v^2}{2\,k\,T}}\,\mathrm{d}v \\ &= \mathrm{Gam}\left(v^2 \middle| \frac{3}{2}, \frac{m}{2\,k\,T}\right)\mathrm{d}v^2.\end{aligned}$$

Example 2.4 Escaping Helium

Molecules can escape the atmosphere if their velocity exceeds the escape velocity and they are pointing in the right direction. This is known as Jeans' escape. The gravitational escape velocity of an object from a mass M at a radius r from the centre of the mass is given by

$$v_e = \sqrt{\frac{2GM}{r}}.$$

For a molecule 500 km above earth this is around 10.75 km/s. The upper level of the atmosphere is known as the exosphere, which starts at the exobase at a height of around 500 km. In the exosphere the mean free path of a gas molecule is sufficiently large that a molecule can easily escape the gravitational pull of earth if it has sufficient velocity. The temperature of atmospheric gas is surprising large at around 1600 K. The velocity for hydrogen and helium is given by the Maxwell–Boltzmann distribution. Figure 2.7 shows the distribution of velocities for both molecular hydrogen and helium. Although small, there is a sufficiently high probability of reaching the escape velocity for the escape of hydrogen and helium to be important. Although hydrogen is lost to space, most of it is retained as it forms molecules with heavy atoms (e.g. water, H_2O), however, helium does not form any molecules and so will, over time, become lost into outer space. The presence of helium in the atmosphere is the result of radioactive alpha decays. The concentration of helium in the atmosphere (5.2 parts per billion) is determined by an equilibrium between its production through alpha decays and its loss from the atmosphere through Jeans' escape.

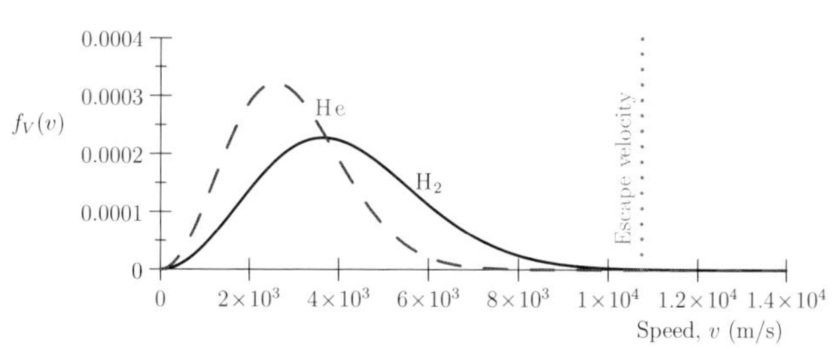

Figure 2.7 Distribution of velocity of hydrogen and helium molecules at 1600 K.

Yet another distribution obtained by making a suitable change of variable is the Weibull distribution. If $Y \sim \text{Exp}(1) = \text{Gam}(1, 1)$ then the random variable $X = \sqrt[k]{\lambda Y}$ is distributed according to a Weibull probability density:

$$\text{Wei}(x|\lambda, k) = \frac{k}{\lambda}\left(\frac{x}{\lambda}\right)^{k-1} \mathrm{e}^{-(x/\lambda)^k}.$$

The mean of the Weibull distribution is $\Gamma(1 + 1/k)$ and the variance is $\lambda^2(\Gamma(1 + 2/k) - \Gamma^2(1 + 1/k))$. Weibull distributions can also be used to fit data involving positive real random variables. It is slightly less convenient than a gamma distribution, although easier enough to fit numerically. It provides slightly differently shaped density profiles to the gamma distribution. This is explored in Exercise 2.6.

Although gamma distributions are not so well known, as these many examples illustrate they deserve to be widely appreciated.

2.2.3 Beta Distribution

The beta distribution is a two-parameter continuous distribution that is defined in the interval $[0, 1]$. It is therefore useful for modelling situations where the random variable lies in a range. It is defined by

$$\text{Bet}(x|a, b) = \frac{x^{a-1}(1-x)^{b-1}}{\text{B}(a, b)} \tag{2.7}$$

where $\text{B}(a, b)$ is the beta function (see Appendix 2.B) given by

$$\text{B}(a, b) = \int_0^1 x^{a-1}(1-x)^{b-1}\text{d}x = \frac{\Gamma(a)\Gamma(b)}{\Gamma(a+b)}.$$

The mean and variance of the beta distribution are $a/(a+b)$ and $a\,b/((a+b)^2(a+b+1))$, respectively. Examples of the beta probability density functions are shown in Figure 2.8.

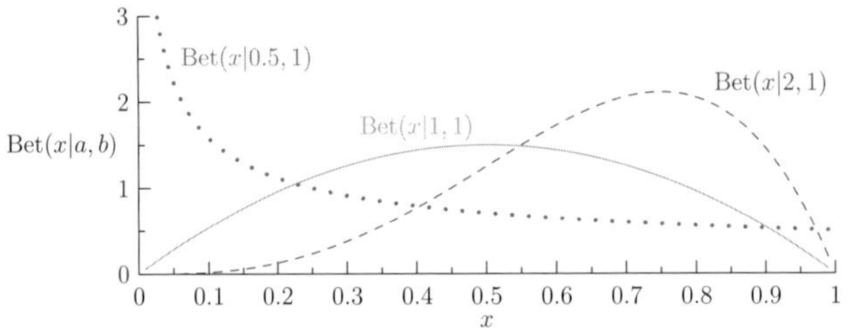

Figure 2.8 Examples of the beta distribution for $(a, b) = (0.5, 1)$ (solid line), $(a.b) = (2, 2)$ (dotted), and $(a, b) = (4, 2)$ (dashed).

A typical application of the beta distribution is to model an unknown probability. The uncertainty might be because you don't know what the value of probability is. For example, you might want to model the probability of a cell dividing in the next hour. In this case, there is some fixed probability p, but you don't know it. To model your uncertainty you can treat p as a random variable that takes some value in the interval $[0, 1]$. Alternatively, you might have different types of cells with different probabilities of dividing. Here, the uncertainty arises because you don't know which type of cell you are looking at. In this case, you are modelling the distribution of p in a population of cells. The beta distribution has a limited parametric form, nevertheless it is sufficiently flexible that it can fit many observed distributions for quantities bounded in an interval quite well, provided the distributions are single peaked (unimodal).

2.2.4 Cauchy Distribution

There are a large number of other continuous distributions, many of which are rather esoteric. However, one type of distribution which you need to be aware of are those with *long tails* – that is, with a significant probability of drawing a sample which is many standard deviations away from the mean. (These are also, perhaps more accurately called *thick-tailed distributions*, since they are usually characterised by a power-law fall-off rather than an exponential fall-off in probability.) A classic example of a distribution with very long (thick) tails is the *Cauchy distribution* (aka Cauchy–Lorentz, Lorentzian, Breit–Wigner), defined through the probability density

$$\mathrm{Cau}(x) = \frac{1}{\pi(1+x^2)}. \tag{2.8}$$

The median and mode of the Cauchy distribution is zero, but rather shockingly the distribution has no mean or variance. That is, if X is drawn from Cau then the improper integrals $\mathbb{E}\left[X\right]$ and $\mathbb{E}\left[X^2\right]$ diverge. We will see that distributions like this behave rather differently to the other distributions we have looked at so far. The Probability Distribution Function (PDF) for the Cauchy distribution is shown in Figure 2.9.

It is tempting to assume the mean must be zero by symmetry. Don't be tempted!

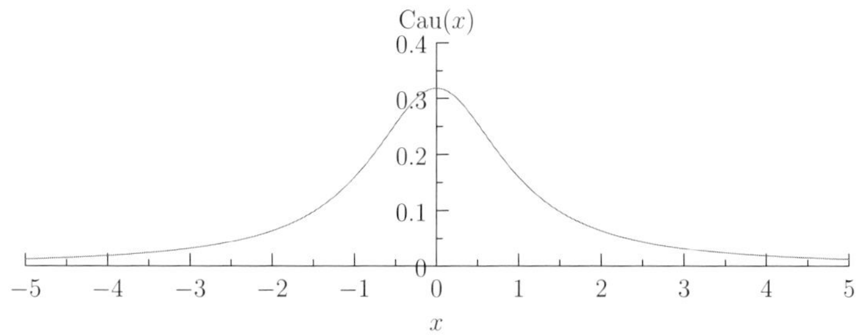

Figure 2.9 The Cauchy distribution.

2.3 Multivariate Distributions

So far we have considered distributions involving a single random variable. Often we have situations where there are many correlated random variables. Distributions that describe more than one random variable are known as *multivariate distributions*, in contrast to distributions of a single variable, which are known as *univariate distributions*. There are multivariate extensions for most univariate distributions, although they often become rather too complex to work with. However, there are three well-known and useful multivariate distributions that are relatively easy to work with: the multinomial distribution, the multivariate normal distribution, and the Dirichlet distribution.

2.3.1 Multinomial Distribution

The *multinomial distribution* is the generalisation of the binomial distribution to more than two classes. We assume that we have k classes. The probability

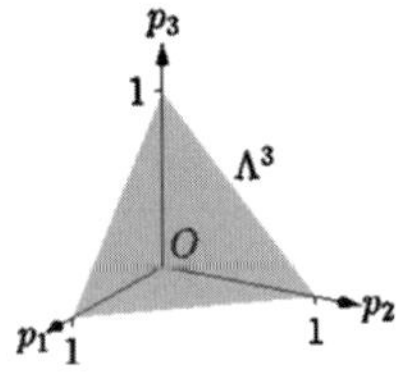

of drawing a sample from class i is given by p_i. Thus the model is described by a vector of probabilities $\boldsymbol{p} = (p_1, p_2, \ldots, p_k)$, with $p_i \geq 0$ for all i and $\sum_i p_i = 1$. The vector of probabilities satisfying these constraints live in the $(k-1)$-*dimensional unit simplex*

$$\Lambda^k = \left\{ \boldsymbol{p} = (p_1, p_2, \ldots, p_k) \,\middle|\, \forall i,\, p_i \geq 0 \text{ and } \sum_{i=1}^{k} p_i = 1 \right\}. \tag{2.9}$$

Note that the simplex is a $(k-1)$-dimensional surface that lives in a k-dimensional space. Suppose we draw a sample of n objects without replacement and we wish to know what is the probability that we have drawn n_1 objects from class 1, n_2 objects from class 2, etc. This probability, $\mathbb{P}\left(\boldsymbol{N} = \boldsymbol{n}\right)$, is given by the multinomial distribution

$$\text{Mult}(\boldsymbol{n}|n, \boldsymbol{p}) = n! \prod_{i=1}^{k} \frac{p_i^{n_i}}{n_i!} \left[\!\left[\boldsymbol{n} \in \Lambda_n^k \right]\!\right] \tag{2.10}$$

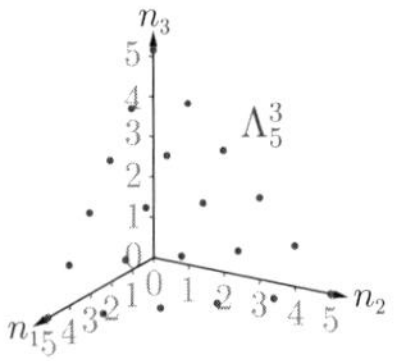

where we use the notation $\left[\!\left[\textit{predicate} \right]\!\right]$ to be the *indicator function* as defined in Section 1.5.1 (note that the bold $\boldsymbol{n}$ signifies a vector with components n_i equal to the number of samples in class i, while the italic n denotes the total number of samples). The set Λ_n^k is the *discrete simplex* defined by

$$\Lambda_n^k = \left\{ \boldsymbol{n} = (n_1, n_2, \ldots n_k) \,\middle|\, \forall i,\, n_i \in \{0, 1, 2, \ldots\} \text{ and } \sum_{i=1}^{k} n_i = n \right\}. \tag{2.11}$$

The mean of the multivariate distribution is $\mathbb{E}\left[\boldsymbol{N}\right] = n\,\boldsymbol{p}$. For multivariate distributions you not only have a variance for each variable $\mathbb{V}\text{ar}\left[N_i\right] = \mathbb{E}\left[N_i^2\right] - \mathbb{E}\left[N_i\right]^2$, you also have a covariance between variables $C_{ij} = \mathbb{E}\left[N_i\, N_j\right] - \mathbb{E}\left[N_i\right] \mathbb{E}\left[N_j\right]$. In general, the *second-order statistics* for a multivariate distribution are described by a *covariance matrix*, $\mathbf{C}$, defined as

$$\mathbf{C} = \mathbb{C}\text{ov}\left[\boldsymbol{N}\right] = \mathbb{E}\left[\boldsymbol{N}\,\boldsymbol{N}^\mathsf{T}\right] - \mathbb{E}\left[\boldsymbol{N}\right] \mathbb{E}\left[\boldsymbol{N}^\mathsf{T}\right].$$

A positive semi-definite matrix has the property that for any vector $\boldsymbol{x}$

$$\boldsymbol{x}^\mathsf{T} \mathbf{C}\, \boldsymbol{x} \geq 0.$$

The covariance matrix is both symmetric and positive semi-definite. For the multinomial distribution the covariance between N_i and N_j is given by

$$\mathbb{C}\text{ov}\left[N_i, N_j\right] = C_{ij} = \mathbb{E}\left[N_i\, N_j\right] - \mathbb{E}\left[N_i\right] \mathbb{E}\left[N_j\right] = n \left[\!\left[i = j \right]\!\right] p_i - n\, p_i\, p_j$$

or in matrix form

$$\mathbf{C} = n\, \left(\text{diag}(\boldsymbol{p}) - \boldsymbol{p}\,\boldsymbol{p}^\mathsf{T}\right),$$

where $\text{diag}(\boldsymbol{p})$ is a diagonal matrix with elements p_i.

The random variables N_i are not independent since their sum adds up to n. A consequence is that each row (or column) of the covariance matrix sums to zero. The multinomial distribution for just two variables only has one degree of freedom (i.e. given p_1 then $p_2 = 1 - p_1$) and in this case the multinomial reduces to the binomial distribution. With three variables, the multinomial is sometimes referred to as the trinomial distribution.

Multinomial distributions are fairly common. Suppose, for example, you had a (possibly biased) dice which you rolled n times. Letting N_i for $i = 1, 2, \ldots, 6$ denote the number of times the dice lands on i, then the probability of the outcome $\boldsymbol{N} = \boldsymbol{n} = (n_1, n_2, \ldots, n_6)$ is given by the multinomial $\mathbb{P}\left(\boldsymbol{N} = \boldsymbol{n}\right) = \text{Mult}(\boldsymbol{n}|n, \boldsymbol{p})$, where the components of the vector $\boldsymbol{p} = (p_1, p_2, \ldots, p_6)$ describe the probability of each possible outcome of a dice roll.

2.3.2 Multivariate Normal Distribution

The most commonly used multivariate distribution for continuous variables is the multivariate normal distribution defined as

$$\mathcal{N}(\boldsymbol{x}|\boldsymbol{\mu}, \boldsymbol{\Sigma}) = \frac{1}{\sqrt{|2\pi\boldsymbol{\Sigma}|}} \mathrm{e}^{-\frac{1}{2}(\boldsymbol{x}-\boldsymbol{\mu})^{\mathsf{T}}\boldsymbol{\Sigma}^{-1}(\boldsymbol{x}-\boldsymbol{\mu})}, \tag{2.12}$$

which has mean vector $\mathbb{E}\left[\boldsymbol{X}\right] = \boldsymbol{\mu}$ and covariance $\boldsymbol{\Sigma}$. A two-dimensional normal distribution is shown in Figure 2.10. Like its univariate counterpart, the multivariate normal (or Gaussian) distribution can be manipulated analytically. This can be a somewhat complicated or awkward business requiring some practice, but it pays off handsomely. A large number of state-of-the-art algorithms from Gaussian processes to Kalman filters rely on being able to manipulate normal distributions analytically. We discuss the multivariate normal distribution in more detail in Section 5.6 on page 96. Applications of multivariate normal distribution reoccur throughout this book.

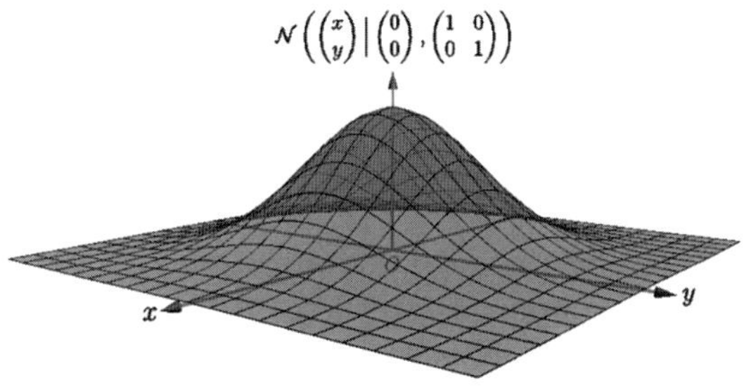

Figure 2.10 A two-dimensional normal distribution.

2.3.3 Dirichlet Distribution

Although the most common multivariate distributions by far are the multinomial and multivariate normal distributions, there exist many others. A particularly convenient distribution for describing a vector of random variables defined on the unit simplex, Λ^k, is the Dirichlet distribution, defined as

$$\text{Dir}(\boldsymbol{x}|\boldsymbol{\alpha}) = \Gamma(\alpha_0) \prod_{i=1}^{k} \frac{x_i^{\alpha_i - 1}}{\Gamma(\alpha_i)}, \tag{2.13}$$

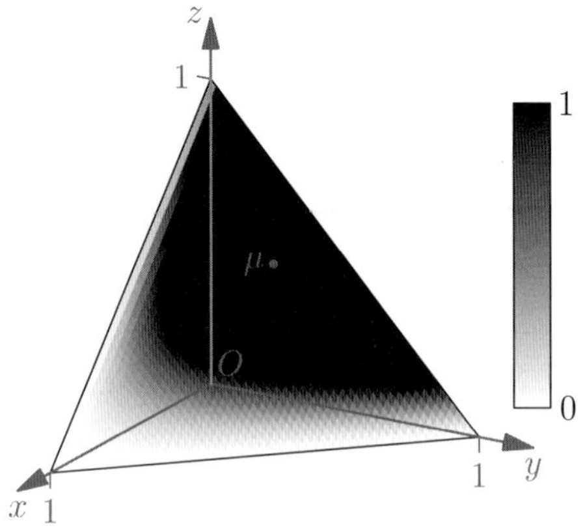

Figure 2.11 A three-dimensional Dirichlet distribution, $\text{Dir}(\boldsymbol{X} = (x, y, z) | \boldsymbol{\alpha} = (1, 2, 3))$. Note that this distribution is defined on the simplex.

where $\alpha_0 = \sum_{i=1}^{n} \alpha_k$. The means are equal to $\mathbb{E}[X_i] = \alpha_i/\alpha_0$ and the covariance

$$C_{i,j} = \mathbb{E}[X_i X_j] - \mathbb{E}[X_i]\,\mathbb{E}[X_j] = \frac{\alpha_i(\alpha_0 [\![i = j]\!] - \alpha_j)}{\alpha_0^2(\alpha_0 + 1)}.$$

An example of a Dirichlet distribution with three variables is shown in Figure 2.11.

Suppose you have a dice and you are not sure whether it is biased. You could model your uncertainty about the probability of rolling any number using a Dirichlet distribution. The random variables X_i, drawn from $\text{Dir}(\boldsymbol{X}, \boldsymbol{\alpha})$, are not all independent as their sum adds up to one. In the two component case there is only one independent variable and the Dirichlet distribution reduces to a beta distribution.

2.4 Exponential Family*

* *Warning, this is more advanced material than the rest of this chapter. It can be skipped.*

Although distributions vary considerably, they also share many properties, sometimes more so than is immediately obvious. One very important family of distributions which share similar properties is the exponential family. These are distributions which can be written in the form

$$f_{\boldsymbol{X}}(\boldsymbol{x}|\boldsymbol{\eta}) = g(\boldsymbol{\eta})\, h(\boldsymbol{x})\, e^{\boldsymbol{\eta}^{\mathsf{T}} \boldsymbol{u}(\boldsymbol{x})}, \tag{2.14}$$

where $\boldsymbol{\eta}$ are *natural parameters* of the distribution, $g(\boldsymbol{\eta})$ and $h(\boldsymbol{x})$ are scalar functions, and $\boldsymbol{u}(\boldsymbol{x})$ is a vector function (i.e. a function for each natural parameter). The distribution can be either for a single random variable or for a random vector, e.g. in the case of a multinomial distribution. The importance of the exponential family is that many properties are known to hold true for distributions belonging to this family. Thus, once we know a distribution belongs to this family we know many of its properties.

It is not immediately obvious, though, which distributions are in the exponential family, because they are often written in ways that don't appear to fit the standard form. Examples of distributions that belong to the exponential family

Distribution	$\boldsymbol{\eta}$	$\boldsymbol{u}(\boldsymbol{x})$	$h(\boldsymbol{x})$	$g(\boldsymbol{\eta})$
$\text{Bern}(x\|\mu) = \mu^x(1-\mu)^{1-x}$	$\log\left(\frac{\mu}{1-\mu}\right)$	x	1	$\frac{1}{1+e^{\eta}}$
$\mathcal{N}(x\|\mu,\sigma) = \frac{1}{\sqrt{2\pi\sigma^2}}\,e^{-\frac{(x-\mu^2)}{2\sigma^2}}$	$\frac{1}{\sigma^2}\begin{pmatrix}\mu \\ -1/2\end{pmatrix}$	$\begin{pmatrix}x \\ x^2\end{pmatrix}$	$\frac{1}{\sqrt{2\pi}}$	$\sqrt{-2\eta_2}\,e^{\eta_1^2/(4\eta_2)}$
$\text{Gam}(x\|a,b) = \frac{b^a\,x^{a-1}\,e^{-b\,x}}{\Gamma(a)}$	$\begin{pmatrix}-b \\ a-1\end{pmatrix}$	$\begin{pmatrix}x \\ \log(x)\end{pmatrix}$	1	$\frac{(-\eta_1)^{\eta_2+1}}{\Gamma(\eta_2+1)}$
$\text{Bet}(x\|a,b) = \frac{x^{a-1}(1-x)^b}{B(a,b)}$	$\begin{pmatrix}a-1 \\ b-1\end{pmatrix}$	$\begin{pmatrix}\log(x) \\ \log(1-x)\end{pmatrix}$	1	$\frac{1}{B(\eta_1+1,\eta_2+1)}$
$\text{Dir}(\boldsymbol{x}\|\boldsymbol{\alpha}) = \Gamma\left(\sum_{i=1}^n \alpha_i\right)\prod_{i=1}^n \frac{x_i^{\alpha_i}}{\Gamma(\alpha_i)}$	$\begin{pmatrix}\alpha_1-1 \\ \vdots \\ \alpha_n-1\end{pmatrix}$	$\begin{pmatrix}\log(x_1) \\ \vdots \\ \log(x_n)\end{pmatrix}$	1	$\frac{\Gamma(n+\sum_{i=1}^n \eta_i)}{\prod_{i=1}^n \Gamma(\eta_i+1)}$

Table 2.1 Examples of distributions belonging to the exponential family. $\text{Bern}(x|\mu)$ is a Bernoulli distribution, which we discuss in Section 4.1.

are shown in Table 2.1. A fuller discussion of the exponential family can be found in Duda et al. (2001).

An important property of members of the exponential family is a relation between the natural parameters and an expectation. To see this we start from the normalisation condition

$$g(\boldsymbol{\eta}) \int h(\boldsymbol{x})\, e^{\boldsymbol{\eta}^{\mathsf{T}} \boldsymbol{u}(\boldsymbol{x})}\, d\boldsymbol{x} = 1.$$

Differentiating both sides with respect to the natural parameters, $\boldsymbol{\eta}$,

$$\boldsymbol{\nabla} g(\boldsymbol{\eta}) \int h(\boldsymbol{x})\, e^{\boldsymbol{\eta}^{\mathsf{T}} \boldsymbol{u}(\boldsymbol{x})}\, d\boldsymbol{x} + g(\boldsymbol{\eta}) \int h(\boldsymbol{x})\, e^{\boldsymbol{\eta}^{\mathsf{T}} \boldsymbol{u}(\boldsymbol{x})}\, \boldsymbol{u}(\boldsymbol{x})\, d\boldsymbol{x} = 0.$$

Rearranging and making use of the normalisation condition we find

$$\frac{-1}{g(\boldsymbol{\eta})} \boldsymbol{\nabla} g(\boldsymbol{\eta}) = g(\boldsymbol{\eta}) \int h(\boldsymbol{x})\, e^{\boldsymbol{\eta}^{\mathsf{T}} \boldsymbol{u}(\boldsymbol{x})}\, \boldsymbol{u}(\boldsymbol{x})\, d\boldsymbol{x}$$

or, equivalently,

$$-\boldsymbol{\nabla} \log\big(g(\boldsymbol{\eta})\big) = \mathbb{E}\left[\boldsymbol{u}(\boldsymbol{x})\right].$$

Thus, the expectation of $\boldsymbol{u}(\boldsymbol{x})$ can be found by taking a derivative of $\log\big(g(\boldsymbol{\eta})\big)$. Furthermore, the covariance and higher order moments can be obtained by taking higher order derivatives of $g(\boldsymbol{\eta})$.

A related property is to do with the maximum likelihood estimate of the natural parameters. Given a collection of independent data points, $\mathcal{D} = (\boldsymbol{x}_1, \boldsymbol{x}_2, \ldots, \boldsymbol{x}_n)$, the likelihood of the data is equal to

$$f(\mathcal{D}|\boldsymbol{\eta}) = \left(\prod_{i=1}^{n} h(\boldsymbol{x}_i)\right) g^n(\boldsymbol{\eta})\, e^{\boldsymbol{\eta}^{\mathsf{T}} \sum_{j=1}^{n} \boldsymbol{u}(\boldsymbol{x}_j)}.$$

The maximum likelihood estimator for the natural parameters, $\hat{\boldsymbol{\eta}}$, satisfies

$$\boldsymbol{\nabla} f(\mathcal{D}|\hat{\boldsymbol{\eta}}) = 0$$

or

$$-\nabla \log\big(g(\hat{\boldsymbol{\eta}})\big) = \frac{1}{n}\sum_{i=1}^{n} \boldsymbol{u}(\boldsymbol{x}_n).$$

From this we can solve for the maximum likelihood estimate $\hat{\boldsymbol{\eta}}$. Thus, the only statistics needed for the maximum likelihood estimate of the natural parameters are the components of the vector $\sum_{i=1}^{n} \boldsymbol{u}(\boldsymbol{x}_n)$. These are therefore *sufficient statistics* for any member of the exponential family. We discuss maximum likelihood estimators and sufficient statistics in more detail in Chapter 4.

Although there are a plethora of probability distributions, a few of them are so common and important that they simply can't be ignored. Of these the binomial and normal (or Gaussian) distributions stand out as particularly important. In the second rank sits the hypergeometric, Poisson, gamma, beta, multinomial, and Dirichlet distributions. You also need to be aware that some distributions can have very long tails and nasty properties. Although not very frequently met in practice, the Cauchy distribution is a particularly pretty example of a long-tailed distribution. We'll meet other long-tailed distributions along the way. Appendix B on page 445 provides tables showing the properties of some of the more commonly encountered distributions.

Additional Reading

A useful table of results for different distributions can be found in the compendium of mathematical formula by Abramowitz and Stegun (1964). If you know which distribution you are interested in, then performing a Google or Wikipedia search on the distribution is a very quick way to find most of the common relationships that you might be interested in.

Exercise for Chapter 2

Exercise 2.1 (answer on page 396)

What distribution might you use to model the following situations:

i. the proportion of the gross national product (GDP) from different sectors of the economy;
ii. the probability of three buses arriving in the next five minutes;
iii. the length of people's stride;
iv. the salary of people;
v. the outcome of a roulette wheel spun many times;
vi. the number of sixes rolled in a fixed number of trials; or
vii. the odds of a particular horse winning the Grand National.

(Note that there is not necessarily a single correct answer.)

Exercise 2.2 (answer on page 397)

Assume that one card is chosen from an ordinary pack of cards. The card is then replaced and the pack shuffled. This is repeated 10 times. What is the chance that an ace is drawn exactly three times?

Exercise 2.3 (answer on page 397)

Assume that a pack of cards is shuffled and 10 cards are dealt. What is the probability that exactly three of the cards are aces?

Exercise 2.4 (answer on page 397)

Show that the hypergeometric distribution, $\text{Hyp}(k|N, m, n)$, converges to a binomial distribution, $\text{Bin}(k|n, p)$ in the limit where N and m go to infinity in such a way that $m/N = p$. Explain why this will happen in words.

Exercise 2.5 (answer on page 398)

In the UK National Lottery players choose six numbers between 1 to 59. On draw day six numbers are chosen and the players who correctly guess two or more of the drawn numbers win a prize. The prizes increase substantially as you guess more of the chosen numbers. Write down the probability of guessing k balls correctly using the hypergeometric distribution and compute the probabilities for k equal 2 to 6.

Exercise 2.6 (answer on page 398)

Show that if $Y \sim \text{Exp}(1)$ then the random variable $X = \lambda \sqrt[k]{Y}$ (or $Y = (X/\lambda)^k$) is distributed according to the Weibull density

$$\text{Wei}(x|\lambda, k) = \frac{k}{\lambda}\left(\frac{x}{\lambda}\right)^{k-1} e^{-(x/\lambda)^k}.$$

Plot the Weibull densities $\text{Wei}(x|1, k)$ and the gamma distribution with the same mean and variance

$$\text{Gam}\left(\frac{\Gamma^2(1+1/k)}{\Gamma(1+2/k) - \Gamma^2(1+1/k)}, \frac{\Gamma(1+2/k)}{\Gamma(1+2/k) - \Gamma^2(1+1/k)}\right)$$

for $k = 1/2, 1, 2,$ and 5.

Appendix 2.A The Gamma Function

The gamma function, $\Gamma(z)$, occurs frequently in probability. For $\Re(z) > 0$ (i.e. the real part of z is positive), the gamma function is defined by the integral

$$\Gamma(z) = \int_0^\infty x^{z-1} e^{-x} dx. \tag{2.15}$$

Using integration by parts (assuming $z > 1$) we obtain the relationship

$$\begin{aligned}\Gamma(z) &= \left[-x^{z-1} e^{-x}\right]_0^\infty + (z-1)\int_0^\infty x^{z-2} e^{-x} dx \\ &= (z-1)\Gamma(z-1).\end{aligned}$$

$$\int_a^b u \frac{dv}{dx} dx = [u\, v]_a^b - \int_a^b v \frac{du}{dx} dx$$

Since

$$\Gamma(1) = \int_0^\infty e^{-x} dx = 1$$

Gauss much more sensibly defined the Pi-function

$$\Pi(z) = \int_0^{\infty} x^z e^{-x} dx$$

so that $\Pi(n) = n!$. Alas, history left us with the gamma function.

we find for integer, n, that $\Gamma(n) = (n-1)\Gamma(n-1) = (n-1)(n-2)\cdots 2 \cdot 1 = (n-1)!$. Thus, the gamma function is intimately related to factorials (although annoyingly with an offset of 1). As the gamma function increases so fast, it will tend to cause overflows or underflows in numerical calculations if used in its raw form. To overcome this it is usual to work with $\log(\Gamma(z))$. In C-based programming languages this function is called `lgamma`. It is also useful for computing factorials. For example, to compute the binomial coefficient $\binom{n}{k}$ we can use

$$\texttt{exp(lgamma(n + 1) − lgamma(k + 1) − lgamma(n − k + 1))}.$$

The gamma distribution is very well approximated by Stirling's approximation

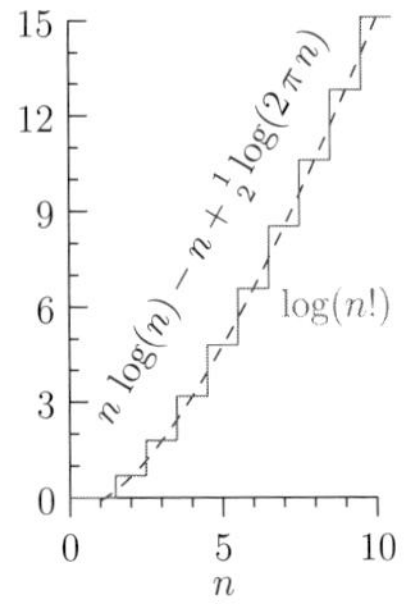

$$\Gamma(z) = \sqrt{\frac{2\pi}{z}} \left(\frac{z}{e}\right)^z \left(1 + O\left(\frac{1}{z}\right)\right).$$

For factorials this is equivalent to

$$n! \approx \left(\frac{n}{e}\right)^n \sqrt{2\pi n}.$$

In proving theorems involving factorials it is occasionally useful to use a bound provided by Stirling's approximation

$$\sqrt{2\pi}\, n^{n+1/2} e^{-n} \le n! \le e\, n^{n+1/2} e^{-n}.$$

Although the integral in Equation (2.15) is only defined for $\Re(z) > 0$, the gamma function can be defined everywhere in the complex plane except at $a = 0, -1, -2, \cdots$, where the function diverges. There are a number of relationships between the gamma function at different values that often help to simplify formulae. We have already seen that $\Gamma(a) = (a-1)\Gamma(a-1)$. Another important relationship is *Euler's reflection formula*

$$\Gamma(1-z)\,\Gamma(z) = \frac{\pi}{\sin(\pi z)}$$

and the *duplication formula*

$$\Gamma(z)\,\Gamma\left(z + \frac{1}{2}\right) = 2^{1-2z}\,\sqrt{\pi}\,\Gamma(2z).$$

For those readers with a more mathematical background, a formula, due to Hermann Hankel, which is occasionally useful is an integral form for the reciprocal of the gamma function in terms of a contour integration

$$\frac{1}{\Gamma(z)} = \frac{1}{2\pi i} \int_C x^{-z} e^x \,dx,$$

where C is a path that starts at $-\infty$ below the branch cut, goes around 0, and returns to $-\infty$ above the branch cut.

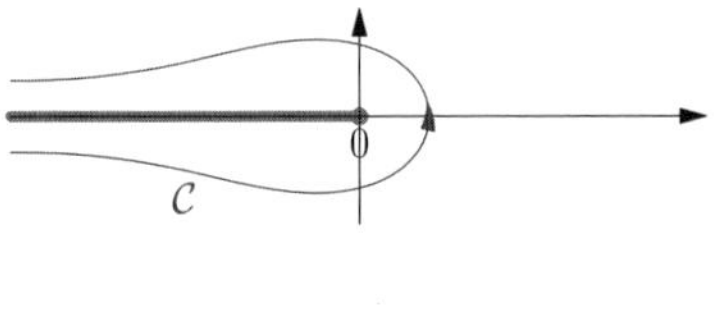

Derivatives of the gamma function arise when computing maximum likelihood estimates for distributions such as the gamma distribution. Rather than using the derivative of the gamma function it is more usual (and often more convenient) to consider the derivative of the (natural) logarithm of the gamma function (throughout this book we use $\log(x)$ to denote the natural logarithm of x). This is known as the *digamma function* which is usually written as

$$\psi(z) = \frac{\mathrm{d}\log(\Gamma(z))}{\mathrm{d}z} = \frac{1}{\Gamma(z)}\frac{\mathrm{d}\Gamma(z)}{\mathrm{d}z}.$$

Although the digamma function is not part of the standard C library, it exists in many numerical packages. The derivative of the digamma function is known as the trigamma function $\psi'(z)$, while higher order derivatives are known as polygamma functions. Like the gamma function the polygamma functions, and particularly the digamma function, have interesting properties that are well documented (e.g. in most tables of mathematical functions as well as in Wikipedia, etc.).

The incomplete gamma functions are defined for $\Re(a) > 0$ by

$$\gamma(a,z) = \int_0^z x^{a-1}\,\mathrm{e}^{-x}\,\mathrm{d}x, \qquad \Gamma(a,z) = \int_z^\infty x^{a-1}\,\mathrm{e}^{-x}\,\mathrm{d}x,$$

with $\gamma(a,z) + \Gamma(a,z) = \Gamma(a)$. The normalised incomplete gamma functions are defined as

$$P(a,z) = \frac{\gamma(a,z)}{\Gamma(a)}, \qquad Q(a,z) = \frac{\Gamma(a,z)}{\Gamma(a)},$$

with $P(a,z) + Q(a,z) = 1$.

Appendix 2.B The Beta Function

The beta function is defined (for $\Re(a), \Re(b) > 0$) through the integral

$$\mathrm{B}(a,b) = \int_0^1 x^{a-1}\,(1-x)^{b-1}\mathrm{d}x.$$

Remarkably, it is related to the gamma function through

$$\mathrm{B}(a,b) = \frac{\Gamma(a)\,\Gamma(b)}{\Gamma(a+b)}.$$

To prove this we start from

$$\Gamma(a)\,\Gamma(b) = \int_0^\infty u^{a-1}\,\mathrm{e}^{-u}\mathrm{d}u \int_0^\infty v^{b-1}\,\mathrm{e}^{-v}\mathrm{d}v = \int_0^\infty\int_0^\infty u^{a-1}\,v^{b-1}\,\mathrm{e}^{-u-v}\mathrm{d}u\,\mathrm{d}v.$$

We make the change of variables $u = z\,t$, $v = z\,(1-t)$, with $t \in [0,1]$ and $z \in [0,\infty]$ (in the u-v plane t determines the angle and z the magnitude). The Jacobian is given by

$$\frac{\partial(u,v)}{\partial(t,z)} = \begin{vmatrix} z & t \\ -z & (1-t) \end{vmatrix} = z\,(1-t) + z\,t = z.$$

With this change of variables we find

$$\Gamma(a)\,\Gamma(b) = \int_0^1 \int_0^\infty (z\,t)^{a-1}\,(z\,(1-t))^{b-1}\,e^{z\,t-z\,(1-t)}\,z\,dt\,dz$$
$$= \int_0^1 t^{a-1}\,(1-t)^{b-1} dt \int_0^\infty z^{a+b-1} e^{-z} dz = \mathrm{B}(a,b)\,\Gamma(a+b).$$

The incomplete beta function is defined as

$$\mathrm{B}_z(a,b) = \int_0^z x^{a-1}\,(1-x)^{b-1}\,dx.$$

The normalised incomplete beta function is defined as $I_z(a,b) = \mathrm{B}_z(a,b)/\mathrm{B}(a,b)$.

3

Monte Carlo

Contents

Modern computers give us the ability to simulate a probabilistic process. By repeating the simulation many times we can calculate many of the properties of interest. Performing multiple runs of probabilistic simulation goes by the name of *Monte Carlo*, drawing on the analogy of the gambler who repeatedly plays the same game (with the forlorn hope that their expected pay-off might be greater than 1). In this chapter, we discuss simple Monte Carlo techniques and ways of generating pseudo-random numbers.

Monte Carlo methods are a recurring theme throughout this book. In particular we revisit Monte Carlo techniques in Chapter 11 when we consider Markov chains.

3.1 Random Deviates

One of the most useful concepts in probabilistic modelling is that of *independent, identically distributed* random variables, know colloquially as iid variables (or even just *iids*). These are assumed to be independent samples drawn from some distribution f_X. If X_1 and X_2 are iid variables then

$$f(X_1, X_2) = f_X(X_1)\, f_X(X_2).$$

The distribution we choose depends on the application. If we are modelling the counts coming from a Geiger counter over different intervals of time, we might model this with a Poisson distribution. If we are modelling the outcome of a series of measurements of a continuous quantity we might choose to model this as *iids* coming from some normal distribution.

To simulate a probabilistic model we would like to be able to generate iid variables computationally. This is a very common requirement. We call such realisations of *iids* either *random deviates* or *random variates*. This chapter describes how such (random) deviates are generated in practice.

Having the ability to generate deviates allows us to simulate probabilistic models, a process known as *Monte Carlo simulation*. Often Monte Carlo simulations are sufficient to solve the problem we are interested in. In some cases it can even be the only method for solving a problem. However, we will frequently want to know more about a system than we can find out through a simulation. We might, for example, want to prove a theorem, or to understand how the system depends on some parameter. Very often there is a parameter that determines the size of the system and we want to know how the system scales with the size. An example of this would be to understand how the fluctuations scale for a sum of random variables as we increase the number of variables in the sum. We might want to obtain a result in some limiting case, for example when the system size grows to infinity (this is often a limit where the behaviour becomes simple). To obtain such results we need mathematical tools, but even in this case simulations can help build up an intuition and provide a sanity check for our theorem. Personally, I often struggle with understanding extremely abstract theorems so building a concrete model (even as a thought experiment) can help to clarify what the theorem tells us.

Before diving into techniques for generating random deviates we consider a couple of applications. These provide a very brief illustration of the use of Monte Carlo methods. We will see many more examples throughout this book.

3.1.1 Estimating Expectations

One of the most common tasks in probabilistic modelling is to compute an expectation of some function of a random variable, $\mathbb{E}\left[g(X)\right]$. The random variable may be discrete, in which case the expectation involves a (possibly infinite) sum, or continuous, in which case it involves an integral. Furthermore, the random variable may be multidimensional so that the expectation would involve either a multiple sum or multiple integral. Except in a few special cases these sums or integrals are intractable and we are forced to resort to numerical estimations. The special cases attract considerable attention, as it is extremely useful to get analytic answers, but they are nevertheless the exception rather than the rule. If the random variables are high dimensional then even numerical estimation of the sums or integrals can become intractable. However, the battle is not lost, for we can approximate any expectation by a sum over random samples

$$\mathbb{E}\left[g(X)\right] \approx \frac{1}{n}\sum_{i=1}^{n} g(X_i^*) \tag{3.1}$$

where the X_i^*s are *random deviates*. Here we struggle with notation, because once we have generated the random deviates they are no longer random variables. In any simulation the right-hand side of Equation (3.1) will give a fixed number, although from simulation to simulation the right-hand side will vary. Conceptually this is not difficult to understand, but using a notation that only distinguishes random versus non-random variables leads to some ambiguity in how to interpret Equation (3.1).

A commonly used notation to denote a random variable drawn from some distribution of $f_X(x)$ is

$$X \sim f_X. \tag{3.2}$$

We use the convention of dropping the variable names in the function argument. So, for example, a normally distributed random variable drawn from $\mathcal{N}(x|\mu, \sigma^2)$ would be written $X \sim \mathcal{N}(\mu, \sigma^2)$.

The accuracy of the approximation in Equation (3.1) depends on the probability distribution and the function $g(X)$. Normally the error in the approximation falls off as $1/\sqrt{n}$. However, this can fail, for example, if the variance of $g(X)$ is infinitely large or the probability distribution has very long/thick tails. Computing expectations in this way is often referred to as *Monte Carlo*, signifying the element of chance that appears in the calculation.

3.1.2 Monte Carlo Integration

Equation 3.1 very naturally replaces the expectation with a sum over random variates. This is often how we think of an expectation – i.e. as the mean result we obtain after very many trials. But if the random variable is continuous, then this expectation can be expressed as an integral. Turning this argument on its head, we can frequently perform an integral by summing over random variates, even when the problem we are interested in has nothing to do with probabilities. This use of probabilistic techniques for solving integrals is known as *Monte Carlo integration*. It is frequently used to perform high-dimensional integrals as the errors decrease because $1/\sqrt{n}$ independently of the dimensionality of the integral.

Example 3.1 Pi

We give a very simple example: to compute the area of a circle of radius 1. We can represent this area by the integral

$$A = \int_{-1}^{1}\int_{-1}^{1} [\![x^2 + y^2 \leq 1]\!] \, dx dy.$$

(Here we are using $[\![x^2 + y^2 \leq 1]\!]$ to denote the indicator function that is equal to 1 if the point (x, y) lies inside a unit circle and 0 otherwise.) To approximate this integral we generate n pairs of random numbers (X_i, Y_i) uniformly distributed in the square, $X_i, Y_i \sim \mathrm{U}(-1, 1)$, (where $U \sim \mathrm{U}(a, b)$ denotes a deviate uniformly distributed in the range from a to b, $\mathrm{U}(x|a, b) = [\![a \leq x \leq b]\!] / (b-a)$). We then count the number of instances such that $X_i^2 + Y_i^2 \leq 1$. The expected probability of landing in the circle is equal to the ratio of the area of a circle to the area of the square. Thus our estimate for A (the area of the unit square) is equal to four times the number of points lying in the circle divided by n. This procedure is illustrated in Figure 3.1. In this simulation, we know that each random deviate either lies in the circle or outside. Thus we can treat the event where a random deviate lies inside the circle as a true–false event (a so-called Bernoulli trial) with a success probability $p = \pi/4$. The probability that m random deviates out of n fall in the unit circle is given by the binomial distribution $\mathrm{Bin}(m|n, p)$. The expected number of random deviates falling in the unit circle is thus $n\,p$ while the standard deviation is $\sqrt{n\,p\,(1-p)}$. Since our estimate for A is $4\,m/n$, the expected value for the area of the circle, A, is $4\,p = \pi$, while the expected magnitude of the error is $4\sqrt{p\,(1-p)/n} \approx 1.642/\sqrt{n}$. Increasing the number of sample points by 4 decreases the size of the expected error by 2.

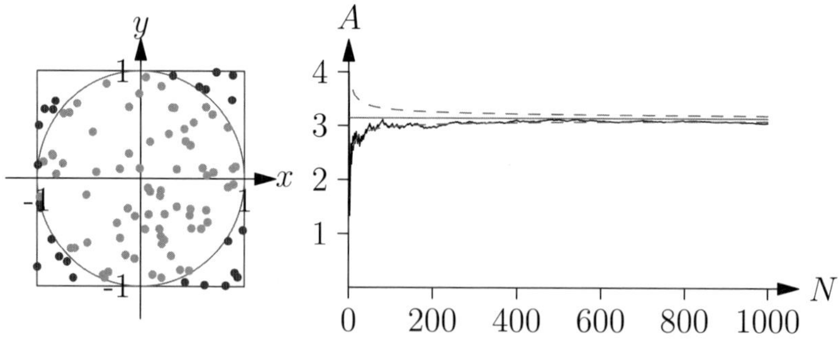

Figure 3.1 Illustration of Monte Carlo integration to compute the area of a circle. On the left we show 100 random points uniformly distributed in a square (points inside the circle are light dots, while those that lie outside the circle are dark). On the right we show our estimate of the area as a function of the number of sample points, n. The dashed lines show the expected errors around the true solution.

The accuracy of Monte Carlo integration is not great. You would certainly obtain higher accuracy much more quickly in one or two dimensions using standard integration algorithms. The disadvantage with standard techniques is that to compute an n-dimensional integral you have to compute the integrand at a number of points, but the computation of each integrand involves computing an $(n - 1)$-dimensional integral. Thus the work increases exponentially with the number of dimensions (halving the distance between sample points in each dimension requires looking at 2^n more points). Although Monte Carlo has a rather poor convergence rate it is independent of the number of dimensions. This is not to say that there isn't a price to pay for working in higher dimensions. You may well spend much longer generating random deviates. Nevertheless, if you want to estimate a multidimensional integral in three dimensions or above, Monte Carlo methods may be very competitive.

For problems such as the integral in Figure 3.1, you can do significantly better using non-random points, so-called *quasi-random numbers*. These are numbers which lie at regular lattice points. Quasi-random numbers can significantly improve the accuracy of Monte Carlo integration, but considerable care is necessary to prevent any bias caused by the underlying lattice. Details on how to use these techniques can be found in *Numerical Recipes* by Press et al. (2007).

3.2 Uniform Random Deviates

'Anyone who attempts to generate random numbers by deterministic means is, of course, living in a state of sin.'
– von Neumann

To perform Monte Carlo calculations we require a generator of random deviates. Alas, there is no algorithm which will generate truly random numbers since the existence of such an algorithm stops these numbers from being truly random. However, there are algorithms which generate very good *pseudo-random numbers*. That is, they generate random deviates with properties very close to those you would expect of real random numbers.

The obvious property which one would expect of a random deviate, X, is that it has the correct distribution. Another way of looking at this is that if we have a set of such random deviates, then a histogram (see Section 6.2) will approximate the probability mass/density distribution and will approach the distribution ever more closely as the data-set size increases. When working with deviates it is always a useful sanity check to plot a histogram of the deviates you are using and compare this with the true probability distribution you are trying to simulate.

However, this is not the only property we require of random deviates. Another important property is that if we draw pairs of deviates X_1 and X_2 then they are independent of each other, and more generally all deviates in a set should be mutually independent. Subtle correlation between deviates can occur when using some random number generators, which can cause systematic biases to occur in simulation results. Computer scientists tend to be much more fearful of this occurring than present-day experience justifies. In the dim and distant past there was a notorious random number generator that had exactly this error, but was nevertheless widely used. Most random number generators around today are much better behaved. Nevertheless, the fast default generators are not perfect and

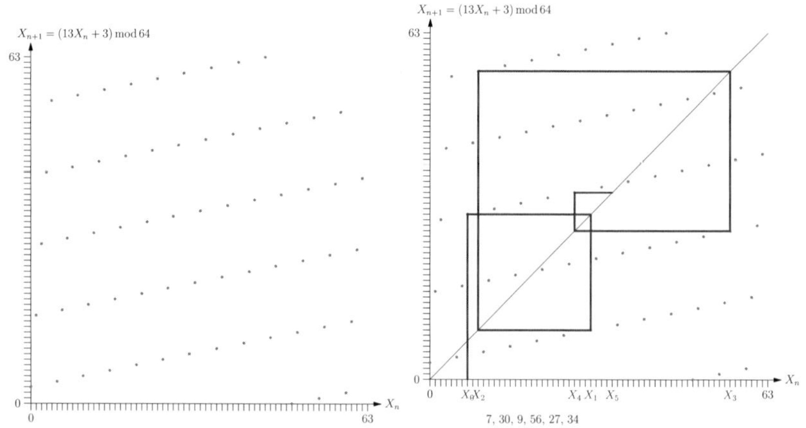

Figure 3.2 A simple linear-congruential update function, $f(X_n) = (13{*}X_n + 3) \bmod 64$. The left-hand graph shows the function. The right-hand graph shows a series of random numbers generated by iterating the update equation starting with $X_0 = 7$.

can lead to inaccuracies. This has only happened to me once in over 30 years of doing probabilistic simulations. In all the many other cases where I've suspected the random number generator, the fault has been a bug in the program or an error in my theory. It is very easy to get hold of superior pseudo-random number generators to the default ones available in C or Java, however, you pay a price in terms of run times. In many simulations the majority of machine time can be spent generating random numbers. Using a simple random number generator can significantly speed up performance.

Linear-Congruential Generators

The basic building block for generating pseudo-random numbers is a uniform random number generator. By far the most popular generators are linear-congruential generators which generate a series of numbers, $X_0, X_1, X_2, \cdots$ using the recursion relation

$$X_{n+1} = f(X_n) = (a\, X_n + c) \bmod m.$$

Usually m is taken to be the greatest representable integer plus 1 (e.g. 2^{31}). An example of the function $f(X_n)$ is shown in Figure 3.2 for a very small m. Note that provided we choose a, c, and m appropriately we should visit all integers between 0 and $m - 1$. As a consequence the distribution is uniform. Note also that this recursion relation is periodic, with a period m. Usually, this is large enough for us not to worry.

The choice of the constants a, c, and m is quite critical. The notoriety of random number generators arose from a routine that was widely used on IBM mainframes and which used a particularly poor choice of constants. Good constants give quite good pseudo-random numbers. These form the basis of most of the default pseudo-random number generators found in most programming languages. Superior pseudo-random number generators exist (i.e. those whose sequences pass much more stringent tests of randomness) and in a few applications it is necessary to resort to these. However, they tend to be considerably slower.

3.3 Non-Uniform Random Deviates

Often we are interested in obtaining random numbers from more complex distributions. Algorithms for generating deviates for most of the standard distributions are readily available. For example, they exist within Matlab or Octave (an open source clone of Matlab). Code is available in Press et al. (2007). A rich source of information on generating random deviates is Knuth (1997b). The GNU scientific library (gsl) provides a very extensive set of random number generators as does the latest version of the C++ standard library (C++11), R and python.

3.3.1 *Transformation Method*

There are a number of different approaches to generating numbers from non-uniform random distributions. The simplest and often fastest methods are the *transformation methods* which transform one type of random variable to another. These almost always start with pseudo-random numbers drawn from a uniform distribution in the interval $[0, 1]$. The transformation methods use the cumulative probability function

$$F_X(x) = \mathbb{P}\left(X \le x\right) = \int_{-\infty}^{x} f_X(t)\,\mathrm{d}t.$$

These functions go from 0 to 1. To obtain a random deviate from the distribution function $f_X(t)$, we can generate a uniform deviate $U \sim \mathrm{U}(0, 1)$ and use it to generate a new random number $X = F^{-1}(U)$. The probability of choosing a U such that the value of X lies in the interval between x and $x + \delta x$ is

$$\begin{aligned}
\mathbb{P}\left(x \le X < x + \delta x\right) &\overset{(1)}{=} \int_0^1 \left[\!\left[x \le F_X^{-1}(u) < x + \delta x \right]\!\right] \mathrm{d}u \\
&\overset{(2)}{=} \int_0^1 \left[\!\left[F_X(x) \le u < F_X(x + \delta x) \right]\!\right] \mathrm{d}u \\
&\overset{(3)}{=} \int_0^1 \left[\!\left[F_X(x) \le u < F_X(x) + f_X(x)\,\delta x + O(\delta x^2) \right]\!\right] \mathrm{d}u \\
&\overset{(4)}{\approx} \int_{F_X(x)}^{F_X(x)+f_X(x)\,\delta x} \mathrm{d}u \overset{(5)}{=} \left[u\right]_{F_X(x)}^{F_X(x)+f_X(x)\,\delta x} \overset{(6)}{=} f_X(x)\,\delta x
\end{aligned}$$

(1) The probability of this event is equal to the range of u where X falls into the required interval.
(2) Rewriting of the indicator function (since the cumulative probability $F_X(x)$ is monotonically increasing it is invertible).
(3) Performing a Taylor expansion of the upper limit (recall that $f_X(x)$ is the derivative of $F_X(x)$).
(4) Absorbing the limits of the integral while discarding terms of order $O\left(\delta x^2\right)$.
(5) Performing the integral.
(6) Substituting in the limits of integration.

Dividing through by δx and taking the limit $\delta x \to 0$ we find

$$\lim_{\delta x \to 0} \frac{\mathbb{P}\left(x \leq X < x + \delta x\right)}{\delta x} = f_X(x)$$

as we would hope (note that the approximation of ignoring terms of order $(\delta x)^2$ becomes exact in this limit). The proof assumes that $F_X(x)$ is differentiable and thus the density function exists. The method is actually more general and works equally well for discrete distributions where $F_X(x)$ changes in steps. This procedure is illustrated schematically in Figure 3.3: we chose a uniform deviate U in the interval $[0, 1]$ and used the inverse of the cumulative probability density to map this to a value X with the desired density.

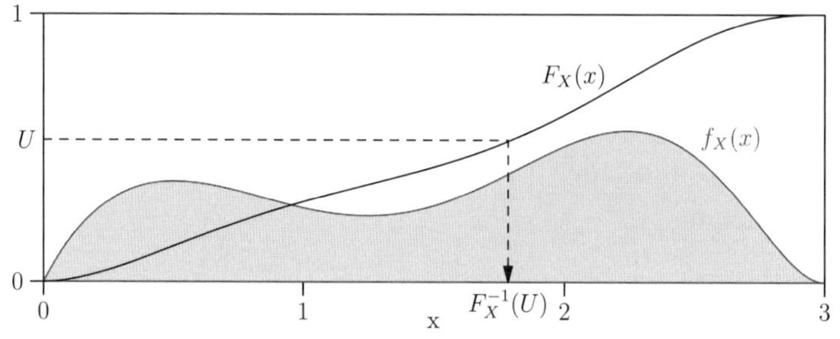

Figure 3.3 Example of using a transformation method for generating a random variate. We show a probability density $f_X(x)$ and the corresponding cumulative probability distribution $F_X(x)$. Given a random variate U from uniform distribution between 0 and 1, we can obtain a random variate of $f_X(x)$ using $F_X^{-1}(U)$.

Example 3.2 Exponential Deviates

As an example, consider drawing a number from the exponential distribution defined, for $x \geq 0$, as

$$\text{Exp}(x|b) = b\,\text{e}^{-b\,x}.$$

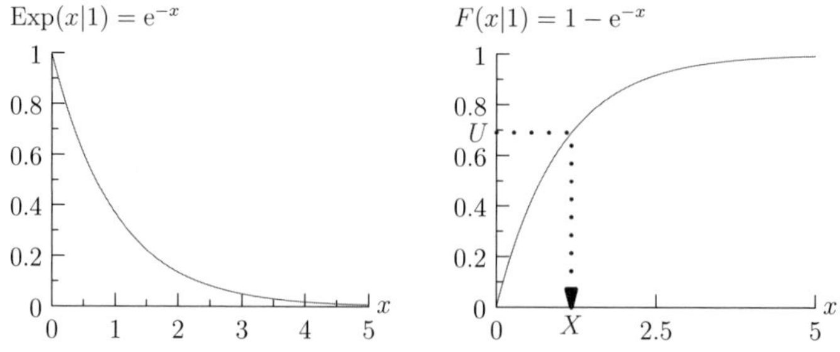

Figure 3.4 Exponential distribution for $b = 1$. The left-hand graph shows the probability density function, while the right-hand graph shows the cumulative probability distribution. We show an example of choosing $U \sim \text{U}(0, 1)$ and using the inverse cumulative distribution function (CDF) to generate a deviate X from an exponential distribution.

The cumulative probability function is $F_X(x|b) = 1 - e^{-bx}$. Inverting this we obtain a function $F_X^{-1}(x|b) = -b^{-1}\log(1-x)$. The density and cumulative probability distribution for the case when $b = 1$ are shown in Figure 3.4. To obtain an exponential deviate, X, from a uniform deviate $U \sim \mathrm{U}(0,1)$, we use $X = -b^{-1}\log(1-U)$. (Mathematically, we could also write $X = -b^{-1}\log(U)$, but in a computer implementation, where U is implemented as `rand()/double(RAND_MAX)`, you have to be wary since `rand()` can return 0 which will crash your program rather unexpectedly. Using $X = -b^{-1}\log(1-U)$ avoids this as U is always strictly less than 1.)

Example 3.3 Normal Deviates

Unfortunately, not all cumulative probability functions are easily invertible. For example, the cumulative distribution for a zero mean unit variance normal distribution is given by the error function

$$\int_{-\infty}^{x} e^{-z^2/2} \frac{dz}{\sqrt{2\pi}} = \frac{1}{2} + \frac{1}{2}\mathrm{erf}\left(\frac{x}{2}\right).$$

We could obtain a normal deviate by inverting the error function (which we could do numerically using Newton's method). However, this is slow. Instead, we use the trick of working in two dimensions. We start by generating two deviates, X_1 and X_2, uniformly distributed over the circle defined by $X_1^2 + X_2^2 \leq 1$. These can be obtained by generating pairs, (X_1, X_2), of uniformly distributed deviates in the interval from -1 to 1 and rejecting all pairs with $X_1^2 + X_2^2 > 1$. We can then construct random deviates $\Theta = \arctan(X_1/X_2)$ and $S = \sqrt{X_1^2 + X_2^2}$, corresponding to the angle made by the vector (X_1, X_2) and the x-axis, and the radius of the point from the origin. These are clearly independent of each other as the angle does not depend on the distance from the origin of (X_1, X_2) (see Figure 3.5). The angle deviate Θ is uniformly distributed between 0 and 2π, while the radius S is distributed according to

$$f_S(s) = 2\,s\,[\![0 \leq s \leq 1]\!]$$

(since the probability of generating a point at radius s is proportional to the circumference of the circle, which is equal to $2\pi s$). Define the random variable $R = 2\sqrt{-\log(S)}$ (or $S = e^{-R^2/4}$) so that

$$f_R(r) = \left|\frac{d\,s(r)}{dr}\right| f_S(s(r)) = \frac{r}{2}e^{-r^2/4} \times 2\,e^{-r^2/4} = r\,e^{-r^2/4}.$$

If we now define the random variables $Y_1 = R\cos(\Theta)$ and $Y_2 = R\sin(\Theta)$ so that $R^2 = Y_1^2 + Y_2^2$, then the probability density of Y_1 and

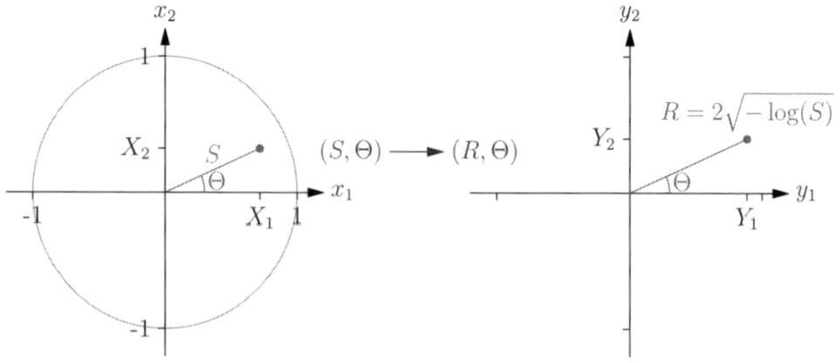

Figure 3.5 Illustration of Box–Muller's algorithm for generating normal deviates. Starting from two uniformly generated deviates lying in the unit circle, we rescale the radius to obtain a normal deviate in two dimensions.

Y_2 are given by

$$f_{Y_1,Y_2}(y_1 = r\cos(\theta), y_2 = r\sin(\theta)) = \left|\frac{\partial(y_1, y_2)}{\partial(r,\theta)}\right|^{-1} f_R(r)\, f_\Theta(\theta),$$

where

$$\left|\frac{\partial(y_1, y_2)}{\partial(r,\theta)}\right| = \begin{vmatrix} \frac{\partial y_1}{\partial r} & \frac{\partial y_1}{\partial \theta} \\ \frac{\partial y_2}{\partial r} & \frac{\partial y_2}{\partial \theta} \end{vmatrix} = \begin{vmatrix} \cos(\theta) & -r\sin(\theta) \\ \sin(\theta) & r\cos(\theta) \end{vmatrix} = r\left(\cos^2(\theta) + \sin^2(\theta)\right) = r$$

is the Jacobian needed to ensure the conservation of probability. Using $f_R(r) = r\,\mathrm{e}^{-r^2/2}$ and $f_\Theta(\theta) = 1/2\pi$ we find

$$f_{Y_1,Y_2}(y_1, y_2) = \frac{1}{2\pi}\mathrm{e}^{-(y_1^2 + y_2^2)/2} = \left(\frac{1}{\sqrt{2\pi}}\mathrm{e}^{-y_1^2/2}\right)\left(\frac{1}{\sqrt{2\pi}}\mathrm{e}^{-y_2^2/2}\right).$$

That is, Y_1 and Y_2 are independent normally distributed variables. Note that $Y_1 = (R/S)\,X_1$ and $Y_2 = (R/S)\,X_2$.

This method for generating random normal deviates is known as the *Box–Muller* method and is widely used. To recap, two uniform deviates X_1, $X_2 \sim \mathrm{U}(0, 1)$ are drawn. If $S^2 = X_1^2 + X_2^2 > 1$ the deviates are rejected and a new pair drawn until a pair is found with $S^2 \leq 1$. A constant $C = R/S = \sqrt{-2\log(S^2)/S^2}$ is computed (note that to speed up the algorithm we don't compute S, which requires taking the square root, but rather work with S^2), and then two normal deviates are computed as $Y_1 = C\,X_1$ and $Y_2 = C\,X_2$.

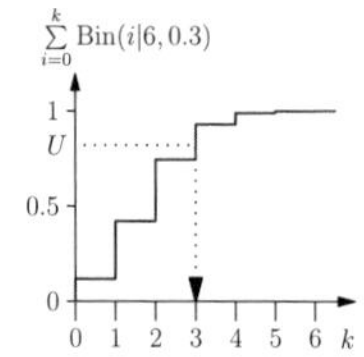

When using the transformation method for a discrete distribution, the CDF consists of a number of steps and its inverse can map many values to the same integer. When there are a small number of discrete values it is relatively efficient to use the transformation method. In a few cases we can invert the CDF efficiently for any number of outcomes, as we explore in Exercise 3.3.

3.3.2 Rejection Sampling

The transformation method is often very efficient, but requires that we can compute the cumulative probability function and invert it efficiently. These are strong restrictions which make the transformation method rather limited. A more general technique is the rejection method. Again we consider generating a random deviate, $X \sim f_X(x)$, but we now perform two steps. We firstly generate a random deviate $Y \sim g_Y(y)$ from and appropriate 'proposal' distribution. We require of the distribution $g_Y(y)$ to be such that for all x

$$c\, g_Y(x) \geq f_X(x)$$

for some known constant $c > 1$. In the second step we accept the random deviate with a probability $f_X(Y)/(c\, g_Y(Y))$, otherwise we reject it. We illustrate this in Figure 3.6. In this example, we want to generate a deviate from the distribution $f_X(x)$ defined in the interval $(0, 3)$. Since, in our example, $f_X(x) < 2\,\mathrm{U}(x|0, 3) = 2/3$, we can generate a random deviate $U \sim \mathrm{U}(0, 3)$ and use a constant $c = 2$. We accept this number with a probability $p = f_X(U)/(c\,\mathrm{U}(U|0, 3)) = 3\, f_X(U)/2$ (which we achieve by generating a second deviate $U' \sim \mathrm{U}(0, 1)$ and accepting $X = U$ if $U' \leq p$), otherwise we reject this number and try again.

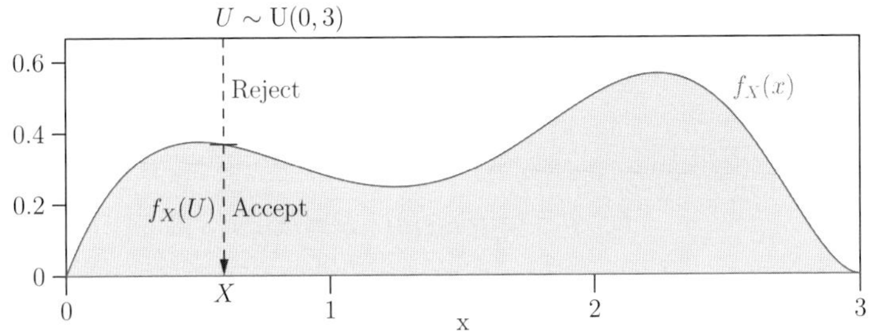

Figure 3.6 Illustration of rejection method. A uniform deviate $U \sim \mathrm{U}(0, 3)$ is generated. This is then accepted with a probability $f_X(U)/(2\,\mathrm{U}(U|0, 3))$. Note that in this example $f_X(X) \leq 2\mathrm{U}(x|0, 3)$ for all x.

The probability density of an accepted deviate is proportional to

$$g_Y(Y) \times \frac{f_X(Y)}{c\, g_Y(Y)} = \frac{f_X(Y)}{c},$$

i.e. the probability density for selecting Y times the probability of accepting Y. The constant of proportionality is equal to c which is the reciprocal of the acceptance probability. Thus, the deviates have the correct probability density. Furthermore, the rejection rate is equal to $1/c$, so by making $g_Y(y)$ as close as possible to $f_X(x)$ (in the sense that we can choose a c as close as possible to 1), we reduce the probability of rejection.

Example 3.4 Normal Deviates Again

As an example of the rejection method, yet another way to generate a normal deviate is to generate a Cauchy deviate, $Y \sim \mathrm{Cau}$, and then reject it with a probability $\mathcal{N}(Y|0, 1)/(1.53\,\mathrm{Cau(Y)})$ – see Exercise 3.1 to see how to generate random Cauchy deviates. We show the rescaled Cauchy distribution and the normal density distribution in Figure 3.7. This method has a slightly higher rejection rate than the Box–Muller method and involves slightly more mathematical operations per deviate. Rejection methods are, however, commonly used for generating gamma, Poisson, and binomial deviates, among many others.

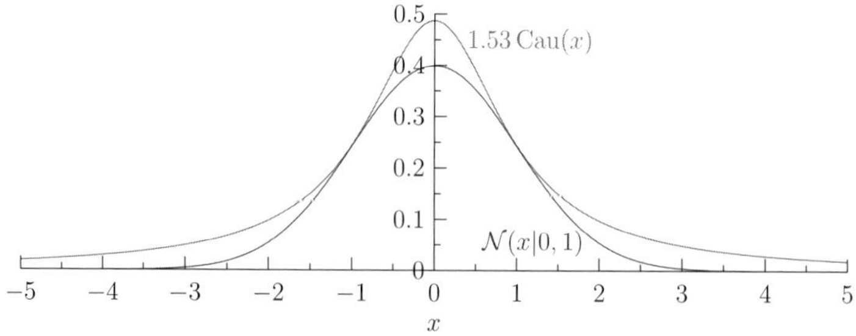

Figure 3.7 Normal distribution and rescaled Cauchy distribution for generating a normal deviate using the rejection method.

Although generally applicable, the rejection method requires a proposal distribution, $g_Y(y)$, which accurately approximates the true function $f_X(x)$. This is not always easy to find. In particular, in high dimensions the proposal distribution $g_{\boldsymbol{Y}}(\boldsymbol{y})$ may be a very poor approximation to the function of interest $f_{\boldsymbol{X}}(\boldsymbol{x})$ and we end up with a very high rejection rate. In these situations a preferred method is the Markov Chain Monte Carlo (MCMC) method. We discuss this in Section 11.2 on page 317.

3.3.3 Multivariate Deviates

Deviates for multivariate distributions are often generated by creating deviates for univariate distributions and transforming them appropriately. For example, to obtain multinomial deviates from a distribution $\boldsymbol{N} \sim \text{Mult}(n, \boldsymbol{p})$ we first generate a binomial deviate $N_1 \sim \text{Bin}(n, p_1)$, then we generate a second binomial deviate $N_2 \sim \text{Bin}(n - N_1, p_2/(1 - p_1))$; this is repeated, each time generating a binomial deviate

$$N_i \sim \text{Bin}\left(n - N_1 - N_2 \cdots - N_{i-1}, \frac{p_i}{1 - p_1 - p_2 \cdots - p_{i-1}}\right).$$

We can stop either when we only have one class left or when we have assigned all n trials available. If there are some classes with a probability of assignment $p_i \ll 1/n$ then it can be worthwhile to reorder the classes so that those with low probability occur at the end. Then it is more likely that we will have assigned all n outcomes before we reach a class with very small p_i, so we can stop the algorithm early.

To generate n-dimensional multivariate normal deviates, $\boldsymbol{X} \sim \mathcal{N}(\boldsymbol{\mu}, \boldsymbol{\Sigma})$, we firstly generate n normal deviates and then transform the deviates, usually by multiplying them by the Cholesky decomposition of the correlation matrix $\boldsymbol{\Sigma}$ and then adding the means $\boldsymbol{\mu}$. Details are given in Section 5.6 on page 96.

Finally, to generate n-dimensional Dirichlet deviates, $\boldsymbol{X} \sim \text{Dir}(\boldsymbol{\alpha})$, we first generate n gamma-distributed deviates $Z_i \sim \text{Gam}(\alpha_i, 1)$ and then normalise them

$$X_i = \frac{Z_i}{\sum_{j=1}^{n} Z_j}.$$

The X_is are Dirichlet deviates. To show this we note that

$$f_Z(z) = \prod_{i=1}^{n} \text{Gam}(z_i|\alpha_i, 1) = \left(\prod_{i=1}^{n} \frac{z_i^{\alpha_i - 1}}{\Gamma(\alpha_i)}\right) \text{e}^{-\sum_{i=1}^{n} z_i}.$$

We denote $\sum_{j=1}^{n} Z_j = Y$ so that $X_i = Z_i/Y$. As a consequence of the normalisation not all X_is are independent. We are free to eliminate one component, $X_n = 1 - \sum_{i=1}^{n-1} X_i$. We can then compute the probability distribution of the random variables $(X_1, \ldots, X_{n-1}, Y)$ using the usual rules for the change of variable. To compute the Jacobian between $\boldsymbol{Z}$ and $(\boldsymbol{X}, Y)$ we note that

$$\frac{\partial z_i}{\partial x_j} = y \, [\![i = j]\!] \qquad\qquad \frac{\partial z_i}{\partial y} = \frac{1}{\frac{\partial y}{\partial z_i}} = 1.$$

The Jacobian is

$$J = \left| \frac{\partial(z_1, \ldots, z_n)}{\partial(x_1, \ldots, x_{n-1}, y)} \right| = \begin{vmatrix} y & 0 & \cdots & 0 & 0 \\ 0 & y & \cdots & 0 & 0 \\ \vdots & \vdots & \ddots & \vdots & \vdots \\ 0 & 0 & \cdots & y & 0 \\ 1 & 1 & \cdots & 1 & 1 \end{vmatrix} = y^{n-1}.$$

Then,

$$\begin{aligned} f_{X,Y}(\boldsymbol{x}, y) = \left| \frac{\partial(z_1, \ldots, z_n)}{\partial(x_1, \ldots, x_{n-1}, y)} \right| f_Z(y\,\boldsymbol{x}) &= y^{n-1} \left(\prod_{i=1}^{n} \frac{(y\, x_i)^{\alpha_i - 1}}{\Gamma(\alpha_i)}\right) \text{e}^{-y} \\ &= \left(\prod_{i=1}^{n} \frac{x_i^{\alpha_i - 1}}{\Gamma(\alpha_i)}\right) y^{\alpha_0 - 1} \, \text{e}^{-y}, \end{aligned}$$

where $\alpha_0 = \sum_{i=1}^{n} \alpha_i$. To compute the distribution of $\boldsymbol{X}$ we marginalise out Y

$$\begin{aligned} f_X(\boldsymbol{x}) &= \int_0^\infty f_{X,Y}(\boldsymbol{x}, y) \, \text{d}y \\ &= \prod_{i=1}^{n} \frac{x_i^{\alpha_i - 1}}{\Gamma(\alpha_i)} \int_0^\infty y^{\alpha_0 - 1} \, \text{e}^{-y} \, \text{d}y = \Gamma(\alpha_0) \prod_{i} \frac{x_i^{\alpha_i - 1}}{\Gamma(\alpha_i)} = \text{Dir}(\boldsymbol{\alpha}). \end{aligned}$$

Monte Carlo techniques have become a staple of scientific computing. The techniques used are not difficult to understand. However, there are a lot of optimisation tricks used to generate good-quality deviates quickly. Lots of people have dedicated a substantial amount of time to writing good generators. The rest of us can enjoy the fruits of their labours. The exception to this is with high-dimensional systems (i.e. those involving a large number of coupled variables) where it is necessary to tailor the code to your specific problem. This is the area of MCMC, which we visit in Chapter 11.

Additional Reading

Donald Knuth provides a good description of generating random numbers in the second volume of his epic *The Art of Computer Programming* (Knuth, 1997b). If you are interested in low-level practical algorithms for generating random deviates, a good starting place is Fishman (1996) or Devroye (1986). A nice introduction to generating random deviates together with practical algorithms for a wide range of distributions is given in Press et al. (2007).

Exercise for Chapter 3

Exercise 3.1 (answer on page 399)

Using the transformation method show how you could generate a Cauchy deviate. The density of the Cauchy distribution is given by

$$\text{Cau}(x) = \frac{1}{\pi(1+x^2)}$$

and the cumulative distribution is

$$F(x) = \int_{-\infty}^{x} \text{Cau}(y)\,\text{d}y = \frac{1}{2} + \frac{\arctan(x)}{\pi}.$$

Exercise 3.2 (answer on page 399)

Implement a normal deviate generator:

i. using the transformation method by numerically inverting the erfc function (see section 5.4 on page 90) using Newton–Raphson's method (alternatively, the third edition of numerical recipes (Press et al., 2007) provides an inverse of the CDF for a normal distribution);
ii. using the standard Box–Muller method;
iii. using the rejection method from Cauchy deviates.

Time how long it takes each implementation to generate 10^6 random deviates.

Exercise 3.3 (answer on page 400)

In simulating a simple model of evolution, one is often faced with mutating each allele in a gene with a small probability $p \ll 1$ (this task also occurs in running a genetic algorithm). It is far quicker to generate a random deviate giving the distance between mutations than to decide for each allele whether to mutate it. The distance between mutations is given by the geometric distribution. Denoting by $K > 0$ the distance to the next allele that is mutated then

$$\mathbb{P}\left(K = k\right) = \text{Geo}(k|p) = p\,(1-p)^{k-1}.$$

Using the transformation method show how to generate a deviate $K \sim \text{Geo}(p)$ starting from a uniform deviate $U \sim \text{U}(0,1)$.

4

Discrete Random Variables

Contents

This chapter focuses on probability distributions for discrete random variables and in particular the binomial, Poisson, and multinomial distributions. On our way, we will come across maximum likelihood estimators, sufficient statistics, moments, cumulants, generating functions, and characteristic functions.

To get the ball rolling, in Chapter 2 I defined a number of distributions and gave their properties. In this and Chapter 5, we derive those properties from first principles, using mathematical tools that have a very wide range of applications.

4.1 Bernoulli Trials

Jacob Bernoulli wrote Ars Conjectandi *(The Art of Conjecture), which was published after his death in 1713. It is widely regarded as laying the foundations of probability and combinatorics.*

A random variable, X, that has two possible outcomes, $\{0, 1\}$, is known as a *Bernoulli trial*, named after Jacob (Jacques) Bernoulli (1654–1705), who was one of the first mathematicians to consider problems involving binary random variables. The probability mass function for a binary random variable is

$$\text{Bern}(X|\mu) = \mu^X (1-\mu)^{1-X} = \begin{cases} \mu & \text{if } X = 1 \\ 1-\mu & \text{if } X = 0. \end{cases} \tag{4.1}$$

The term $\mu^X (1-\mu)^{1-X}$ looks rather intimidating, but as $X \in \{0, 1\}$ it really just says that the probability of $X = 1$ is μ while the probability of $X = 0$ is $1 - \mu$. The mean and variance are given by

$$\mathbb{E}_X\left[X\right] = 1 \times \mu + 0 \times (1-\mu) = \mu$$
$$\mathbb{V}\text{ar}_X\left[X\right] = \mathbb{E}_X\left[X^2\right] - \mathbb{E}_X\left[X\right]^2 = \left(1^2 \times \mu + 0^2 \times (1-\mu)\right) - \mu^2 = \mu(1-\mu).$$

Examples of Bernoulli trials abound. For example, in tossing a coin, if we assign 1 to the event 'heads' and 0 to the event 'tails' we have a Bernoulli trial. Bernoulli trials where there is an equal probability of either outcome and where the random variables take values of 1 and -1 are sometimes referred to as Rademacher variables.

Suppose we have an experiment with two possible outcomes: success or failure. We can assign 1 to the event that the experiment was successful and 0 otherwise. This is a Bernoulli trial. The success probability, μ, provides a measure of performance of the experiment. We often want to measure the performance, μ, assuming we are given some data $\mathcal{D} = (X_1, X_2, \ldots, X_n)$ describing n independent trials where $X_i \in \{0, 1\}$. The collection of data is often called a 'data set', although it is more properly described as a multiset (i.e. a collection where order doesn't matter, but where we take into account repetitions). How can we estimate the success probability from our data? One way to answer this question is to choose the value of μ which maximises the likelihood of the data. This is known as the *maximum likelihood estimator*. The likelihood of the data given μ is

$$\mathbb{P}\left(\mathcal{D}|\mu\right) = \prod_{i=1}^{n} \text{Bern}(X_i|\mu) = \prod_{i=1}^{n} \mu^{X_i} (1-\mu)^{1-X_i}.$$

Here we are making the assumption that the data is independent – in most well-designed experiments we would hope this to be a very good approximation. This equation looks rather difficult to deal with. However, maximising this quantity is equivalent to maximising the logarithm of this quantity, because the logarithm is a monotonically increasing function. Taking logarithms we have

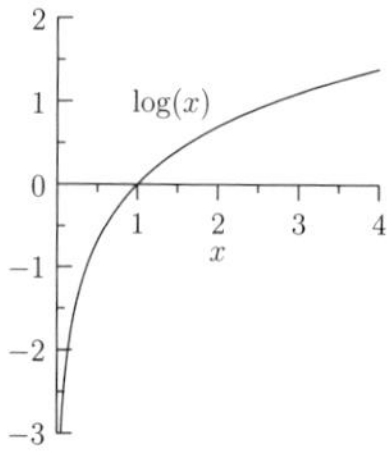

$$\begin{aligned}\log(\mathbb{P}\left(\mathcal{D}|\mu\right)) &= \sum_{i=1}^{n} \left(X_i \log(\mu) + (1-X_i)\log(1-\mu)\right) \\ &= K\log(\mu) + (n-K)\log(1-\mu)\end{aligned}$$

where $K = \sum_{i=1}^{n} X_i$ is the number of successes. To find the maximum likelihood, we set the derivative of the log likelihood with respect to μ equal to zero

$$\left.\frac{\mathrm{d}\log(\mathbb{P}(\mathcal{D}|\mu))}{\mathrm{d}\mu}\right|_{\mu=\hat{\mu}} = \frac{K}{\hat{\mu}} - \frac{n-K}{1-\hat{\mu}} = 0.$$

Solving for $\hat{\mu}$, we find $\hat{\mu} = K/n$. This is the intuitive answer, it is the empirical success probability given a sample of size n. We follow a frequently used convention of denoting an estimator of a quantity with a hat.

The maximum likelihood estimator depends only on the number of successful trials $K = \sum_{i=1}^{n} X_i$. It does not depend, for example, on the order of successful trials (this is hardly surprising given our assumption that the trials were all independent). This sum, K, is an example of a *sufficient statistic*. That is, it is a statistic of the data that captures all the information in a data set required to determine the maximum likelihood estimators of the parameters of the distribution. The estimator $\hat{\mu} = K/n$ is said to be a *consistent estimator* because, in the limit $n \to \infty$, it approaches the true value of μ. It is also an *unbiased estimator* because its expected value is equal to μ

$$\mathbb{E}\left[\hat{\mu}\right] = \frac{1}{n}\mathbb{E}\left[K\right] = \frac{1}{n} n\,\mu = \mu.$$

(Note that until we make a measurements our estimators are random variables – we are our breaking our self-imposed convention by not writing these as capitals.) You might think that all estimators are going to be consistent and unbiased, but that is not always true, particularly when trying to estimate non-linear functions of a random variable (e.g. its skewness).

The maximum likelihood estimate seems to have all the properties you might wish of an estimator, but it has one deficiency. If μ is very small (or very large) then for small samples we are likely to find $K = 0$ (or $K = n$), leading to an estimate of $\hat{\mu} = 0$ (or $\hat{\mu} = 1$). In many ways this is not a bad estimate, but sometimes there is a significant difference between an event occurring rarely and an event not occurring at all. If we have a prior belief that the event might occur then we can use a Bayesian approach to obtain a probability distribution for μ. We pursue this approach in Chapter 8.

4.2 Binomial Distribution

In the previous section we obtained an estimator for μ based on the number of successful Bernoulli trials. We can take another viewpoint of the experiment and ask what is the probability of K successes in n Bernoulli trials given a success probability of p (note that in the Bernoulli distribution, $\text{Bern}(X|\mu)$, the parameter of the distribution coincided with the mean so it makes sense to call it μ, while in the binomial distribution, $\text{Bin}(k|n, p)$, the mean number of successes is $n\,p$, therefore we have given the parameter a different name). Treating the experiment as a sequence of trials, we can represent the space of all possible events, Ω, as a tree where each branch represents a possible outcome. This is

Figure 4.1 Example of the possible results arising from a series of Bernoulli trials together with the probability of each outcome. Note that the frequency of occurrence of $p^i\,(1-p)^{n-i}$ after n trials is given by the binomial coefficient $\binom{n}{i}$.

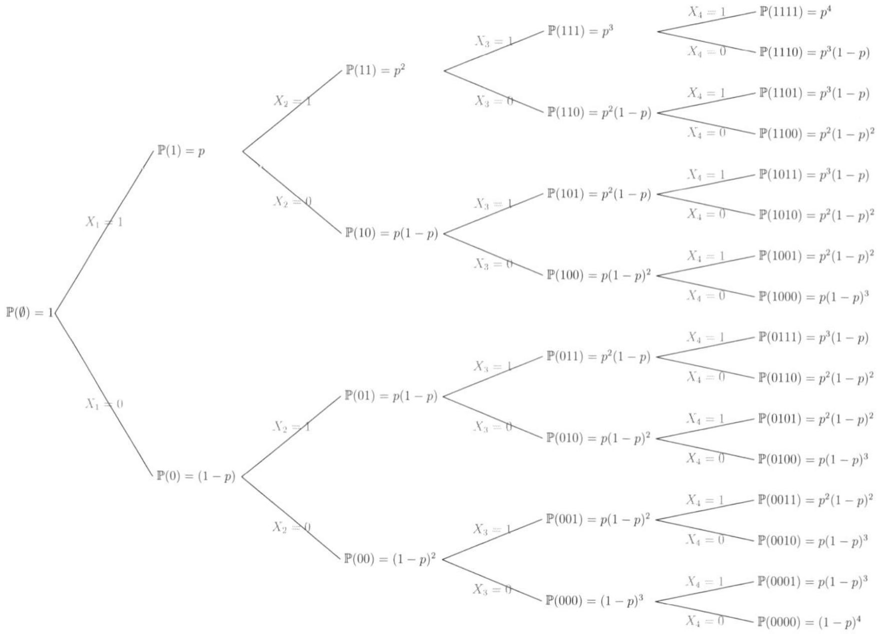

shown in Figure 4.1 for an experiment with four binary trials. The number of trials that end up with k successes out of n trials is given by the binomial coefficient $\binom{n}{k}$ ('n choose k'), which gives rise to the binomial distribution

$$\text{Bin}(k|n,p) = \binom{n}{k} p^k\,(1-p)^{n-k}.$$

This and the normal distribution are the most frequently occurring distributions in probability theory.

4.2.1 Statistical Properties of the Binomial Distribution

The binomial distribution appears awkward to work with. It involves both combinatorial factors and powers. However, there is a simple way to compute many of its statistical properties. That is, to use a *generating function*. We start by considering the *moment generating function*. The mth *moment*, μ_m, of a random variable K is defined as

$$\mu_m = \mathbb{E}_K\left[K^m\right].$$

The first moment is therefore the mean. We define the moment generating function

$$M(l) = \mathbb{E}_K\left[e^{K\,l}\right] = \mathbb{E}_K\left[1 + K\,l + \frac{(K\,l)^2}{2!} + \frac{(K\,l)^2}{2!} + \cdots\right]$$

where the last inequality we obtain from a Taylor expansion of the exponential. The name, *moment generating function*, comes from the fact that we can compute the *mth-order moment* by computing the mth derivative of the moment generating function and setting l to zero

$$\mu_m = \mathbb{E}_K\left[K^m\right] = \left.\frac{\mathrm{d}^m M(l)}{\mathrm{d}\, l^m}\right|_{l=0} = M^{(m)}(0),$$

which follows from taking derivatives of the exponential. Note that the moments are the coefficients of the Taylor expansion of $M(l)$. This approach is powerful because we can obtain a convenient closed form expression for the moment generating function of the binomial distribution

$$\begin{aligned} M(l) = \mathbb{E}_K\left[\mathrm{e}^{K\,l}\right] &= \sum_{k=0}^{n} \mathrm{Bin}(k|n,p)\,\mathrm{e}^{k\,l} \\ &= \sum_{k=0}^{n} \binom{n}{k} p^k\,(1-p)^{n-k}\,\mathrm{e}^{k\,l}. \end{aligned}$$

To compute this sum we recall the binomial expansion

$$(a+b)^n = \sum_{k=0}^{n} \binom{n}{k} a^k\, b^{n-k}.$$

According to Arthur Conan Doyle, Sherlock Holmes' arch-rival Professor Moriarty wrote A Treatise on the Binomial Theorem *at the age of 21. It was, of course, Newton who discovered the binomial theorem.*

Identifying $a = p\,\mathrm{e}^l$ and $b = 1-p$, we can write the moment generating function as

$$M(l) = \left(1 + p\,(\mathrm{e}^l - 1)\right)^n = U^n(l), \tag{4.2}$$

where $U(l) = \left(1 + p(\mathrm{e}^l - 1)\right)$. Now,

$$\begin{aligned} M'(l) &= n\,p\,\mathrm{e}^l U^{n-1}(l), & M'(0) &= n\,p, \\ M''(l) &= n\,p\,\mathrm{e}^l U^{n-1}(l) + n\,(n-1)p^2\mathrm{e}^{2l}U^{(n-2)}(l), & M''(0) &= n\,p + n\,(n-1)\,p^2, \end{aligned}$$

where we have used $U(0) = 1$. Thus,

$$\begin{aligned} \mathbb{E}_K\left[K\right] &= n\,p \\ \mathbb{V}\mathrm{ar}_K\left[K\right] &= \mathbb{E}_K\left[K^2\right] - \left(\mathbb{E}_K\left[K\right]\right)^2 = n\,p\,(1-p) \end{aligned}$$

as advertised in Section 2.1.1.

The moment generating function is intimately related to the *characteristic function* which is defined as the Fourier transform of the probability distribution or

$$\phi(\omega) = \mathbb{E}_X\left[\mathrm{e}^{\mathrm{i}\,\omega\,X}\right],$$

where $\mathrm{i} = \sqrt{-1}$ (known to many engineers as j). For the binomial distributions,

$$\phi(\omega) = M(\mathrm{i}\,\omega) = \left(1 + p\,(\mathrm{e}^{\mathrm{i}\,\omega} - 1)\right)^n.$$

Generally, $\phi(\omega) = M(\mathrm{i}\,\omega)$, provided the moment generating function exists. The moment generating function is more straightforward to use simply because it doesn't contain the constant i (the square root of -1). However, the characteristic function has one important advantage. Namely, it exists for all probability distributions. This follows trivially from the inequality

$$\left|\mathbb{E}_X\left[e^{i\omega X}\right]\right| \leq \mathbb{E}_X\left[\left|e^{i\omega X}\right|\right] = \mathbb{E}_X\left[1\right] = 1.$$

Note that

$$\left|\sum_i a_i\right| \leq \sum_i |a_i| \qquad \text{and} \qquad \left|\int f(x)\mathrm{d}x\right| \leq \int |f(x)|\,\mathrm{d}x,$$

from this it follows (since $p(x)$ is real and non-negative) that

$$\left|\mathbb{E}_X\left[f(X)\right]\right| \leq \mathbb{E}\left[|f(X)|\right].$$

The moment generating function exists only for functions with finite support (i.e. defined in a finite region) or those that fall off suitably quickly (at least exponentially fast). Distributions such as the Cauchy distribution (see Section 2.2.4) don't have a moment generating function but do have a characteristic function (the characteristic function of the Cauchy distribution is $e^{-|\omega|}$, see Appendix 5.B on page 107).

4.2.2 Cumulants

Moments are not the most useful statistic. For example, rather than the second moment, we are usually more interested in the *variance* of a random variable defined as

$$\mathbb{V}\mathrm{ar}_X\left[X\right] = \mathbb{E}_X\left[(X-\mu)^2\right] = \mu_2 - \mu^2$$

where $\mu = \mu_1$ is the mean and μ_2 is the second moment. This is an example of a *central moment* defined as

$$\bar{\mu}_m = \mathbb{E}_X\left[(X-\mu)^m\right].$$

However, central moments are not the most useful statistics when considering high-order statistics. There exists another set of statistics which have more interesting properties: these are the *cumulants*. The first three cumulants are the same as the first three central moments, but thereafter they differ. The cumulants are most easily defined through the *cumulant generating function* (cumulant generation function). For a random variable X, the cumulant generation function is defined as

$$G(l) = \log\big(M(l)\big) = \log\big(\mathbb{E}_X\left[e^{lX}\right]\big).$$

The cumulants are defined as the derivative of the CGF $\kappa_m = G^{(m)}(0)$. Thus,

$$G'(l) = \frac{M'(l)}{M(l)} \qquad \kappa_1 = G'(0) = \mu_1$$

$$G''(l) = \frac{M''(l)}{M(l)} - \frac{(M'(l))^2}{M^2(l)} \qquad \kappa_2 = G''(0) = \mu_2 - \mu_1^2$$

$$G'''(l) = \frac{M'''(l)}{M(l)} - 3\frac{M'(l)M''(l)}{M^2(l)} + 2\frac{(M'(l))^3}{M^3(l)} \qquad \kappa_3 = G'''(0) = \mu_3 - 3\mu_2\mu_1 + 2\mu_1^3.$$

The algebra gets more complicated but the fourth cumulant is equal to

$$\kappa_4 = \mathbb{E}\left[(x-\kappa_1)^4\right] - 3\,\kappa_2^2 = \mu_4 - 4\,\mu_3\,\mu_1 - 3\,\mu_2^2 + 12\,\mu_2\,\mu_1^2 - 6\,\mu_1^4.$$

We can obtain the cumulant generation function for the binomial distribution from its moment generating function, Equation (4.2),

$$G(l) = \log(M(l)) = n\,\log\bigl(1 + p\,(\mathrm{e}^l - 1)\bigr)\,.$$

It is proportional to the number of trials n. As a consequence all the cumulants are also proportional to n. This is not true of the high-order moments or central moments. It is a direct consequence of the fact that a binomial random variable, K, is the sum of n independent Bernoulli trials $K = \sum_{i=1}^{n} X_i$. Thus

$$\mathrm{Bin}(k|n,p) = \sum_{\boldsymbol{x}\in\{0,1\}^n} \left(\prod_{i=1}^{n} \mathrm{Bern}(x_i|p)\right) \left[\!\!\left[k = \sum_{i=1}^{N} x_i \right]\!\!\right];$$

multiply by $\mathrm{e}^{l\,k}$ and summing over k we find

$$\begin{aligned}
M_K(l) &= \sum_{k=0}^{n} \mathrm{e}^{l\,k}\,\mathrm{Bin}(k|n,p) \overset{(1)}{=} \sum_{k=0}^{n} \mathrm{e}^{l\,k} \sum_{\boldsymbol{x}\in\{0,1\}^n} \left(\prod_{i=1}^{n} \mathrm{Bern}(x_i|p)\right) \left[\!\!\left[k = \sum_{i=1}^{N} x_i \right]\!\!\right] \\
&\overset{(2)}{=} \sum_{\boldsymbol{x}\in\{0,1\}^n} \left(\prod_{i=1}^{n} \mathrm{Bern}(x_i|p)\right) \mathrm{e}^{l\,\sum_{i=1}^{N} x_i} \overset{(3)}{=} \prod_{i=1}^{n} \left(\sum_{x_i\in\{0,1\}} \mathrm{Bern}(x_i|p)\mathrm{e}^{l\,x_i} \right) \\
&\overset{(4)}{=} M_X(l)^n
\end{aligned}$$

(1) Using the definition for $\mathrm{Bin}(k|n,p)$ given above.
(2) Changing the order of summation and summing over k (note the power of using the indicator function).
(3) Using the fact that $\sum_{\boldsymbol{x}} = \prod_i \sum_{x_i}$ and $\exp(\sum_i a_i) = \prod_i \mathrm{e}^{a_i}$ to reorder the sum and product.
(4) Identifying the sum as the moment generating function of the Bernoulli distribution.

The CGF for the binomial distribution is therefore given by

$$G_K(l) = \log\bigl(M_K(l)\bigr) = \log\bigl(M_X^n(l)\bigr) = n\,G_X(l),$$

where

$$\begin{aligned}
G_X(l) &= \log\left(\sum_{x_i\in\{0,1\}} \mathrm{Bern}(x_i|p)\mathrm{e}^{l\,x_i} \right) = \log\bigl(p\,\mathrm{e}^l + (1-p)\bigr) \\
&= \log\bigl(1 + p(\mathrm{e}^l - 1)\bigr)\,.
\end{aligned}$$

Whenever a random variable is the sum of many independent random variables then its cumulant generation function will be the sum of the cumulant generation function of its component parts. One consequence of this is that the cumulants will all be proportional to the number of component parts. This is also the basis for the central limit theorem, which we will discuss in more detail in Chapter 5.

Returning to the CGF for the binomial distribution, we can perform a Taylor expansion:

$$\begin{aligned} G_K(l) &= n \log\big(1 + p\,(e^l - 1)\big) \\ &= l\,n\,p + \frac{l^2}{2!} n\,p\,(1-p) + \frac{l^3}{3!} n\,p\,(1-p)\,(1-2p) \\ &\quad + \frac{l^4}{4!} n\,p\,(1-p)\,(1-6p+6p^2) + O(l^5). \end{aligned}$$

From this we can read off the cumulants

$$\kappa_1 = n\,p \qquad\qquad \kappa_2 = n\,p\,(1-p)$$

$$\kappa_3 = n\,p\,(1-p)\,(1-2p) \qquad\qquad \kappa_4 = n\,p\,(1-p)\,(1-6p+6p^2)$$

This may seem a long-winded method for calculating the mean and variance, but it is widely applicable and often saves considerable work, particularly when calculating higher order statistics. It is also easy to write a program in a symbolic manipulation language such as Mathematica to do all the tedious algebra.

4.3 Beyond the Binomial Distribution

4.3.1 *Large* n *Limit*

In the large n limit, a binomial distribution 'converges' to a normal distribution. This can be proved directly, for example, by using Stirling's approximation for factorials $r! \sim \sqrt{2\pi r}\, r^r e^{-r}$ (here we use $\sim$ to denote that this is an asymptotic expansion), and collecting together all terms of order n^j and expanding. We will, however, avoid the details as the calculation is rather involved. However, this result is a special example of the *central limit theorem*, which states that the sum of random variables will very often converge to a normal distribution. There is a caveat, which is that this can fail for random variables drawn from distributions which have long tails – although this caveat does not apply here. We have already seen that the random variable for a binomial is equal to the sum of n Bernoulli random variables so we would expect the central limit theorem to apply here. We will discuss the central limit theorem in its general form in Section 5.3.

In Figure 4.2 we show different binomial distributions overlaid with a normal distribution with the same mean and variance. That is, given a binomial $\text{Bin}(k|n,p)$, we overlay the normal distribution $\mathcal{N}(x|n\,p,\, n\,p\,(1-p))$. As we increase n we see that the binomial distribution more closely approximates a normal distribution.

The difference between the cumulative distribution function (CDF) for the binomial and normal distribution is bounded by the Berry–Esseen formula. If $X \sim \text{Bin}(n,p)$ and $Y \sim \mathcal{N}(n\,p, n\,p\,(1-p))$ then

$$\left|\mathbb{P}\left(X \le x\right) - \mathbb{P}\left(Y \le x\right)\right| \le \frac{4\,(1-2p\,(1-p))}{5\,\sqrt{n\,p\,(1-p)}}.$$

In Figure 4.3 we plot $\mathbb{P}\left(X \le x\right) - \mathbb{P}\left(Y \le x\right)$ in the case $n = 50$ and $p = 0.5$. We see that the Berry–Esseen bound is not very tight, particularly away from

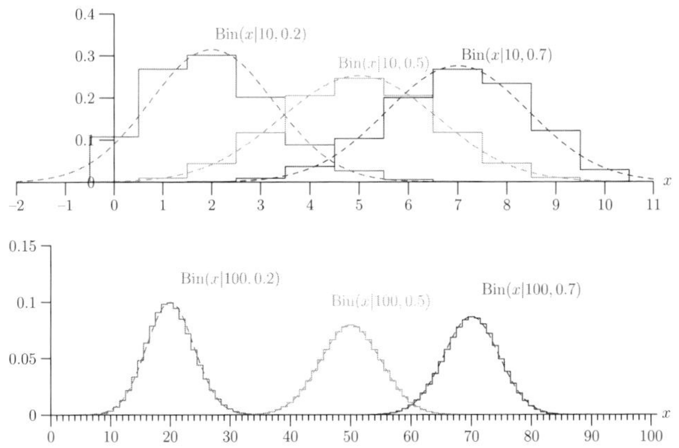

Figure 4.2 Examples of binomial distributions with a Gaussian having the same mean and variance overlaid. In the top set of graphs $n = 10$, while in the bottom set $n = 100$.

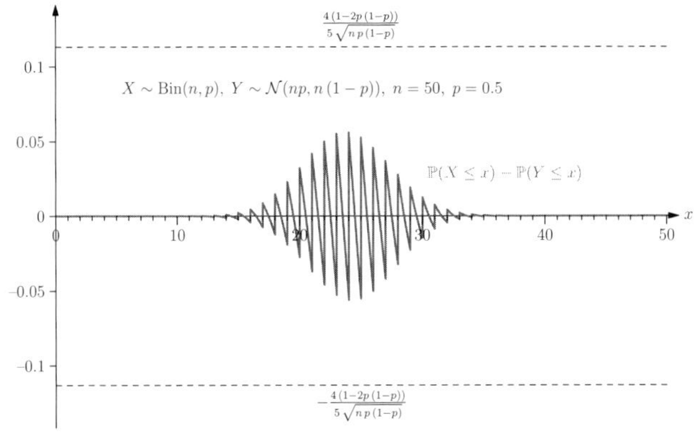

Figure 4.3 $\mathbb{P}\left(X \leq x\right) - \mathbb{P}\left(Y \leq x\right)$ versus x for $X \sim \text{Bin}(50, 0.5)$ and $Y \sim \mathcal{N}(25, 12.5)$. Also shown are the Berry–Esseen bounds.

the mean. However, it demonstrates that the difference between the cumulative distribution functions are of order $O(1/\sqrt{n})$. We discuss a method of obtaining tighter bounds for the tails of the distribution in Section 7.2.4.

4.3.2 Poisson Distribution

In the previous subsection we showed that the limit of the binomial distribution as $n \to \infty$ is a normal distribution. However, there is a special case when we simultaneously take the limits $n \to \infty$ and $p \to 0$, such that $p \times n \to \mu$. As we showed in the Section 2.1.3, this limit leads to the Poisson distribution

$$\text{Poi}(k|\mu) = \frac{\mu^k}{k!}\,\text{e}^{-\mu}.$$

This is a very commonly occurring distribution. For example, the number of excitations in a Geiger counter caused by background radiation or the number of cars passing a traffic light between 12:00 and 12:01 are closely approximated by a Poisson distribution. We return to a discussion of this in Section 12.3.1 when we discuss Poisson processes. Poisson distributions are also a little easier to work with than binomial distributions and so they are often used as approximations to binomials where n is large and p is small.

The moment generating function for a Poisson distribution is

$$M(l) = \mathbb{E}_K\left[\mathrm{e}^{l\,K}\right] = \sum_{k=0} \frac{(\mu\,\mathrm{e}^{l})^k}{k!}\mathrm{e}^{-\mu}$$
$$= \mathrm{e}^{\mu(\mathrm{e}^{l}-1)}.$$

The CGF is thus

$$G(l) = \log\bigl(M(l)\bigr) = \mu(\mathrm{e}^{l} - 1)$$

and all the cumulants are simply $\kappa_n = \mu$. In particular both the mean and variance are equal to μ.

4.3.3 Multinomial Distribution

The generalisation of the Bernoulli trial is a k-way trial where the probability of outcome $i \in \{1, 2, \ldots, k\}$ is given by p_i. This probability distribution is sometimes referred to as the *categorical distribution*

$$\begin{pmatrix}1\\0\\0\\0\end{pmatrix}\begin{pmatrix}0\\1\\0\\0\end{pmatrix}\begin{pmatrix}0\\0\\1\\0\end{pmatrix}\begin{pmatrix}0\\0\\0\\1\end{pmatrix}$$

One-hot 4-D vectors.

$$\mathrm{Cat}(X|\boldsymbol{p}) = \sum_{i=1}^{k} p_i \left[\!\left[X = i\right]\!\right]$$

where $\boldsymbol{p} = (p_1, p_2, \ldots, p_k)^{\mathsf{T}}$. Sometimes, the outcome is encoded as a "one-hot" vector $\boldsymbol{X}$ of length k with one element equal to 1 and all other elements equal to 0.

Just as the binomial distribution describes the result of a series of n independent Bernoulli trials, the multinomial distribution describes the result of a series of n independent k-way trials

$$\mathrm{Mult}(\boldsymbol{n}|n, \boldsymbol{p}) = n! \prod_{i=1}^{k} \frac{p_i^{n_i}}{n_i!} \left[\!\left[\boldsymbol{n} \in \Lambda_n^k\right]\!\right] \tag{4.3}$$

where

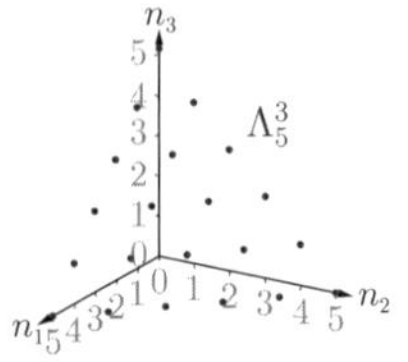

$$\Lambda_n^k = \left\{ \boldsymbol{n} = (n_1, n_2, \ldots n_k) \,\middle|\, \forall i,\, n_i \in \{0, 1, 2, \ldots\} \text{ and } \sum_{i=1}^{k} n_i = n \right\}$$

(see Equation (2.10) on page 36) – take care to distinguish between the vector of integers $\boldsymbol{n}$ and the total number of trials n. Note, that if we assume $\boldsymbol{n} \in \Lambda_n^k$, then the indicator function is not necessary in Equation (4.3). Similarly, if we assume the variables n_i are positive integers of 0 then it is sufficient to impose the constraint that $\sum_i n_i = n$. As a consequence the multinomial distribution is often expressed slightly differently in different texts.

The multinomial distribution looks rather painful to work with; however, the moment generating function is

$$M(\boldsymbol{l}) = \mathbb{E}_{\boldsymbol{N}}\left[\exp\left(\sum_{i=1}^{k} l_i\, N_i\right)\right] = n! \sum_{\boldsymbol{n}\in\Lambda_n^k} \prod_{i=1}^{k} \frac{(p_i\,\mathrm{e}^{l_i})^{n_i}}{n_i!} = \left(\sum_{i=1}^{k} p_i\,\mathrm{e}^{l_i}\right)^n$$

where we have used the multinomial expansion

$$\left(\sum_{i=1}^{k} a_i\right)^n = n! \sum_{\boldsymbol{n} \in \Lambda_n^k} \prod_{i=1}^{k} \frac{a_i^{n_i}}{n_i!}.$$

It follows that the cumulant generating function is

$$G(\boldsymbol{l}) - n \log\left(\sum_{i=1}^{k} p_i \,\mathrm{e}^{l_i}\right).$$

From this it is rather simple to obtain the cumulants. For example,

$$\left.\frac{\partial G(\boldsymbol{l})}{\partial l_i}\right|_{\boldsymbol{l}=0} = \left.\frac{p_i \,\mathrm{e}^{l_i}}{\sum_k p_k \,\mathrm{e}^{l_k}}\right|_{\boldsymbol{l}=0} = p_i$$

$$\left.\frac{\partial^2 G(\boldsymbol{l})}{\partial l_i \partial l_j}\right|_{\boldsymbol{l}=0} = \frac{p_i \,\mathrm{e}^{l_i} [\![i = j]\!]}{\sum_k p_k \,\mathrm{e}^{l_k}} - \left.\frac{p_i \,\mathrm{e}^{l_i} \, p_j \,\mathrm{e}^{l_j}}{\left(\sum_k p_k \,\mathrm{e}^{l_k}\right)^2}\right|_{\boldsymbol{l}=0} = p_i \, [\![i == j]\!] - p_i \, p_j.$$

Note that, in a similar manner to the binomial, the CGF for the multinomial is simple because the random variable $\boldsymbol{N} = (N_1, N_2, \ldots, N_k)$ can be thought of as a sum of n random variables $\boldsymbol{X}(i) = (X_1(i), X_2(i), \ldots, X_k(i))$ where $\boldsymbol{X}(i)$ is a one-hot vector. Thus the cumulant generating function for the multinomial is just equal to n times the cumulant generating function for the categorical distribution, which is trivially given by

$$G(\boldsymbol{l}) = \log\left(\sum_{i=1}^{k} p_i \,\mathrm{e}^{l_i}\right).$$

Obtaining moments and cumulants for the multinomial distribution is relatively easy, but for many other kinds of calculations the multinomial distribution can be awkward to work with. There are a couple of useful tricks that can help out. They both remove the constraint that appears in the sum. The first goes by the name of *Poissonisation*. Here's the trick. Consider drawing a deviate N from a Poisson distribution $\mathrm{Poi}(\lambda)$ and then drawing a vector of integers $\boldsymbol{N} = (N_1, N_2, \ldots, N_k)$ from the multinomial $\mathrm{Mult}(N, \boldsymbol{p})$, where N is the integer we draw from a Poisson distribution. Then the probability of drawing a vector of integers $\boldsymbol{N} = \boldsymbol{n}$ is

$$\begin{aligned}
\mathbb{P}\left(\boldsymbol{N} = \boldsymbol{n}\right) &\stackrel{(1)}{=} \sum_{n=0}^{\infty} \mathrm{Mult}(\boldsymbol{n}|n, \boldsymbol{p}) \mathrm{Poi}(n|\lambda) \\
&\stackrel{(2)}{=} \sum_{n=0}^{\infty} n! \left(\prod_{i=1}^{k} \frac{p_i^{n_i}}{n_i!}\right) \left[\!\!\left[\sum_{i=1}^{k} n_i = n \right]\!\!\right] \frac{\lambda^n}{n!} \,\mathrm{e}^{-\lambda} \\
&\stackrel{(3)}{=} \left(\prod_{i=1}^{k} \frac{p_i^{n_i}}{n_i!}\right) \mathrm{e}^{-\lambda} \sum_{n=0}^{\infty} \lambda^n \left[\!\!\left[\sum_{i=1}^{k} n_i = n \right]\!\!\right] \\
&\stackrel{(4)}{=} \left(\prod_{i=1}^{k} \frac{p_i^{n_i}}{n_i!}\right) \mathrm{e}^{-\lambda} \lambda^{\sum_{i=1}^{k} n_i} \stackrel{(5)}{=} \mathrm{e}^{-\lambda} \prod_{i=1}^{k} \frac{(\lambda p_i)^{n_i}}{n_i!}
\end{aligned}$$

(1) Assuming we draw n from a Poisson distribution and $\boldsymbol{N} = \boldsymbol{n}$ from a multinomial distribution.
(2) Writing out the distributions.
(3) Taking terms out of the sum that don't depend on n.
(4) Taking the sum over n and using $\sum_n f(n) \llbracket n = a \rrbracket = f(a)$. In this case a is a sum.
(5) Using $\lambda^{\sum_i n_i} = \prod_i \lambda^{n_i}$ and rearranging.

However, as $\sum_i p_i = 1$ we can write $e^{-\lambda} = \prod_i e^{-\lambda p_i}$ so that

$$\mathbb{P}\left(\boldsymbol{N} = \boldsymbol{n}\right) = \prod_{i=1}^{k} \frac{(\lambda p_i)^{n_i}}{n_i!} e^{-\lambda p_i} = \prod_{i=1}^{k} \text{Poi}(\lambda\, p_i).$$

From this we see that the variables n_i are distributed as if they were independently drawn from a Poisson distribution, $\text{Poi}(\lambda\, p_i)$.

Why does this help? Dealing with independent variables is easier than dealing with variables that sum to n, but usually n is fixed, so one has to go through some contortions to actually use Poissonisation. However, as the following example shows, Poissonisation can help.

Example 4.1 No Empty Boxes

Consider putting n balls randomly into k boxes. What is the probability that no box is empty? Rather than tackle this directly, we consider the case when $n = N$ is drawn from a Poisson distribution $\text{Poi}(\lambda)$, in which case the probability of N_i balls in box i is just $\text{Poi}(\lambda\, p_i) = \text{Poi}(\lambda/k)$ (assuming all boxes are equally likely to be visited). Then, using Poissonisation to treat N_i as independent, we get

$$\mathbb{P}\left(N_1 > 0,\, N_2 > 0, \ldots N_k > 0\right) = \prod_{i=1}^{n} \mathbb{P}\left(N_i > 0\right) = \left(1 - e^{-\lambda/k}\right)^k$$

since $P(N_i = 0) = e^{-\lambda/k}$. By expanding terms and carefully rearranging, we obtain

$$\begin{aligned}
&\mathbb{P}\left(N_1 > 0,\, N_2 > 0,\, \ldots,\, N_k > 0\right) \\
&= \left(1 - e^{-\lambda/k}\right)^k \\
&\overset{(1)}{=} \sum_{i=0}^{k} \binom{k}{i} (-1)^i e^{-i\,\lambda/k} \\
&\overset{(2)}{=} \sum_{i=0}^{k} \binom{k}{i} (-1)^i e^{\lambda\,(1-i/k)}\, e^{-\lambda} \\
&\overset{(3)}{=} e^{-\lambda} \sum_{i=0}^{k} \binom{k}{i} (-1)^i \sum_{n=0}^{\infty} \frac{1}{n!} \lambda^n \left(1 - \frac{i}{k}\right)^n
\end{aligned}$$

$$\stackrel{(4)}{=} \sum_{n=0}^{\infty} \frac{e^{-\lambda}}{n!} \lambda^n \sum_{i=0}^{k} \binom{k}{i} (-1)^i \left(1 - \frac{i}{k}\right)^n$$

$$\stackrel{(5)}{=} \sum_{n=0}^{\infty} \text{Poi}(n|\lambda) f(n)$$

where

$$f(n) = \sum_{i=0}^{k} \binom{k}{i} (-1)^i \left(1 - \frac{i}{k}\right)^n.$$

(1) Using the binomial expansion $(1 + a)^k = \sum_{i=0}^{k} \binom{k}{i} a^k$ with $a = -e^{-\lambda/k}$.
(2) Taking out $e^{-\lambda}$ which we want to use later.
(3) Taylor expanding $e^{\lambda(1-i/k)}$.
(4) Changing the order of summation.
(5) Recognising the first terms as the Poisson distribution and naming the second summation as $f(n)$.

Because we assumed that the number of balls was chosen according to the distribution $\mathbb{P}(N = n) = \text{Poi}(n|\lambda)$, then the probability of there being no empty boxes given n balls is just $f(n)$. Here we use the Poissonisation result in reverse. Since our choice of λ is arbitrary this is the only value $f(n)$ can take. Although the expression for $f(n)$ doesn't look very pretty, it is much easier to compute than the original sum over all $\boldsymbol{n}$ lying in the integer simplex.

For my taste, Poissonisation is awkward and I prefer a trick used heavily in statistical physics where the indicator function is replaced with an integral representation. This takes some getting used to, but it is a very general trick (we can also use it with the Dirac delta function discussed in Chapter 5). I won't go into it here as I don't want to be distracted from our main purpose of surveying probability rather than discussing the esoterics of combinatorics. However, for those interested, in Exercise 4.6 I repeat Example 4.1 using the integral representation of an indicator function rather than Poissonisation.

The binomial, Poisson, and multinomial distributions arise very frequently in modelling the world. They are not the most convenient functions to work with directly, but their statistical properties can be readily computed through the through the cumulative generation function (CGF). This has a simple form because these distributions arise as a sum of iid random binary variables (in the case of the multinomial it is a binary vector). Generating functions are widely used in studying probabilistic models.

Additional Reading

An outstanding book on combinatorial manipulations, working with generating functions, and much else besides, is *Concrete Mathematics* by Graham et al. (1989). A volume that explores generating functions in very great depth is Dobrushkin (2010).

Exercise for Chapter 4

Exercise 4.1 (answer on page 401)

Consider a succession of Bernoulli trials with a success probability of p. The probability that the first success occurs after k trials is given by the *geometric distribution*

$$\mathbb{P}\left(\text{First success after } k \text{ trial}\right) = \text{Geo}(k|p) = p\,(1-p)^{k-1}.$$

(The geometric distribution is sometimes defined to be the probability of success after k failures so is defined as $p\,(1-p)^k$ where $k = 0, 1, 2, \ldots$) Compute the CGF for the geometric distribution and from this compute the mean and variance. If the probability of winning a lottery is 10^{-6}, how many attempts do you have to make on average before you win?

Exercise 4.2 (answer on page 402)

Show that the probability of k successes until r failures have occurred in a sequence of Bernoulli trials with success probability p is given by the *negative binomial distribution*

$$\text{NegBin}(k|r,p) = \binom{k+r-1}{k} p^k\,(1-p)^r$$

(note that the last trial will have been a failure, hence the combinatorial factor). Compute the CGF for the negative binomial and compute the mean and variance. If you are playing a game of pure chance with four friends how many times do you expect to win before you lose nine times? (Note that $\text{NegBin}(k|r,p)$ is a properly normalised distribution so that

$$\sum_{k=0}^{\infty} \binom{k+r-1}{k} p^k = \frac{1}{(1-p)^r}$$

which is true for any p. This is exactly the sum we need to compute the CGF.)

Exercise 4.3 (answer on page 403)

Consider a series of Bernoulli trials occurring at a time interval δt with a success probability $\mu\,\delta t$ (so the expected number of successes in one second is μ). Show that, in the limit $\delta t \to 0$, the probability of the first successful event occurring at time t is distributed according to the *exponential distribution*

$$f(t) = \text{Exp}(t|\mu) = \mu\,\text{e}^{-\mu\,t}.$$

(This is the waiting time between Poisson distributed events. It is used, for example, in simulating chemical reactions at a molecular level – see Section 12.3.2.)

Exercise 4.4 (answer on page 403)

The cumulative probability mass function for a Poisson distribution $\text{Poi}(\mu)$ is given by

$$\mathbb{P}\left(\text{number of success} < k\right) = \sum_{i=0}^{k-1} \frac{\mu^i}{i!} e^{-\mu}.$$

Starting from the normalised incomplete gamma function (see Appendix 2.A) defined by

$$Q(k,\mu) = \frac{1}{(k-1)!} \int_\mu^\infty z^{k-1} e^{-z} \, dz$$

obtain a relationship between $Q(k,\mu)$ and $Q(k-1,\mu)$ using integration by part and hence prove that

$$\mathbb{P}\left(\text{number of success} < k\right) = Q(k,\mu).$$

Exercise 4.5 (answer on page 404)

The probability of k or more success in n Bernoulli trials with a success probability of p is given by

$$\mathbb{P}\left(\text{number of success} \geq k\right) = \sum_{i=k}^{n} \binom{n}{i} p^i (1-p)^{n-i} = I_p(k, n-k+1).$$

Starting from the normalised incomplete beta function (see Appendix 2.B) defined by

$$I_p(k, n-k+1) = \frac{n!}{(k-1)!\,(n-k)!} \int_0^p z^{k-1} (1-z)^{n-k} \, dz$$

obtain a relationship between $I_p(k, n-k+1)$ and $I_p(k-1, n-k+2)$ using integration by part and hence prove

$$\mathbb{P}\left(\text{number of success} \geq k\right) = I_p(k, n-k+1).$$

Exercise 4.6 (answer on page 405)

In Example 4.1 on page 70 we calculated the probability that there are no empty boxes when we randomly place n balls in k boxes. This probability was

$$p_{neb} = \left(\prod_{j=1}^{k} \sum_{n_j > 0}\right) \text{Mult}(\boldsymbol{n}|n, \boldsymbol{p}) \left[\!\!\left[\sum_{j=1}^{k} n_j = n \right]\!\!\right]$$

where $p_j = 1/k$ for, $j = 1, 2, \ldots, k$. Perform this sum using the integral representation of the indicator function

$$\left[\!\!\left[\sum_{j=1}^{k} n_j = n \right]\!\!\right] = \int_{-\infty}^{\infty} e^{i\omega\left(\sum_j n_j - n\right)} \frac{d\omega}{2\pi}. \tag{4.4}$$

Note that the indicator function ensures that the total number of balls equals n, so we can treat the sums as unconstrained. Having performed the sums expand out the expression similar to what was done in Example 4.1 and then use Equation (4.4) to get a simplified answer? (Be warned the answer is not in an identical form to that given in Example 4.1, although it can be made so by a change of variables.)

5

The Normal Distribution

Contents

The Gaussian or normal distribution is the bedrock for a huge chunk of probability and statistics. Its popularity is due to a number of different reasons. It arises naturally due to the central limit theorem and also through maximum entropy arguments (although we don't cover this until Chapter 9). The normal distribution has nice analytic properties, which means that closed-form solutions to many problems can often be obtained. However, working with normal distribution is fiddly and non-intuitive. In this chapter we familiarise ourselves with techniques for manipulating normal distribution. We look at its moments and cumulants, which sets us up for understanding the central limit theorem. The central limit theorem is discussed and we point out its limitations. In passing we discuss the log-normal distribution. We then consider the cumulative distribution function (CDF) for the normal distribution (aka error functions). We finish by discussing the multivariate normal distribution and its properties.

5.1 Ubiquitous Normals

Many quantities that we meet are approximately normally distributed. For example, the distribution of people's height or the velocity of particles in a gas. Normal distributions arise as the limit of many other distributions. We illustrate this in Figure 5.1, where we show how the binomial, gamma, and beta distribution come closer to a normal distribution as we increase the parameter values. Similarly, the multinomial and Dirichlet distributions also converge towards a multivariate Gaussian in the limit when their parameters increase.

It is not, of course, inevitable that a distribution will tend towards a normal distribution as the parameters increase. We have, for example, seen for a binomial distribution that, if $p \to 0$ as $n \to \infty$ such that $p\,n$ remains finite, then we tend to a Poisson distribution rather than a normal distribution (although when the product $p\,n$ becomes large the Poisson distribution again converges towards a normal distribution).

One reason why normal distributions arise so commonly as a limit distribution is that many distributions can be written as

$$f(x) = \mathrm{e}^{b\,g(x)}$$

where b is a parameter which typically grows with the system size and $g(x)$ is some, often complicated, function of x. In this case, if $g(x)$ reaches some maximum value around x^* then Taylor expanding around this maximum we find

$$\begin{aligned} g(x) &= g(x^*) + (x - x^*)\,g'(x^*) + \frac{1}{2}(x - x^*)^2\,g''(x^*) + O\left((x - x^*)^3\right) \\ &= g(x^*) + \frac{1}{2}(x - x^*)^2\,g''(x^*) + O\left((x - x^*)^3\right) \end{aligned}$$

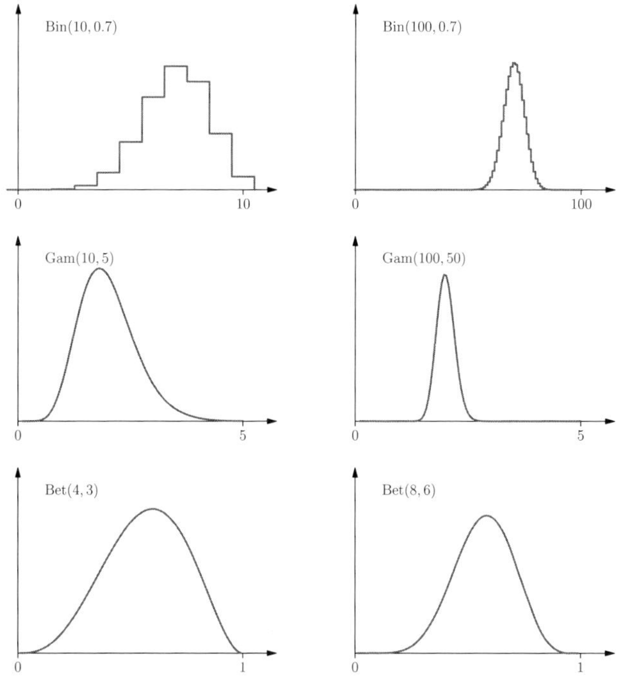

Figure 5.1 Examples of how the binomial, gamma, and beta distribution tend towards a normal distribution in certain limits.

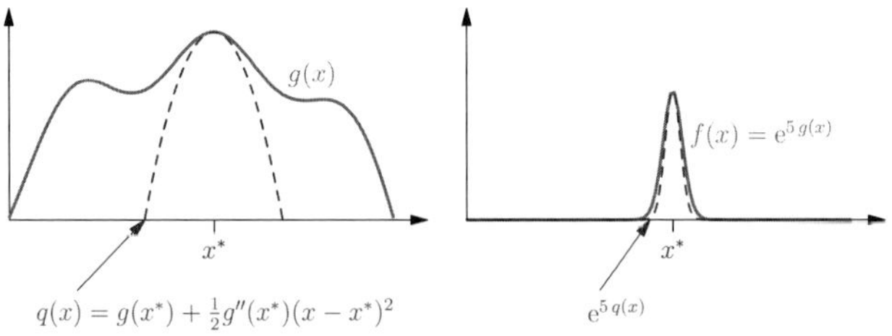

Figure 5.2 Left figure shows a typical exponent, $g(x)$, and its quadratic approximation, $q(x)$. Right figure shows $\exp(b\,g(x))$ and the quadratic approximation, $\exp(b\,q(x))$, for the case $b = 5$. The approximation becomes more accurate as b increases.

since $g'(x^*) = 0$. Close to x^* the term $O\left((x - x^*)^3\right)$ is small compared to the $(x - x^*)^2$ term so that

$$f(x) \approx c\,\mathrm{e}^{(x-x^*)^2\,b\,g''(x^*)/2}.$$

As we are at a maximum, $g''(x^*) < 0$ (see Figure 5.2). The function $f(x)$ therefore resembles a normal distribution with mean x^* and variance $-1/(b\,g''(x^*))$. As b increases the variance decreases so that most of the probability mass is around x^*. The same argument applies in higher dimensions. This relies on the fact that $g''(x)$ doesn't vanish at $x = x^*$, but, unless there is some special reason for both the first and second derivative to vanish at the same point, this will very rarely happen. One way for this argument to break down is if x^* is bounded and the maximum occurs on the boundary.

Although we have seen that many random variables naturally become normally distributed in some limit, the argument tells us nothing about how the tails of the distribution look. Thus, although nearly normal distributions abound, it is a common experience to find outliers occurring more frequently than would be predicted by a normal distribution.

5.2 Basic Properties

In this section we compute the normalisation constant for the normal distribution and its moments and cumulants. Becoming familiar with manipulating normal distributions pays off handsomely, but it does require practice and often many sheets of paper.

5.2.1 Integrating Gaussians

One of the key attractions of the normal or Gaussian distribution is that it is relatively easy to work with analytically. It is thus surprising that the Gaussian integral

$$I_0 = \int_{-\infty}^{\infty} \mathrm{e}^{-x^2/2}\,\mathrm{d}x$$

is challenging. If you haven't been shown how to calculate this integral then it's tough. Although you rarely actually have to calculate it, it is such an important building block that it's worth seeing how to do it. One approach is to start with the simpler integral

$$I_1 = \int_0^\infty x\,e^{-x^2/2}\,dx$$

which can be solved by a change of variables. Let $u = x^2/2$ then $du = x\,dx$, thus

$$I_1 = \int_0^\infty x\,e^{-x^2/2}\,dx = \int_0^\infty e^{-u}\,du = \left[-e^{-u}\right]_0^\infty = 1.$$

But how does that help us with out initial integral? The answer is a bit of a trick. We work in two dimensions, then

$$I_0^2 = \int_{-\infty}^\infty e^{-x^2/2}\,dx \int_{-\infty}^\infty e^{-y^2/2}\,dy.$$

Now we make a change of variables to polar coordinates so that $x = r\cos(\theta)$, $y = r\sin(\theta)$, and $x^2 + y^2 = r^2(\cos^2(\theta) + \sin^2(\theta)) = r^2$. We note that the Jacobian is given by

$$\frac{\partial\,(x, y)}{\partial\,(r, \theta)} = \begin{vmatrix} \frac{\partial\, r\,\cos(\theta)}{\partial\, r} & \frac{\partial\, r\,\cos(\theta)}{\partial\,\theta} \\ \frac{\partial\, r\,\sin(\theta)}{\partial\, r} & \frac{\partial\, r\,\sin(\theta)}{\partial\,\theta} \end{vmatrix} = \begin{vmatrix} \cos(\theta) & -r\,\sin(\theta) \\ \sin(\theta) & r\,\cos(\theta) \end{vmatrix} = r(\cos^2(\theta) + \sin^2(\theta)) = r.$$

We can also understand this by considering the area swept out by making an infinitesimal change in the coordinates (see Figure 5.3). Using this change of variable we find

$$\begin{aligned} I_0^2 &= \int_{-\infty}^\infty \int_{-\infty}^\infty e^{-(x^2+y^2)/2}\,dx\,dy \\ &= \int_0^{2\pi} d\theta \int_0^\infty r\,e^{-r^2/2}\,dr = 2\pi\,I_1 = 2\pi \end{aligned}$$

or $I_0 = \sqrt{2\pi}$.

The general form of a Gaussian integral (i.e. an integral of an exponential of a quadratic function) is

$$I_3(a, b) = \int_{-\infty}^\infty e^{-a\,x^2/2 + b\,x}\,dx.$$

When faced with an integral like this, the first step is to *complete the square* in the exponent

$$-\frac{a\,x^2}{2} + b\,x = -\frac{a}{2}\left(x - \frac{b}{a}\right)^2 + \frac{b^2}{2\,a}.$$

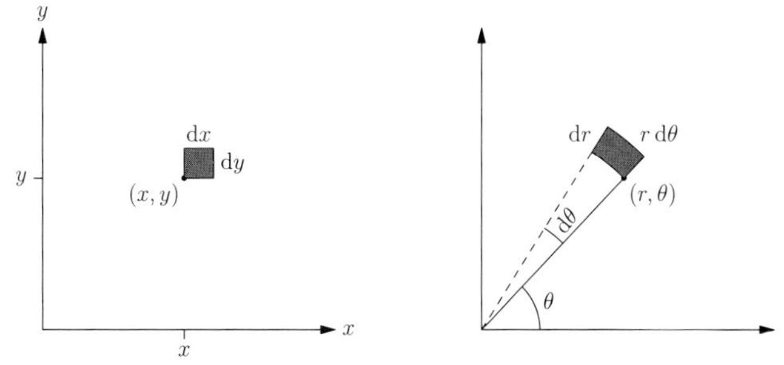

Figure 5.3 Change of coordinates. The area produced by an infinitesimal change in the coordinates is $dx\,dy$ in Cartesian coordinates and $r\,dr\,d\theta$ in polar coordinates.

We then substitute this into the integral and make the change of variables $y = \sqrt{a}(x - b/a)$. Noting that $\mathrm{d}y = \sqrt{a}\,\mathrm{d}x$, we have

$$I_3(a,b) = \frac{\mathrm{e}^{b^2/(2a)}}{\sqrt{a}} \int_{-\infty}^{\infty} \mathrm{e}^{-y^2/2}\,\mathrm{d}y = \sqrt{\frac{2\pi}{a}}\,\mathrm{e}^{b^2/(2a)}.$$

5.2.2 Moments and Cumulants

The normal distribution is defined as

$$\mathcal{N}(x|\mu,\sigma^2) = \frac{1}{\sqrt{2\pi\sigma^2}}\mathrm{e}^{-(x-\mu)^2/(2\sigma^2)}.$$

You should be able to use the integrals we have already computed to verify that the normalisation term is correct. The moment generating function is

$$M(l) = \mathbb{E}_X\left[\mathrm{e}^{lX}\right] = \int_{-\infty}^{\infty} \mathrm{e}^{lx}\,\mathcal{N}(x|\mu,\sigma^2)\,\mathrm{d}x.$$

We can compute this by completing the square. To do this we note

$$-\frac{a}{2}x^2 + bx - \frac{c}{2} = -\frac{1}{2}a\left(x - \frac{b}{a}\right)^2 + \frac{b^2}{2a} - \frac{c}{2}. \tag{5.1}$$

Thus,

$$\mathrm{e}^{lx}\,\mathcal{N}(x|\mu,\sigma^2) = \frac{1}{\sqrt{2\pi}\,\sigma}\mathrm{e}^{-\frac{(x-\mu)^2}{2\sigma^2}+lx} = \frac{1}{\sqrt{2\pi}\,\sigma}\mathrm{e}^{-\frac{x^2}{2\sigma^2}+\left(l+\frac{\mu}{\sigma^2}\right)x-\frac{\mu^2}{2\sigma^2}}.$$

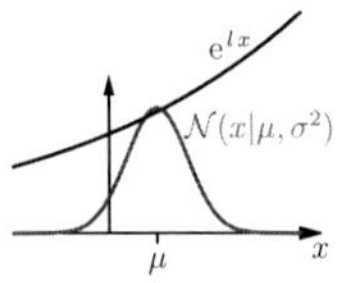

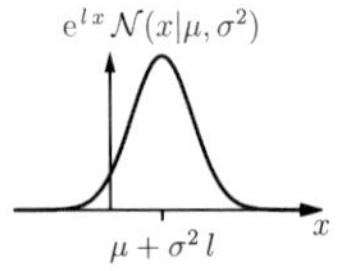

Identifying $a = 1/\sigma^2$, $b = l + \mu/\sigma^2$, and $c = \mu^2/\sigma^2$, then using Equation (5.1), we can rewrite the exponent of $\mathrm{e}^{lx}\,\mathcal{N}(x|\mu,\sigma^2)$ as

$$\begin{aligned}
-\frac{x^2}{2\sigma^2} + \left(l + \frac{\mu}{\sigma^2}\right)x - \frac{\mu^2}{2\sigma^2} &= -\frac{1}{2\sigma^2}(x - \mu - \sigma^2 l)^2 + \frac{\sigma^2}{2}\left(\frac{\mu}{\sigma^2} + l\right)^2 - \frac{\mu^2}{2\sigma^2} \\
&= -\frac{1}{2\sigma^2}(x - \mu - \sigma^2 l)^2 + \mu l + \frac{\sigma^2 l^2}{2}
\end{aligned}$$

so that

$$\mathrm{e}^{lx}\,\mathcal{N}(x|\mu,\sigma^2) = \mathrm{e}^{l\mu+\sigma^2 l^2/2}\,\mathcal{N}(x|\mu+\sigma^2 l,\sigma^2).$$

Using this,

$$M(l) = \mathbb{E}_X\left[\mathrm{e}^{lX}\right] = \mathrm{e}^{l\mu+\sigma^2 l^2/2}\int_{-\infty}^{\infty}\mathcal{N}(x|\mu+\sigma^2 l,\sigma^2)\mathrm{d}x = \mathrm{e}^{l\mu+\sigma^2 l^2/2}.$$

The cumulant generating function (CGF) is thus given by

$$G(l) = \log\bigl(M(l)\bigr) = l\,\mu + \sigma^2 l^2/2.$$

From this we can read off the cumulants trivially: the mean is $\kappa_1 = \mu$, the variance is $\kappa_2 = \sigma^2$, and all higher cumulants are zero.

We can interpret the higher cumulants of a distribution as a measure of how far that distribution differs from a normal distribution. However, the absolute size of the cumulants of a distribution tell us little about how much the distribution deviates from a normal distribution, as the cumulants depend on the units in which we measure our random variables. That is, if we have a distribution $f_X(x)$ and we rescale x by a so that $Y = a\,X$, then the cumulants of $f_Y(y)$ will be $\kappa_n^Y = a^n\,\kappa_n^X$. To obtain a scale invariant measurement of the deviation away from a normal distribution we rescale by $1/\sqrt{\kappa_2}$ so that we compare all distributions with unit variance. The third cumulant of the rescaled distribution is known as the *skewness*, which in terms of the cumulants of the original distribution is equal to $\kappa_3/\kappa_2^{3/2}$. The skewness measures the lopsidedness of a distribution (see upper graphs in Figure 5.4). The fourth cumulant of the rescaled distribution is known as the *kurtosis*, which in terms of the cumulants for the original distribution is given by κ_4/κ_2^2. This measures whether the tails fall off faster or slower than a normal distribution (see lower graphs in Figure 5.4).

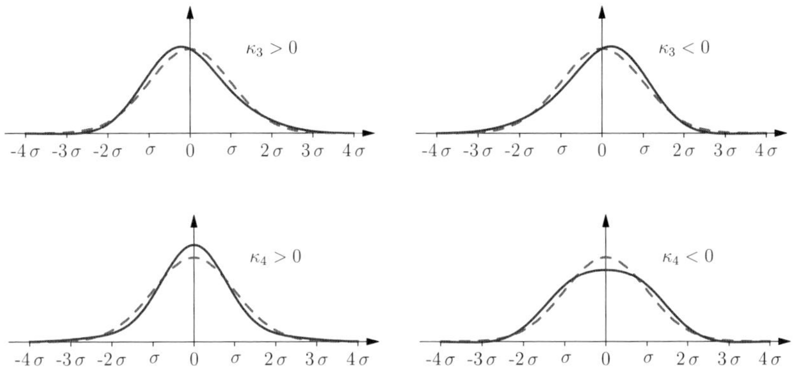

Figure 5.4 The upper two graphs show examples of skewed distributions while the lower graphs show examples of distributions with zero skewness but a positive and negative kurtosis. Also shown is a Gaussian with the same mean and variance (dashed lines).

Incidentally, there is another way to compute integrals of the form

$$I_4(n) = \int_0^\infty x^n\,\mathrm{e}^{-x^2/2}\mathrm{d}x,$$

which is to make the change of variables $u = x^2/2$ so that $\mathrm{d}u = x\,\mathrm{d}x$, $x = \sqrt{2\,u}$, and

$$I_4(n) = 2^{(n-1)/2}\int_0^\infty u^{(n-1)/2}\,\mathrm{e}^{-u}\,\mathrm{d}u = 2^{(n-1)/2}\,\Gamma\left(\frac{n+1}{2}\right),$$

where $\Gamma(z)$ is the gamma function (see Appendix 2.A on page 41). To compute the moments we note that $\Gamma(1/2) = \sqrt{\pi}$ (which is proved by noting that $I_4(0) = \Gamma(1/2)/\sqrt{2} = I_0/2 = \sqrt{\pi/2}$). Now the odd moments of the unit normal distribution $\mathcal{N}(0,1)$ are $\mu_{2n+1} = 0$ because the integrand, $x^{2n+1}\mathcal{N}(x|0,1)$, is odd. The even moments are given by

$$\begin{aligned}\mu_{2n} &= \frac{1}{\sqrt{2\pi}}\int_{-\infty}^\infty x^{2n}\,\mathrm{e}^{-x^2/2}\mathrm{d}x = \frac{2}{\sqrt{2\pi}}\,I_4(2n)\\ &= \frac{2^n}{\sqrt{\pi}}\,\Gamma\left(\frac{2n+1}{2}\right).\end{aligned}$$

Using the property $\Gamma(x) = (x-1)\,\Gamma(x-1)$ we have

$$\Gamma\left(\frac{2n+1}{2}\right) = \left(\frac{2n-1}{2}\right)\left(\frac{2n-3}{2}\right)\cdots\left(\frac{1}{2}\right)\Gamma\left(\frac{1}{2}\right)$$
$$= \frac{1}{2^n}\sqrt{\pi}\,(2n-1)\,(2n-3)\,\cdots\,1.$$

Thus,

$$\mu_{2n} = (2n-1)\,(2n-3)\,\cdots\,1 = (2n-1)!!$$

The double factorial, $(2n-1)!!$, is a fairly standard notation signifying the product of every odd integer from $2n-1$ down to 1. We can rewrite the double factorial in terms of normal factorials

$$(2n-1)!! = (2n-1)\times(2n-3)\times\cdots\times 1$$
$$= \frac{(2n)\times(2n-1)\times(2n-2)\times(2n-3)\times\cdots\times 2\times 1}{(2n)\times(2n-2)\times\cdots\times 4\times 2}$$
$$= \frac{(2n)!}{2^n\times n\times(n-1)\times\cdots\times 2\times 1} = \frac{(2n)!}{2^n\,(n!)}.$$

Note that $\mu_2 = 1$, $\mu_4 = 3$, $\mu_6 = 15$, etc.

Sometimes it is easier to work with the characteristic function of the normal distribution. The characteristic function of a normal distribution, $\mathcal{N}(x|\mu,\sigma^2)$, is

$$\phi(\omega) = \tilde{\mathcal{N}}(\omega|\mu,\sigma^2) = \int_{-\infty}^{\infty} \mathrm{e}^{\mathrm{i}\,\omega\,x}\,\mathcal{N}(x|\mu,\sigma^2)\,\mathrm{d}x = \mathrm{e}^{\mathrm{i}\,\mu\,\omega-\frac{1}{2}\sigma^2\,\omega^2}$$

where using the standard inverse Fourier transform

$$\mathcal{N}(x|\mu,\sigma^2) = \int_{-\infty}^{\infty} \mathrm{e}^{-\mathrm{i}\,\omega\,x}\,\tilde{\mathcal{N}}(\omega|\mu,\sigma^2)\,\frac{\mathrm{d}\omega}{2\,\pi}.$$

Example 5.1 Convolution of Normals

Consider the convolution of two normal distributions

$$I_6(y) = \int_{-\infty}^{\infty} \mathcal{N}(x|\mu_1,\sigma_1^2)\,\mathcal{N}(y-x|\mu_2,\sigma_2^2)\,\mathrm{d}x.$$

We can compute this by completing the squares and performing the integral, however, completing the square is very fiddly. Instead, we can express the normal distributions in terms of their characteristic function

$$I_6(y) = \int_{-\infty}^{\infty}\int_{-\infty}^{\infty}\int_{-\infty}^{\infty} \mathrm{e}^{-\mathrm{i}\,\omega_1\,x-\mathrm{i}\,\omega_2\,(y-x)}\,\tilde{\mathcal{N}}(\omega_1|\mu_1,\sigma_1^2)\,\tilde{\mathcal{N}}(\omega_2|\mu_2,\sigma_2^2)$$
$$\frac{\mathrm{d}\omega_1}{2\,\pi}\,\frac{\mathrm{d}\omega_2}{2\,\pi}\,\mathrm{d}x.$$

Changing the order of integration and using the standard result

$$\int_{-\infty}^{\infty} \mathrm{e}^{-\mathrm{i}\,x\,(\omega_1-\omega_2)}\,\mathrm{d}x = 2\,\pi\,\delta(\omega_1-\omega_2),$$

where $\delta(\omega_1 - \omega_2)$ is the Dirac delta function (see Appendix 5.A), gives us

$$I_6(y) = 2\pi \int_{-\infty}^{\infty} \int_{-\infty}^{\infty} \delta(\omega_1 - \omega_2)\, \mathrm{e}^{-\mathrm{i}\omega_2 y}\, \tilde{\mathcal{N}}(\omega_1|\mu_1, \sigma_1^2)\, \tilde{\mathcal{N}}(\omega_2|\mu_2, \sigma_2^2) \, \frac{\mathrm{d}\omega_1}{2\pi} \frac{\mathrm{d}\omega_2}{2\pi}$$
$$= \int_{-\infty}^{\infty} \mathrm{e}^{-\mathrm{i}\omega_1 y}\, \tilde{\mathcal{N}}(\omega_1|\mu_1, \sigma_1^2)\, \tilde{\mathcal{N}}(\omega_1|\mu_2, \sigma_2^2)\, \frac{\mathrm{d}\omega_1}{2\pi}.$$

But,

$$\begin{aligned}\tilde{\mathcal{N}}(\omega_1|\mu_1, \sigma_1^2)\, \tilde{\mathcal{N}}(\omega_1|\mu_2, \sigma_2^2) &= \mathrm{e}^{\mathrm{i}\mu_1\omega_1 - \frac{1}{2}\sigma_1^2\omega_1^2}\, \mathrm{e}^{\mathrm{i}\mu_2\omega_1 - \frac{1}{2}\sigma_2^2\omega_1^2} \\ &= \mathrm{e}^{\mathrm{i}(\mu_1+\mu_2)\omega_1 - \frac{1}{2}(\sigma_1^2+\sigma_2^2)\omega_1^2} \\ &= \tilde{\mathcal{N}}(\omega_1|\mu_1 + \mu_2, \sigma_1^2 + \sigma_2^2)\end{aligned}$$

so that

$$\begin{aligned}I_6(y) &= \int_{-\infty}^{\infty} \mathrm{e}^{-\mathrm{i}\omega_1 y}\, \tilde{\mathcal{N}}(\omega_1|\mu_1 + \mu_2, \sigma_1^2 + \sigma_2^2)\, \frac{\mathrm{d}\omega_1}{2\pi} \\ &= \mathcal{N}(y|\mu_1 + \mu_2, \sigma_1^2 + \sigma_2^2).\end{aligned}$$

That is, we have shown that

$$\int_{-\infty}^{\infty} \mathcal{N}(x|\mu_1, \sigma_1^2)\, \mathcal{N}(y - x|\mu_2, \sigma_2^2)\, \mathrm{d}x = \mathcal{N}(y|\mu_1 + \mu_2, \sigma_1^2 + \sigma_2^2),$$

which can be a useful result.

5.3 Central Limit Theorem

The central limit theorem concerns the distribution of a sum of independent random variables $S = \sum_{i=1}^{n} X_i$. The distribution of S is given by

$$f_S(s) = \mathbb{E}_{X_1, X_2, \ldots, X_n}\left[\delta\left(s - \sum_{i=1}^{n} X_i\right)\right]$$

where $\delta(\cdots)$ is the Dirac delta function, which has the property that

$$\int_{-\infty}^{\infty} f(x)\, \delta(x - y)\, \mathrm{d}x = f(y).$$

We cover the Dirac delta function in Appendix 5.A on page 103. The moment generating function for S is defined as

$$M_S(l) = \mathbb{E}_S\left[\mathrm{e}^{l S}\right] = \mathbb{E}_{X_1, X_2, \ldots, X_n}\left[\mathrm{e}^{l \sum_{i=1}^{n} X_i}\right] = \mathbb{E}_{X_1, X_2, \ldots, X_n}\left[\prod_{i=1}^{n} \mathrm{e}^{l X_i}\right]$$
$$= \prod_{i=1}^{n} \mathbb{E}_{X_i}\left[\mathrm{e}^{l X_i}\right] = \prod_{i=1}^{n} M_{X_i}(l)$$

and the cumulant generating function is given by

$$G_S(l) = \log\bigl(M_S(l)\bigr)$$
$$= \sum_{i=1}^{n} \log\bigl(M_{X_i}(l)\bigr) = \sum_{i=1}^{n} G_{X_i}(l).$$

This is an important property of cumulative generationg function (CGF). That is,

> *the cumulant generating function for a sum of independent variables is equal to the sum of the cumulant generating function for each variable.*

This is the same property we discussed when we examined the cumulant generating function for the binomial and multinomial functions. Expanding both sides of $G_S(l) = \sum_i G_{X_i}(l)$ in powers of l (recall that for cumulant generating functions, $\kappa_n = G^{(n)}(0)$ so that the Taylor expansion of $G(l)$ is $\sum_i \kappa_i\, l^i/i!$) we find

$$l\, \kappa_1^S + \frac{l^2}{2}\kappa_2^S + \cdots = l \sum_i \kappa_1^{X_i} + \frac{l^2}{2} \sum_i \kappa_2^{X_i} + \cdots$$

Equating coefficients in l, we find the kth cumulant is given by

$$\kappa_k^S = \sum_{i=1}^{n} \kappa_k^{X_i}.$$

Thus the cumulants of any random variable, that is, the sum of independent random variables, are just equal to the sum of the constituent cumulants. This is one of the important properties that makes cumulants of considerable interest.

We notice that if the X_is are normally distributed (so have $\kappa_n = 0$ for $n \geq 3$) then X will have no higher cumulants and so will also be normally distributed! This is again sufficiently important to deserve highlighting:

> *the sum of independent normally distributed random variables is itself normally distributed.*

This is true even when the variables have different means and variances. A consequence of this is that you cannot 'deconvolve' random Gaussian noise into its initial components – this has important implications, for example in independent component analysis.

Example 5.2 Example 5.1 Revisited

In Example 5.1 we considered the convolution of two normal distributions

$$I_6(y) = \int_{-\infty}^{\infty} \mathcal{N}(x|\mu_1, \sigma_1^2)\, \mathcal{N}(y - x|\mu_2, \sigma_2^2)\, \mathrm{d}x.$$

We can rewrite this as

$$I_6(y) = \int_{-\infty}^{\infty} \int_{-\infty}^{\infty} \delta(y - x - z)\, \mathcal{N}(x|\mu_1, \sigma_1^2)\, \mathcal{N}(z|\mu_2, \sigma_2^2)\, \mathrm{d}x\, \mathrm{d}y$$

where we see that we can interpret $I_6(y)$ as the density function for a random variable $Y = X+Z$ where $X \sim \mathcal{N}(\mu_1, \sigma_1^2)$ and $Z \sim \mathcal{N}(\mu_2, \sigma_2^2)$. As a consequence Y will be normally distributed with mean $\mu_1 + \mu_2$ and variance $\sigma_1^2 + \sigma_2^2$. A fact which we proved through a rather long-winded chain of algebra in Example 5.1.

Suppose that S is the sum of *independent, identically distributed* random variables – such variables are colloquially known as *iids*. If the X_is are iid then as well as being independent they are also drawn from the same distribution $f_X(x)$. The cumulants of the sum of n *iids* is equal to $\kappa_k^S = n\, \kappa_k^X$. Consider the variable

$$Y = \frac{S - n\, \kappa_1^X}{\sqrt{n\, \kappa_2^X}}$$

then we find that $\kappa_1^Y = 0$, $\kappa_2^Y = 1$ and for $k > 2$

$$\kappa_k^Y = \frac{\kappa_k^X}{n^{k/2-1}\, (\kappa_2^X)^{k/2}}.$$

As $n \to \infty$, we find for $k > 2$ that $\kappa_k^Y = O(n^{1-k/2}) \to 0$ and so $Y \sim \mathcal{N}(0, 1)$. This result is known as the *central limit theorem*. This derivation of the central limit theorem also provides the corrections (i.e. the size of the higher cumulants when n is not infinite) which can be useful in practice.

To recap, the kth cumulant, κ_k^S, for a sum of n *iids* is equal to $n\, \kappa_k^X$, where κ_k^X is the kth cumulant of each individual *iid*. If we rescale the distribution for the sum so that the variance is equal to 1 then the rescaled cumulants are proportional to $n^{1-k/2}$. In the limit $n \to 0$, the rescaled higher cumulants vanish. Thus, the distribution more closely resembles a normal distribution.

The central limit theorem is, in fact, more general, in that provided the higher cumulants of the component distributions are well behaved then the sum of any set of independent random variables will tend to a normal distribution as the number of components increases.

It is sometimes said that normal distributions are ubiquitous because most observables can be considered as the sum of many independent uncertainties. This may occasionally be true, but is actually a big assumption. On the other hand, many other people will tell you that observed quantities are rarely normally distributed and, in particular, outliers are much more frequent than a normal distribution predicts. In a way both these views are valid. Many random variables have approximately bell-shaped distributions which are reasonably approximated by normal distributions, but their tails are rarely well described by a normal distribution. This may or may not matter, depending on whether you are interested in the great bulk of the population or in the outliers.

The derivation of the central limit theorem given assumes that the CGF exists. This, as we have said, is not guaranteed. After all, e^{lx} grows exponentially with x so its expectation will only exist if the distribution falls off sufficiently fast as $x \to \infty$. This certainly is not the case for the Cauchy distribution

$$\text{Cau}(x) = \frac{1}{\pi(1 + x^2)}$$

for which neither the mean nor variance is defined. Although the moment generating function is not defined, the characteristic function is. If X is Cauchy distributed then

$$\phi_X(\omega) = \mathbb{E}_X\left[e^{i\omega X}\right] = e^{-|\omega|}$$

(see Appendix 5.B). Note that this is not differentiable at $\omega = 0$, reflecting the fact that the Cauchy distribution has no moments. Defining $Y = \frac{1}{n}\sum_{i=1}^{n} X_i$, where X_i is Cauchy distributed, then

$$\begin{aligned}\phi_Y(\omega) &= \mathbb{E}_Y\left[e^{i\omega Y}\right] = \mathbb{E}_{X_1,X_2,\ldots,X_n}\left[e^{i\omega \sum_{i=1}^{n} X_i/n}\right] = \prod_{i=1}^{n} \mathbb{E}_{X_i}\left[e^{i\omega X_i/n}\right] = \phi_X^n(\omega/n)\\ &= \left(e^{-|\omega/n|}\right)^n = e^{-|\omega|}.\end{aligned}$$

Thus Y is also Cauchy distributed! This is rather remarkable: the mean of n Cauchy random variables, rather than becoming closer to a normal distribution as n increases, remains Cauchy distributed. In general, *the central limit theorem only applies to variables with a finite mean and variance*.

Another limitation of the central limit theorem is that it really only tells you about the distribution a few standard deviations around the mean. It reveals very little about the tails of the distribution. You should not assume that if S_n is the sum of n independent variables that the tails for the distribution of S_n fall off as fast as a normal distribution as this will generally not be true. In Chapter 7 we discuss Chernoff's inequality, which provides a bound on the probability of a large deviation from the mean for a sum of random variables.

Example 5.3 Log-Normal Distribution

Before leaving the central limit theorem, we consider another application of it in a slightly different scenario: a random variable which arises as a product of random variables

$$X = \prod_{i=1}^{n} Y_i.$$

An example of this would be if we dropped a rod of glass. We assume that the rod breaks into two and then each half breaks into two and this carries on n times. Each time it breaks, it does so at a random point. This process is illustrated in Figure 5.5. We consider a fragment of glass and ask how long it is. Assume that the initial rod

Figure 5.5 An example of a process giving rise to a log-normal distribution. A glass rod breaks into two n times. Each break happens at a random position. Notice that the final lengths vary very considerably.

of glass had length 1. Let Y_i be the fraction along the glass that the ith break occurs. Thus, after the first break the length of the fragment will be Y_1, after the second break it will be $Y_1\,Y_2$. After n breaks the length of the fragment will be given by X defined above.

To understand how X is distributed we consider the random variable $L = \log(X)$, then

$$L = \sum_{i=1}^{n} \log(Y_i),$$

where

$$\mathbb{E}_{Y_i}\left[\log(Y_i)\right] = \int_0^1 \log(y)\,\mathrm{d}y = -1,$$
$$\mathbb{E}_{Y_i}\left[\log^2(Y_i)\right] - \left(\mathbb{E}_{Y_i}\left[\log(Y_i)\right]\right)^2 = 1.$$

Thus, for large n (invoking the central limit theorem)

$$f_L(l) \approx \frac{1}{\sqrt{2\pi}} \mathrm{e}^{-(l+n)^2/(2\,n)}.$$

Making the change of variables $X = \mathrm{e}^L$ (see Example 1.7 on page 14 for details on the change of variables) we find the distribution of lengths is given by

$$f_X(x) \approx \frac{1}{x\sqrt{2\,\pi\,n}} \mathrm{e}^{-(\log(x)+n)^2/(2\,n)}.$$

This is an example of a *log-normal distribution* defined by

$$\mathrm{LogNorm}(x|\mu,\sigma^2) = \frac{1}{x\,\sigma\sqrt{2\pi}} \mathrm{e}^{-(\log(x)-\mu)^2/(2\sigma^2)}. \qquad (5.2)$$

This has mean $\kappa_1 = \mathrm{e}^{\mu+\sigma^2/2}$, variance $\kappa_2 = (\mathrm{e}^{\sigma^2} - 1)\,\mathrm{e}^{2\mu+\sigma^2}$, and skewness $(\mathrm{e}^{\sigma^2} + 2)\sqrt{\mathrm{e}^{\sigma^2} - 1}$. In our example $\mu = -n$ and $\sigma^2 = n$, which would give a $\kappa_1 = \mathrm{e}^{-n/2}$, with a variance $\kappa_2 \approx 1$ and a third cumulant $\kappa_3 \approx \mathrm{e}^{3n/2}$. Thus the distribution is massively skewed to the right. Log-normal distributions typically have a very long tail (see Figure 5.6).

The derivation we gave of the log-normal distribution has some minor blemishes. The obvious one is that when breaking a rod the sizes of the pieces are not independent. We discuss this later in this

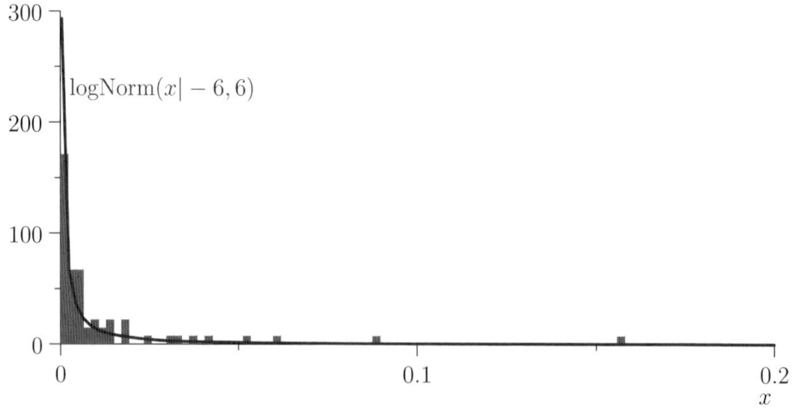

Figure 5.6 Illustration of a log-normal distribution LogNorm(x| − 6, 6) together with a histogram of length from Example 5.3. Note the very long tail, with many short pieces and a few much longer pieces. The curve for the long-normal distribution actually goes back down to zero at $x = 0$ – that is it has a very high narrow peak at e^{-6} as well as a long tail for large values of x.

section. There is a second subtle problem. The central limit theorem tells us about convergence in distribution, which is a rather weak form of convergence (see Section 7.1). In particular, it does not guarantee much about the tails. The change of variables back to $X = e^{L}$ makes the tails very important. As a consequence, the mean of X as predicted by the log-normal distribution is $e^{-n/2}$ while the actual mean of X is 2^{-n}, which follows trivially from the fact that we considered a unit length rod to break in two sections successively n times. This discrepancy shows that considerable care is necessary when invoking the central limit theorem – it tells us about the bulk of the distribution, but not the tails.

Applying the central limit theorem to problems with long tails is dangerous even when it is feasible. For example, log-normal deviates satisfy the condition for the central limit theorem in that they have a well-defined first and second cumulant. However, consider the random variable $S = \sum_{i=1}^{N} X_i$ where the X_is are *iid* log-normal deviates with $\sigma^2 = 10$ (not unreasonable in Example 5.3). Then the skewness of the X_i is 3.27×10^6 and the skewness of S is $3.27 \times 10^6/\sqrt{N}$. Therefore, we need N to be very large (e.g. $N \approx 10^{13}$) before the skewness becomes small (order 1). Even if a mathematician tells you that the central limit theorem applies as the distribution has a well-defined mean and variance, be on your guard.

We should finally make a few comments about another key assumption concerning the central limit, namely that the variables are independent. When the variables aren't independent then the sum of many random variables can be distributed in much more complicated ways. However, if the dependence is 'sufficiently weak' the central limit theorem may still apply. The problem arises of deciding how weak is sufficiently weak? – a question which is very difficult to answer in general. In our example the pieces all sum to 1 so we cannot assume the pieces to be independent. However, as shown in Figure 5.6, the length of the pieces are well approximated by a log-normal distribution.

Long-tailed (aka thick, heavy, or fat-tailed) distributions, such as the log-normal, occur quite frequently in the real world. In such cases, the typical behaviour (that which is found most often) is associated with the mode of the distribution, which can be very different from the mean. To find this typical behaviour it is common (at least amongst physicists) to consider the mean of the logarithm of random variables with long tails. This reduces the influence of very rare outliers compared to the samples around the mode of the distribution.

Example 5.4 Betting to Win!

Historically one of the driving forces behind developing the theory of probability was to understand betting. This motivation led to the rather sterile conclusion that you aren't going to win in any game where the odds are against you. You're not even going to win in a game where the odds are even, unless you are (a) very lucky or (b) richer than the person you're betting against. But what if you are betting in a game where the odds are with you? Admittedly, there are not that many bookies who will offer you such odds, although arguably financial investment might provide this kind of situation. The question then is how should you bet in a series of games when the odds are in your favour?

The surprising answer to this was supplied by John L. Kelly in a seminal paper (Kelly, 1956). Suppose we have a series of games with two possible outcomes. Denote the outcome of game i by the Bernoulli random variable X_i, and let the return on winning a bet of one pound in round i be r_i pounds. Let C_i be your capital after round i and f_i be the fraction of the capital that you bet. Then your capital after round k is

$$C_k = \prod_{i=1}^{k} \left(1 - f_i + X_i\, r_i\, f_i\right) C_0$$

where C_0 is your initial capital. The log-gain is thus

$$\log\left(\frac{C_k}{C_0}\right) = \sum_{i=1}^{k} \log\left(1 - f_i + X_i\, r_i\, f_i\right).$$

This is a sum of random variables, by which the central limit theorem will be approximately normally distributed. The typical return is the mode of this distribution and we maximise our typical winning by maximising the expected log-gain

$$\mathbb{E}\left[\log\left(\frac{C_k}{C_0}\right)\right] = \sum_{i=1}^{k} \mathbb{E}\left[\log(1 - f_i + X_i\, r_i\, f_i)\right]$$
$$= \sum_{i=1}^{k} \left(p_i \log(1 - f_i + r_i\, f_i) + (1 - p_i)\log(1 - f_i)\right)$$

where $p_i = \mathbb{P}\left(X_i = 1\right)$. To maximise this we set the derivative of the log-gain with respect to f_i to zero

$$\frac{\partial}{\partial f_i}\mathbb{E}\left[\log\left(\frac{C_k}{C_0}\right)\right] = \frac{p_i(r_i - 1)}{1 - f_i + r_i\, f_i} - \frac{1 - p_i}{1 - f_i} = 0$$

or $f_i = (p_i\, r_i - 1)/(r_i - 1)$. Note that we are usually only allowed to place positive bets so we require $p_i\, r_i > 1$ (if you want to win in expectation, it clearly makes sense to only make a bet when the probability of winning times the size of the reward exceeds the cost of the bet). The expected winnings in each round are $(p_i\, r_i - 1)\, f_i$, so this will be positive only if $p_i\, r_i > 1$.

The surprising part of this strategy is that it does not maximise the expected winnings. The expected winnings are

$$\mathbb{E}\left[C_k\right] = \prod_{i=1}^{k} \left(p_i\,(1 - f_i + r_i\, f_i) + (1 - p_i)(1 - f_i)\right) C_0$$
$$= \prod_{i=1}^{k} \left(1 + (p_i\, r_i - 1)\, f_i\right) C_0.$$

Provided the expected winnings in each round satisfy $p_i\, r_i > 1$, then the total expected winnings are maximised when $f_i = 1$. For example, suppose that $p_i = 1/2$ (you are betting on a fair coin) and that $r_i = 4$. In other words you are playing a game with your fairy godmother where she tosses a fair coin 10 times. You bet on the outcome of each toss. If you win the toss you get four times what you bet, otherwise you lose your stake. If you want to maximise your expected winnings then you should bet your entire capital at each round. In this case you only win if you are right on every toss of the coin, in which case you win $4^{10}\, C_0 \approx 10^6\, C_0$. Assuming your initial capital $C_0 = £1$, then you would will win just over 1 million pounds. Of course, the chance of winning is only 1/1024, otherwise you lose everything. Your expected winnings would be £1024. However, if you follow Kelly, then you would stake only one third of your capital each round and your expected winnings would be $(4/3)^{10} \times C_0 \approx 17.8\, C_0$. In this scenario there are 10 possible outcomes (since the amount you win does not depend on the order in which you win). The probability of these outcomes and the log-gain in your capital are shown in Figure 5.7. Following Kelly, if you are lucky and win all 10 rounds then you win £1024; on the other hand, you have over 80% chance of winning something. If you think of life as a long-distance race rather than a sprint, then playing Kelly's strategy is a smart move.

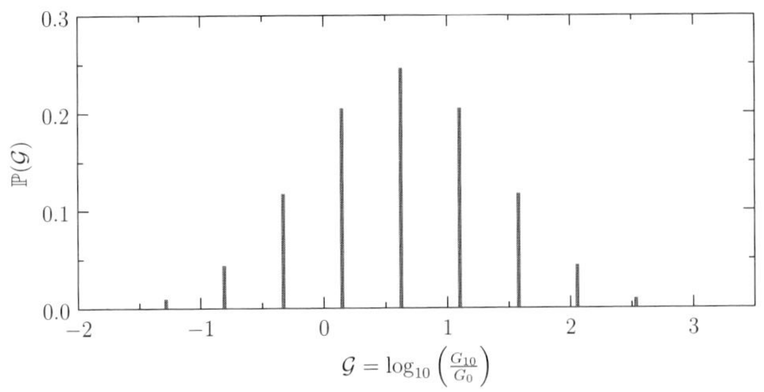

Figure 5.7 Log-gain, $\mathcal{G}$, in your capital after 10 rounds of playing a game biased in your favour, using Kelly's criterion for choosing your bid.

In this example, if you are a gambler you may well decide that it's worth taking the high-risk bet to become a millionaire rather than be content with, at most, only winning £1024. However, consider playing the same game 1000 times. In this case you have a probability of $2^{-1000} \approx 10^{-333}$ of becoming rich beyond your imagination. On the other hand, following Kelly you have a chance of greater than 0.999 of winning over £10^{38}. Call me risk adverse, but I would follow Kelly in this case.

Of course, Kelly comes with its drawbacks. Firstly you need to have an unbiased estimate of the probability of success. Because you select good odds to bet on, they tend to be biased (you are much more likely to bet on events whose probabilities you've overestimated than events whose probabilities you've underestimated). Thus using Kelly may not be optimal. The second drawback of using Kelly to become rich is that you need to find a bookie who gives you good odds. A nice (though somewhat irrelevant) story is that Kelly worked with Claude Shannon (the inventor of information theory, which we cover in Chapter 9) and the mathematics professor Ed Thorp (a proponent of Kelly's strategy in financial investments). To bias the odds in their favour they developed the world's first wearable computer that predicted the likely outcome of roulette by observing the state of the ball just before the roulette wheel stopped spinning. Shannon, Thorp, and their wives tried this out in Las Vegas. This is now illegal and was probably always dangerous (winning too much in a casino is not necessarily healthy) so I would not recommend this, but it certainly provides a different appreciation of Claude Shannon than you might get from working through his theorems.

5.4 Cumulative Distribution Function of a Normal Distribution

The CDF for a normally distributed random variable, $X \sim \mathcal{N}(\mu, \sigma^2)$, is defined as

$$\mathbb{P}\left(X \le x | \mu, \sigma^2\right) = \int_{-\infty}^{x} e^{-(y-\mu)^2/(2\sigma^2)} \frac{dy}{\sqrt{2\pi\sigma^2}}.$$

Alas, this integral is not expressible in terms of nice well-known functions (although, see Exercise 5.3). Instead, we have to define a function. The most convenient such function is

$$\Phi(x) = \int_{-\infty}^{x} e^{-y^2/2} \frac{dy}{\sqrt{2\pi}}$$

for which

$$\mathbb{P}\left(X \le x | \mu, \sigma^2\right) = \Phi\left(\frac{x-\mu}{\sigma}\right).$$

Writing the CDF for a normal using the notation Φ (i.e. the Greek letter capital phi) is fairly common. The CDF, $\Phi(x)$, denotes the area under the curve of normal distribution from $-\infty$ to x, as illustrated in Figure 5.8.

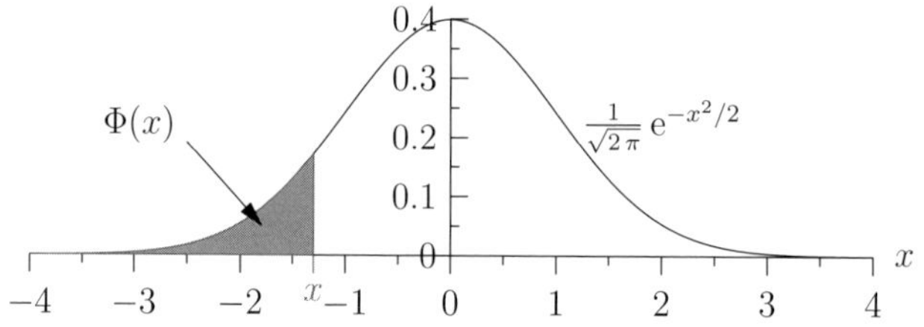

Figure 5.8 Area under the probability distribution function (PDF) for a normal distribution is equal to the CDF, $\Phi(x)$.

For historical reasons the standard function related to the CDF of the normal distribution is the *error function*, $\text{erf}(x)$, defined as

$$\text{erf}(x) = \frac{2}{\sqrt{\pi}} \int_0^x e^{-t^2} dt,$$

which is related to our function by

$$\Phi(x) = \frac{1}{2} + \frac{1}{2}\text{erf}\left(\frac{x}{\sqrt{2}}\right).$$

The error function is available in most computer languages and mathematical packages. In the limit $x \to -\infty$ the error function approaches -1 so that $\Phi(x)$ approaches zero. However, using the definition above, $\Phi(x)$ would be computed by subtracting a number very close to a half from a half. This can make the numerical estimate rather imprecise for large negative x. To overcome this the complementary error function, $\text{erfc}(x) = 1 - \text{erf}(x)$, is also available in most computer environments. Using this we can write

$$\Phi(x) = \frac{1}{2}\text{erfc}\left(\frac{-x}{\sqrt{2}}\right).$$

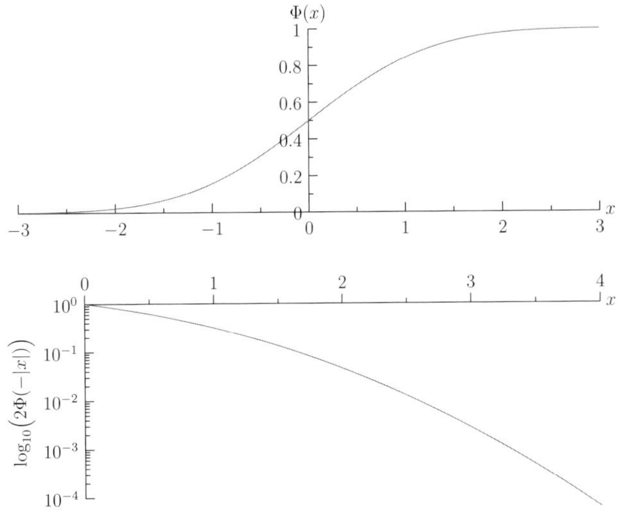

Figure 5.9 CDF for the normal distribution with mean 0 and variance 1. Also shown is $\log_{10}\left(2\Phi(-|x|)\right)$, which shows the behaviour of $\Phi(x)$ as $-x$ increases.

The top graph in Figure 5.9 shows $\Phi(x)$ versus x. The probability of observing a normally distributed random variable at least z standard deviations from the mean is given by

$$\mathbb{P}\left(|X-\mu| > z\,\sigma\right) = 2\Phi\left(-|z|\right).$$

The function $\Phi(x)$ falls very rapidly to zero as x becomes negative. The bottom graph in Figure 5.9 shows the log-probability of being at least x standard deviations away from the mean. If you have data consisting of normally distributed variables then you expect 68% of it to lie within one standard deviation of the mean, 95% of it to lie within two standard deviations of the mean, and 99.7% of it to lie within three standard deviations of the mean. To find an analytic approximation for $\Phi(x)$ for large negative x we use integration by parts

$$\begin{aligned}\Phi(x) &= \int_{-\infty}^{x} \frac{-1}{z}\left(-z\frac{\mathrm{e}^{-z^2/2}}{\sqrt{2\pi}}\right)\mathrm{d}z = \left[-\frac{1}{z}\frac{\mathrm{e}^{-z^2/2}}{\sqrt{2\pi}}\right]_{z=-\infty}^{z=x} - \int_{-\infty}^{x}\frac{1}{z^2}\frac{\mathrm{e}^{-z^2/2}}{\sqrt{2\pi}}\mathrm{d}z \\ &= -\frac{\mathrm{e}^{-x^2/2}}{\sqrt{2\pi}\,x} - \int_{-\infty}^{x}\frac{1}{z^2}\frac{\mathrm{e}^{-z^2/2}}{\sqrt{2\pi}}\mathrm{d}z.\end{aligned}$$

The integral on the right-hand side is less than $\Phi(x)/x^2$ so that $\Phi(x) \approx -\mathrm{e}^{-x^2/2}/(\sqrt{2\pi}\,x)$ as $x \to -\infty$. We can sandwich $\Phi(x)$ between lower and upper bounds which are close when x is large.

$$-\frac{x}{x^2+1}\frac{\mathrm{e}^{-x^2/2}}{\sqrt{2\pi}} < \Phi(-x) < -\frac{1}{x}\frac{\mathrm{e}^{-x^2/2}}{\sqrt{2\pi}}.$$

To obtain more accuracy we can derive a full asymptotic expansion by applying integration by parts iteratively, although a better approximation of $\Phi(x)$ for large negative x is given by

$$\Phi(x) \approx \frac{(\sqrt{4+x^2}+x)}{2} \frac{\mathrm{e}^{-x^2/2}}{\sqrt{2\pi}}.$$

For large x we note that, because $\mathcal{N}(x|0,1)$ is symmetric, $\Phi(x) = 1 - \Phi(-x)$; thus we can use the same approximations in the limit $x \to \infty$.

Occasionally you have to integrate over products of Gaussians and error functions. There are a surprising number of these integrals you can compute in closed form (although they tend not to be that simple). These integrals are not well known or easy to find so I list some of them below.

$$\begin{aligned}
\int_{-\infty}^{\infty} \mathrm{e}^{-x^2/2}\Phi(a\,x+b)\,\frac{\mathrm{d}x}{\sqrt{2\pi}} &= \Phi\left(\frac{b}{\sqrt{1+a^2}}\right)\\
\int_{-\infty}^{\infty} \mathrm{e}^{-x^2/2+b\,x}\Phi(a\,x)\,\frac{\mathrm{d}x}{\sqrt{2\pi}} &= \mathrm{e}^{b^2/2}\,\Phi\left(\frac{a\,b}{\sqrt{1+a^2}}\right)\\
\int_{-\infty}^{0} \mathrm{e}^{-x^2/2}\Phi(-a\,x)\,\frac{\mathrm{d}x}{\sqrt{2\pi}} &= \frac{1}{2}+\frac{1}{\pi}\arcsin\left(\frac{a}{\sqrt{1+a^2}}\right)\\
\int_{-\infty}^{\infty} \mathrm{e}^{-x^2/2}\Phi(a\,x)\,\Phi(b\,x)\,\frac{\mathrm{d}x}{\sqrt{2\pi}} &= \frac{2}{\pi}\arccos\left(\frac{-a\,b}{\sqrt{1+a^2+b^2+a^2\,b^2}}\right)\\
\int_{-\infty}^{\infty} \mathrm{e}^{-x^2/2}\Phi^2(a\,x)\,\frac{\mathrm{d}x}{\sqrt{2\pi}} &= \frac{2}{\pi}\arccos\left(\frac{-a^2}{1+a^2}\right).
\end{aligned}$$

At the time of writing a more comprehensive list, including many indefinite integrals, can be found on Wikipedia under the title *List of integrals of Gaussian functions*.

5.5 Best of *n*

Given a set of n *iid* continuous random variable $\mathcal{S}_n = \{X_1, X_2, \ldots, X_n\}$, where each element is drawn from a distribution f_X, what is the distribution of the largest element, L_n, in $\mathcal{S}_n$? This question arises in many different application areas (we give one example at the end of this section). The CDF for L_n is given by

$$F_{L_n}(x) = \mathbb{P}\left(L_n \leq x\right) = \prod_{i=1}^{n} \mathbb{P}\left(X_i \leq x\right) = F_X^n(x).$$

If $F_X(x)$ is differentiable then the PDF for L_n is simply

$$f_{L_n}(x) = \frac{\mathrm{d}\,F_{L_n}(x)}{\mathrm{d}x} = \frac{\mathrm{d}\,F_X^n(x)}{\mathrm{d}x} = n\,f_X(x)\,F_X^{n-1}(x)$$

since the PDF, $f_X(x)$, is the derivative of the CDF, $F_X(x)$.

When the *iid*s are drawn from a normal distribution

$$f_{L_n}(x) = n\,\mathcal{N}(x|\mu,\sigma^2)\,\Phi\left(\frac{x-\mu}{\sigma}\right)^{n-1}.$$

Examples of this are shown in Figure 5.10.

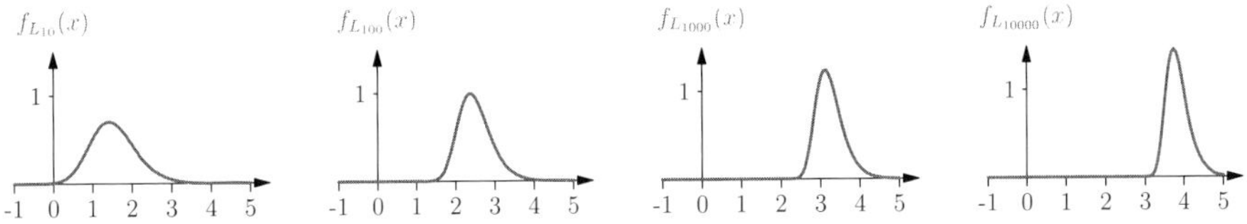

Figure 5.10 Distributions of the largest element of the set $\mathcal{S}_n = \{X_1, X_2, \dots, X_n\}$ where $X_i \sim \mathcal{N}(0, 1)$ for $n = 10$, 100, 1000, and 10 000.

For discrete random variables, $K_i \sim f_K$, we still have that $F_{L_n}(k) = F_K^n(k)$, following the same logic as for continuous random variables. Assuming the random variables take integer values, the PDF is given by

$$f_{L_n}(k) = F_{L_n}(k) - F_{L_n}(k-1) = F_K^n(k) - F_K^n(k-1).$$

This expression, however, doesn't simplify any further.

Extreme Value Distribution

For very large n it turns out that, for continuous variables, $f_{L_n}(x)$ converges to a distribution known as an *extreme value distribution*. There are a small number of universal extreme value distribution families that are shared by many distributions: f_X. Which extreme value distribution family we converge to depends on the fall off of the right-hand tail of the distribution of the X_is.

Consider the case where $X_i \sim \mathcal{N}(0, 1)$. The largest member of the set $\mathcal{S}_n$ will be in the right-hand tail of the distribution when n becomes large. Furthermore, as n increases the distribution of L_n becomes narrower (we see both these effects in Figure 5.10). Consider a point x^* where $f_L(x^*)$ is relatively large (i.e. close to its mean, median, or mode). Taylor expanding x^2 around x^* we find

$$x^2 = (x^*)^2 + 2(x - x^*)\,x^* + (x - x^*)^2 = -(x^*)^2 + 2\,x\,x^* + (x - x^*)^2$$

so that

$$\mathcal{N}(x|0,1) = \frac{1}{\sqrt{2\pi}} e^{-x^2/2} = \frac{1}{\sqrt{2\pi}} e^{(x^*)^2/2 - x\,x^* - (x-x^*)^2/2}.$$

If we consider x very close to x^*, the term $(x - x^*)^2$ will be small and (setting $(x - x^*)^2 = 0$)

$$\mathcal{N}(x|0,1) \approx c\,e^{-x\,x^*}$$

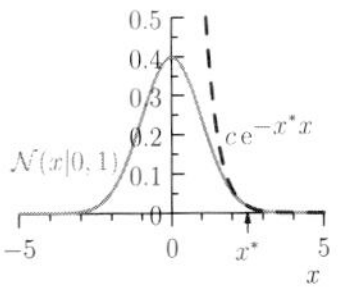

where $c = e^{(x^*)^2/2}/\sqrt{2\pi}$. Using this approximation, the corresponding CDF is given by

$$F_X(x) = 1 - \mathbb{P}(X > x) \approx 1 - \int_x^\infty c\,e^{-x\,x^*} dx = 1 - \frac{c}{x^*} e^{-x\,x^*}.$$

Thus

$$F_{L_n}(x) \overset{(1)}{\approx} \left(1 - \frac{c}{x^*} e^{-x\,x^*}\right)^n \overset{(2)}{=} e^{n \log\left(1 - \frac{c}{x^*} e^{-x\,x^*}\right)}$$
$$\overset{(3)}{\approx} e^{-\frac{n\,c}{x^*} e^{-x\,x^*}} \overset{(4)}{=} e^{-e^{-(x-\mu)\,x^*}}$$

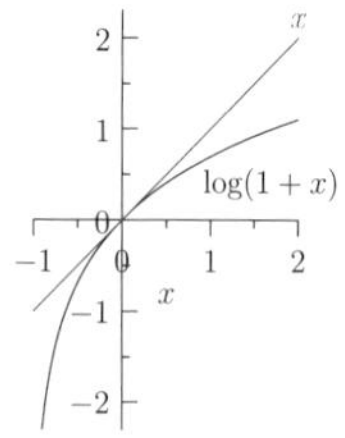

(1) Using $F_{L_n}(x) = F_X^n(x)$ and the approximation for $F_X(x)$.
(2) Using $a^n = \mathrm{e}^{n\,\log(a)}$.
(3) Using the approximation $\log(1+x) \approx x$ when $|x| \ll 1$.
(4) Defining $\frac{n\,c}{x^*} = \mathrm{e}^{\mu\,x^*}$ or $\mu = \frac{1}{x^*}\log\left(\frac{n\,c}{x^*}\right)$.

Note that as n increases, L_n will typically become larger so that $F_X(x^*)$ becomes closer to 1. As a consequence, the approximation made in step (3) becomes more accurate. Differentiating $F_{L_n}(x)$ we obtain the PDF for the extreme value distribution

$$f_{L_n}(x) \approx x^*\,\mathrm{e}^{-(x-\mu)\,x^* - \mathrm{e}^{-(x-\mu)\,x^*}}.$$

This limiting distribution is known as the Gumbel distribution or the *generalised extreme value distribution type I*. Often this distribution is written with $x^* = 1/\beta$. The distribution is the limit distribution for a large number of distributions, $f_X(x)$, where the tail of $f_X(x)$ can be approximated by an exponential fall-off. This includes the family of gamma distribution. Other families of extreme value distributions occur when the tail of f_X has different properties, such as falling off as a power law.

We have still to determine x^* (a point where $f_{L_n}(x)$ is relatively large). This is not as critical as it may seem (it is the point around which we approximate a normal distribution by an exponential – this approximation does not vary greatly in the region where $f_{L_n}(x)$ is large). Since we are dealing with a set of normally distributed variables

$$F_{L_n}(x) = \Phi^n(x) = \big(1 - \Phi(-x)\big)^n = \mathrm{e}^{n\,\log\left(1-\Phi(-x)\right)} \approx \mathrm{e}^{-n\,\Phi(-x)}.$$

When $n\,\Phi(-x) \gg 1$ then $F_{L_n}(x) \approx 0$ and when $n\,\Phi(-x) \ll 1$ then $F_{L_n}(x) \approx 1$. Thus a good place to choose x^* is where $\Phi(-x) = 1/n$. Using the asymptotic expansion for $\Phi(-x)$ as $x \to \infty$ given by $\Phi(-x) \approx \exp(-x^2/2)/(\sqrt{2\pi}\,x)$, then

$$\frac{\mathrm{e}^{-(x^*)^2/2}}{\sqrt{2\pi}\,x^*} = \frac{1}{n}.$$

There is no closed-form solution to this. We could solve it numerically (although we might just as easily solve $\Phi(-x^*) = 1/n$ in that case). However, for large n it turns out $n \gg x^*$ so that

$$\frac{\mathrm{e}^{-(x^*)^2/2}}{\sqrt{2\pi}} \approx \frac{1}{n}$$

or

$$x^* \approx \sqrt{2\log(n) - \log(2\pi)}.$$

With more work we can compute corrections to this of order $\log(\log(n))$, although when n is very large we can get away with the even simpler approximation $x^* \approx \sqrt{2\log(n)}$. The quality of the approximation is reasonable for large n, as illustrated in Figure 5.11. However, I am yet to come across an application where using the extreme value distribution rather than the exact distribution for L_n

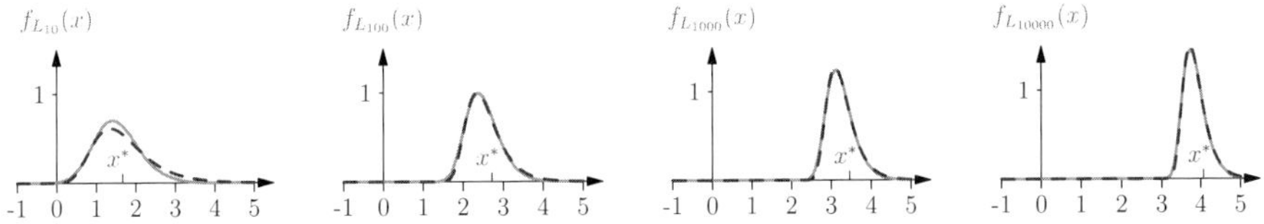

Figure 5.11 Distributions of the largest element of the set $\mathcal{S}_n = \{X_1, X_2, \ldots, X_n\}$ where $X_i \sim \mathcal{N}(0, 1)$ for $n = 10$, 100, 1000, and 10 000. We also show the approximate extreme value distribution (dashed curve) and the point $x^* = \sqrt{2\log(n) - \log(2\pi)}$.

is worthwhile. The Gumbel distribution has a slightly cleaner functional form (it involves elementary functions rather than $\Phi(x)$), but by the time you have computed x^* and μ it doesn't seem to save any work.

Example 5.5 Significant Sequence Alignments

A very important application in bioinformatics is sequence alignment. This may be aligning DNA sequences or protein sequences. Its importance comes from the fact that different species tend to have proteins and DNA that are clearly related both in terms of sequence similarity and function. In sequence alignment we have two sequences, e.g.

$$S_1 = KKASKPKKAASKAPTKKPKATPVK$$
$$S_2 = KKAAKPKKAASKAPSKKPKATPVK.$$

In this example S_1 is part of the Histone H1 protein in humans, while S_2 is part of the Histone H1 protein in mice. In these two subsequences there are two sites where the sequences differ.

Sequence alignment is a classic and well-studied problem in computer science (which refers to it as inexact string matching). Given two sequences, there is an efficient algorithm for finding the 'best alignment' (in fact, there are a number of different algorithms depending, for example, on whether you want to match the whole string or find the best subsequences of the two strings to match). To determine what we mean by 'best alignment,' biologists have empirically estimated the probability of any letter in the sequence (amino acid in the case of proteins or bases in the case of DNA) being substituted by another letter. In addition, elements in either sequence can be deleted or new elements inserted, again with some small probability. By taking the product of these probabilities (or more conveniently the sum of the log-probabilities), we can compute a score giving the probability of a particular alignment. The number of possible ways of matching two sequences is exponential in the length of the strings, and for all but the shortest strings it would be computationally infeasible to try all possible matches. Fortunately, using an algorithmic method known as dynamic programming it is nevertheless possible to find the best alignment very efficiently.

An important scientific question is whether the best alignment that is found is biologically significant (i.e. due to an evolutionary shared inheritance). There are an exponential number of possible alignments and the algorithm has found the best. However, because we have found the best of a large number of alignments, the score we obtain will be much higher than chance. To determine whether the alignment we have found is likely to be biologically significant we can estimate the distribution of scores for a random alignment involving the same number of elements as the alignment we have found. We denote the score of a random alignment by X and its CDF by $F_X(x)$. The score we obtain is typically a sum of the log-probabilities for all the elements that have been aligned so that X will typically be approximately normally distributed. We also need to compute the total number, N, of possible alignments of this length that we could have found. This is a complicated combinatorial problem, but typically N will be very large. The CDF for the best alignments that would occur by chance is $F_X^N(x)$. If we find an alignment with a score X_{best} then the probability that we could have found a score that is as good as, or better than, X_{best}, given that we are aligning random sequences, is $F_X^N(X_{\text{best}})$. If this is very small then we can be confident that the alignment found is biologically significant.

5.6 Multivariate Normal Distributions

The normal distribution naturally extends to the multivariate case; furthermore, because the exponent is quadratic it is relatively easy to compute expectations analytically. This can be very important in high dimensions as the only other feasible approach may be Monte Carlo, which is slow. There is a very well-developed field of multivariate statistics which builds on the multi-dimensional normal distribution.

Computing high-dimensional Gaussian integrals requires some knowledge of linear algebra. You can accept these formulas on faith, but if you work constantly with multivariate normal distributions then it is worth understanding how they are obtained. The multivariate normal distribution is

We are being a bit sneaky. For any constant c the determinant $|c\,\mathsf{M}|$ of an $n \times n$ matrix, M, is equal to $c^n|\mathsf{M}|$. Thus our normalisation could also be written $(2\pi)^{-n/2}/\sqrt{|\boldsymbol{\Sigma}|}$.

$$\mathcal{N}(\boldsymbol{x}|\boldsymbol{\mu}, \boldsymbol{\Sigma}) = \frac{1}{\sqrt{|2\pi\boldsymbol{\Sigma}|}} \mathrm{e}^{-\frac{1}{2}(\boldsymbol{x}-\boldsymbol{\mu})^{\mathsf{T}}\boldsymbol{\Sigma}^{-1}(\boldsymbol{x}-\boldsymbol{\mu})}$$

where $\boldsymbol{\mu}$ is the mean (now in a multidimensional space), $\boldsymbol{\Sigma}$ is the covariance matrix, and $|2\pi\boldsymbol{\Sigma}|$ denotes the determinant of the matrix $2\pi\boldsymbol{\Sigma}$. The covariance matrix must be symmetric and, furthermore, positive definite. Symmetry just means that $\Sigma_{ij} = \Sigma_{ji}$. Positive definite means that for all non-zero vectors $\boldsymbol{u}$ we have $\boldsymbol{u}^{\mathsf{T}}\boldsymbol{\Sigma}\boldsymbol{u} > 0$, which is required to ensure that the distribution goes to zero (rather than infinity) as we move away from the mean. An entirely equivalent condition for a matrix to be positive definite is that all of its eigenvalues are positive.

Example 5.6 Two-Dimensional Normal Distribution
Figure 5.12 plots the multivariate Gaussian

$$\mathcal{N}\left(\begin{pmatrix} x \\ y \end{pmatrix} \middle| \begin{pmatrix} 1 \\ 1 \end{pmatrix}, \begin{pmatrix} 2 & -1 \\ 1 & 1 \end{pmatrix}\right) = \frac{1}{2\pi} \mathrm{e}^{-\frac{1}{2}\left(x^2+2\,x\,y+2\,y^2-4\,x-6\,y+5\right)}.$$

Note that the eigenvalues of $\mathbf{\Sigma} = \begin{pmatrix} 2 & -1 \\ -1 & 1 \end{pmatrix}$ are $(3+\sqrt{5})/2$ and $(3-\sqrt{5})/2$ and the corresponding eigenvectors are $\begin{pmatrix} -0.85065 \\ 0.52573 \end{pmatrix}$ and $\begin{pmatrix} 0.52573 \\ 0.85065 \end{pmatrix}$. The determinant of $\mathbf{\Sigma}$ is equal to 1 and the inverse is $\mathbf{\Sigma}^{-1} = \begin{pmatrix} 1 & 1 \\ 1 & 2 \end{pmatrix}$.

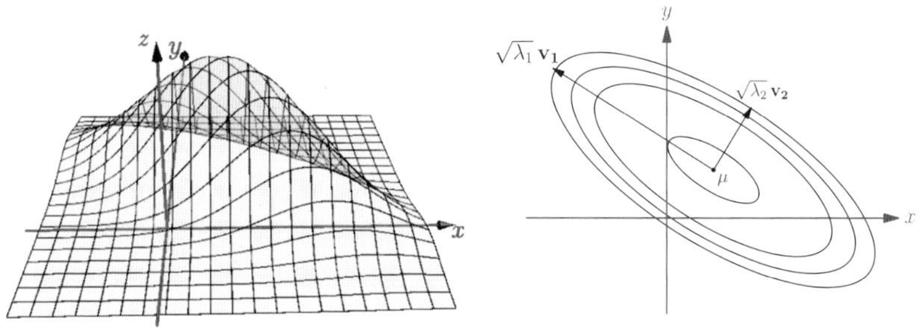

Figure 5.12 Illustration of a two-dimensional Gaussian with covariance $\mathbf{\Sigma} = \begin{pmatrix} 2 & -1 \\ -1 & 1 \end{pmatrix}$ and mean $\boldsymbol{\mu} = (1, 1)$. It shows the relationship between the eigenvalues and eigenvectors of $\mathbf{\Sigma}$ and the shape of the distribution.

To show that $\mathcal{N}(\boldsymbol{x}|\boldsymbol{\mu}, \mathbf{\Sigma})$ is normalised (i.e. to compute the integral) it is necessary to diagonalise the covariance matrix. For any symmetric matrix $\mathbf{\Sigma}$ it is possible to find an orthogonal matrix $\mathbf{V}$ such that

$$\mathbf{V}^{\mathsf{T}}\mathbf{\Sigma}\mathbf{V} = \mathbf{\Lambda}$$

where $\mathbf{\Lambda}$ is a diagonal matrix. The diagonal elements $\Lambda_{ii} = \lambda_i$ are the eigenvalues of $\mathbf{\Sigma}$ while the columns of the orthogonal matrix $\mathbf{V}$ are the eigenvectors of $\mathbf{\Sigma}$. The eigenvectors of a symmetric matrix are orthogonal to each other, which implies that $\mathbf{V}^{\mathsf{T}}\mathbf{V} = \mathbf{I}$ (the identity matrix), so that $\mathbf{V}^{-1} = \mathbf{V}^{\mathsf{T}}$ – a defining property of an orthogonal matrix. Multiplying the formula above on the left by $\mathbf{V}$ and on the right by $\mathbf{V}^{\mathsf{T}}$ we find

$$\mathbf{\Sigma} = \mathbf{V}\mathbf{\Lambda}\mathbf{V}^{\mathsf{T}}.$$

Using the fact that $(\mathbf{A}\,\mathbf{B}\,\mathbf{C})^{-1} = \mathbf{C}^{-1}\mathbf{B}^{-1}\mathbf{A}^{-1}$ (which is easily verified, by showing when you multiple by $\mathbf{A}\,\mathbf{B}\,\mathbf{C}$ you obtain the identity matrix) we find

$$\mathbf{\Sigma}^{-1} = (\mathbf{V}^{\mathsf{T}})^{-1}\mathbf{\Lambda}^{-1}\mathbf{V}^{-1} = \mathbf{V}\mathbf{\Lambda}^{-1}\mathbf{V}^{\mathsf{T}}$$

where $\mathbf{\Lambda}^{-1}$ is a diagonal matrix with elements $\Lambda_{ii}^{-1} = \lambda_i^{-1}$. Thus, the eigenvectors of a matrix are also eigenvectors of the inverse. We can diagonalise the inverse matrix using the same similarity transform as the original matrix

$$\mathbf{V}^{\mathsf{T}}\mathbf{\Sigma}^{-1}\mathbf{V} = \mathbf{\Lambda}^{-1}.$$

To check the normalisation of the normal distribution we change variables $\boldsymbol{y} = \mathbf{V}\,(\boldsymbol{x} - \boldsymbol{\mu})$, which gives

$$\int_{-\infty}^{\infty} \mathcal{N}(\boldsymbol{x}|\boldsymbol{\mu}, \boldsymbol{\Sigma})\,\mathrm{d}\boldsymbol{x} = \frac{1}{\sqrt{|2\pi\boldsymbol{\Sigma}|}} \int_{-\infty}^{\infty} \mathrm{e}^{-\boldsymbol{y}\,\mathbf{V}^{\mathsf{T}}\,\boldsymbol{\Sigma}^{-1}\,\mathbf{V}\,\boldsymbol{y}/2}\, J \prod_{i=1}^{n} \mathrm{d}y_i$$

where J is the Jacobian defined by

$$J = \begin{vmatrix} \frac{\partial x_1}{\partial y_1} & \frac{\partial x_1}{\partial y_2} & \cdots & \frac{\partial x_1}{\partial y_n} \\ \frac{\partial x_2}{\partial y_1} & \frac{\partial x_2}{\partial y_2} & \cdots & \frac{\partial x_2}{\partial y_n} \\ \vdots & \vdots & \ddots & \vdots \\ \frac{\partial x_n}{\partial y_1} & \frac{\partial x_n}{\partial y_2} & \cdots & \frac{\partial x_n}{\partial y_n}. \end{vmatrix}$$

Since $\boldsymbol{x} = \mathbf{V}^{\mathsf{T}}\boldsymbol{y} + \boldsymbol{\mu}$ we find $J = |\mathbf{V}^{\mathsf{T}}|$. The determinant of an orthogonal matrix is should be ± 1. A matrix can be thought of as mapping points from one space to another space. If we map all points in some volume in the first space to the second space, then the ratio of volumes is given by the determinant (up to a sign describing whether that volume suffers a reflection or not). Because an orthogonal matrix just corresponds to a rotation and a possible reflection the determinant is ± 1. Returning to the integral and using the diagonalisation formula we find

$$\begin{aligned}\int_{-\infty}^{\infty} \mathcal{N}(\boldsymbol{x}|\boldsymbol{\mu}, \boldsymbol{\Sigma})\,\mathrm{d}\boldsymbol{x} &= \frac{1}{\sqrt{|2\pi\boldsymbol{\Sigma}|}} \int_{-\infty}^{\infty} \mathrm{e}^{-\boldsymbol{y}\,\boldsymbol{\Lambda}^{-1}\,\boldsymbol{y}/2} \prod_{i=1}^{n} \mathrm{d}y_i \\ &= \frac{1}{\sqrt{|2\pi\boldsymbol{\Sigma}|}} \prod_{i=1}^{n} \int_{-\infty}^{\infty} \mathrm{e}^{-y_i^2/(2\lambda_i)}\,\mathrm{d}y_i = \frac{1}{\sqrt{|2\pi\boldsymbol{\Sigma}|}} \prod_{i=1}^{n} \sqrt{2\pi\,\lambda_i} \\ &= \frac{(2\pi)^{n/2}\sqrt{\prod_{i=1}^{n}\lambda_i}}{\sqrt{|2\pi\boldsymbol{\Sigma}|}}.\end{aligned}$$

If $\mathbf{M}$ is an $n \times n$ matrix then $|a\,\mathbf{M}| = a^n\,|\mathbf{M}|$. Thus the factors of 2π cancel each other out. Using the property of the determinant $|\mathbf{A}\,\mathbf{B}| = |\mathbf{A}| \times |\mathbf{B}|$ (which immediately implies $|\mathbf{A}\,\mathbf{B}\,\mathbf{C}| = |\mathbf{A}| \times |\mathbf{B}| \times |\mathbf{C}|$), $|\boldsymbol{\Sigma}| = |\mathbf{V}\boldsymbol{\Lambda}\mathbf{V}^{\mathsf{T}}| = |\mathbf{V}| \times |\boldsymbol{\Lambda}| \times |\mathbf{V}^{\mathsf{T}}| = |\boldsymbol{\Lambda}|$, since the determinants of the orthogonal matrix are equal to ± 1. However, the determinant of a diagonal matrix is equal to the product of its diagonal elements so that $|\boldsymbol{\Lambda}| = \prod_{i=1}^{n} \lambda_i$, which completes our verification of the normalisation of the normal distribution.

It follows from this derivation that if $\boldsymbol{X} \sim \mathcal{N}(\boldsymbol{\mu}, \boldsymbol{\Sigma})$ where $\boldsymbol{\Sigma} = \mathbf{V}\boldsymbol{\Lambda}\mathbf{V}^{\mathsf{T}}$ then $\boldsymbol{Y} = \boldsymbol{\Lambda}^{-1/2}\mathbf{V}(\boldsymbol{X} - \boldsymbol{\mu})$ will be distributed according to $\mathcal{N}(\mathbf{0}, \mathbf{I})$. That is, the components of $\boldsymbol{Y}$ are independent normally distributed variables with zero mean and unit variance. The transformation from $\boldsymbol{X}$ to $\boldsymbol{Y}$ involves a shift by $\boldsymbol{\mu}$, followed by a rotation and a possible reflection given by $\mathbf{V}$, followed by a rescaling of each component by $\lambda_i^{-1/2}$. These are all affine (i.e. linear) transformations of the variables. Thus,

> *up to an affine transformation any multivariate normal distribution is equivalent to distribution for n independent normal variables.*

That is, given any multivariate normal distribution, by means of a translation, rotation, and rescaling of the axes we can obtain the distribution $\mathcal{N}(\mathbf{0}, \mathbf{I})$. You should be able to convince yourself that this is credible for the distribution shown in Figure 5.12.

By inverting this transformation we can obtain a multivariate normal deviate in n dimensions starting from n independent normally distributed variables. In practice, however, it is more efficient to compute the Cholesky decomposition of the covariance matrix $\mathbf{L}$, such that $\mathbf{L}\mathbf{L}^\mathsf{T} = \boldsymbol{\Sigma}$ where $\mathbf{L}$ is a lower triangular matrix. Then to generate a multivariate normal deviate we first generate $\boldsymbol{Y} \sim \mathcal{N}(0, \mathbf{I})$ and use $\boldsymbol{X} = \mathbf{L}\boldsymbol{Y} + \boldsymbol{\mu}$. Its expectation is

$$\mathbb{E}\left[\boldsymbol{X}\right] = \mathbb{E}\left[\mathbf{L}\boldsymbol{Y} + \boldsymbol{\mu}\right] = \mathbf{L}\mathbb{E}\left[\boldsymbol{Y}\right] + \boldsymbol{\mu} = \boldsymbol{\mu}.$$

The covariance is given by

$$\begin{aligned}\mathbb{C}\mathrm{ov}\left[\boldsymbol{X}, \boldsymbol{X}\right] &= \mathbb{E}\left[(\boldsymbol{X} - \boldsymbol{\mu})\,(\boldsymbol{X} - \boldsymbol{\mu})^\mathsf{T}\right] \\ &= \mathbb{E}\left[\mathbf{L}\,\boldsymbol{Y}\,\boldsymbol{Y}^\mathsf{T}\,\mathbf{L}^\mathsf{T}\right] = \mathbf{L}\,\mathbb{E}\left[\boldsymbol{Y}\,\boldsymbol{Y}^\mathsf{T}\right]\,\mathbf{L}^\mathsf{T} = \mathbf{L}\,\mathbf{I}\,\mathbf{L}^\mathsf{T} = \mathbf{L}\,\mathbf{L}^\mathsf{T} = \boldsymbol{\Sigma}.\end{aligned}$$

As the components of $\boldsymbol{X}$ are sums of normally distributed variables (plus a constant) they will be normally distributed. Thus $\boldsymbol{X} \sim \mathcal{N}(\boldsymbol{\mu}, \boldsymbol{\Sigma})$.

Moments and Cumulants. Fortunately, most expectations can be obtained easily from the moment generating function

$$M(\boldsymbol{l}) = \mathbb{E}_{\boldsymbol{X}}\left[\mathrm{e}^{\boldsymbol{l}^\mathsf{T}\boldsymbol{X}}\right] = \int_{-\infty}^{\infty} \mathrm{e}^{\boldsymbol{l}^\mathsf{T}\boldsymbol{x}}\,\mathcal{N}(\boldsymbol{x}|\boldsymbol{\mu}, \boldsymbol{\Sigma})\mathrm{d}\boldsymbol{x}.$$

To compute this we complete the square in the exponent

$$\begin{aligned}&\boldsymbol{l}^\mathsf{T}\boldsymbol{x} - \frac{1}{2}(\boldsymbol{x} - \boldsymbol{\mu})^\mathsf{T}\boldsymbol{\Sigma}^{-1}(\boldsymbol{x} - \boldsymbol{\mu}) \\ &= \boldsymbol{l}^\mathsf{T}\boldsymbol{\mu} + \frac{1}{2}\boldsymbol{l}^\mathsf{T}\boldsymbol{\Sigma}\,\boldsymbol{l} - \frac{1}{2}(\boldsymbol{x} - \boldsymbol{\mu} - \boldsymbol{\Sigma}\,\boldsymbol{l})^\mathsf{T}\boldsymbol{\Sigma}^{-1}(\boldsymbol{x} - \boldsymbol{\mu} - \boldsymbol{\Sigma}\,\boldsymbol{l}).\end{aligned}$$

Substituting this into the moment generating function we find

$$\begin{aligned}M(\boldsymbol{l}) &= \mathrm{e}^{\boldsymbol{l}^\mathsf{T}\boldsymbol{\mu} + \frac{1}{2}\boldsymbol{l}^\mathsf{T}\boldsymbol{\Sigma}\,\boldsymbol{l}} \int_{-\infty}^{\infty} \mathcal{N}(\boldsymbol{x}|\boldsymbol{\mu} - \boldsymbol{\Sigma}\,\boldsymbol{l}, \boldsymbol{\Sigma})\mathrm{d}\boldsymbol{x} \\ &= \mathrm{e}^{\boldsymbol{l}^\mathsf{T}\boldsymbol{\mu} + \frac{1}{2}\boldsymbol{l}^\mathsf{T}\boldsymbol{\Sigma}\,\boldsymbol{l}}.\end{aligned}$$

In a similar way to the one-dimensional case we can compute the mean by taking a partial derivative of the moment generating function and setting $\boldsymbol{l}$ to zero

$$\mathbb{E}_{\boldsymbol{X}}\left[\boldsymbol{X}\right] = \boldsymbol{\nabla} M(\mathbf{0}) = \boldsymbol{\mu}.$$

The second moments are given by

$$\mathbb{E}_{\boldsymbol{X}}\left[X_i X_j\right] = \frac{\partial M(\boldsymbol{0})}{\partial l_i \partial l_j} = \Sigma_{ij} + \mu_i \mu_j.$$

The multidimensional equivalent of the variance is the covariance defined by

$$\mathbb{C}\mathrm{ov}\left[X_i, X_j\right] = \mathbb{E}_{\boldsymbol{X}}\left[X_i X_j\right] - \mathbb{E}_{\boldsymbol{X}}\left[X_i\right] \mathbb{E}_{\boldsymbol{X}}\left[X_j\right] = \Sigma_{ij}.$$

This can be written in matrix form as

$$\mathbb{C}\mathrm{ov}\left[\boldsymbol{X}\right] = \mathbb{E}_{\boldsymbol{X}}\left[\boldsymbol{X}\boldsymbol{X}^{\mathsf{T}}\right] - \mathbb{E}_{\boldsymbol{X}}\left[\boldsymbol{X}\right] \mathbb{E}_{\boldsymbol{X}}\left[\boldsymbol{X}^{\mathsf{T}}\right] = \boldsymbol{\Sigma}$$

where we use the notation that $\mathsf{M} = \boldsymbol{a}\,\boldsymbol{b}^{\mathsf{T}}$ is the outer-product matrix with components $M_{ij} = a_i\, b_j$. Note that we can also define a multidimensional CGF

$$G(\boldsymbol{l}) = \log\bigl(M(\boldsymbol{l})\bigr) = \boldsymbol{l}^{\mathsf{T}} \boldsymbol{\mu} + \frac{1}{2}\boldsymbol{l}^{\mathsf{T}} \boldsymbol{\Sigma}\, \boldsymbol{l}$$

where $\boldsymbol{\nabla} G(0)$ gives us the mean and the second derivative the covariance. The diagonal elements of the covariance matrix are just

$$\Sigma_{ii} = \mathbb{E}_{\boldsymbol{X}}\left[X_i^2\right] - \mathbb{E}_{\boldsymbol{X}}\left[X_i\right]^2$$

which we recognise as the variance of X_i. The off-diagonal elements encode the correlation between the components. Because the covariance matrix is readily observable, we usually try to work with the this rather than its inverse.

An important property often known as the *marginalisation property* of a normal distribution is that if we marginalise a normal distribution $\mathcal{N}(\boldsymbol{x}|\boldsymbol{\mu}, \boldsymbol{\Sigma})$ where

$$\boldsymbol{\mu} = \begin{pmatrix} \mu_1 \\ \mu_2 \\ \vdots \\ \mu_n \end{pmatrix} \qquad \boldsymbol{\Sigma} = \begin{pmatrix} \Sigma_{11} & \Sigma_{12} & \cdots & \Sigma_{1n} \\ \Sigma_{21} & \Sigma_{22} & \cdots & \Sigma_{2n} \\ \vdots & \vdots & \ddots & \vdots \\ \Sigma_{n1} & \Sigma_{n2} & \cdots & \Sigma_{nn} \end{pmatrix}$$

with respect to x_n, then

$$\int_{-\infty}^{\infty} \mathcal{N}(\boldsymbol{x}|\boldsymbol{\mu}, \boldsymbol{\Sigma})\, \mathrm{d}x_n = \mathcal{N}(\boldsymbol{x}|\boldsymbol{\mu}', \boldsymbol{\Sigma}')$$

where

$$\boldsymbol{\mu}' = \begin{pmatrix} \mu_1 \\ \mu_2 \\ \vdots \\ \mu_{n-1} \end{pmatrix} \qquad \boldsymbol{\Sigma}' = \begin{pmatrix} \Sigma_{11} & \Sigma_{12} & \cdots & \Sigma_{1\,n-1} \\ \Sigma_{21} & \Sigma_{22} & \cdots & \Sigma_{2\,n-1} \\ \vdots & \vdots & \ddots & \vdots \\ \Sigma_{n-1\,1} & \Sigma_{n-1\,2} & \cdots & \Sigma_{n-1\,n-1} \end{pmatrix}.$$

Thus neglecting one dimension does not change the means and covariances in the other directions. However, it is much harder to see if we try to compute the integrals explicitly, as the exponent $-\frac{1}{2}(\boldsymbol{x} - \boldsymbol{\mu})^{\mathsf{T}}\boldsymbol{\Sigma}^{-1}(\boldsymbol{x} - \boldsymbol{\mu})$ couples all the components (see Exercise 5.4).

Schur Complement

One of the complications of working with multivariate normal distributions is the need to shift between the covariance matrix $\boldsymbol{\Sigma}$ and its inverse $\boldsymbol{\Sigma}^{-1}$. This is easy enough numerically, but it is tedious to do analytically. Often rather convoluted derivations are used to avoid going backwards and forwards between the covariance matrix and its inverse. Sometimes it is sufficient to break up the random variables into two groups (e.g. when marginalising over some set of nuisance variables). This often leads to inverting a partitioning of the matrix. We can rewrite any matrix $\mathbf{M}$ into the product of three matrices

The purpose of this section is just to point out that identities exist. They aren't beautiful, at least not to me, but occasionally they are useful. You can always look them up when necessary.

$$\underbrace{\begin{pmatrix} \mathbf{A} & \mathbf{B} \\ \mathbf{C} & \mathbf{D} \end{pmatrix}}_{\mathbf{M}} = \underbrace{\begin{pmatrix} \mathbf{I} & \mathbf{B}\mathbf{D}^{-1} \\ \mathbf{0} & \mathbf{I} \end{pmatrix}}_{\mathbf{U}} \underbrace{\begin{pmatrix} \mathbf{A} - \mathbf{B}\mathbf{D}^{-1}\mathbf{C} & \mathbf{0} \\ \mathbf{0} & \mathbf{D} \end{pmatrix}}_{\mathbf{W}} \underbrace{\begin{pmatrix} \mathbf{I} & \mathbf{0} \\ \mathbf{D}^{-1}\mathbf{C} & \mathbf{I} \end{pmatrix}}_{\mathbf{L}}$$

where $\mathbf{A}$, $\mathbf{B}$, $\mathbf{C}$, and $\mathbf{D}$ are matrices. This decomposition is easily proved by simply multiplying out the matrix product. We denote $\hat{\mathbf{A}} = \mathbf{A} - \mathbf{B}\mathbf{D}^{-1}\mathbf{C}$ as the *Schur complement* of $\mathbf{M}$. The first advantage of this form is that we can quickly determine the determinant as $|\mathbf{A}| = |\mathbf{U}| \times |\mathbf{W}| \times |\mathbf{L}|$. But because of their structure $|\mathbf{U}| = |\mathbf{L}| = 1$, so that $|\mathbf{A}| = |\mathbf{W}| = |\hat{\mathbf{A}}| \times |\mathbf{D}|$.

This form also allows us to compute the matrix inverse relatively easily

$$\begin{aligned} \underbrace{\begin{pmatrix} \mathbf{A} & \mathbf{B} \\ \mathbf{C} & \mathbf{D} \end{pmatrix}^{-1}}_{\mathbf{M}^{-1}} &= \underbrace{\begin{pmatrix} \mathbf{I} & \mathbf{0} \\ -\mathbf{D}^{-1}\mathbf{C} & \mathbf{I} \end{pmatrix}}_{\mathbf{L}^{-1}} \underbrace{\begin{pmatrix} \hat{\mathbf{A}}^{-1} & \mathbf{0} \\ \mathbf{0} & \mathbf{D}^{-1} \end{pmatrix}}_{\mathbf{W}^{-1}} \underbrace{\begin{pmatrix} \mathbf{I} & -\mathbf{B}\mathbf{D}^{-1} \\ \mathbf{0} & \mathbf{I} \end{pmatrix}}_{\mathbf{U}^{-1}} \\ &= \begin{pmatrix} \hat{\mathbf{A}}^{-1} & -\hat{\mathbf{A}}^{-1}\mathbf{B}\mathbf{D}^{-1} \\ -\mathbf{D}^{-1}\mathbf{C}\hat{\mathbf{A}}^{-1} & \mathbf{D}^{-1} + \mathbf{D}^{-1}\mathbf{C}\hat{\mathbf{A}}^{-1}\mathbf{B}\mathbf{D}^{-1} \end{pmatrix}. \end{aligned} \tag{5.3}$$

The decomposition we used is not unique (in particular, $\mathbf{A}$ and $\mathbf{D}$ are treated differently). If we swap the role of $\mathbf{A}$ and $\mathbf{D}$ then we can also define a second Schur complement of $\hat{\mathbf{D}} = \mathbf{D} - \mathbf{C}\mathbf{A}^{-1}\mathbf{B}$, such that

$$\mathbf{M}^{-1} = \begin{pmatrix} \mathbf{A}^{-1} + \mathbf{A}^{-1}\mathbf{B}\hat{\mathbf{D}}^{-1}\mathbf{C}\mathbf{A}^{-1} & -\mathbf{A}^{-1}\mathbf{B}\hat{\mathbf{D}}^{-1} \\ -\hat{\mathbf{D}}^{-1}\mathbf{C}\mathbf{A}^{-1} & \hat{\mathbf{D}}^{-1} \end{pmatrix}. \tag{5.4}$$

The determinant of the matrix $\mathbf{M}$ is also equal to $|\mathbf{M}| = |\mathbf{A}|\,|\hat{\mathbf{D}}|$.

Although Equations (5.3) and (5.4) look different, they must be the same, which implies

$$\hat{\mathbf{A}}^{-1} = (\mathbf{A} - \mathbf{B}\mathbf{D}^{-1}\mathbf{C})^{-1} = \mathbf{A}^{-1} + \mathbf{A}^{-1}\mathbf{B}\hat{\mathbf{D}}^{-1}\mathbf{C}\mathbf{A}^{-1}. \tag{5.5}$$

This identity is known as the Woodbury matrix identity and has many uses. It can be proved directly by multiplying $\hat{\mathbf{A}} = \mathbf{A} - \mathbf{B}\mathbf{D}^{-1}\mathbf{C}$ by $\mathbf{A}^{-1} + \mathbf{A}^{-1}\mathbf{B}\hat{\mathbf{D}}^{-1}\mathbf{C}\mathbf{A}^{-1}$ and showing that you obtain the identity matrix.

For a symmetric matrix

$$\boldsymbol{\Sigma} = \begin{pmatrix} \mathbf{A} & \mathbf{B} \\ \mathbf{B}^\mathsf{T} & \mathbf{C} \end{pmatrix}$$

the inverse is given by

$$\boldsymbol{\Sigma}^{-1} = \begin{pmatrix} \mathbf{A} & \mathbf{B} \\ \mathbf{B}^\mathsf{T} & \mathbf{C} \end{pmatrix}^{-1} = \begin{pmatrix} (\mathbf{A} - \mathbf{B}\mathbf{C}^{-1}\mathbf{B}^\mathsf{T})^{-1} & -(\mathbf{A} - \mathbf{B}\mathbf{C}^{-1}\mathbf{B}^\mathsf{T})^{-1}\mathbf{B}\mathbf{C}^{-1} \\ -\mathbf{C}^{-1}\mathbf{B}^\mathsf{T}(\mathbf{A} - \mathbf{B}\mathbf{C}^{-1}\mathbf{B}^\mathsf{T})^{-1} & (\mathbf{C} - \mathbf{B}\mathbf{A}^{-1}\mathbf{B}^\mathsf{T})^{-1} \end{pmatrix} \tag{5.6}$$

(note there are a number of other ways of writing this). Since $|\boldsymbol{\Sigma}| = |\hat{\mathbf{A}}| \times |\mathbf{C}|$ we see that $\boldsymbol{\Sigma}$ is positive definite if and only if both $\mathbf{C}$ and the Schur complement, $\hat{\mathbf{A}} = \mathbf{A} - \mathbf{B}\mathbf{C}\mathbf{B}^\mathsf{T}$, are positive definite.

Another non-obvious but useful set of identities is

$$(\mathbf{A}^{-1} + \mathbf{B}^{-1})^{-1} = \mathbf{A} - \mathbf{A}\,(\mathbf{A} + \mathbf{B})^{-1}\,\mathbf{A} = \mathbf{B} - \mathbf{B}\,(\mathbf{A} + \mathbf{B})^{-1}\,\mathbf{B}.$$

This can be proved by direct multiplication, i.e.

$$\begin{aligned}
(\mathbf{A}^{-1} + \mathbf{B}^{-1})\,(\mathbf{A}^{-1} + \mathbf{B}^{-1})^{-1} &= (\mathbf{A}^{-1} + \mathbf{B}^{-1})\left(\mathbf{A} - \mathbf{A}\,(\mathbf{A} + \mathbf{B})^{-1}\,\mathbf{A}\right) \\
&= \mathbf{I} - (\mathbf{A} + \mathbf{B})^{-1}\,\mathbf{A} + \mathbf{B}^{-1}(\mathbf{A} - \mathbf{A}\,(\mathbf{A} + \mathbf{B})^{-1}\,\mathbf{A}) \\
&= \mathbf{I} - \mathbf{B}^{-1}\,\mathbf{B}\,(\mathbf{A} + \mathbf{B})^{-1}\,\mathbf{A} + \mathbf{B}^{-1}(\mathbf{A} - \mathbf{A}\,(\mathbf{A} + \mathbf{B})^{-1}\,\mathbf{A}) \\
&= \mathbf{I} - \mathbf{B}^{-1}\,(\mathbf{I} - (\mathbf{A} + \mathbf{B})\,(\mathbf{A} + \mathbf{B})^{-1})\,\mathbf{A} = \mathbf{I}.
\end{aligned}$$

The second identity $(\mathbf{A}^{-1} + \mathbf{B}^{-1})^{-1} = \mathbf{B} - \mathbf{B}\,(\mathbf{A} + \mathbf{B})^{-1}\,\mathbf{B}$ follows by symmetry. These identities are useful when multivariate normal PDFs are multiplied together. For example, we can use this identity to show

$$\mathcal{N}(\boldsymbol{x}|\boldsymbol{a}, \mathbf{A})\,\mathcal{N}(\boldsymbol{x}|\boldsymbol{b}, \mathbf{B}) = \mathcal{N}(\boldsymbol{x}|\boldsymbol{c}, \mathbf{C})\,\mathcal{N}(\boldsymbol{a} - \boldsymbol{b}|\mathbf{0}, \mathbf{A} + \mathbf{B})$$

where $\mathbf{C} = (\mathbf{A}^{-1} + \mathbf{B}^{-1})^{-1}$ and $\boldsymbol{c} = \mathbf{C}\,(\mathbf{A}^{-1}\boldsymbol{a} + \mathbf{B}^{-1}\boldsymbol{b})$.

Normal distributions are tremendously useful because they are analytically tractable. They form the basis of techniques such as multivariate statistics, Gaussian processes (in machine learning), and Kalman filters. Working with them is, however, fiddly. You often have to complete squares and invert matrices. Furthermore, the CDF is sometimes awkward to work with. All that said, don't be put off by the algebra: normal distributions are usually much easier to work with than the alternatives.

Exercise for Chapter 5

Exercise 5.1 (answer on page 407)

Let

$$S_n = \frac{1}{\sqrt{n/12}} \sum_{i=1}^{n} (U_i - 1/2)$$

where $U_i \sim \mathrm{U}(0,1)$. That is, S_n is the sum of n uniform deviates normalised so that it has zero mean and unit variance. Generate a histogram of deviates S_5 and compare it to the density $\mathcal{N}(x|0,1)$ (see Section 6.2 on page 115 for details on generating histograms). Repeat this for deviates S_{10}.

Exercise 5.2 (answer on page 408)

Let

$$T_n = \frac{1}{n}\sum_{i=1}^{n} C_i$$

where $C_i \sim \mathrm{Cau}$ are Cauchy deviates. Generate a histogram T_{10} and compare this with the density $\mathrm{Cau}(x)$.

Exercise 5.3 (answer on page 408)

Show that the CDF for normal random variables,

$$\Phi(x) = \int_{-\infty}^{x} \mathrm{e}^{y^2/2} \frac{\mathrm{d}y}{\sqrt{2\pi}}$$

can be written in terms of an incomplete gamma function by making the change of variables $y^2/2 = t$.

Exercise 5.4 (answer on page 409)

Marginalise out the variables $\boldsymbol{y}$ from the multivariate normal distribution

$$\begin{pmatrix} \boldsymbol{x} \\ \boldsymbol{y} \end{pmatrix} \sim \mathcal{N}\left(\begin{pmatrix} \boldsymbol{\mu}_x \\ \boldsymbol{\mu}_y \end{pmatrix}, \begin{pmatrix} \mathsf{A} & \mathsf{B} \\ \mathsf{B}^\mathsf{T} & \mathsf{C} \end{pmatrix}^{-1} \right)$$

$$= (2\pi)^{n/2} \begin{vmatrix} \mathsf{A} & \mathsf{B} \\ \mathsf{B}^\mathsf{T} & \mathsf{C} \end{vmatrix}^{-1/2} \mathrm{e}^{-\frac{1}{2}(\boldsymbol{x}-\boldsymbol{\mu}_x)^\mathsf{T}\mathsf{A}(\boldsymbol{x}-\boldsymbol{\mu}_x)-(\boldsymbol{x}-\boldsymbol{\mu}_x)^\mathsf{T}\mathsf{B}(\boldsymbol{y}-\boldsymbol{\mu}_y)-\frac{1}{2}(\boldsymbol{y}-\boldsymbol{\mu}_y)^\mathsf{T}\mathsf{C}(\boldsymbol{y}-\boldsymbol{\mu}_y)}$$

to obtain a density for the variables $\boldsymbol{x}$.

Exercise 5.5 (answer on page 411)

Let $U = S^2 = \sum_{i=1}^{n} X_i^2$ where $X_i \sim \mathcal{N}(0,1)$, show that U is chi-squared distributed

$$f_U(u) = \chi^2(n) = \mathrm{Gam}\left(u \middle| \frac{n}{2}, \frac{1}{2}\right) = \frac{u^{n/2-1}\,\mathrm{e}^{-u/2}}{2^{n/2}\,\Gamma(n/2)}.$$

Dirac's delta function was immediately criticised by hardened mathematicians such as von Neumann who rightly pointed out that no such function can exist. It took mathematicians 20 years to catch up and finally develop a theory of distributions which put the delta function on a rigorous footing.

Appendix 5.A Dirac Delta

The Dirac delta 'function' is a useful device in computing probabilities. Strictly it is not a function as it does not take a well-defined value when its argument becomes zero. We can think of it as a limit of a family of functions, $\delta_i(x)$, such that

$$\int_{-\infty}^{\infty} \delta_i(x)\,\mathrm{d}x = 1$$

and which become closer to zero for $x \neq 0$ as $i \to \infty$. There are many such families of functions. For example, the family of Gaussian

$$\delta_i(x) = \mathcal{N}\left(x \middle| 0, \frac{1}{i^2}\right).$$

Alternatively, we can think of the function as $\delta(x) = \lim_{\epsilon \to 0} \mathcal{N}(x|0, \epsilon)$. The Dirac delta is sometimes referred to as a *generalised function* or a *distribution* rather than a function. Although Dirac deltas are awkward objects on their own, they become invaluable inside integrals. Furthermore, they are very easy to use. Their defining equation is

$$\int_{-\infty}^{\infty} f(x)\, \delta(x - y)\, \mathrm{d}\, x = f(y).$$

Dirac delta functions have one rather surprising property, namely

$$\begin{aligned}\int_{-\infty}^{\infty} \delta(a\, x)\, \mathrm{d}\, x &= \frac{1}{a} \int_{-\infty}^{\infty} \delta(a\, x)\, \mathrm{d}\, (a\, x) \\ &= \frac{1}{a} \int_{-\infty/a}^{\infty/a} \delta(y)\, \mathrm{d}\, y \\ &= \frac{\mathrm{sign}(a)}{a} = \frac{1}{|a|}.\end{aligned}$$

The $\mathrm{sign}(a)$ appears because if $a < 0$ the order of the limits change giving an extra -1. If we have an integral

$$I = \int_{-\infty}^{\infty} f(x)\, \delta(y - g(x))\, \mathrm{d}\, x$$

we can Taylor expand $g(x)$ around the point x_i^* where $y - g(x_i^*) = 0$. Doing this we get

$$\begin{aligned}I &= \int_{-\infty}^{\infty} f(x) \sum_i \delta\left(y - g(x_i^*) + (x - x_i^*) g'(x_i^*) + O\left((x - x_i^*)^2\right)\right) \mathrm{d}\, x \\ &= \sum_i \int_{-\infty}^{\infty} f(x)\, \delta\left((x - x_i^*) g'(x_i^*) + O\left((x - x_i^*)^2\right)\right) \mathrm{d}\, x \\ &= \sum_i \frac{f(x_i^*)}{|g'(x_i^*)|}.\end{aligned}$$

In the case where $g(x)$ is invertible so that $x^* = g^{-1}(y)$ is unique, we obtain

$$\int_{-\infty}^{\infty} f(x)\, \delta(y - g(x))\, \mathrm{d}\, x = \frac{f(g^{-1}(y))}{|g'(g^{-1}(y))|}.$$

However, we observe that $g(g^{-1}(y)) = y$, so differentiating both sides we find

$$\frac{\mathrm{d} g(g^{-1}(y))}{\mathrm{d} y} = g'(g^{-1}(y))\, \frac{\mathrm{d} g^{-1}(y)}{\mathrm{d} y} = 1$$

or

$$\frac{\mathrm{d}g^{-1}(y)}{\mathrm{d}y} = \frac{1}{g'(g^{-1}(y))},$$

thus

$$\int_{-\infty}^{\infty} f(x)\,\delta(y-g(x))\,\mathrm{d}\,x = \left|\frac{\mathrm{d}g^{-1}(y)}{\mathrm{d}y}\right| f(g^{-1}(y)).$$

The behaviour of the delta function should not be too surprising as it can be viewed as a (limit of a) probability density. As we saw in Section 1.4, densities pick up the Jacobian under changes of variables.

Dirac delta functions can also be defined in higher dimensions

$$\int_{-\infty}^{\infty} f(\boldsymbol{x})\,\delta(\boldsymbol{x}-\boldsymbol{y})\,\mathrm{d}\,\boldsymbol{x} = f(\boldsymbol{y})$$

where the integral is now a multidimensional integral. We can interpret the multidimensional Dirac delta as a product of one-dimensional Dirac deltas

$$\delta(\boldsymbol{x}-\boldsymbol{y}) = \prod_{i=1}^{n} \delta(x_i - y_i).$$

Given a transformation of the coordinates from $\boldsymbol{x}$ to $\boldsymbol{y}(\boldsymbol{x})$ the Dirac delta function transforms as

$$\delta(\boldsymbol{x}-\boldsymbol{x}^*) = |J|\,\delta(\boldsymbol{y}-\boldsymbol{y}^*)$$

where $\boldsymbol{y}^* = \boldsymbol{y}(\boldsymbol{x}^*)$ and J is the Jacobian $\det(\mathbf{T})$ where $\mathbf{T}$ is a matrix with components

$$T_{ij} = \frac{\partial y_i(\boldsymbol{x})}{\partial x_i}.$$

Multidimensional Dirac deltas make many mathematicians go weak at the knees, but physicists have been happily using them for years with little ill consequences. They take some getting used to, but they often make tricky calculations transparent.

The Dirac delta is often useful when treating probabilities. In particular, under a transformation of variables $Y = g(X)$, the probability distribution for Y is given by

$$f_Y(y) = \int_{-\infty}^{\infty} \delta(y-g(x))\,f_X(x)\,\mathrm{d}\,x$$

which we have just learnt how to compute. If $g(x)$ is invertible we find

$$f_Y(y) = \left|\frac{\mathrm{d}g^{-1}(y)}{\mathrm{d}y}\right| f_X(g^{-1}(y))$$

(cf. Section 1.4 on page 12).

The Heaviside function is the step function $[\![x \geq y]\!]$ (sometimes written as $\Theta(x-y)$). Note that

$$\int_{-\infty}^{\infty} f(x)\,[\![x \geq y]\!]\,\mathrm{d}x = \int_{y}^{\infty} f(x)\,\mathrm{d}x$$

so taking derivatives with respect to y

$$\frac{\mathrm{d}}{\mathrm{d}y}\int_{-\infty}^{\infty} f(x)\,[\![x \ge y]\!]\,\mathrm{d}x = \frac{\mathrm{d}}{\mathrm{d}y}\int_{y}^{\infty} f(x)\,\mathrm{d}x = -f(y).$$

Being rather rash we could conclude (by taking the derivative inside the integral) that

$$\frac{\mathrm{d}[\![x \ge y]\!]}{\mathrm{d}y} = \frac{\mathrm{d}\Theta(x-y)}{\mathrm{d}y} = -\delta(x-y)$$

or $\Theta'(x) = \delta(x)$. Of course, any self-respecting mathematician would deplore such a derivation. The Heaviside function is non-differentiable and it is quite incorrect to take the derivative inside the integral sign. However, we can make this identification kosher, for example, by defining a function

$$\Theta_\epsilon(x) = \Phi\left(\frac{x}{\epsilon}\right)$$

where $\Phi(x/\epsilon)$ is the CDF of the normal distribution $\mathcal{N}(0, \epsilon^2)$. In the limit $\epsilon \to 0$ we see that

$$\lim_{\epsilon\to 0}\Theta_\epsilon(x) = \Theta(x) \qquad\qquad \lim_{\epsilon\to 0}\Theta'_\epsilon(x) = \mathcal{N}(0, \epsilon^2) = \delta(x).$$

In this sense we can relate the Heaviside function and the Dirac delta function.

One of the more useful representations of a Dirac delta is

$$\delta(x) = \int_{-\infty}^{\infty} \mathrm{e}^{\mathrm{i}\,\omega\,x}\,\frac{\mathrm{d}\omega}{2\pi}.$$

This is not entirely obvious (as $\delta(x)$ is not a well-defined mathematical object it is not entirely true except as some limiting process). One way to understand this identity is as a limit of the sum of discrete Fourier components

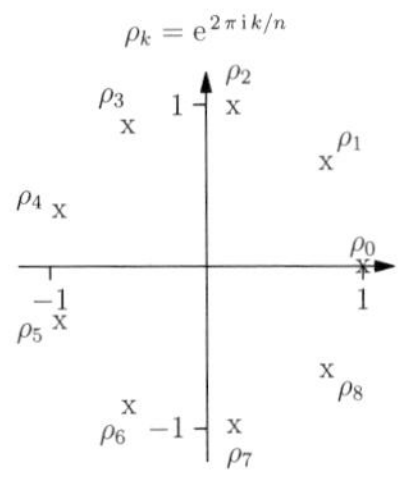

$$[\![k = 0]\!] = \frac{1}{n}\sum_{l=-n/2+1}^{n/2} \mathrm{e}^{\frac{2\pi\mathrm{i}kl}{n}}$$

which follows for integer k as the sum consists of roots of 1 which sum to 0, except when $k = 0$, where the exponential is just equal to 1. We can consider $[\![k = 0]\!]$ as a discrete impulse. To turn this into a Dirac delta function in the limit of large n, we define $x_k = k/\sqrt{n}$ and $\omega_l = 2\pi l/\sqrt{n}$ so that $\Delta\omega = \omega_l - \omega_{l-1} = 2\pi/\sqrt{n}$ and

$$\sqrt{n}\,[\![\sqrt{n}\,x_k = 0]\!] = \frac{1}{\sqrt{n}}\sum_{l=-n/2+1}^{n/2} \mathrm{e}^{\frac{2\pi\mathrm{i}kl}{n}} = \frac{1}{2\pi}\sum_{l=-n/2+1}^{n/2} \mathrm{e}^{\mathrm{i}x_k\omega_l}\,\Delta\omega$$

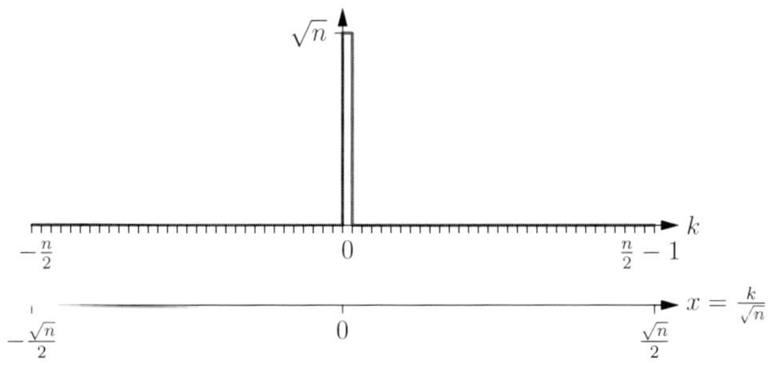

Taking the limit $n \to \infty$ the sum goes over to an integral at the same time the indicator function becomes closer to a Dirac delta function.

A more direct way to obtain this result is to define a limiting process where the integral becomes well defined

$$\int_{-\infty}^{\infty} \mathrm{e}^{\mathrm{i}\,\omega\,x}\,\frac{\mathrm{d}\omega}{2\,\pi} \overset{(1)}{=} \lim_{\epsilon\to 0}\int_{-\infty}^{\infty} \mathrm{e}^{-\epsilon\,\omega^2/2}\mathrm{e}^{\mathrm{i}\,\omega\,x}\,\frac{\mathrm{d}\omega}{2\,\pi} \overset{(2)}{=} \lim_{\epsilon\to 0}\int_{-\infty}^{\infty} \mathrm{e}^{-\frac{\epsilon}{2}(\omega-\frac{\mathrm{i}\,x}{\epsilon})^2-\frac{x^2}{2\,\epsilon}}\,\frac{\mathrm{d}\omega}{2\,\pi}$$

$$\overset{(3)}{=} \lim_{\epsilon\to 0}\frac{1}{\sqrt{2\,\pi\,\epsilon}}\mathrm{e}^{-\frac{x^2}{2\,\epsilon}} = \lim_{\epsilon\to 0}\mathcal{N}(x|0,\epsilon) = \delta(x)$$

(1) Using $\lim_{\epsilon\to 0}\mathrm{e}^{-\epsilon\,\omega^2/2} = 1$ (we have recklessly exchanged the order of the limit and integral).
(2) Follows from completing the square.
(3) Calculating the integral.

Appendix 5.B Characteristic Function for the Cauchy Distribution

This is provided solely out of a wish not to pull results out of a hat. The reader should feel free to skip this section if it is of no interest or you are unfamiliar with contour integration.

We derive the characteristic function for the Cauchy distribution using contour integration. It is beyond the scope of this text to explain contour integration. The characteristic function of the Cauchy distribution is defined as

$$\phi(\omega) = \mathbb{E}\left[\mathrm{e}^{\mathrm{i}\,\omega\,X}\right] = \int_{-\infty}^{\infty}\frac{\mathrm{e}^{\mathrm{i}\,\omega\,x}}{\pi(1+x^2)}\,\mathrm{d}x.$$

To evaluate the integral for $\omega \geq 0$ we consider the contour integral

$$I_1(R) = \oint_{\mathcal{C}_+(R)}\frac{\mathrm{e}^{\mathrm{i}\omega z}}{\pi(1+z^2)}\mathrm{d}z$$

where $\mathcal{C}_+(R)$ is given by

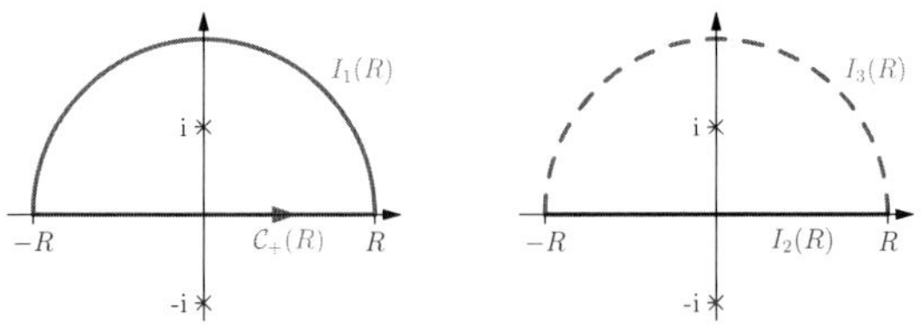

Dividing the contour integral into two sections we find $I_1(R) = I_2(R) + I_3(R)$ where

$$I_2(R) = \int_{-R}^{R} \frac{e^{i\omega x}}{\pi\,(1+x^2)} dx, \qquad I_3(R) = \int_0^{\pi} \frac{e^{i\omega R\cos(\theta) - \omega R\sin(\theta)}}{\pi\,(1+R^2 e^{2i\theta})} d\theta.$$

In the limit $R \to \infty$ we find $I_2(R) \to \phi(\omega)$ and $I_3(R) \to 0$ (since $-\omega R \sin(\theta) < 0$ for $0 \le \theta \le \pi$ so the integrand goes to zero as $R \to \infty$). Thus $\phi(\omega) = \lim_{R\to\infty} I_1(R)$. To evaluate $I_1(R)$ we use the residue theorem. We note that $I_1(R)$ is equal to

$$I_1(R) = \oint_{C(R)} \frac{e^{i\omega z}}{\pi\,(z+i)\,(z-i)} dz$$

so that it has poles at $\pm$i. By the residue theorem

$$I_1(R) = 2\pi i \sum \text{residues}$$

where the residues are computed at the poles that lie within the contour $\mathcal{C}(R)$. Provided $R > 1$, there is one simple pole at $z = i$ with a residue of (denoting the integrand of I_1 by $f_1(z)$ so that $I_1(R) = \oint f_1(z)\,dz$):

$$\lim_{z\to i}(z-i) f_1(z) = \frac{e^{-\omega}}{2\pi i}.$$

Thus, for $R > 1$ we find $I_1(R) = e^{-\omega}$, so that $\phi(\omega) = e^{-\omega}$ if $\omega \ge 0$.

If $\omega < 0$ the integral $I_3(R)$ involves an integrand that diverges as $R \to \infty$ so instead we consider a second integral

$$I_4(R) = \oint_{C_-(R)} \frac{e^{i\omega z}}{\pi\,(1+z^2)} dz$$

where $C_-(R)$ is given by

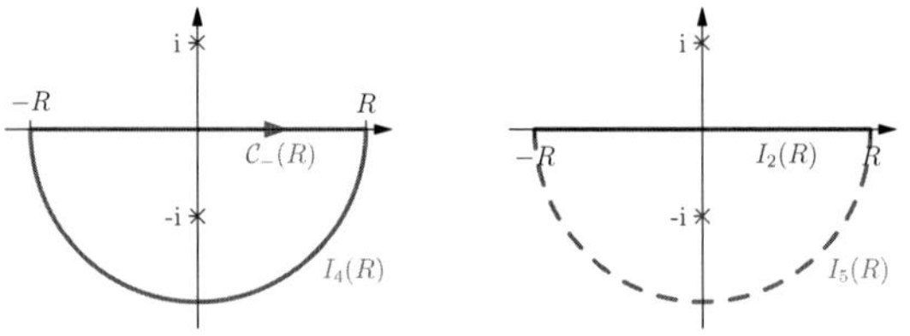

Once again dividing the contour integral into two sections we find $I_4(R) = I_2(R) + I_5(R)$ where $I_2(R)$ is given above and

$$I_5(R) = \int_0^{\pi} \frac{e^{i\omega R\cos(\theta) + \omega R\sin(\theta)}}{\pi(1+R^2 e^{-2i\theta})} d\theta.$$

Again in the limit $R \to \infty$, we find $I_2(R) \to \phi(\omega)$ and (assuming $\omega < 0$) $I_5(R) \to 0$. Thus, for $\omega < 0$, we find $\phi(\omega) = \lim_{R\to\infty} I_4(R)$. Applying the residue theorem (noting that we get an extra factor of -1 because we are going around the pole clockwise) we find a residue of

$$\lim_{z \to -\mathrm{i}} -(z + \mathrm{i}) f_4(z) = \frac{\mathrm{e}^{\omega}}{2\pi\mathrm{i}}$$

(to obtain this we note that $I_4(R) = \oint f_4(z)\,\mathrm{d}z$ where $f_4(z) = f_1(z)$ – the integral $I_4(R)$ differs from $I_1(R)$ only in the contour being traversed). Thus, for $R > 1$ we find $I_4(R) = \mathrm{e}^{\omega}$. Combining this result with the previous result valid for $\omega \geq 0$ we find $\phi(\omega) = \mathrm{e}^{-|\omega|}$ as announced in the text.

6

Handling Experimental Data

Contents

This chapter dips its toes into the often murky waters of *statistics*. The field can on occasion cause heated passion as academics argue about the validity of a particular statistical test. For others the field generates little or no passion. For many researchers, the application of common sense with a passing understanding of some of the issues involved in statistics is sufficient for a lifetime in science or engineering. This chapter attempts to fulfil this need. For others, whose research pivots on the correct interpretation of scarce and expensive data, a more complete immersion into the maelstrom of statistics may be necessary. For those readers I can only direct you towards a wealth of literature covering many different aspects of statistics and catering for many varied tastes.

Experimental data usually comes with errors. We can treat the data points as random variables and apply our knowledge of probability when treating them. If you collect any data then you really need to know how big your errors are to interpret your results properly. If you don't do this yourself, you will find that referees of your papers are likely to impose this on you. This chapter discusses errors and how to treat them and discusses common methods for extracting statistical information from raw data.

6.1 Estimating the Error in the Mean

The problem of estimating quantities from empirical data is the domain of statistics. This chapter deals with classical notions of statistics. In this context, a statistic is some quantity that can be inferred from the observed data. Thus, if we have random variables $X \sim \mathcal{N}(\mu, \sigma^2)$, μ and σ^2 are not statistics in this sense. Estimates of μ and σ^2 from data would be considered statistics. Probability has a wider remit, but when dealing with experimental data an understanding of statistics is useful and sometimes essential.

When making some measurements there is usually some random component which means that the quantity that you are measuring changes each time you make the measurement. We can treat the quantity you are measuring as a random variable, X. The underlying quantity of interest is often the expectation $\mathbb{E}\left[X\right]$ (assuming the errors are not systematic). To estimate the expected value accurately we repeat the measurement many times to get a collection of measurements $\mathcal{D} = (X_1, X_2, \ldots, X_n)$. We assume that these measurements are independent – that is, the uncertainty we are measuring results in different errors each time we make a measurement. An estimate of the expected value is given by the empirical mean

$$\hat{\mu} = \frac{1}{n} \sum_{i=1}^{n} X_i.$$

Since $\mathbb{E}\left[X_i\right] = \mu$ then $\mathbb{E}\left[\hat{\mu}\right] = \mu$, which implies that this estimator is unbiased. However, knowledge of $\hat{\mu}$ is of little use without some estimate of its uncertainty. That is, we also want an estimate of the possible error in the mean. This is usually written $\mu = \hat{\mu} \pm \Delta$, but how should we estimate Δ? Intuitively this should have to do with the variance in X. So how can we estimate the variance?

An unbiased estimator for the variance is

$$\hat{\sigma}^2 = \frac{1}{n-1} \sum_{i=1}^{n} (X_i - \hat{\mu})^2. \tag{6.1}$$

This definition of the estimator for the variance seems at first sight perplexing. Why divide by $n - 1$ rather than n? The reason why dividing by n is wrong is because we are using the same data set to estimate both $\hat{\mu}$ and $\hat{\sigma}$, but the reason why the correction is $n/(n-1)$ is harder to see. However, we can verify by direct calculation that this estimator is unbiased

$$\begin{aligned}
\mathbb{E}\left[\hat{\sigma}^2\right] &\overset{(1)}{=} \frac{1}{n-1} \mathbb{E}\left[\sum_{i=1}^{n} \left(X_i - \frac{1}{n}\sum_{j=1}^{n} X_j\right)^2\right] \\
&\overset{(2)}{=} \frac{1}{n-1} \mathbb{E}\left[\sum_{i=1}^{n} \left(X_i^2 - \frac{2}{n} X_i \sum_{j=1}^{n} X_j + \frac{1}{n^2}\left(\sum_{j=1}^{n} X_j\right)\left(\sum_{k=1}^{n} X_k\right)\right)\right] \\
&\overset{(3)}{=} \frac{1}{n-1} \mathbb{E}\left[\sum_{i=1}^{n} X_i^2 - \frac{1}{n}\left(\sum_{j=1}^{n} X_j\right)\left(\sum_{k=1}^{n} X_k\right)\right]
\end{aligned}$$

$$\stackrel{(4)}{=} \frac{1}{n-1}\mathbb{E}\left[\left(1-\frac{1}{n}\right)\sum_{i=1}^{n} X_i^2 - \frac{1}{n}\sum_{\substack{j=1,k=1\\k\neq j}}^{n} X_j\, X_k\right]$$

$$\stackrel{(5)}{=} \frac{1}{n-1}\left(\left(1-\frac{1}{n}\right) n\,\mu_2 - \frac{1}{n}\, n\,(n-1)\,\mu_1^2\right) \stackrel{(6)}{=} \frac{n-1}{n-1}\left(\mu_2-\mu_1^2\right) \stackrel{(7)}{=} \sigma^2.$$

(1) From Equation (6.1).
(2) Expanding the square.
(3) Summing over i and cancelling terms.
(4) Separating out the terms $k = j$.
(5) Taking the expectation over each term in the sum and using the definition of moments $\mathbb{E}\left[X_i^2\right] = \mu_2$, $\mathbb{E}\left[X_i\right] = \mu_1$, and the independence of X_i and X_j.
(6) Taking out the common term $(n-1)$.
(7) Cancelling and using the definition of the variance $\sigma^2 = \mu_2 - \mu_1^2$.

Dividing $\sum_{i=1}^{n}(X_i - \hat{\mu})^2$ by n would give us a consistent, but biased estimator (it is consistent in the sense that in the limit $n \to \infty$ it converges to the true variance). Unbiased estimators for higher cumulants are even more complicated and many books get them wrong. They are known as Fisher statistics, after Ronald Fisher, who wrote a paper deriving the first 10. The unbiased estimator for the third and fourth cumulants are

$$\hat{\kappa}_3 = \frac{n}{(n-1)\,(n-2)}\sum_{i=1}^{n}(X_i-\hat{\mu})^3$$

$$\hat{\kappa}_4 = \frac{n^2}{(n-1)\,(n^2-6\,n+6)}\sum_{i=1}^{n}(X_i-\hat{\mu})^4 - \frac{(3\,n-4)\,n}{n^2-6\,n+6}\left(\frac{1}{n-1}\sum_{i=1}^{n}(X_i-\hat{\mu})^2\right)^2.$$

I'm unaware of any unbiased estimators for the skewness or kurtosis (these are complicated by involving ratios of powers of cumulants).

Returning to the issue of estimating the error in the estimated mean, the estimated mean is

$$\hat{\mu} = \frac{1}{n}\sum_{i=1}^{n} X_i.$$

To calculate the variance we observe that the X_is are independent with variance σ^2 so

$$\mathbb{V}\text{ar}\left[\hat{\mu}\right] \stackrel{(1)}{=} \sum_{i=1}^{n}\mathbb{V}\text{ar}\left[\frac{X_i}{n}\right] \stackrel{(2)}{=} \sum_{i=1}^{n}\frac{\sigma^2}{n^2} \stackrel{(3)}{=} \frac{\sigma^2}{n}.$$

(1) Using the definition of $\hat{\mu}$ and that for independent random variables $\mathbb{V}\text{ar}\left[\sum_i X_i\right] = \sum_i \mathbb{V}\text{ar}\left[X_i\right]$.
(2) Using $\mathbb{V}\text{ar}\left[a\,X_i\right] = a^2\,\mathbb{V}\text{ar}\left[X_i\right]$ and the fact that X_i are all independent, identically distributed (*iid*) so that $\mathbb{V}\text{ar}\left[X_i\right] = \sigma^2$.
(3) Using $\sum_{i=1}^{n} c = n\,c$.

The uncertainty in the estimated mean is the standard deviation, $\sqrt{\mathbb{V}\mathrm{ar}\left[\hat{\mu}\right]}$, divided by $\sqrt{n}$, $\Delta = \sigma/\sqrt{n}$. This is known as the *error in the mean*. Of course, we don't know σ, however we have an unbiased estimate for it, namely $\hat{\sigma}$. Thus, the *estimated error in the mean* is equal to $\hat{\Delta} = \hat{\sigma}/\sqrt{n}$. When plotting graphs, the error bars show by convention the estimated errors in the mean, $\hat{\Delta}$, *not* the estimated standard deviation $\hat{\sigma}$. An extremely common mistake in scientific papers is showing the standard deviation. The error bars decrease as $1/\sqrt{n}$ so to reduce the error bars by two we have to use four times as much data – your estimate of the standard deviation doesn't change much as you increase the sample size. In the margin figure we show some histograms of data sampled from $\mathcal{N}(0, 1)$ for sample sizes of 4, 16, 64, and 256. We observe that the estimated standard deviation remains much the same, but the estimated error in the mean reduces by approximately two when we increase the sample size by four.

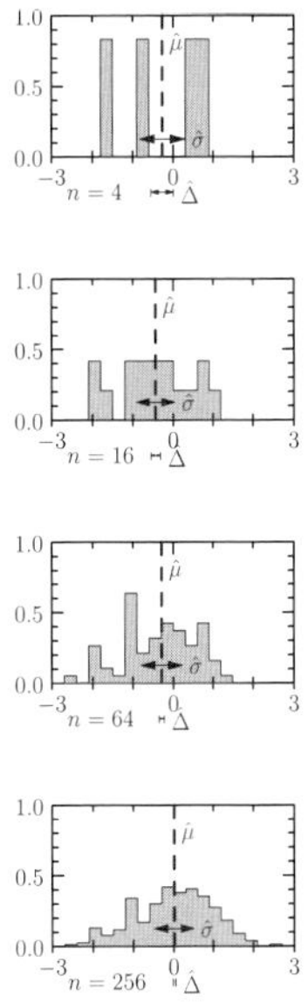

If a curve is drawn through a set of data points with error bars then only around 60% of the time should it pass through the error bar. Figure 6.1 illustrates a graph with correctly drawn error bars and with the standard deviation used as error bars. Notice that using standard deviation gives no indication of expected errors in the mean. Occasionally, people use error bars that show 95% confidence intervals (i.e. with 95% confidence the mean should be within the shown error bars). In this case, rather than plotting $\sigma/\sqrt{n}$, you need to plot $2\sigma/\sqrt{n}$. This is common practice, for example, in much of biology.

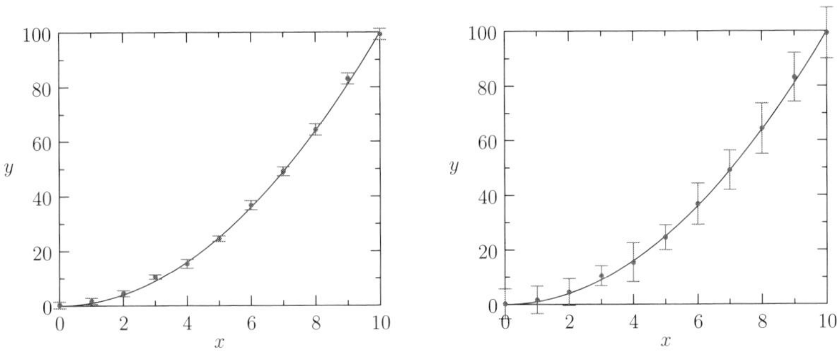

Figure 6.1 The eye-ball test – have you got the right error bars? We suppose we are empirically trying to determine some function $y(x)$, where we have some noisy measurements of y together with estimated errors in the mean. The left-hand plot shows error bars with height equal to the estimated error in the mean. The right-hand plot shows one standard deviation in the data – these should NOT be drawn as error bars.

If you are really unlucky, you might be measuring a variable without a variance or even a mean. We saw that Cauchy distributed variables don't have a well-defined mean or variance. This would be an unusual situation. If your estimates for the mean and variance seem to vary wildly from sample to sample then plotting a histogram, as described in Section 6.2, might help you see what is happening. If you really have a random variable whose distribution appears to have extremely large tails then you might consider looking at the statistics of its logarithm (although this only works if the random variables are strictly positive).

6.1.1 Computing Means with Errors

Every time you collect data you should automatically compute the errors. The easy way to do this is to write a routine that does this for you automatically. In object-oriented languages you can write a class to do this. The class should become part of a library of routines which gets included automatically. You need only keep a count of the number of data points n and an estimate for the mean and squared difference defined as

$$\hat{\mu}_n = \frac{1}{n} \sum_{i=1}^{n} X_i, \qquad \hat{S}_n = \sum_{i=1}^{n} (X_i - \hat{\mu}_n)^2.$$

Straightforward algebra shows that these quantities obey the recursion relations

$$\hat{\mu}_n = \hat{\mu}_{n-1} + \frac{X_n - \hat{\mu}_{n-1}}{n}$$

$$\hat{S}_n = \hat{S}_{n-1} + \left(1 - \frac{1}{n}\right)(X_n - \hat{\mu}_{n-1})^2.$$

The estimated variance in the data is $\hat{\sigma}^2 = \hat{S}_n/(n-1)$ and the estimated error in the mean is $\hat{\Delta} = \hat{\sigma}/\sqrt{n}$. If you deal with data, but don't yet have a class that computes this, then create one now – it will pay you many times over.

6.1.2 Bernoulli Trials

There is one situation when you don't need to estimate the variance separately, namely when you are dealing with Bernoulli trials (i.e. when your data falls into two classes). We have seen that the maximum likelihood estimator for the success rate of a Bernoulli trial is $\hat{\mu} = K/n$, where K is the number of successes in the sample.

Suppose you develop a new method for doing a task and you find it has an estimated success rate of $\hat{\mu} = 0.78$ compared with $\hat{\mu} = 0.75$ for someone else's method. You want to express how confident you are that your method is really better than theirs. One approach is to perform a statistical significance test, but a good engineer should have a gut intuition of the significance. Fortunately, for Bernoulli trials this is easy to do, at least approximately.

The outcomes of n independent Bernoulli trials with success probability μ is distributed according to a binomial distribution, $K \sim \text{Bin}(n, \mu)$, where the expected number of successes is $n\,\mu$ and the variance is $\sigma^2 = n\,\mu\,(1-\mu)$. Our estimate of μ would then be $\hat{\mu} \pm \Delta$ where

$$\Delta = \sqrt{\frac{\mu\,(1-\mu)}{n}}.$$

This is our estimated error in the mean. Of course, in practice we don't know μ. There are several approaches to overcoming this. The worst-case error occurs when $\mu = 1/2$, in which case $\sigma = 1/(2\sqrt{n})$ – this provides an upper bound on your error in the mean. If, for example, you tested your method on 100 samples then the size of the error is unlikely to be more than 5% while if you tested your method on 10,000 samples the error is unlikely to be more than 0.5% (and in the example above a success rate of 0.78 is significantly better than 0.75 if measured on this number of samples). The problem with this worse case is that it substantially overestimates the size of the error when μ is close to 0 or 1. An improved error estimate is obtained by plugging in the estimated error in the mean, which is easily found to be $\sqrt{\hat{\mu}(1-\hat{\mu})/(n-1)}$. Unfortunately, this can give you an underestimation of the uncertainty. In particular, if you have seen no successes or failures then the uncertainty estimate is zero, which is definitely an underestimate of the true uncertainty.

If we require a more accurate estimate of the error, we can resort to Bayes rule, Equation (1.4), which we can write as

$$f(\mu|k,n) = \frac{\mathbb{P}\left(k|n,\mu\right) f(\mu)}{\mathbb{P}\left(k|n\right)}$$

where $f(\mu)$ is our prior density distribution for μ and

$$\mathbb{P}\left(k|n\right) = \int_0^1 \mathbb{P}\left(k|n,\mu\right) f(\mu)\, \mathrm{d}\mu.$$

Assuming that every value of μ is equally likely (following the maximum likelihood philosophy) so that $f(\mu) = 1$, then

$$f(\mu|k,n) = \frac{\mathrm{Bin}(k|n,\mu)}{\int_0^1 \mathrm{Bin}(k|n,\mu)\, \mathrm{d}\mu}.$$

We can use this to obtain tight error bounds, although it is usually rather awkward numerically to do this. Various researchers have proposed approximations to this. As an example, suppose we have a 100% success rate so that $k = n$. In this case $\mathrm{Bin}(k = n|n,\mu) = \mu^n$ so that $f(\mu|k,n) = (n+1)\mu^n$. The expected value of μ is

$$\mathbb{E}\left[\mu\right] = \int_0^1 \mu\, f(\mu|k,n)\, \mathrm{d}\mu = 1 - \frac{1}{n+2}$$

while the variance is (following a similar calculation) $\sigma^2 = 1/((n+2)(n+3))$. Thus the expected error is approximately $1/n$, which is, of course, much smaller than the previous bound of $1/(2\sqrt{n})$. Rather surprisingly, the assumption that all values of μ are equally likely is not the most uninformative prior, but we delay a discussion of uninformative priors until Chapter 8.

6.2 Histogram

Sometimes you are not just interested in the mean of the data you have collected, but are also interested in the distribution. To visualise this distribution you can draw a histogram. Given a collection of data $\mathcal{D} = (X_1, X_2, \ldots, X_n)$ which lies in some range x_{min} to x_{max}, you divide up the range into a set of bins. You then count the number of data points that lie in each bin. It is traditional to plot your data so that the area under the curve sums up to 1. Thus in plotting a histogram

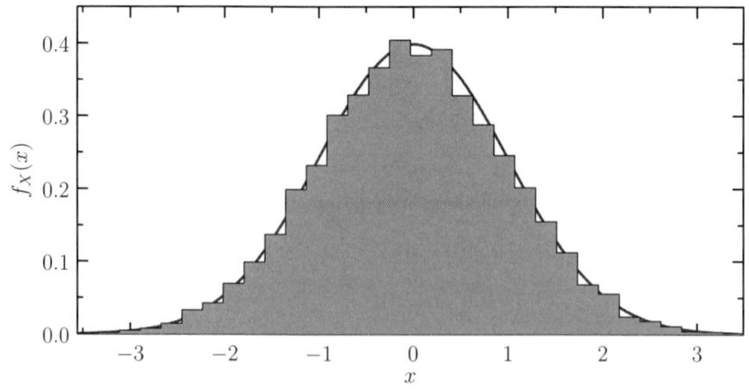

Figure 6.2 Example of a histogram showing 10, 000 normal deviates, together with the probability distribution function (PDF) for a normally distributed random variable.

you divide the count of points in each bin by the total number of data points, but also by the width of the bin. This allows you to compare your histogram to the probability density functions. Figure 6.2 illustrates a histogram of normal deviates. The size of the bin affects the quality of the plot. If too small then the uncertainty in the count will be large and the fluctuations very large. If too large you lose resolution on the x-axis. There are 'optimal' strategies for choosing the bin size, but usually trial and error is sufficient. There are lots of packages for drawing histograms or you can write one yourself, but it's worthwhile to have a package at hand so you can visualise distributions easily.

I have on occasion torn my hair out trying to understand some complex process where the mean doesn't behave as I expected. After days of agonising, I've finally got around to plotting a histogram of my data, which showed that the quantity I was studying had thick tails and its typical behaviour was not well captured by the means. Histograms have saved me from premature baldness, so I wholeheartedly recommend them. I also use them routinely as a sanity check every time I perform some Monte Carlo-based simulations.

6.3 Significance Tests

Is my algorithm better than hers? Suppose I want to show this empirically from two sets of data $\mathcal{D}_X = (X_1, X_2, \ldots, X_n)$ and $\mathcal{D}_Y = (Y_1, Y_2, \ldots, Y_m)$ collected on the two algorithms, where I assume that the data has some uncertainty. From empirical data alone it is impossible to *prove* one algorithm is better than the other. There is always a (possibly very small) chance of the difference being due to luck. What you can do though is give a probability that one result is better than another. This is precisely what a *statistical significance test* does.

Before you do a statistical significance test you should have a reasonable idea of whether the means differ significantly just by looking at the error bars (provided you have calculated your error bars correctly). You should also be aware that a *statistically significant difference* is not always the same as a *significant difference*. For example, I can compute an empirical estimate of the means of my two data sets. I might find $\hat{\mu}_Y > \hat{\mu}_X$ and doing a significance test I might prove with high probability that the true mean of the process that generated $\mathcal{D}_Y$ is higher than that for $\mathcal{D}_X$. They are (statistically) significantly different, but the difference might be very small – smaller perhaps than anyone would care about. Of course, to determine if something is statistically significantly when the means differ by only a small amount requires a lot of data. Conversely, if I have a little data then there could be a large difference in the true means, but the data might not be statistically significant. Often the need to do a significance test is a sign that the difference is not so interesting. If my algorithm was so much better than hers then it would be easy to see that the difference was much larger than the estimated error in the estimator so a significance test might be superfluous. This is not to say that significance tests shouldn't be done, but personally, I'm always on my guard when I see one.

So when should you do a significance test? Firstly, if you have little data then your intuition might be a false guide and a significance test can be helpful. If you are tackling a very competitive problem then any improvement might be of interest so you might want to show that it is significant. You might have an extremely professional attitude and so want to present your data with as much information as possible. These are all good reasons for performing significant tests.

What significance tests should you perform? There are quite a lot of tests out there depending on what you want to show. However, you need to read the small print. Many of the classical tests assume that your data is normally distributed. If the distribution of your data has significant tails then many of these tests are likely to give you overly confident results. You can first test whether your data is, at least approximately, normally distributed, e.g. by using the Shapiro–Wilk or Kolmogorov–Smirnov test (the former specifically tests for normality while the latter provides a more general measure between two distributions; details can be found on Wikipedia). There are also tests that make no assumption about your data, although they might give you overly conservative estimates of significance. You should also be aware that if you have two data sets that are tested on the same set of examples then by examining the data in pairs you might find significant difference in the samples that may otherwise be hidden by fluctuations between samples. We discuss 'paired significance tests' after describing the classical *t*-test.

T-tests

The granddaddy of all significance tests is Student's t-test. Student was a pseudonym of William Sealy Gosset, a statistician working for the Guinness brewery in Dublin – he wasn't allowed to publish under his own name. Student's t-test tests for the *null hypothesis* that the data comes from two normal distributions with the same mean. It also make an assumption that the two distributions have the same variance. Thus, if we perform a t-test and obtain a very small probability for the null hypothesis we can conclude that the means are probably distinct, provided the underlying assumptions are true. If the variances of the two distributions are different then we can use a variation of the t-test, officially known as Welch's t-test.

Independent two-sample t-test. To compute the t-test for two collections of data $\mathcal{D}_X = (X_1, X_2, \ldots, X_n)$ and $\mathcal{D}_Y = (Y_1, Y_2, \ldots, Y_m)$ where you believe the data has the same variance, you first compute the t-value

$$t = \frac{\hat{\mu}_X - \hat{\mu}_Y}{\hat{\sigma}_{X+Y}}$$

where $\hat{\sigma}^2_{X+Y}$ is a kind of combined estimated error in the mean given by

$$\hat{\sigma}^2_{X+Y} = \frac{(n-1)\,\hat{\sigma}^2_X + (m-1)\,\hat{\sigma}^2_Y}{n+m-2}\left(\frac{1}{n} + \frac{1}{m}\right)$$

and $\hat{\mu}_X$, $\hat{\mu}_Y$, $\hat{\sigma}^2_X$, and $\hat{\sigma}^2_Y$ are unbiased estimates for the mean and variance of the two data sets. The number of degrees of freedom is taken to be $\nu = n + m - 2$. If the data sets have different variances then use

$$\hat{\sigma}_{X+Y} = \sqrt{\frac{\hat{\sigma}^2_X}{n} + \frac{\hat{\sigma}^2_Y}{m}}$$

and take the number of degrees of freedom to be

$$\nu = \frac{\left(\hat{\sigma}^2_X/n + \hat{\sigma}^2_Y/m\right)^2}{(\hat{\sigma}^2_X/n)^2/(n-1) + (\hat{\sigma}^2_Y/m)^2/(m-1)}.$$

In both cases the probability of the null hypothesis (the p-value) is given by the cumulative distribution function (CDF) of the t-distribution

$$p = 1 - \int_{-t}^{t} \mathrm{T}(t|\nu)\mathrm{d}t = 1 - \frac{1}{\sqrt{\nu}\,B(1/2,\,\nu/2)} \int_{-t}^{t} \left(1 + \frac{x^2}{\nu}\right)^{-\frac{\nu+1}{2}} \mathrm{d}\,x$$

where $B(1/2, \nu/2)$ is the standard beta function. For readers who take nothing on trust, or who just enjoy ploughing through algebra, I give a derivation of Student's t-test in Appendix 6.A. There are many computer packages that will compute this. If you have written a class to automatically accumulate the mean and estimated error in the mean you might want to add the ability to perform a t-test between two such sets of data.

Example 6.1 T-test

As an example, we generate two sets of random deviates $X \sim \mathcal{N}(0.4, 1)$ and $Y \sim \mathcal{N}(0.6, 0.25)$ each of size n. In Figure 6.3 we show the p-value as a function of n. We also show the error bars. The p-value measures the probability that the two samples come from normal distributions of the same mean. Thus a small p-value of say less than 0.01 gives pretty firm evidence that the underlying distribution has a different mean.

Note that t-tests use almost the same information that is shown by error bars (the p-values depend on the number of samples, which is not shown in the error bars); their additional contribution is that they give you the probability that

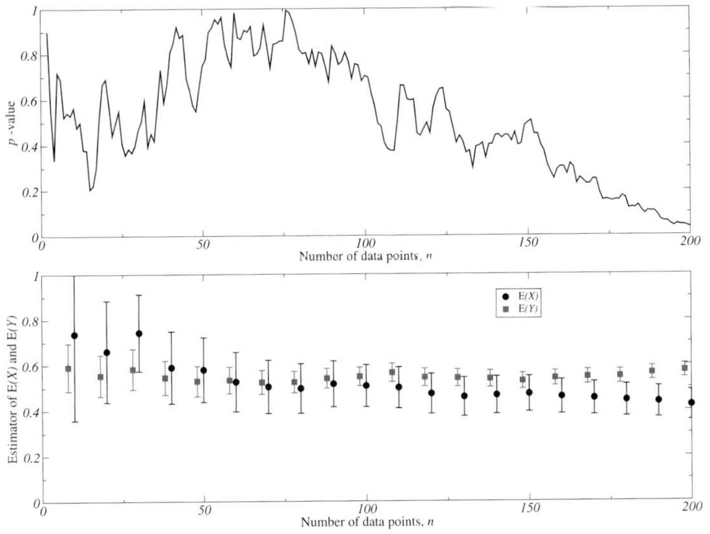

Figure 6.3 (a) Plot of p-values versus number of data points, and (b) the error bars in the estimator for X and Y.

such error bars could arise by chance if the data actually came from the same distribution. It is very easy to obtain low p-values by being selective about your data. Don't take anything under a p-value of 0.01 as being significant, it is just too easy to get p-values around 0.1 even when you think you are being honest. If you can't get your p-values low enough it is probably a sign that you need to collect more data or that your samples just aren't significantly different.

To get a feel for the t-test let us assume that two sets of data, $\mathcal{D}_X$ and $\mathcal{D}_Y$, are drawn from a normal distribution and that empirically they have the same observed variance $\hat{\sigma}_X^2 = \hat{\sigma}_Y^2 = \hat{\sigma}^2$. The size of both data sets we assume to be n. If we measure the difference between the means of our two samples in units of the estimated error in the mean, $\Delta = \hat{\sigma}/\sqrt{n}$, then

$$\hat{\mu}_X - \hat{\mu}_Y = \frac{k\,\hat{\sigma}}{\sqrt{n}};$$

i.e. k is the number of error bars separating the sample means. By straightforward substitution we find $\hat{\sigma}_{X+Y}^2 = 2\,\hat{\sigma}^2/n$, so $t = (\hat{\mu}_X - \mu_Y)/\hat{\sigma}_{X+Y} = k/\sqrt{2}$, and the number of degrees of freedom is $\nu = 2n - 2$. Figure 6.4 shows the p-value for $k = 2$ and $k = 3$. The dependency on n arises because of the uncertainty in the estimate of the variance. We notice that when the means differ by twice the estimated error in the mean then the difference is never significant (the p-value is around 0.045 for reasonable size n). When the difference reaches three times the estimated error in the mean then the p-value becomes significant (i.e. below 0.01) for $n \geq 8$. For $n > 20$ with $|\hat{\mu}_X - \mu_Y| > 3\,\Delta$ then $p \approx 0.0027$.

Paired t-test. There are times when the difference between the means of two sets of data can be significantly masked by some other source of stochasticity. An example of this would be if we were comparing the performance of two random algorithms on different problem instances. For example, you might be interested in the quality of the solution found by two random search algorithms

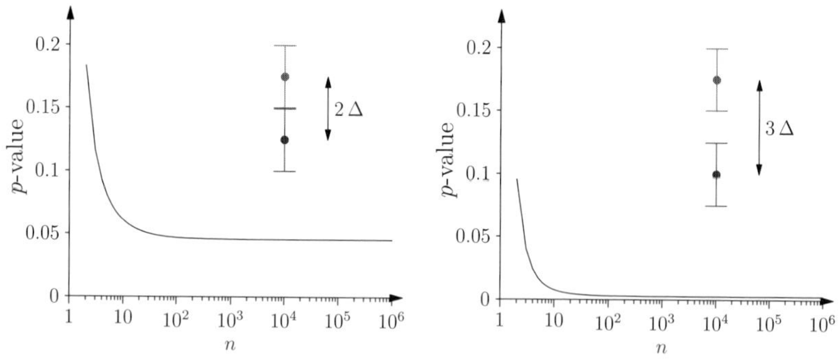

Figure 6.4 The p-value is plotted against n (the number of observations) for data points differing by two times the estimated error in the mean and three times the estimated error in the mean.

on a combinatorial optimisation problem such as graph-colouring, given a fixed computational budget. In this case, there are two possible sources of stochasticity. The first is from the problem instance being tested and the second is from the algorithms. In this situation a 'paired' statistical test can be useful in identifying whether the performance is significantly different. In these families of tests we would generate our data in pairs. In our example, we would first generate the problem instance and then test the two algorithms on that instance (thus having a data pair). This procedure is repeated to obtain a set of pairs. Assuming the data is normally distributed (which we should test), we could use a 'paired t-test'. This would remove the stochasticity due to the problem. Such a test is more likely to show a significant difference in performance of the two algorithms, if such a difference exists, than a non-paired t-test.

To perform a paired t-test you compute the combined variances as

$$\hat{\sigma}^2_{X+Y} = \frac{\hat{\sigma}^2_X + \hat{\sigma}^2_Y - \hat{c}_{X,Y}}{n}$$

where

$$\hat{c}_{X,Y} = \frac{1}{n-1} \sum_{i=1}^{n} (X_i - \hat{\mu}_X)\,(Y_i - \hat{\mu}_Y)$$

is the unbiased estimator for the covariance. The number of degrees of freedom is taken to be $\nu = n - 1$. Note that if the data is strongly correlated then $\hat{\sigma}^2_{X+Y}$ for the paired test can be significantly smaller than that for unpaired test, giving you much higher confidence.

Example 6.2 Paired T-test

As an illustration we consider two random variables $X = Z + X_1$ and $Y = Z + Y_1$ where $Z \sim \mathcal{N}(0, 0.8)$, $X_1 \sim \mathcal{N}(0.4, 0.2)$, and $Y_1 \sim \mathcal{N}(0.6, 0.2)$. Thus, $X \sim \mathcal{N}(0.2, 1)$ and $Y \sim \mathcal{N}(0, 1)$, however, X and Y are strongly correlated. As a consequence, the paired t-test gives a much higher significance (smaller p-value) than a standard t-test. The p-values for both the paired and unpaired t-test are shown in Figure 6.5. We also show the error bars for estimated errors in the mean for $\hat{\mu}_X$, $\hat{\mu}_Y$ and $\hat{\mu}_{Y-X}$. Although it would be difficult to

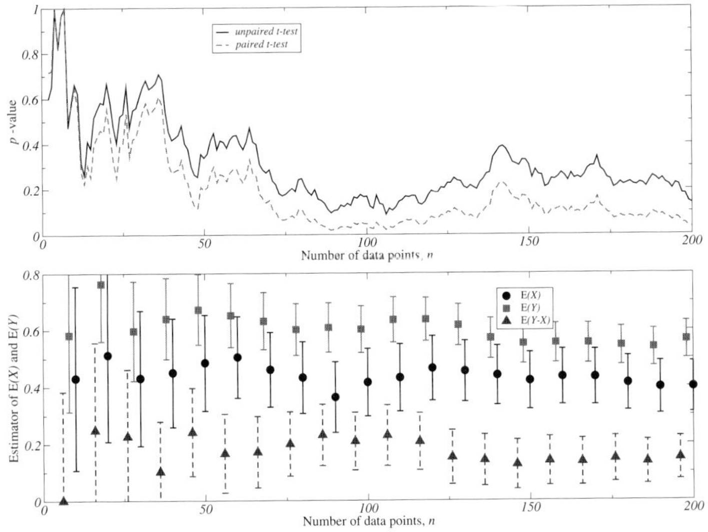

Figure 6.5 (a) Plot of p-values versus number of data points computed using a t-test and a paired t-test, and (b) error bars for X, Y, and $Y - X$.

conclude from the error bars for $\hat{\mu}_X$ and $\hat{\mu}_Y$ that X and Y come from different distributions there is quite strong evidence that $\hat{\mu}_{Y-X}$ has a mean different from zero.

It is worth stressing the need to apply confidence tests scrupulously. If, for example, you test 100 different systems against some baseline and then you choose the best, then it is fairly likely that that system will have a p-value of less than 0.01, even when all the systems were identical with the baseline system. That is what the p-value is telling you: the chances of the null hypothesis happening (i.e. the systems have the same expected value) is p, so in expectation you would see a p-value of less that or equal to p in $1/p$ trials. To do the confidence tests fairly, you would need to perform a t-test (assuming your data is normally distributed) on the baseline and best system using an independent set of measurements. The take-home message is that being honest and applying some common sense is far more important than pushing your data into a statistics package and obtaining some confidence value that you don't really understand.

There are many other tests. For example Fisher's F-test is used to determine the equality of variances for two data sets assuming a normal distribution. We first compute the F-ratio, which is the ratio of the empirical variance. The test gives the probability of this F-ratio occurring by chance under the null hypothesis that both sets of data have the same variance. The F-ratio can be used in other contexts. For example, in regression (fitting observed data depending on some independent variables by a line or surface) we can use the F-ratio to measure the ratio between the variance explained by the regression model (i.e. the reduction in variance due to the model compared with the variance without the model) to the unexplained variance (i.e. the remaining variance in the data given the model). Under the assumption that the noise is normal, the F-test gives a measure of the probability that the fit happened by chance. F-ratios are sometimes used in machine learning to weight the input features, so that features that better explain the data receive a higher weighting.

There are also many tests which don't make assumptions of normality. For testing the equality of means for independent samples, the Mann–Whitney U-test is used, while for related samples, either the binomial test or the Wilcoxon signed-rank test are used. Other tests for measuring the equality of variance include Levene's test, Bartlett's test, the Brown–Forsythe test, or the O'Brien test.

For more complex experiments where you want to compare more than one distribution then performing multiple t-tests become less reliable. Consider a set of field trials where there are k known independent variables (external features that can be changed). If we perform binary t-tests between every group where some of the independent variables are set and the rest aren't then there are $k\,(k-1)/2$ separate t-tests to carry out. However, this grows rapidly so there is a reasonable chance that a significant p-value occurs in one of the binary tests by chance. Instead of using t-tests there is a whole body of tools known as *analysis of variance* or ANOVO that provides a more robust test to see whether different batches come from the same distribution or from different distributions. For example, suppose you have three textbooks and you want to decide which is the best to use in a class. If you had the grades of students who used different books then you can ask the question: Did the textbook used make any difference? The null hypothesis would be that the textbook was irrelevant and the difference in scores between the different groups of students was entirely due to chance. To test this hypothesis we see whether the variance between the three groups of students is compatible with the variance within the groups of student – this uses Fisher's F-test. You can see this as a regression problem: if M_i is the grade of student i then we can fit a linear model

$$M_i = \sum_{k=1}^{3} w_k \left[\!\left[\text{student } i \text{ used book } k \right]\!\right] + \epsilon_i$$

where ϵ_i is the error that is not explained by the model. To find the best parameters, w_i, that fit the data we can minimise the squared error

$$\sum_{i=1}^{n} \epsilon_i^2 = \sum_{i=1}^{n} \left(M_i - \sum_{k=1}^{3} w_k \left[\!\left[\text{student } i \text{ used book } k \right]\!\right] \right).$$

This is a standard procedure for performing regression. In this case we obtain a very simple solution that w_k is the mean grade for the students who used book k. To determine the goodness of fit for each parameter we compute the reduction in the variance caused by setting w_k to the mean grade, compared to the variance in the grades with no model. We can then compute the F-ratio by dividing this reduction in variance with the remaining variance given the model. We can compute the probability of obtaining this F-ratio by chance (this depends on the number of observations, n, and the number of parameters, k). ANOVA and its many variants have been highly developed. It is used in many disciplines including social sciences. In complex disciplines where there are many possible explanations of the data, the complexity of ANOVA provides an incredibly useful tool. In contrast, it tends to get little used in fields where the uncertainty in the model is smaller or where we seek to make predictions (where classic machine learning dominates) rather than try to explain the data. I will curtail any further discussion of ANOVA as the subject fills many textbooks.

A good source of statistical algorithms is Press et al. (2007). Many of these tests are also built into packages such as Matlab and Octave, and statistical programming languages such as R. Packages such as SPSS allow the application of tests like ANOVA with relatively little pain, although I dread to think of the numerous sins of statistics that have resulted.

6.4 Maximum Likelihood Estimate

Given some *iid* data $\mathcal{D} = (X_1, X_2, \ldots X_n)$ from a probability distribution $f(\boldsymbol{x}|\boldsymbol{\theta})$ that depends on parameters $\boldsymbol{\theta}$, then the maximum likelihood estimate of the parameters are those parameters $\hat{\boldsymbol{\theta}}$ that maximise the likelihood

$$\mathbb{P}\left(\mathcal{D}|\boldsymbol{\theta}\right) = \prod_{i=1}^{n} f(\boldsymbol{X}_i|\boldsymbol{\theta}).$$

Maximum likelihood estimates can be useful for obtaining approximate parameters of a distribution from empirical measurements. This approach was introduced by Ronald Fisher and became an established part of classical statistics (much sooner than it took for Bayesian statistics to become respectable). For distributions that are differentiable functions of their parameters (which include all the common distributions), we can find the maximum likelihood distribution by a (usually) straightforward optimisation of their parameters. Most simple distributions are convex functions of their parameters so they have a unique extremal value corresponding to the maximum likelihood solution – this is not commonly true when the likelihood is a complicated function. As the likelihood function is a product of the probability distributions it is nearly always easier to maximise the log-likelihood (because the logarithm is a monotonically increasing function, the parameters that maximise the log-likelihood are the same as those that maximise the likelihood). Thus the maximum likelihood estimates are the solution to the equations

$$\frac{\partial \log\left(\mathbb{P}\left(\mathcal{D}|\boldsymbol{\theta}\right)\right)}{\partial \theta_i} = \sum_{i=1}^{n} \frac{\partial \log\left(f(\boldsymbol{X}_i|\boldsymbol{\theta})\right)}{\partial \theta_i} = 0.$$

We consider a couple of simple examples.

Consider estimating the mean and variance of normally distributed data. Now

$$\log\left(f(\boldsymbol{X}_i|\mu,\sigma)\right) = -\frac{1}{2}\log\left(2\,\pi\,\sigma^2\right) - \frac{(X_i-\mu)^2}{2\,\sigma^2}$$

so that, on taking derivatives with respect to μ and σ, and setting them to zero,

$$\sum_{i=1}^{n} -\frac{X_i - \mu}{2\,\sigma^2} = 0, \qquad \sum_{i=1}^{n} \left(-\frac{1}{\sigma} + \frac{(X_i - \mu)^2}{\sigma^3} \right) = 0.$$

The first equation gives us an estimate of the mean

$$\hat{\mu}_{ML} = \frac{1}{n} \sum_{i=1}^{n} X_i$$

while the second equation give us an estimate of the variance

$$\hat{\sigma}^2_{ML} = \frac{1}{n} \sum_{i=1}^{n} (X_i - \hat{\mu}_{ML})^2 = \frac{1}{n} \sum_{i=1}^{n} X_i^2 - \hat{\mu}^2_{ML}.$$

Note that to estimate the mean and variance we need only know $\sum_{i=1}^{n} X_i$, and $\sum_{i=1}^{n} X_i^2$. These are known as sufficient statistics (sufficient in the sense that they are all you need to obtain maximum likelihood statistics). We observe in passing that the maximum likelihood estimator for the variance of a normal distribution is a biased estimator (it equals $n/(n-1)$ times the unbiased estimator).

Things get more complicated for most other distributions. For example, for the gamma distribution where

$$f_X(x) = \text{Gam}(x|a, b) = \frac{b^a\, x^{a-1}\, \mathrm{e}^{-b\,x}}{\Gamma(a)}$$

the log-likelihood of the data is given by

$$\mathcal{L} = n\, a\, \log(b) + (a-1) \sum_{i=1}^{n} \log(X_i) - b \sum_{i=1}^{n} X_i - \log(\Gamma(a)).$$

Taking derivatives

$$\frac{\partial \mathcal{L}}{\partial a} = n\, \log(b) + \sum_{i=1}^{n} \log(X_i) - \psi(a)$$

$$\frac{\partial \mathcal{L}}{\partial b} = \frac{n\,a}{b} - \sum_{i=1}^{n} X_i$$

where $\psi(a)$ is the diagamma function (see Section 2.A on page 41). Setting the partial derivatives to zero we find

$$\frac{a}{b} = \frac{1}{n} \sum_{i=1}^{n} X_i, \qquad \psi(a) - \log(a) = \frac{1}{n} \sum_{i=1}^{n} \log(X_i) - \log\left(\frac{1}{n} \sum_{i=1}^{n} X_i \right).$$

This can be solved numerically, for example, using Newton–Raphson's method to find a and then use the first equation to find b. Notice that the sufficient statistics are $\sum_{i=1}^{n} X_i$ and $\sum_{i=1}^{n} \log(X_i)$.

With work we can also compute confidence intervals telling us that the maximum likelihood parameters fall within a given interval with some probability (say 95%). Technically this is non-trivial although far from impossible to compute.

Confidence intervals often provide an alternative to hypothesis testings (e.g. t-tests). When we know that two sets of random variables have non-overlapping confident intervals for their empirical means then with overwhelming likelihood we can say that one set has a higher value than the other. The standard error in the mean we computed earlier are a poor man's confidence interval in this sense. Whereas $\hat{\mu} \pm 2\Delta = \hat{\mu} \pm 2\sigma/\sqrt{n}$ provides a good estimate of a 95% confidence interval for a normally distributed random variable, when we use an estimate of the standard deviation, $\hat{\sigma}$, for a small sample of data, then we may substantially underestimate the true standard deviation. Thus, for small samples, the 95% confidence interval is greater than $\hat{\mu} \pm 2\hat{\Delta} = \hat{\mu} \pm 2\hat{\sigma}/\sqrt{n}$, with the size of the confidence interval having a subtle dependence on the size of the sample. The interested reader should consult a more general text on statistics such as Williams (2001).

Maximum likelihood estimators are part of classical statistics, but they are not always particularly good estimators (e.g. they are often biased and the value you obtain, for example, depends on the representation of the probability distribution). We return to this in Chapter 8 where we consider Bayesian estimators.

Handling data usually just requires a bit of common sense. My top tip when handling real data is to compute your estimated error in the mean of all the quantities you are trying to estimate. Without an estimate of your error you really have no idea what the data means. The person you're most likely to cheat by not understanding your errors is yourself. If you underestimate your errors you often end up chasing phantom patterns which are just fluctuations that arise by chance. When your data is confusing, examining its histogram can be very useful; sometimes the distribution of your data can have very large tails which may throw your intuition into confusion. Significance tests are beloved by pedants and anonymous referees. They have a role when performed carefully. However, be wary: significance tests are frequently abused. Understanding the errors in your data (e.g. by always showing your error bars) is the best guard against being misled by random fluctuations.

Additional Reading

There are a huge number of books on statistics covering many different aspects and catering for different tastes. An informal but comprehensive guide to statistics is by Field et al. (2012). A comprehensive mathematical treatment of statistics is given by Williams (2001).

Exercise for Chapter 6

Exercise 6.1 (answer on page 411)

Generate 50 deviates $X_i \sim \mathcal{N}(0, 1)$ and 50 deviates $Y_i \sim \mathcal{N}(0, 4)$ and compute the empirical means and variances for the X and Y variables. From this compute the t-

value and the corresponding p-value using Welch's t-test. Repeat this $n = 10^6$ times recording the p-values. Sort the p-values such that the pair (t_r, p_r) has

$$t_1 \le t_2 \le t_3 \le \cdots \le t_r \le \cdots \le t_n.$$

Note that the p-value is a monotonic function of the t-value so it is sufficient to sort out the t-values. Recall that the p-value is defined such that, if the null hypothesis is true (i.e. the two sets of deviates have the same mean, which in our case is true), then $\mathbb{P}\left(|T| \ge t_r\right) = p_r$. Thus we would expect that a proportion p_r of the sample have t-values greater than t_r. Empirically this proportion is $(n-r)/n$. Thus, if we plot p_r versus $1 - r/n$ where r is the rank of the t-value we should get, roughly, a straight line. Plot this (it helps to do this on a log-log plot).

Exercise 6.2 (answer on page 412)

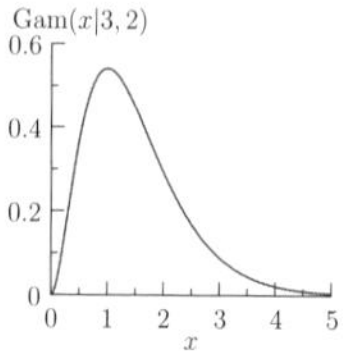

Repeat the experiment described in Exercise 6.1, but for $X_i, Y_i \sim \text{Gam}(3,2)$. As our random variables are not normally distributed we have no right to believe in the p-value for the t-test. Interestingly, the p-value is not a bad estimate of the probability of the t-value occurring. This gives some confidence that even when the data is not exactly normal the p-values of the t-test are not always too misleading. Of course, it is always possible to cook up examples where the p-values are significantly underestimated, particularly for thick-tailed distributions.

Exercise 6.3 (answer on page 413)

Recall in Section 2.4 on page 38 we observed that many distributions, including the normal and gamma distributions, could be written in the form

$$f_{\boldsymbol{X}}(\boldsymbol{x}|\boldsymbol{\eta}) = g(\boldsymbol{\eta})\, h(\boldsymbol{x})\, \mathrm{e}^{\boldsymbol{\eta}^{\mathsf{T}} \boldsymbol{u}(\boldsymbol{x})}$$

where $\boldsymbol{u}(\boldsymbol{x})$ are some functions of our random variables and $\boldsymbol{\eta}$ are 'natural parameters'. We showed that the maximum likelihood estimators for the natural parameters satisfy

$$-\boldsymbol{\nabla} \log\Big(g(\boldsymbol{\eta})\Big) = \frac{1}{n}\sum_{i=1}^{n} \boldsymbol{u}(\boldsymbol{x}_n).$$

Compute $\boldsymbol{\nabla} \log\Big(g(\boldsymbol{\eta})\Big)$ and $\boldsymbol{u}(\boldsymbol{x}_n)$ for the normal and gamma distributions and show that they are consistent with the maximum likelihood estimators we obtained in Section 6.4.

Appendix 6.A Deriving the T-Distribution

The derivation of Student's t-test takes time, but has its rewards along the way. Recall in the t-test, we consider the random variable

$$T = \frac{\hat{\mu}_X - \hat{\mu}_Y}{\hat{\sigma}_{X+Y}}$$

where

$$\hat{\sigma}^2_{X+Y} = \frac{(n-1)\,\hat{\sigma}^2_X + (m-1)\,\hat{\sigma}^2_Y}{n+m-2}\left(\frac{1}{n} + \frac{1}{m}\right)$$

and

$$\hat{\mu}_X = \frac{1}{n}\sum_{i=1}^{n} X_i, \qquad \hat{\mu}_Y = \frac{1}{m}\sum_{i=1}^{m} Y_i,$$

$$\hat{\sigma}_X^2 = \frac{1}{n-1}\sum_{i=1}^{n}(X_i - \hat{\mu}_X)^2, \qquad \hat{\sigma}_Y^2 = \frac{1}{m-1}\sum_{i=1}^{m}(Y_i - \hat{\mu}_Y)^2.$$

Now we assume that $X_i, Y_i \sim \mathcal{N}(\mu, \sigma^2)$. That is, our random variables come from the same *normal* distribution. This is our null hypothesis. We then ask what is the probability that $|T| \geq t$. If $p = \mathbb{P}\left(|T| \geq t\right)$ is very small where t is our measured t-value then we have a confidence of $1 - p$ that our data has different means (although this is only true if our data has the same variance σ^2 and is normally distributed).

We first observe that T is independent of the actual value of the true mean and variance μ and σ. In particular, we can make the change of variables $X_i' = (X_i - \mu)/\sigma$ and $Y_i' = (Y_i - \mu)/\sigma$ so that $X_i', Y_i' \sim \mathcal{N}(0,1)$. Under this change of variables

$$\hat{\mu}_X = \frac{1}{n}\sum_{i=1}^{n} X_i = \frac{1}{n}\sum_{i=1}^{n} \sigma\, X_i' + \mu = \sigma\, \hat{\mu}_{X'} + \mu \qquad \text{where} \qquad \hat{\mu}_{X'} = \frac{1}{n}\sum_{i=1}^{n} X_i'$$

and similarly $\hat{\mu}_Y = \sigma\, \hat{\mu}_{Y'} + \mu$. Note that $\hat{\mu}_X - \hat{\mu}_Y = \sigma\left(\hat{\mu}_{X'} - \hat{\mu}_{Y'}\right)$. Also we note that

$$\hat{\sigma}_X^2 = \frac{1}{n-1}\sum_{i=1}^{n}(X_i - \hat{\mu}_X)^2 = \frac{1}{n-1}\sum_{i=1}^{n}(\sigma\, X_i' + \mu - \hat{\mu}_X)^2$$

but $\hat{\mu}_X - \mu = \sigma\, \hat{\mu}_{X'}$, so

$$\hat{\sigma}_X^2 = \frac{\sigma^2}{n-1}\sum_{i=1}^{n}(X' - \hat{\mu}_{X'})^2 = \sigma^2 \hat{\sigma}_{X'}^2$$

and similarly $\hat{\sigma}_Y^2 = \sigma^2 \hat{\sigma}_{Y'}^2$. Thus, we see that $\hat{\sigma}_{X+Y}^2 = \sigma^2\, \hat{\sigma}_{X'+Y'}^2$ and the T value for X_i, $Y_i \sim \mathcal{N}(\mu, \sigma^2)$ is the same as the T value for $X_i' = (X_i - \mu)/\sigma$ and $Y_i' = (Y_i - \mu)/\sigma$ where now X_i', $Y_i' \sim \mathcal{N}(0,1)$. We can thus assume without loss of generality that X_i, $Y_i \sim \mathcal{N}(0,1)$.

The next observation is that

$$\hat{\mu}_X = \frac{1}{n}\sum_{i=1}^{n} X_i \qquad \text{and} \qquad \hat{\sigma}_X^2 = \frac{1}{n-1}\sum_{i=1}^{n}(X_i - \hat{\mu}_X)^2$$

are independent random variables. This is far from obvious. To prove this we note that if $\boldsymbol{X} \sim \mathcal{N}(0, \mathbf{I})$ then so is any linear combination of the X_is that corresponds to a simple rotation. That is, if $\boldsymbol{Z} = \mathbf{V}\,\boldsymbol{X}$ where $\mathbf{V}$ is an orthogonal matrix (i.e. a rotation matrix with the property $\mathbf{V}^{-1} = \mathbf{V}^{\mathsf{T}}$) then $\boldsymbol{Z} \sim \mathcal{N}(0, \mathbf{I})$. This follows trivially from the observation that

Recall that $(\mathbf{M}\,\boldsymbol{x})^{\mathsf{T}} = \boldsymbol{x}^{\mathsf{T}}\,\mathbf{M}^{\mathsf{T}}$

$$\sum_i Z_i^2 = |\boldsymbol{Z}|^2 = \boldsymbol{Z}^{\mathsf{T}}\boldsymbol{Z} = (\boldsymbol{X}^{\mathsf{T}}\,\mathbf{V}^{\mathsf{T}})(\mathbf{V}\,\boldsymbol{X}) = \boldsymbol{X}^{\mathsf{T}}\mathbf{I}\,\boldsymbol{X} = |\boldsymbol{X}|^2 = \sum_i X_i^2. \qquad (6.2)$$

Using the usual rule for a change of variables, the distribution of $\boldsymbol{Z}$ is given by $f(\boldsymbol{Z}) = |J|\, f(\boldsymbol{X}(\boldsymbol{Z}))$ where $\boldsymbol{X}(\boldsymbol{Z}) = \mathsf{V}^{\mathsf{T}}\boldsymbol{Z}$ and J is the Jacobian, which in this case $|\mathsf{V}^{\mathsf{T}}| = 1$ (note the determinant of an orthogonal matrix is 1). Thus

$$f(\boldsymbol{Z}) = \frac{1}{(2\pi)^{n/2}} \mathrm{e}^{-|\boldsymbol{X}(\boldsymbol{Z})|^2/2} = \frac{1}{(2\pi)^{n/2}} \mathrm{e}^{-|\boldsymbol{Z}|^2/2} = \mathcal{N}(0, \boldsymbol{I}).$$

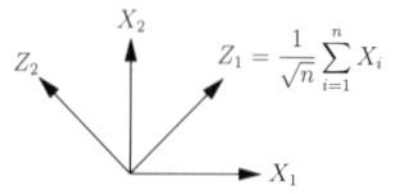

We can see this intuitively as $f(\boldsymbol{X}) = \mathcal{N}(0, \boldsymbol{I})$ is spherically symmetric so any rotation of the axes will have the same distribution. We choose

$$Z_1 = \frac{1}{\sqrt{n}} \sum_{i=1}^{n} X_i \tag{6.3}$$

since Z_1 is a sum of normally distributed variables it will itself be normally distributed. Now $\mathbb{E}\left[Z_1\right] = 0$ since $\mathbb{E}\left[X_i\right] = 0$. Under the assumption that X_is are independent then

$$\mathbb{V}\mathrm{ar}\left[Z_1\right] = \sum_{i=1}^{n} \mathbb{V}\mathrm{ar}\left[\frac{X_i}{\sqrt{n}}\right] = \frac{1}{n}\sum_{i=1}^{n} \mathbb{V}\mathrm{ar}\left[X_i\right] = \frac{1}{n}\sum_{i=1}^{n} 1 = 1$$

(where we used $X_i \sim \mathcal{N}(0,1)$ so $\mathbb{V}\mathrm{ar}\left[X_i\right] = 1$). Since Z_1 is normally distributed with mean zero and variance one we have $Z_1 \sim \mathcal{N}(0,1)$. The variables $\boldsymbol{Z}$ are distributed according to

$$f(\boldsymbol{Z}) = \frac{1}{(2\pi)^{n/2}} \mathrm{e}^{-|\boldsymbol{Z}|^2/2} = \frac{1}{(2\pi)^{n/2}} \mathrm{e}^{-Z_1^2/2 - \sum_{i=2}^{n} Z_i^2/2}$$

but

$$\sum_{i=2}^{n} Z_i^2 = \sum_{i=1}^{n} Z_i^2 - Z_1^2 = \sum_{i=1}^{n} X_i^2 - \frac{1}{n}\left(\sum_{i=1}^{n} X_i\right)^2$$

where we have used Equations (6.2) and (6.3). However,

$$\begin{aligned} S_X^2 &= \sum_{i=1}^{n} (X_i - \hat{\mu}_X)^2 = \sum_{i=1}^{n} X_i^2 - 2\hat{\mu}_X \sum_{i=1}^{n} X_i + \hat{\mu}_X^2 \sum_{i=1}^{n} 1 \\ &= \sum_{i=1}^{n} X_i^2 - 2n\hat{\mu}_X^2 + n\hat{\mu}_X^2 = \sum_{i=1}^{n} X_i^2 - n\hat{\mu}_X^2 \\ &= \sum_{i=1}^{n} X_i^2 - \frac{1}{n}\left(\sum_{i=1}^{n} X_i\right)^2 \end{aligned}$$

but this is just equal to $\sum_{i=2}^{n} Z_i^2$. As an aside, observe that by subtracting the estimator for the mean, S_X^2 is equal to the sum of $n-1$ normally distributed variables rather than n. We have effectively lost a *'degree of freedom'*. This is the reason why in the unbiased estimator for the variance we divide by $n-1$ rather than n. Returning to the derivation of the t-test, we see that $\hat{\mu}_X = Z_1/\sqrt{n}$, but since Z_1 is independent of Z_i for $i > 1$, then $\hat{\mu}_X$ is independent of S_X^2 and hence $\hat{\sigma}_X^2$. We have just shown that S_X^2 can be written as

$$S_X^2 = \sum_{i=2}^{n} Z_i^2$$

where $Z_i \sim \mathcal{N}(0,1)$ but we showed in Question 5.5 (see page 411 for the solution to this question) that such a sum will be distributed according to the chi-squared distribution $\chi^2(n-1)$.

We are now on a downhill run. We observe that

$$\hat{\mu}_X - \hat{\mu}_Y = \sum_{i=1}^{n} \frac{X_i}{n} - \sum_{i=1}^{m} \frac{Y_i}{m}$$

is a sum of normally distributed variables so will be normally distributed with mean 0 and variance (again using the assumption of independence)

$$\begin{aligned}
\mathbb{V}\text{ar}\left[\hat{\mu}_X - \hat{\mu}_Y\right] &\overset{(1)}{=} \sum_{i=1}^{n} \mathbb{V}\text{ar}\left[\frac{X_i}{n}\right] + \sum_{i=1}^{m} \mathbb{V}\text{ar}\left[\frac{-Y_i}{m}\right] \\
&\overset{(2)}{=} \frac{1}{n^2} \sum_{i=1}^{n} \mathbb{V}\text{ar}\left[X_i\right] + \frac{1}{m^2} \sum_{i=1}^{m} \mathbb{V}\text{ar}\left[-Y_i\right] \\
&\overset{(3)}{=} \frac{1}{n^2} \sum_{i=1}^{n} 1 + \frac{1}{m^2} \sum_{i=1}^{m} 1 \overset{(4)}{=} \frac{1}{n} + \frac{1}{m}
\end{aligned}$$

(1) From the definition of $\hat{\mu}_X - \hat{\mu}_Y$ given above.
(2) Using $\mathbb{V}\text{ar}\left[cX\right] = c^2 \mathbb{V}\text{ar}\left[X\right]$.
(3) Using the fact that $X_i, Y_i \sim \mathcal{N}(0,1)$, so that $\mathbb{V}\text{ar}\left[X_i\right] = \mathbb{V}\text{ar}\left[Y_i\right] = 1$.
(4) Performing the sums.

Using this result and noting that since $X_i, Y_i \sim \mathcal{N}(0,1)$ the sum will be normally distributed, so that

$$\frac{\hat{\mu}_X - \hat{\mu}_Y}{\sqrt{\frac{1}{n} + \frac{1}{m}}} \sim \mathcal{N}(0,1).$$

We have also shown that

$$S_X^2 = \sum_{i=1}^{n} (X_i^2 - \hat{\mu}_X)^2 \sim \chi^2(n-1) \qquad S_Y^2 = \sum_{i=1}^{m} (Y_i^2 - \hat{\mu}_Y)^2 \sim \chi^2(m-1).$$

Recall that if $Z_i \sim \mathcal{N}(0,1)$ then

$$\sum_{i=1}^{N} Z_i^2 \sim \chi^2(n).$$

We know that if W_i are *iid* random variables drawn from $\mathcal{N}(0,1)$ with $Q_1 = \sum_{i=1}^{k} W_i^2$ and $Q_2 = \sum_{i=k+1}^{k+l} W_i^2$, then $Q_1 \sim \chi^2(k)$ and $Q_2 \sim \chi^2(l)$, but then

$$Q_1^2 + Q_2^2 = \sum_{i=1}^{k+l} W_i^2 \sim \chi^2(k+l).$$

We see therefore that $S_X^2 + S_Y^2 \sim \chi^2(n+m-2)$. Defining

$$U = \frac{\hat{\mu}_X - \hat{\mu}_Y}{\sqrt{\frac{1}{n} + \frac{1}{m}}} \qquad\qquad V = S_X^2 + S_Y^2$$

then $T = \sqrt{n+m-2}\, U/\sqrt{V}$ with $U \sim \mathcal{N}(0,1)$ and $V \sim \chi^2(n+m-2)$, where U and V are independent.

Defining the number of degrees of freedom by $\nu = n + m - 2$ then the distribution of T is thus given by

$$\begin{aligned}
f_T(t) &\overset{(1)}{=} \int_0^\infty \int_{-\infty}^\infty \delta\left(t - \frac{\sqrt{\nu}\, u}{\sqrt{v}}\right) f_U(u)\, f_V(v)\, du\, dv \\
&\overset{(2)}{=} \int_0^\infty \int_{-\infty}^\infty \frac{\sqrt{v}}{\sqrt{\nu}}\, \delta\left(u - \frac{\sqrt{v}\, t}{\sqrt{\nu}}\right) \mathcal{N}(u|0,1)\, \chi^2(v|\nu)\, du\, dv \\
&\overset{(3)}{=} \int_0^\infty \int_{-\infty}^\infty \frac{\sqrt{v}}{\sqrt{\nu}}\, \delta\left(u - \frac{\sqrt{v}\, t}{\sqrt{\nu}}\right) \frac{e^{-u^2/2}}{\sqrt{2\pi}} \frac{v^{\nu/2-1}\, e^{-v/2}}{2^{\nu/2}\, \Gamma(\nu/2)}\, du\, dv \\
&\overset{(4)}{=} \frac{1}{\sqrt{\pi\, \nu}\, 2^{(\nu+1)/2}} \int_0^\infty e^{-v(1+t^2/\nu)/2}\, v^{(\nu+1)/2-1}\, dv \\
&\overset{(5)}{=} \frac{\Gamma\left(\frac{\nu+1}{2}\right)}{\sqrt{\pi\,\nu}\,\Gamma\left(\frac{\nu}{2}\right)} \left(1 + \frac{t^2}{\nu}\right)^{-\frac{\nu+1}{2}} \overset{(6)}{=} \mathrm{T}(t|\nu).
\end{aligned}$$

(1) Using $T = \sqrt{\nu} U/\sqrt{V}$.
(2) Using $\delta(x) = |a|\, \delta(a\, x)$ while $f_U(u) = \mathcal{N}(u|0,1)$ and $f_V(v) = \chi^2(v|\nu)$.
(3) Writing out $\mathcal{N}(u|0,1)$ and $\chi^2(v|\nu)$.
(4) Performing the integral over u and using $\int \delta(u-x)\, f(u)\, du = f(x)$.
(5) From a change ov variables $w = v(1+t^2/\nu)/2$ and performing the integral (which is equal to $\Gamma\left(\frac{\nu+1}{2}\right)$).
(6) By definition of Student's t-distribution $\mathrm{T}(t|\nu)$.

Finally, we find

$$\begin{aligned}
\mathbb{P}\left(|T| \geq t\right) = 1 - \mathbb{P}\left(|T| < t\right) &\overset{(1)}{=} 1 - \frac{\Gamma\left(\frac{\nu+1}{2}\right)}{\sqrt{\pi\,\nu}\,\Gamma\left(\frac{\nu}{2}\right)} \int_{-t}^{t} \left(1 + \frac{\tau^2}{\nu}\right)^{-\frac{\nu+1}{2}} d\tau \\
&\overset{(2)}{=} 1 - \frac{1}{B(\nu/2, 1/2)} \int_{\frac{\nu}{\nu+t^2}}^1 x^{\nu/2-1}\, (1-x)^{-1/2}\, dx \\
&\overset{(3)}{=} \frac{1}{B(\nu/2, 1/2)} \int_0^{\frac{\nu}{\nu+t^2}} x^{\nu/2-1}\, (1-x)^{-1/2}\, dx \overset{(4)}{=} I_{\frac{\nu}{\nu+t^2}}(\nu/2, 1/2)
\end{aligned}$$

(1) Using the definition for the density that we obtained above.
(2) Writing $x = (1+\tau^2/\nu)^{-1}$ so that $\tau = \sqrt{\nu} x^{-1/2}\, (1-x)^{1/2}$ then $d\tau = -\sqrt{\nu} x^{-3/2}\, (1-x)^{-1/2}\, dx/2$. We also use that $\sqrt{\pi} = \Gamma(1/2)$ and $B(a,b) = \Gamma(a)\,\Gamma(b)/\Gamma(a+b)$.
(3) Using $\int_{-1}^{1} x^{\nu/2-1}\, (1-x)^{-1/2}\, dx = B(\nu/2, 1/2)$.

(4) Where by definition $I_x(a, b)$ is the incomplete beta function (see Appendix 2.B on page 43).

As mentioned, this took a long time to calculate. Let's recap the argument

- The first observation was that because of the way we chose the statistic t we would get the same result if our observation came from $\mathcal{N}(\mathbf{0}, \mathbf{I})$ rather than from any other normal distribution. We therefore assume that $X_i, Y_i \sim \mathcal{N}(\mathbf{0}, \mathbf{I})$.
- The derivation relies on the fact that if $\mathcal{D} = \{X_1, X_2, \ldots, X_n\}$ is a data set with $X_i \sim \mathcal{N}(0, 1)$ then $\hat{\mu}$ and $\hat{\sigma}^2$ are independent of each other. This is far from obvious, but we could prove this by considering a rotated vector $\boldsymbol{Z} = (Z_1, Z_2, \ldots, Z_n) = \mathbf{V}\boldsymbol{X}$ where $\mathbf{V}$ is a rotation matrix and $\boldsymbol{X} = (X_1, X_2, \ldots, X_n)$ is a vector whose elements are the random variables that we observe. Because $\boldsymbol{X} \sim \mathcal{N}(\mathbf{0}, \mathbf{I})$ it follows that $\boldsymbol{Z} \sim \mathcal{N}(\mathbf{0}, \mathbf{I})$. If we now choose the rotation matrix so that $Z_1 = \sum_i X_i/\sqrt{n} = \sqrt{n}\,\hat{\mu}$ then $\hat{\sigma}^2 = \sum_{i=2}^n Z_i^2$. As every component of $\boldsymbol{Z}$ is independent this proves $\hat{\mu}$ and $\hat{\sigma}^2$ are independent of each other. Furthermore, we find that because $\hat{\mu} = Z_1/\sqrt{n}$ then $\hat{\mu} \sim \mathcal{N}(0, 1/n)$. Similarly, because $\hat{\sigma}^2 = \sum_{i=2}^n Z_i^2$ then $\hat{\sigma}^2 \sim \chi^2(n-1)$. This will be true both for our X_is and Y_i's.
- Now our T-statistic is given by $T = (\hat{\mu}_X - \hat{\mu}_Y)/\hat{\sigma}_{X+Y}$ with $\hat{\sigma}_{X+Y} = a\,\hat{\sigma}_X + b\,\hat{\sigma}_Y$ where a and b are carefully chosen so that $T = \sqrt{\nu}\,U/\sqrt{V}$ and where $\nu = m+n-2$ is known as the number of degrees of freedom (it is the total number of data points in the two data sets minus 2), $U \sim \mathcal{N}(0, 1)$ and $V \sim \chi^2(\nu)$.
- The rest is algebra. Rather remarkably we get a probability density in closed form. When integrated we get the solution in terms of an incomplete beta function.

7

Mathematics of Random Variables

Contents

Proving results in probability requires a specialised tool set that forms the subject of this chapter. We start with the sometimes dry and confusing subject of convergence. We next discuss the laws of large numbers, which are convergence results for sums of random variables. Martingales are briefly introduced. One of their primary uses is in obtaining convergence results.

We then visit some inequalities that can be very useful for proving results involving random variables. For example, a very common problem is to prove that something happens (or perhaps doesn't happen) with high probability. These proofs are often simplified by using inequalities. We review some of the most commonly encountered inequalities used in probability.

7.1 Convergence

Convergence is a useful concept in mathematics. We are happy to say that $1/n$ converges to zero as n goes to infinity, or that the function $\Phi(x/\epsilon)$ (where Φ is the cumulative density function for a normal distribution) converges towards a unit step function at zero as ϵ goes to 0.

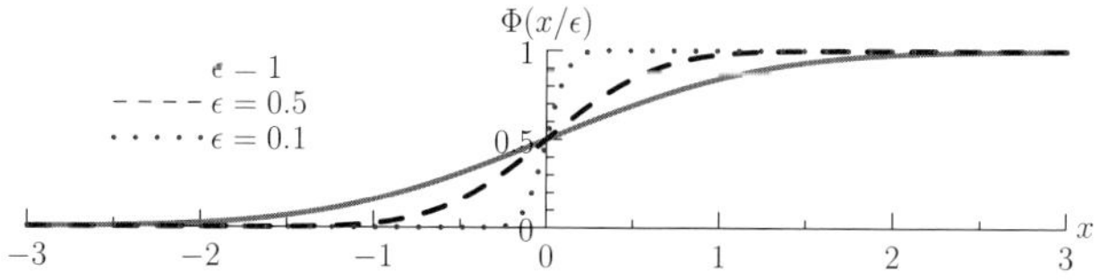

Applying the concept of convergence to random variables, however, turns out to be non-trivial. Consider, for example, a set of coin flips and define the random variable

$$X_n = \frac{1}{n}\sum_{i=1}^{n} B_i$$

where B_i is the result of the ith coin flip and is equal to 1 if the coin ends up heads and 0 otherwise. Assuming that the coin is fair, and that flips are independent, we would intuitively expect X_n to 'converge' to 0.5. But what does this mean? In real analysis there are different notions of convergence (e.g. point-wise convergence and uniform convergence). In probability we must be a little more careful for even the weak condition of point-wise convergence is too strong to be of use. Let us recall what convergence means. The classical notion of convergence is that, given some sequence $a_1, a_2, a_3, \ldots$, we say the sequence converges to a limit a if for any $\epsilon > 0$ there exists an integer n such that $|a_m - a| < \epsilon$ for all $m \geq n$. We cannot apply this notion of convergence directly to our sequence of random variables, X_n, as they don't have a single value but are distributed according to some probability distribution.

We can, however, think of a random variable as a mapping of an elementary event, ω, to a real number $X(\omega)$. Thus, a good starting place is to examine what convergence means for a sequence of mappings or functions. Point-wise convergence of a sequence of functions $f_n(x)$ for $n = 1, 2, 3, \ldots$ to a function $f(x)$ means that, for sufficiently large n, we can make the distance between $f_n(x)$ and $f(x)$ as small as we like *at all points* x – the choice of n will depend on how close we want $f_n(x)$ to be to $f(x)$ and on x. For example,

$$\lim_{n\to\infty} \cos\left(x + \frac{x}{n}\right) = \cos(x).$$

That is, for an arbitrary value of x and any $\epsilon > 0$ we can choose an n so that $|\cos(x + x/m) - \cos(x)| \leq \epsilon$ for all $m \geq n$. The stronger notion of convergence is uniform convergence that states that for any $\epsilon > 0$ we can find an n such that, for all $m \geq n$, $|f_m(x) - f(x)| \leq \epsilon$ at all points x. That is, we can find an n that depends on ϵ but is independent of x. Note, in the example above we cannot do this. That is, for $0 < \epsilon < 2$ and any fixed n, I can always choose an x which will give an error greater than ϵ (e.g. by choosing $x = n\pi$). In contrast, the function $\cos(x + 1/n)$ converges to $\cos(x)$ uniformly.

Recall that a random variable is a mapping from the set of elementary events to a real number ($X : \Omega \to \mathbb{R}$). Thus the natural interpretation of point-wise convergence would be to require that, for all $\omega \in \Omega$, we have

$$\lim_{n\to\infty} X_n(\omega) \to X(\omega).$$

However, this is not a particularly useful definition of convergence in the field of probability. In the example given above concerning the coin tosses, the event $\omega = (T, T, T, T, \ldots)$ has the property that $X_n(\omega) = 1$, which is not arbitrarily close to 1/2. Nor is this elementary event rarer than any other (for a fair coin all elementary events – i.e. sequences of heads and tails – are equally probable). Of course, point-wise convergence does not include any notion of probability so it is not surprising that it is not of much interest when considering random variables. There are four definitions of convergence commonly used in probability which try to fill the gap.

Convergence in distribution. Often written as $X_n \overset{D}{\to} X$ if

$$\lim_{n\to\infty} F_{X_n}(x) = F_X(x)$$

for each point of continuity x. Thus we require that the cumulative distribution function (CDF) converges to some function at each point x. The central limit theorem was an example of convergence in distribution. Recall that it states that if the Y_is are random variable with a finite mean and variance then

$$X_n = \frac{\sum_{i=1}^{n} Y_i - n\,\mathbb{E}\left[Y_i\right]}{\sqrt{n\,\mathbb{V}\mathrm{ar}\left[Y_i\right]}}$$

will converge to $\mathcal{N}(0, 1)$ in the limit $n \to \infty$. This, however, is a rather weak definition of convergence. One reason for this is that it says nothing about the ratio of $F_{X_n}(x)$ to $F_X(x)$. Both quantities become small in the tails, but their ratio may be very different from 1.

Convergence in probability. A stronger form of convergence is *convergence in probability*, which is denoted by $X_n \overset{P}{\to} X$, and defined such that for any $\epsilon > 0$

$$\lim_{n\to\infty} \mathbb{P}\left(|X_n - X| < \epsilon\right) = 1.$$

An illustration of this is shown in Figure 7.1.

If we consider the space of all elementary events, Ω, then convergence in probability requires that there exists an n such that, for $m \geq n$, all but a proportion at most δ of such events must be within ϵ of X. Both ϵ and δ can be arbitrarily small (although strictly greater than 0), with n dependent on our choice of ϵ and δ. Convergence in probability implies convergence in distribution, but not the other way around. However, this does not say that the difference

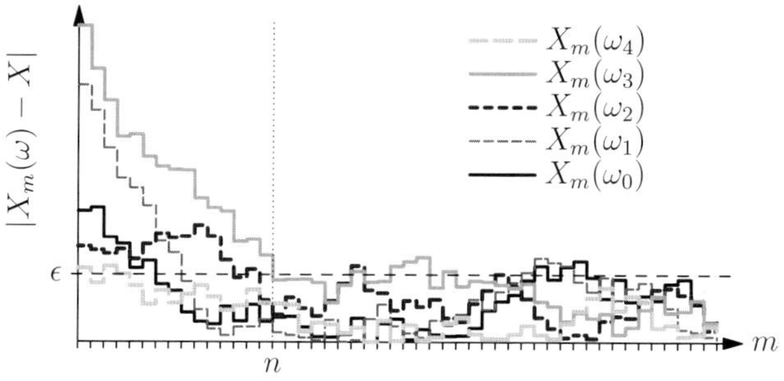

Figure 7.1 Illustration of $|X_m(\omega) - X|$ for a few different elementary events ω. For convergence in probability we require for all $m \geq n$, the probability of having a deviation $|X_m(\omega) - X| > \epsilon$ is less than δ.

between X_n and X is necessarily small, but rather that the probability of the difference being small converges to 1. That is, with overwhelming probability $(1 - \delta)$, the random variable X_n differs from X by, at most, a very small amount (ϵ).

Almost sure convergence. A yet stronger form of convergence is *convergence almost surely*, which is denoted by $X_n \xrightarrow{a.s.} X$, or sometimes as $X_n \to X$ (a.s.), and is defined by

$$\mathbb{P}\left(\left\{\omega \,\middle|\, \lim_{n\to\infty} X_n(\omega) \to X(\omega)\right\}\right) = 1.$$

Although this looks similar to convergence in probability it is subtly stronger. We now require all but a proportion, δ, of elementary events, ω, to be within ϵ of X for *all* $m \geq n$. The examples below clarify the difference between this and convergence in probability.

Convergence in moments. The fourth form of convergence is *convergence in the p-moment*, which is denoted by $X_n \xrightarrow{L^p} X$. It is defined for $p \geq 1$ by

$$\mathbb{E}\left[|X_n - X|^p\right] \to 0, \qquad \text{as } n \to \infty.$$

The most commonly used moments are the first moment giving $X_n \xrightarrow{L^1} X$, *convergence in mean*, and the second moment giving $X_n \xrightarrow{L^2} X$, *convergence in mean squared*. The larger the value of p the stronger the convergence condition. Convergence in moments is a separate condition to convergence almost surely, although it implies convergence in probability. It is not, however, implied by convergence in probability.

Example 7.1 Converges in Mean But Not Almost Surely

Consider a sequence of independent random variables drawn from the distribution

$$X_n = \begin{cases} 1 & \text{with probability } n^{-1} \\ 0 & \text{with probability } 1 - n^{-1}. \end{cases} \tag{7.1}$$

Three random samples, $\omega \in \Omega$, for the sequence defined in Equation (7.1) are shown in Figure 7.2.

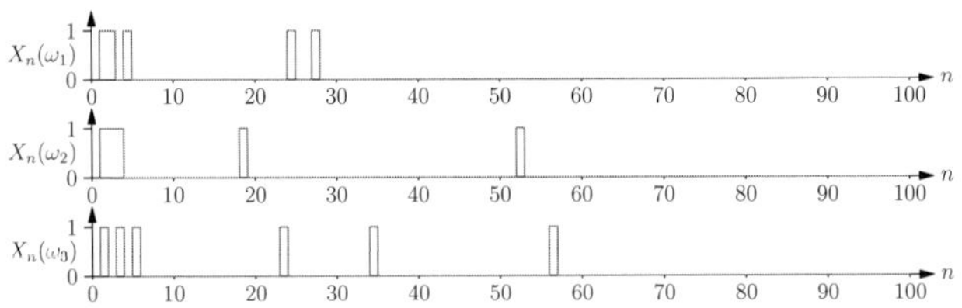

Figure 7.2 Illustration of three samples of the sequence X_n defined by Equation (7.1) for n up to 100.

We see that $X_n \xrightarrow{P} 0$, since the probability that $|X_n - 0| > \epsilon$ (for $0 < \epsilon < 1$) is n^{-1}, which converges to zero. But, for convergence almost surely we require that there exists an n such that for *all* $m \geq n$ the random variable $X_m(\omega)$ is arbitrarily close to X. However, for the example above this doesn't happen. In this example, almost sure convergence requires that for any δ there exists an n such that $\mathbb{P}\left(X_m = 0 \text{ for all } m \geq n\right) > 1 - \delta$, but we can show that this is not the case:

$$
\begin{aligned}
\mathbb{P}\left(X_m = 0 \text{ for all } m \geq n\right) &\overset{(1)}{=} \lim_{r\to\infty} \mathbb{P}\left(X_m = 0 \text{ for all } n \leq m \leq r\right) \\
&\overset{(2)}{=} \lim_{r\to\infty} \left(1 - \frac{1}{n}\right)\left(1 - \frac{1}{n+1}\right)\left(1 - \frac{1}{n+2}\right)\cdots\left(1 - \frac{1}{r}\right) \\
&\overset{(3)}{=} \lim_{r\to\infty} \left(\frac{n-1}{n}\right)\left(\frac{n}{n+1}\right)\left(\frac{n+1}{n+2}\right)\cdots\left(\frac{r-1}{r}\right) \\
&\overset{(4)}{=} \lim_{r\to\infty} \left(\frac{n-1}{r}\right) = 0
\end{aligned}
$$

(1) When we take the limit $r \to \infty$ these are identical. It is easier to reason about a finite product.
(2) Using $\mathbb{P}\left(X_m = 0 \text{ for all } n \leq m \leq r\right) = \prod_{m=n}^{r} \mathbb{P}\left(X_m = 0\right)$ and $\mathbb{P}\left(X_m = 0\right) = 1 - 1/m$.
(3) Putting each term over a common denominator.
(4) Cancelling the numerator of each term with the denominator of the preceding term.

Thus, for this example, almost no elementary event converges. However,

$$\mathbb{E}\left[\,|X_n - 0|\,\right] = \frac{1}{n}$$

so X_n converges in mean, i.e. $X_n \xrightarrow{L^1} 0$.

Although this is a concrete example it is often difficult to gain intuition without embedding the example in the real world. We therefore cook up a rather silly example. Suppose we are modelling the accidents of drivers. As the drivers become more experienced the probability of an accident reduces (by the reciprocal of the number of journeys they make). We may want to choose an ϵ, 0.01 say, as an accident probability which we can tolerate. Thus, if we give licences after 100 journeys we are assured (in our rather simplified model of the world) that the probability of any licensed driver having an accident in their next journey is less than or equal to 0.01 (due to convergence in probability). Nevertheless, however experienced a driver is, they will, with probability 1, have an accident some time in the future (due to the lack of almost sure convergence). In fact, our drivers will have accidents infinitely often in the future (the future is a long time). If our drivers pay an insurance premium in proportion to their risk per journey, then as they become more experienced their insurance premium will converge to zero (due to convergence in mean).

We now consider the reverse scenario.

Example 7.2 Almost Sure Convergence But Not in Mean
Consider an independent sequence of random variables

$$X_n = \begin{cases} n^3 & \text{with probability } n^{-2} \\ 0 & \text{with probability } 1 - n^{-2}. \end{cases}$$

A silly real-world scenario would be to think of stockbrokers trading for a large bank. As they become more experienced the probability of them making a mistake reduces as n^{-2}, however, with experience they are trusted with stock of value n^3, so that each mistake becomes considerably more expensive.

Since $\mathbb{P}\left(|X_n - 0| > \epsilon\right) = n^{-2}$ goes to zero as $n \to 0$, we have $X_n \xrightarrow{P} 0$. In this case we have almost sure convergence since, for any $0 < \epsilon < 1$,

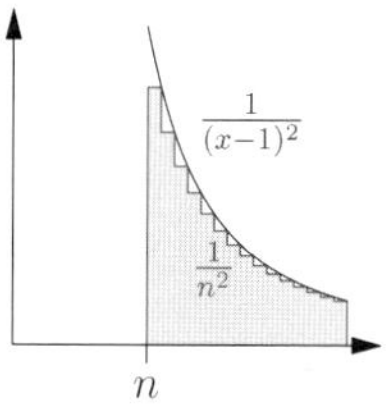

$$\mathbb{P}\left(\forall m \geq n,\ |X_m - 0| > \epsilon\right) = \sum_{m=n}^{\infty} \frac{1}{m^2} \leq \int_n^{\infty} \frac{1}{(x-1)^2}\,\mathrm{d}x = \frac{2}{n-1}$$

which converges to 0 as $n \to \infty$ so that $X_n \xrightarrow{a.s.} 0$. Thus, in our stockbroker scenario, after 201 transactions the probability of a trader making another error is less than 1%. Furthermore, for any δ we can choose $n = \lceil 1/\delta \rceil + 1$, such that the probability of making an error any time in the future is less than δ. However, the expected value of $|X_n - 0|$ is $\mathbb{E}\left[|X_n - 0|\right] = n^3 \times n^{-2} = n$, which actually diverges. Thus X_n does not converge in mean to 0. If you were insuring these stockbrokers, although after a certain time most traders would never make another error, the cost of those few who did would be so high that it could bankrupt a nation. (Maybe this isn't such a silly model after all.)

The hierarchy of convergence results is shown below.

Strong Convergence				*Weak Convergence*		
		Convergence almost surely (a.s.)	$\Rightarrow$	Convergence in probability (P)	$\Rightarrow$	Convergence in distribution (D)
Convergence in mean squared (L^2)	$\Rightarrow$	Convergence in mean (L^1)	$\Rightarrow$	Convergence in probability (P)		

Although the definition allows convergence to a random variable, in most instances convergence in probability, almost surely, and in moments is used to describe sequences which converge to a constant value, while convergence in distribution usually describes situations when the sequence of random variables converges to a distribution. As Examples 7.1 and 7.2 illustrate, despite a random variable converging in probability to a constant, this does not guarantee that the random variable won't be some distance away from zero some time in the future or that the expected value of the random variable will be equal to that constant. Whether this matters to you depends on the situation you find yourself in – you may be a driver wondering whether to insure yourself or an insurance company deciding whether to insure a bank. It may seem that these examples are so pathological that they will never occur in practice, but long- (or thick-) tailed distributions are quite common and this can lead to seemingly counter-intuitive results precisely because the variables don't converge in moment or almost surely.

Convergence results are very worthy and can be important, but I must confess that occasionally I lose patience with them. The limit $n \to \infty$ is a luxury that is often not afforded to those of us modelling real-world systems. I've seen a few too many examples of convergence proofs that don't capture the essence of the problem that they purport to address. Of course, there are times when finite-sized systems are close enough to the limit $n \to \infty$ for the convergence results to be entirely appropriate. But this cannot be taken for granted. For many systems quite large n can be very different from the limit value. The important question in these cases is the rate of convergence, and to address this issue the inequalities discussed in Section 7.2 are often more appropriate than theorems about convergence.

7.1.1 Laws of Large Numbers

The law of large numbers addresses the question of what happens to the sum of n random variables in the limit of large n

$$S_n = \sum_{i=1}^{n} X_i .$$

There are different versions of this law. One of the simplest describes what happens when X_i are independent, identically distributed random variables (*iid*s). Intuitively we expect that if $\mathbb{E}\left[X_i\right] = \mu$ then S_n/n will converge to μ. This is actually the question we started this section with, where we considered the proportion of coin flips that land on heads. The question is does this convergence happen and how strong is it? There are many possible results concerning the laws of large numbers depending on the assumptions we make about the random variable. If $\mathbb{E}\left[X_i^2\right] < \infty$ then S_n will converge to μ almost surely and in mean squared. The proof of convergence in mean squared is easy. We consider

$$\mathbb{E}\left[\left(\frac{S_n}{n} - \mu\right)^2\right] \overset{(1)}{=} \frac{1}{n^2}\mathbb{E}\left[\left(S_n - \mathbb{E}\left[S_n\right]\right)^2\right] \overset{(2)}{=} \frac{1}{n^2}\mathbb{V}\mathrm{ar}\left[S_n\right] \overset{(3)}{=} \frac{1}{n^2}\mathbb{V}\mathrm{ar}\left[\sum_{i=1}^{n} X_i\right]$$

$$\overset{(4)}{=} \frac{1}{n^2}\sum_{i=1}^{n}\mathbb{V}\mathrm{ar}\left[X_i\right] \overset{(5)}{=} \frac{1}{n}\mathbb{V}\mathrm{ar}\left[X\right] \overset{(6)}{\to} 0 \quad \text{as } n \to 0$$

(1) Using $\mu = \mathbb{E}\left[S_n\right]/n$ and taking out a factor $1/n$ from the square.
(2) Identifying the expectation as the definition of the variance.
(3) Using $S_n = \sum_{i=1}^{n} X_i$.
(4) As for independent variables $\mathbb{V}\mathrm{ar}\left[\sum_i X_i\right] = \sum_i \mathbb{V}\mathrm{ar}\left[X_i\right]$.
(5) Because the X_is are *iid* then $\mathbb{V}\mathrm{ar}\left[X_i\right] = \mathbb{V}\mathrm{ar}\left[X\right]$.
(6) As $\mathbb{V}\mathrm{ar}\left[X\right]$ is assumed finite.

Proof of almost sure convergence is trickier and we suggest the interested reader consults a standard probability text, such as Grimmett and Stirzaker (2001a). In fact, almost sure convergence only requires the weaker condition that $\mathbb{E}\left[|X_i|\right] < \infty$. This is known as a strong law of large numbers as a strong form of convergence is guaranteed. Under weaker assumptions we often still get convergence, but only convergence in distribution or in probability. These are known as weak laws of large numbers. The central limit theorem is an example of a weak law of large numbers, where $(S_n - n\,\mu)/(\sigma\sqrt{n})$ converges in distribution to $\mathcal{N}(0, 1)$.

The law of large numbers may seem self-evident, but we have already seen that if S_n is a sum of Cauchy distributed numbers, i.e. $X_i \sim \text{Cau}$ where

$$\text{Cau}(x) = \frac{1}{\pi(1 + x^2)}$$

then S_n/n is itself Cauchy distributed. Thus, the sum S_n does not converge to a number. Of course, both $\mathbb{E}\left[X^2\right]$ and $\mathbb{E}\left[|X|\right]$ are non-existent (i.e. they diverge to ∞) for Cauchy distributed variables so this does not contradict the statements made above.

7.1.2 Martingales

The name martingale comes from a bidding strategy involving doubling your bet every time you lose (not a strategy I would recommend).

A martingale is a sequence of random variable $X_1, X_2, \ldots$ such that the expected value of X_n is equal to X_{n-1}. For example, X_n might be your winnings in a fair casino (a mythical place known only to mathematicians). The idea of martingales can be extended to random variables $Y_n = f(X_n)$ as follows: if $\mathbb{E}\left[|Y_n|\right] < \infty$ and $\mathbb{E}\left[Y_{n+1}|X_1, X_2, \ldots X_n\right] = Y_n$ then we can consider the sequence of Y_ns to be a martingale. This extension is useful as the process we are interested in may not be a martingale, but we may be able to construct some function that is a martingale.

The martingale's tool set is a bunch of theorems that remind me of a set of odd-shaped spanners which don't fit any nuts. Exponents of martingales cleverly add extra arguments so that they can put to work their martingale theorems. For example, one of the primary theorems of martingales is known as the *optimal stopping theorem*. The theorem is a 'no free lunch' type result. It tells you that if you are a gambler playing a fair game you cannot come up with a stopping strategy which will allow you to win in expectation. More precisely, it says that if the gambler does not have infinite resources (of time or money) then she/he isn't going to win on average. Formally, for a particular martingale, $(X_i|i \geq 0)$, we define a stopping time, τ, which is an integer that depends on the value X_i (and with $\mathbb{P}\left(\tau < \infty\right) = 1$), then (provided $\mathbb{E}\left[|X_\tau|\right] < \infty$) for some $n > 0$ let $\tau_n = \min(\tau, n)$ and provided $\lim_{n\to\infty} \mathbb{E}\left[X_{\tau_n}\right] = \mathbb{E}\left[X_\tau\right]$ we have

$$\mathbb{E}\left[X_\tau\right] = \mathbb{E}\left[X_0\right].$$

That is, irrespective of our stopping strategy the martingale property still holds. (The optimal stopping theorem is often expressed in terms of submartingales – sequences where $\mathbb{E}\left[X_{n+1}\right] \geq X_n$, but we leave this generalisation for the interested reader to pursue independently.) There are a lot of mainly technical caveats, although understanding these is often important in using the optimal stopping theorem.

To see how to apply the optimal stopping theorem consider an honest game where a gambler starts with $£a$ and at each round she either wins $£1$ or loses $£1$ with equal probability. The games stops when the gambler has won $£b$ or has lost all her money. Let S_n be the wealth of the gambler after n rounds of the game. As the game is fair, $\mathbb{E}\left[S_{n+1}\right] = S_n$, so we have a martingale. In this case all the small print is satisfied so we can apply the theorem. We find that

$$\mathbb{E}\left[S_\tau\right] = 0 \times \mathbb{P}\left(S_\tau = 0\right) + b \times \mathbb{P}\left(S_\tau = b\right) = \mathbb{E}\left[S_0\right] = a,$$

thus $\mathbb{P}\left(S_\tau = b\right) = a/b$. That is, the probability of reaching b rather than going bankrupt is a/b. The optimal stopping theorem provides a simple way to get this answer, but it is a rather devious route getting there.

Sometimes, applying the optimal stopping theorem is very subtle. Suppose we bet on the outcome of a coin flip. At each step we bet $£1$. We play in phases. We carry on the phase until we win $£1$ which we pocket and then start a new phase. Often, we will be in debt, but eventually our winning streak will return and we should reach the state when our winnings are $£1$. This is a strategy which seems to break the optimal stopping theorem – we always win! – so somewhere we must have violated the small print. This game is equivalent to a symmetric random walk for which it is known that the probability of reaching the goal in less than n steps is $\mathbb{P}\left(\tau < n\right) = 1 - O(1/\sqrt{n})$. This converges to 1 so that $\mathbb{P}\left(\tau < \infty\right) = 1$. However, in this case, the condition $\lim_{n\to\infty} \mathbb{E}\left[S_{\tau_n}\right] = \mathbb{E}\left[S_\tau\right] (= 1)$ is *not* satisfied. For any finite number of steps, n, we will have reached our stopping criterion $S_n = 1$ with a probability $1-O(1/\sqrt{n})$, but with probability $O(1/\sqrt{n})$ the walker won't have reached the state 1, and for these walkers $\mathbb{E}\left[S_n\right] = -O(\sqrt{n})$, so that

We use 'big-O' notation to mean of that order. More formally this should be expressed by 'big-Θ', but we follow the less precise, but more common, usage.

$$\mathbb{E}\left[S_{\tau_n}\right] \approx 1 \times \left(1 - \frac{1}{\sqrt{n}}\right) - \sqrt{n} \times \frac{1}{\sqrt{n}} = 0.$$

Thus, even in the limit $n \to \infty$ we have that $\mathbb{E}\left[S_{\tau_n}\right] \neq 1$. In other words, although by making n sufficiently large we can ensure that we will win with a probability at least $1 - \epsilon = 1 - O(1/\sqrt{n})$, for those few ($\epsilon$) proportion of cases where we don't win we will have lost so much that in expectation our winnings will be 0.

The condition $\lim_{n\to\infty} \mathbb{E}\left[S_{\tau_n}\right] = \mathbb{E}\left[S_\tau\right]$ is often difficult to check – sometimes it seems to me that the theorem is applicable except where it does not work. To apply the optimal stopping theorem we want some sufficient conditions that are easier to verify. One such condition is that if our step size is finite and $\mathbb{E}\left[X_0\right]$ is finite then the optimal stopping condition holds if $\mathbb{E}\left[\tau\right] < \infty$. In the example above, we know that the step is finite as is $\mathbb{E}\left[X_0\right]$ so that $\mathbb{E}\left[\tau\right] = \infty$, otherwise the optimal stopping theorem would be violated (and we are not allowed to violate theorems). That is, the expected time to win £1 is infinite. Of course, most of the time we will win £1 quite frequently, but with a probability of $O(1/\sqrt{n})$ the expected time to reach £1 will be n and in summing all these terms we end up with an infinite expected winning time (in Exercise 10.1 we ask you to empirically determine the distribution of times for a random walk, where the walker returns to his/her starting position).

The optimal stopping theorem therefore gives us a short cut showing that the expected time for a one-dimensional random walk to reach a position right of our starting point is infinite, even though the probability of this happening converges to 1. To me this is a long-winded way of obtaining this result, which leaves me with little intuition of what is happening, but then I'm not a trained mathematician.

Another property of martingales is that they satisfy a strong convergence theorem. That is, if the sequence $(X_n | n = 1, 2, \ldots)$ is a martingale that also satisfies $\mathbb{E}\left[X_n^2\right] < M < \infty$ for some M and all n, then there exists a random variable X such that X_n converges to X almost surely and in mean squared. This provides a useful short cut to proving strong convergence. As this is of great interest to mathematicians, many books on probability (written by mathematicians) spend a lot of time on martingales.

For my taste martingales seem a convoluted way of solving problems. They have a certain elegance for proving theorems as all you have to do is map your problem onto an appropriate martingale and pull the appropriate theorem out of your toolbox (making sure that you have properly understood the small print). They are useful for proving convergence results and laws of large numbers, which is of great interest to mathematicians (whose job is to prove theorems).

However, for engineers and scientists, convergence in the limit $n \to \infty$ is not always that interesting. Often the rate of convergence is more important. That is, we might have a finite sum of random variables and we want to know the probability that they deviate from their expected value. To answer these kinds of questions inequalities are often more useful than theorems about what happens in the limit of infinitely large n. I can only give you my own experience. This is often inadequate. I regularly discover that mathematics that I once thought of as having academic use only, turn out to be of great practical use for solving some real-world problem. It may well be that one day I will discover this for martingales, but for me that day has not yet come.

7.2 Inequalities

We often need to prove some property of a random system. These proofs are frequently accomplished by using a few standard inequalities. For example, we can prove many properties if we know the mean or variance of a random variable – these quantities are often much simpler to compute than the full distribution. The inequalities are not particularly difficult to prove. We briefly cover some of the most commonly used inequalities.

7.2.1 Cauchy–Schwarz Inequality

Cauchy invented and proved the inequality. Schwarz produced a nice proof some 50 years later.

The most useful inequality in mathematics is due to Cauchy and is known as either the Cauchy, Cauchy–Schwarz, or the Schwarz inequality. It applies both to sums and integrals, but has a nice form in terms of expectations

$$\mathbb{E}\left[X\,Y\right]^2 \le \mathbb{E}\left[X^2\right]\,\mathbb{E}\left[Y^2\right].$$

There are many proofs of this – one of the most elegant being due to Schwarz. In the context of expectation, we consider the function

$$f(t) = \mathbb{E}\left[(t\,X + Y)^2\right] = \mathbb{E}\left[X^2\right]\,t^2 + 2\,\mathbb{E}\left[X\,Y\right]\,t + \mathbb{E}\left[Y^2\right].$$

Since $(t\,X+Y)^2 \ge 0$ we have that $f(t) \ge 0$. So, either $f(t) = 0$ for some t, in which case $t\,X$ equals $-Y$ (with probability 1). If this is the case then it is easy to see that the equality holds. Otherwise, $f(t) > 0$ for all t, which means the equation $f(t) = 0$ has no solution. For a quadratic equation $a\,t^2 + 2\,b\,t + c = 0$ to have no real solution implies that $b^2 < a\,c$ (in which case the equation has only imaginary solutions). But this condition is the Cauchy–Schwarz inequality. We can sharpen the inequality by considering $\mathbb{E}\left[(t\,|X| + |Y|)^2\right]$, which gives rise to the inequality

$$\mathbb{E}\left[|X\,Y|\right]^2 \le \mathbb{E}\left[X^2\right]\,\mathbb{E}\left[Y^2\right]$$

which is stricter in the sense that $\mathbb{E}\left[X\,Y\right]^2 \le \mathbb{E}\left[|X\,Y|\right]^2 \le \mathbb{E}\left[X^2\right]\,\mathbb{E}\left[Y^2\right]$.

To get inequalities to work for you, it is often necessary that you use them close to where the equality condition holds. (When going through a proof that involves an inequality you can often get a lot of intuition by considering where the inequality would be tight.) It therefore pays to be aware of when this occurs. The equality condition for Cauchy–Schwarz is that X and Y are proportional to each other.

Example 7.3 Positivity of Variances
As a simple application we can show that variances are always non-negative. Let X be a random variable and $Y = 1$ (i.e. it is not a random variable at all). Then using the Cauchy–Schwarz inequality

$$\mathbb{E}\left[X\right]^2 = \mathbb{E}\left[X\,1\right]^2 \le \mathbb{E}\left[X^2\right]\mathbb{E}\left[1^2\right] = \mathbb{E}\left[X^2\right].$$

Thus $\mathbb{V}\mathrm{ar}\left[X\right] = \mathbb{E}\left[X^2\right] - \mathbb{E}\left[X\right]^2 \ge 0$. Note that for the equality to hold (i.e. $\mathbb{E}\left[X\,1\right]^2 = \mathbb{E}\left[X^2\right]\mathbb{E}\left[1^2\right]$) X is required to be proportional to 1; that is, X is a constant, c. In this case $\mathbb{V}\mathrm{ar}\left[c\right] = 0$.

Example 7.4 Mean Absolute Error
The mean absolute error is defined as $\mathbb{E}\left[|E|\right]$ where E is an error measurement. Using the Cauchy–Schwarz inequality we find

$$\mathbb{E}\left[|E|\right] = \mathbb{E}\left[|E| \times 1\right] \le \sqrt{\mathbb{E}\left[|E|^2\right]\mathbb{E}\left[1^2\right]} = \sqrt{\mathbb{E}\left[E^2\right]}.$$

Thus the mean absolute error is always less than or equal to the root mean squared (RMS) error.

One of the difficulties with applying inequalities is to get them the right way around. It is useful to have a quick example in your head to remember which way around the inequality has to be. One such example is to consider X to be 1 if a coin toss gives heads and 0 otherwise, contrariwise we can chose Y to be 1 if a coin toss is tails and 0 otherwise. Thus $\mathbb{E}\left[X\,Y\right] = 0$, while (for an honest coin) $\mathbb{E}\left[X^2\right] = \mathbb{E}\left[Y^2\right] = 1/2 > 0$. Consequently, we can have a situation were $\mathbb{E}\left[X\,Y\right] = 0$ and $\mathbb{E}\left[X^2\right] = \mathbb{E}\left[Y^2\right] > 0$, but not the other way around.

7.2.2 Markov's Inequality

A useful inequality when we only know the mean of a distribution is *Markov's inequality*. It states that for any random variable X

$$\mathbb{P}\left(|X| \ge t\right) \le \frac{\mathbb{E}\left[|X|\right]}{t}, \quad \text{for any } t > 0.$$

Its proof is straightforward. Consider the random variables, X and $Z = t\left[\!\left[|X| \ge t\right]\!\right]$. Z is a function of X as shown

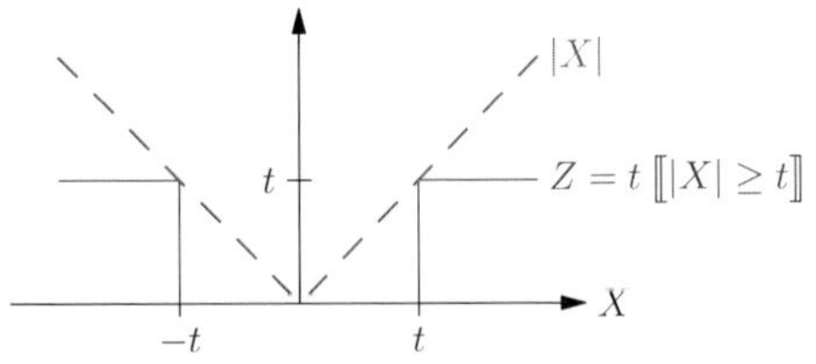

Clearly, $|X| \geq Z$ so that

$$\mathbb{E}\left[|X|\right] \overset{(1)}{\geq} \mathbb{E}\left[Z\right] \overset{(2)}{=} \mathbb{E}\left[t\left[\!\left[|X| \geq t\right]\!\right]\right] \overset{(3)}{=} t\,\mathbb{E}\left[\left[\!\left[|X| \geq t\right]\!\right]\right] \overset{(4)}{=} t\,\mathbb{P}\left(|X| \geq t\right)$$

(1) Since $|X| \geq Z$.
(2) Putting in the definition for Z.
(3) Since t is a constant it can be taken out of the expectation.
(4) Because the expectation of an indicator function for an event is just the probability of that event.

Dividing both sides by t gives Markov's inequality. If $X \geq 0$, then Markov's inequality becomes

$$\mathbb{P}\left(X \geq t\right) \leq \frac{\mathbb{E}\left[X\right]}{t}, \qquad \text{(where } X \geq 0\text{)}.$$

Example 7.5 Tail Distributions

Markov's inequality is often extremely conservative as it has to hold for all distributions. In Figure 7.3 we show the tail distributions for $X \sim \text{Poi}(1)$, $X \sim \mathcal{N}(0, \pi/2)$, $X \sim \mathit{Gam}(\epsilon, \epsilon)$ with $\epsilon = 0.001$, $X \sim \text{LogNorm}(-2, 4)$, and $X \sim \text{Cau}$. In the first four cases $\mathbb{E}\left[|X|\right] = 1$, so by Markov's inequality $\mathbb{P}\left(|X| \geq t\right) \leq 1/t$.

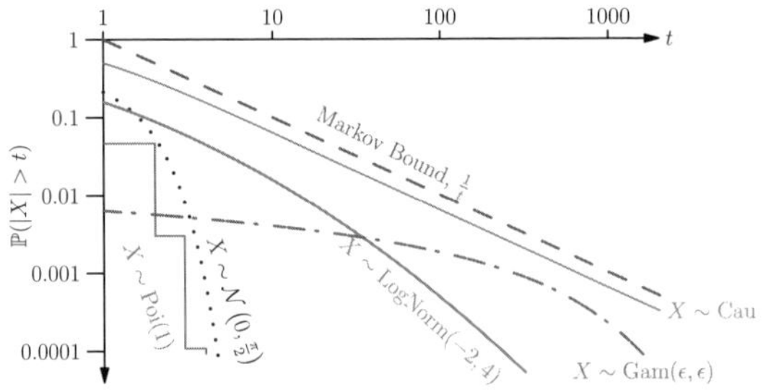

Figure 7.3 $\mathbb{P}\left(X \geq t\right)$ versus t plotted on a log-log scale for a Poisson, normal, gamma ($\epsilon = 0.001$), log-normal, and Cauchy random variable. Also shown is Markov bound for a random variable with $\mathbb{E}\left[|X|\right] = 1$.

To compute these we need to know the CDFs for each distribution.

1. For the Poisson distribution we can use the result of Exercise 4.4 that the CDF $\mathbb{P}\left(X < k\right) = Q(k, 1)$ where $Q(k, \mu)$ is the normalised incomplete gamma function (see Appendix 2.A). In fact we want the complement $\mathbb{P}\left(X \geq k\right) = 1 - Q(k, 1) = P(k, 1)$. However, we

can compute the first few terms directly from the series

$$\mathbb{P}\left(X \geq k\right) = 1 - \mathrm{e}^{-1}\left(1 + \frac{1}{2!} + \cdots + \frac{1}{k!}\right).$$

2. For a normal deviate $X \sim \mathcal{N}(\mu, \sigma)$ the probability $\mathbb{P}\left(|X| \geq t\right) = 2\,\Phi((\mu - t)/\sigma)$, so in our case

$$\mathbb{P}\left(|X| \geq t\right) = 2\,\Phi\left(\frac{-t}{\sqrt{\pi/2}}\right) = \mathrm{erfc}\left(\frac{t}{\sqrt{\pi}}\right).$$

3. For a gamma deviate, $X \sim \mathit{Gam}(a, b)$,

$$\mathbb{P}\left(|X| \geq t\right) = Q(a,\, b\,t)$$

   where $Q(a, x)$ is the normalised incomplete gamma function (see Appendix 2.A).
4. For a log-normal distribution, $\mathrm{LogNorm}(\mu, \sigma^2)$, the mean is $\mathrm{e}^{\mu+\sigma^2/2}$ so if $\mu = -\sigma^2/2$ the mean will equal 1. The CDF is, from a simple change of variables, $\mathbb{P}\left(X \leq t\right) = \Phi((\log(t) - \mu)/\sigma)$ and

$$\mathbb{P}\left(X \geq t\right) = 1 - \Phi\left(\frac{\log(t) - \mu}{\sigma}\right) = \mathrm{erfc}\left(\frac{\log(t) - \mu}{\sqrt{2}\sigma}\right).$$

   We show the case when $\mu = -2$ and $\sigma = 2$. The asymptotic fall-off of the tail decreases as we increase σ.
5. Finally, for a Cauchy deviate,

$$\mathbb{P}\left(|X| \geq t\right) = \frac{1}{\pi}\arctan(Y).$$

For the Cauchy deviate, $X \sim \mathrm{Cau}$, the expectation, $\mathbb{E}\left[|X|\right] = \infty$, and Markov's inequality would be trivial. However, as we can see from Figure 7.3, $\mathbb{P}\left(X \leq t\right) < 1/t$ even for Cauchy deviates.

For very many random variables Markov's inequality is a poor approximation. However, some log-normal distributions can have very significant probability mass in their tails. Similarly, extreme gamma distributions can have significant tails. We will see Chapter 8 that such gamma distributions occur when we wish to be very non-committal about our prior belief in quantities such as the standard deviation.

One common use of Markov's inequality is with non-negative integer valued random variables. Here we often take $t = 1$, so that

$$\mathbb{P}\left(X \neq 0\right) \overset{(1)}{=} \mathbb{P}\left(X \geq 1\right) \overset{(2)}{\leq} \mathbb{E}\left[X\right], \quad \text{where } X \in \{0, 1, 2, 3, \ldots\}$$

(1) Since X takes non-negative integer values.
(2) Using the Markov inequality with $t = 1$.

In a typical application we might want to show that $X = 0$ with high probability. Since $\mathbb{P}\left(X=0\right) = 1 - \mathbb{P}\left(X \geq 1\right) \geq 1 - \mathbb{E}\left[X\right]$, we can show that $X = 0$ with high probability by showing $\mathbb{E}\left[X\right] \to 0$. This is sometimes referred to as the first moment method.

Example 7.6 Chromatic Number of Random Graphs
Consider random graphs of n vertices (nodes) where each edge occurs with a probability p. This collection of graphs is sometimes known as an Erdős–Rényi ensemble and is denoted $\mathcal{G}(n, p)$. We wish to colour the vertices so that no edge has its two vertices the same colour. A question of considerable interest is how few colours are necessary on average to cover all the vertices of a graph drawn from $\mathcal{G}(n, p)$ with no colour conflicts (i.e. no edges exist whose vertices have the same colour).

We can use the first moment method to obtain a lower bound on the expected chromatic number. We let k denote the number of colours. An assignment of colours to the vertices correspond to a partitioning of the n vertices into k partitions. We denote a partition of the vertices as $\mathcal{P} = \{\mathcal{V}_1, \mathcal{V}_2, \ldots, \mathcal{V}_k\}$, where the number of nodes in each partition is $|\mathcal{V}_i| = n_i$. Now we generate a random graph from $\mathcal{G}(n, p)$ by drawing a set of edges, $\mathcal{E}$, where we include a possible edge, (i, j), with a probability p. The probability of a randomly chosen graph, G, being coloured by a partition, $\mathcal{P}$, is:

Partitioning of vertices

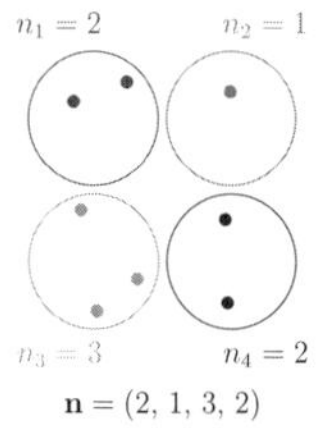

Example of edge set, $\mathcal{E}$, giving a legal colouring

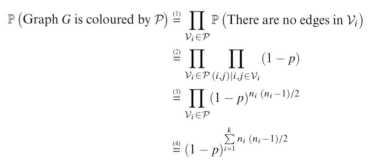

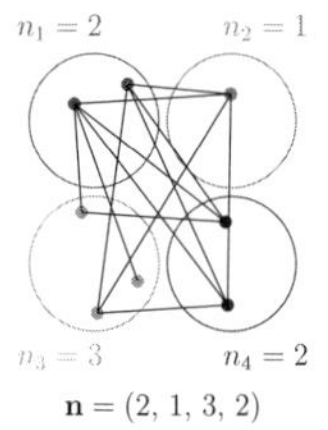

$$
\begin{aligned}
\mathbb{P}\left(\text{Graph } G \text{ is coloured by } \mathcal{P}\right) &\overset{(1)}{=} \prod_{\mathcal{V}_i \in \mathcal{P}} \mathbb{P}\left(\text{There are no edges in } \mathcal{V}_i\right) \\
&\overset{(2)}{=} \prod_{\mathcal{V}_i \in \mathcal{P}} \prod_{(i,j) | i,j \in \mathcal{V}_i} (1-p) \\
&\overset{(3)}{=} \prod_{\mathcal{V}_i \in \mathcal{P}} (1-p)^{n_i\,(n_i-1)/2} \\
&\overset{(4)}{=} (1-p)^{\sum_{i=1}^{k} n_i\,(n_i-1)/2}
\end{aligned}
$$

(1) G is coloured if there are not edges in any partition.
(2) The probability of there being no edges between any two vertices is $1 - p$.
(3) There are $n_i\,(n_i - 1)/2$ edges in a partition with $|\mathcal{V}_i| = n_i$ vertices.
(4) There are k partitions. And $\prod_i \mathrm{e}^{f_i} = \mathrm{e}^{\sum_i f_i}$.

This probability only depends on the number of vertices in a partition. We denote the set of partition sizes by $\boldsymbol{n} = (n_1, n_2, \ldots, n_k)$. The set of possible values that $\boldsymbol{n}$ can take is Λ_n^k (this is the integer simplex defined in Equation (2.11) on page 36).

We call an assignment of colours to the vertices that produces no colour conflicts a *colouring* of the graph. Define the random variable

$X(G)$ to be the number of different colourings of the graph $G(\mathcal{V}, \mathcal{E})$. Then the expected number of colourings will be

$$\begin{aligned}
\mathbb{E}_G\left[X(G)\right] &= \sum_{\boldsymbol{n} \in \Lambda_k^n} N(\boldsymbol{n}) \times \mathbb{P}\left(G \text{ is coloured by } \boldsymbol{n}\right) \\
&= \sum_{\boldsymbol{n} \in \Lambda_k^n} n! \prod_{i=1}^{n} \frac{1}{n_i!} (1-p)^{n_i\,(n_i-1)/2}
\end{aligned}$$

where $N(\boldsymbol{n}) = n!/(\prod_{i=1}^{k} n_i!)$ counts the number of ways of partitioning vertices into groups with sizes $\boldsymbol{n}$.

We are now in a position to use Markov's inequality. Recall that for non-negative random variables

$$\mathbb{P}\left(X(G) \geq t\right) \leq \frac{\mathbb{E}_G\left[X(G)\right]}{t}.$$

In graph colouring with k colours there exists a $k!$-fold permutation symmetry of the colours. In consequence, if there exists a colouring of a graph then there must be at least $k!$ colourings of the graph. The probability of a graph being colourable is therefore equal to

$$\mathbb{P}\left(G \text{ is colourable}\right) = \mathbb{P}\left(X(G) \geq k!\right) \leq \frac{\mathbb{E}_G\left[X(G)\right]}{k!}.$$

When $\mathbb{P}\left(G \text{ is colourable}\right) \leq 1/2$ we know that most of the graphs are not colourable. The value of k when this happens provides a lower bound on the minimum number of colours required to colour a random graph – this is known as the chromatic number of a graph. The reason why Markov's inequality just gives a lower bound is that a few graphs may have a large number of colourings while most graphs may have no colourings.

In Figure 7.4 we show a plot of the lower bound on the chromatic number for random graphs for $p = 1/2$ obtained from this inequality. We also show the smallest number of colours necessary to colour some randomly chosen graphs, as found by a state-of-the-art heuristic search algorithm. The algorithm finds an upper bound on the chromatic number for a particular graph (this is a hard problem where the heuristic search algorithm may fail to find the best colourings). We see, at least, for small n that the lower bound seems reasonably tight. We also note that for large n the quantity $\mathbb{E}_G\left[X(G)\right]/k!$ falls very rapidly from a very large number to a number much less than 1. This suggests that the chromatic number marks a very strong phase transition from nearly all random graphs being colourable for $k > \chi$ (the chromatic number) to almost all random graphs being uncolourable for $k < \chi$. Empirically, it is found that the k-value where a search algorithm first succeeds in finding a

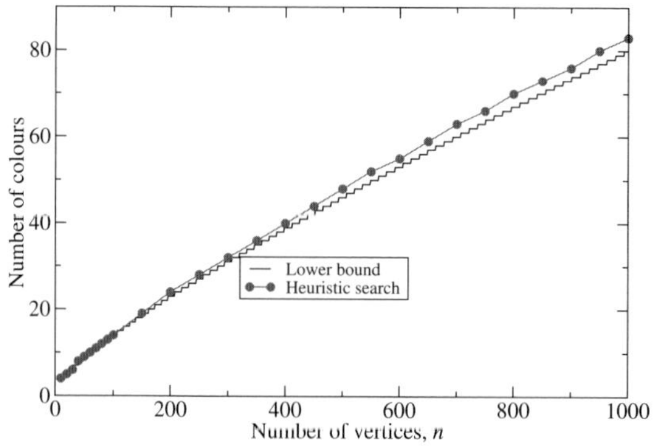

Figure 7.4 Illustration of the lower bound on the chromatic number of random graphs with edge probability $p = 1/2$ (we show where the bound guarantees that the probability of the graph being colourable is less than 0.5). Also shown are the empirically obtained chromatic number of random graphs found using the best available heuristic search algorithm.

colouring seems to be the same for almost all randomly generated graphs.

Although we have an exact result for $\mathbb{E}_G\left[X(G)\right]$ it involves a sum over the integer simplex Λ_n^k. For moderate-sized n the sum can be efficiently evaluated using dynamic programming. For large n, computing $\mathbb{E}_G\left[X(G)\right]$ is intractable. However, we can obtain a bound for $\mathbb{E}_G\left[X(G)\right]$ which is reasonably close to the exact value. We start from the identity

$$\sum_{i=1}^{k}\left(n_i - \frac{n}{k}\right)^2 = \sum_{i=1}^{k}\left(n_i^2 - 2\,n_i\,\frac{n}{k} + \left(\frac{n}{k}\right)^2\right) = -\frac{n^2}{k} + \sum_{i=1}^{k} n_i^2$$

which we can use to show

$$\frac{1}{2}\sum_{i=1}^{k} n_i\,(n_i - 1) = \frac{n^2}{2k} - \frac{n}{2} + \frac{1}{2}\sum_{i=1}^{k}\left(n_i - \frac{n}{k}\right)^2$$

thus

$$\mathbb{E}_G\left[X(G)\right] = (1-p)^{\frac{n^2}{2k}-\frac{n}{2}} \sum_{\boldsymbol{n}\in\Lambda_k^n} n! \prod_{i=1}^{n} \frac{1}{n_i!}(1-p)^{\left(n_i - \frac{n}{k}\right)^2/2}.$$

But $(1-p)^{\left(n_i-\frac{n}{k}\right)^2/2} \leq 1$, so an upper bound on the expected number of colourings of a graph is given by

$$\mathbb{E}_G\left[X(G)\right] \leq (1-p)^{\frac{n^2}{2k}-\frac{n}{2}} \sum_{\boldsymbol{n}\in\Lambda_k^n} n! \prod_{i=1}^{n} \frac{1}{n_i!} = k^n\,(1-p)^{\frac{n^2}{2k}-\frac{n}{2}}.$$

This example also provides an illustration of the dangers of blindly accepting the large n limit. A famous result of Bollobás is that the chromatic number, χ, of a random graph is given by

$$\chi = \frac{1}{2}\left(1 + o(1)\right)\log\left(\frac{1}{1-p}\right)\frac{n}{\log(n)}.$$

The term $o(1)$ indicates a function that goes to zero as n becomes infinite. In many asymptotic analyses such terms are fairly innocuous,

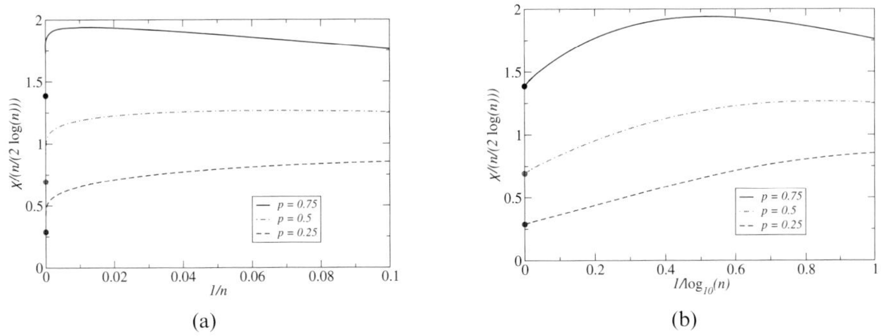

Figure 7.5 Lower bound for the chromatic number divided by $n/(2\log(n))$ for random graphs $\mathcal{G}(n,p)$ with $p = 0.75$, 0.5, and 0.25 versus (a) $1/n$ and (b) $1/\log(n)$. The asymptotic result is shown by the point on the y-axis.

as, for reasonable-sized problems, the corrections are usually negligible. However, Figure 7.5 shows the lower bound for the chromatic number (defined as the number of colours when the probability of a colouring becomes smaller than 0.5) plotted against (a) $1/n$ and (b) $1/\log(n)$. The asymptotic result is shown by a point on the y-axis. Alas, the asymptotic result (setting $o(n)$ to zero) provides no guidance for the chromatic number for the typical size of graph that most people are interested in. Even with $n = 10^{10}$ the asymptotic result is less than the lower bound by approximately 20%. Asymptotic results are often extremely useful ways of summarising the behaviour of large systems, but beware, this is not always the case.

7.2.3 Chebyshev's Inequality

If we know the variance of a random variable as well as the mean then we can use Chebyshev's inequality

$$\mathbb{P}\left(|X - \mathbb{E}\left[X\right]| \geq t\right) \leq \frac{\mathbb{V}\mathrm{ar}\left[X\right]}{t^2}.$$

This follows from considering the random variable $Y = (X - \mathbb{E}\left[X\right])^2$ which being a non-negative random variable satisfies Markov's inequality $\mathbb{P}\left(Y \geq s\right) \leq \mathbb{E}\left[Y\right]/s$. Therefore,

$$\mathbb{P}\left(|X - \mathbb{E}\left[X\right]| \geq t\right) = \mathbb{P}\left((X - \mathbb{E}\left[X\right])^2 \geq t^2\right) \leq \frac{\mathbb{E}\left[(X - \mathbb{E}\left[X\right])^2\right]}{t^2} = \frac{\mathbb{V}\mathrm{ar}\left[X\right]}{t^2}.$$

We show an example of Chebyshev's inequality where we denote $\mathbb{E}\left[X\right] = \mu$ and $\mathbb{V}\mathrm{ar}\left[X\right] = \sigma^2$ in Figure 7.6.

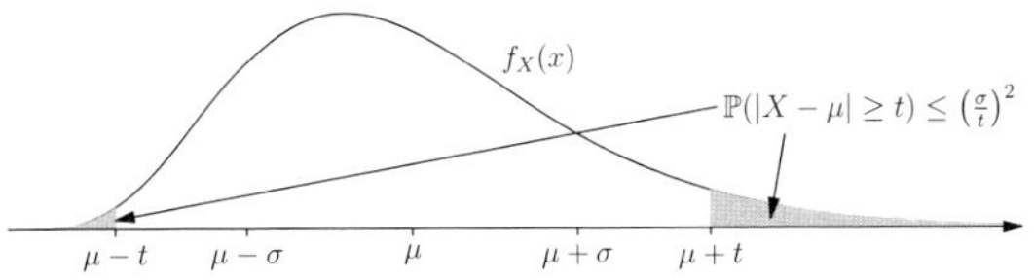

Figure 7.6 Illustration of Chebyshev's inequality.

Example 7.7 Tail Distributions Revisited
In Figure 7.7 we show the tail distributions for $X \sim \text{Poi}(1)$, $X \sim \mathcal{N}(0,1)$, $X \sim \mathit{Gam}(\epsilon, \sqrt{\epsilon})$ with $\epsilon = 0.0001$, and $X \sim \text{LogNorm}(-\log(2)/2, \log(2))$. In each case $\mathbb{V}\text{ar}\left[X\right] = 1$, so according to Chebyshev's inequality $\mathbb{P}\left(|X - \mu| \geq t\right) \leq 1/t^2$. In addition we plot the tail distribution for a random variable X with density

$$f_X(x) = \frac{3\sqrt{3}}{2\pi\,(1 + x^3)}$$

defined for $x \geq 0$. This has mean 1, but its variance diverges. Again there is no reason for it to satisfy a Chebyshev bound with $\mathbb{V}\text{ar}\left[X\right] = 1$, although it does. It has the same asymptotic behaviour as the Chebyshev bound.

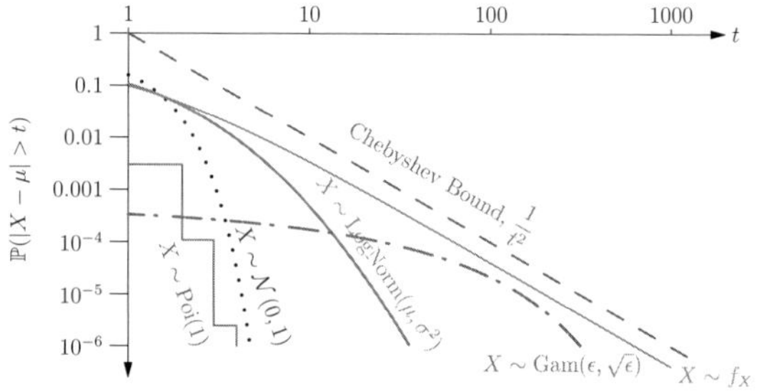

Figure 7.7 $\mathbb{P}\left(|X - \mu| \geq t\right)$ versus t plotted on a log-log scale for a Poisson, normal, gamma ($\epsilon = 0.0001$), and log-normal, all with variance of 1. Also shown is Chebyshev's inequality random variables with $\mathbb{V}\text{ar}\left[X\right] = 1$. We also show tail bounds for a variable $X \sim f_X$ with a heavy tail when the variance diverges.

If we choose $t = |\mathbb{E}\left[X\right]|$ in Chebyshev's inequality then

$$\mathbb{P}\left(\left|X - \mathbb{E}\left[X\right]\right| \geq \left|\mathbb{E}\left[X\right]\right|\right) \leq \frac{\mathbb{V}\text{ar}\left[X\right]}{\mathbb{E}\left[X\right]^2}.$$

We note that if $X = 0$ then $\left|X - \mathbb{E}\left[X\right]\right| = \left|\mathbb{E}\left[X\right]\right|$ so that

$$\mathbb{P}\left(X = 0\right) \leq \mathbb{P}\left(\left|X - \mathbb{E}\left[X\right]\right| \geq \left|\mathbb{E}\left[X\right]\right|\right) \leq \frac{\mathbb{V}\text{ar}\left[X\right]}{\mathbb{E}\left[X\right]^2}.$$

If the mean is large compared to the standard deviation then $\mathbb{P}\left(X = 0\right)$ is very small. This is often referred to as the second moment method. Notice that, in comparison with the first moment method obtained from the Markov inequality, this inequality provides a bound in the other direction.

A tighter bound can be obtained for the case when the random variable takes non-negative values $X \geq 0$. This is known as *Shepp's inequality*

$$\mathbb{P}(X=0) \le \frac{\mathbb{V}\mathrm{ar}[X]}{\mathbb{E}[X^2]}.$$

This follows from the observation that for non-negative random variables

$$\mathbb{E}[X]^2 = \mathbb{E}\left[0 [\![X=0]\!] + X [\![X>0]\!]\right]^2 = \mathbb{E}\left[X [\![X>0]\!]\right]^2.$$

Applying the Cauchy–Schwarz inequality gives

$$\begin{aligned}\mathbb{E}[X]^2 &= \mathbb{E}\left[X [\![X>0]\!]\right]^2 \le \mathbb{E}\left[X^2\right] \mathbb{E}\left[[\![X>0]\!]^2\right] \\ &= \mathbb{E}\left[X^2\right] \mathbb{E}\left[[\![X>0]\!]\right] = \mathbb{E}\left[X^2\right] (1-\mathbb{P}(X=0)),\end{aligned}$$

rearranging we get

$$\mathbb{P}(X=0)\,\mathbb{E}\left[X^2\right] \le \mathbb{E}\left[X^2\right] - \mathbb{E}[X]^2 = \mathbb{V}\mathrm{ar}[X],$$

which gives us Shepp's inequality. Since $\mathbb{E}[X^2] \ge \mathbb{E}[X]^2$ (with equality only when $\mathbb{V}\mathrm{ar}[X] = 0$) we note that

$$\mathbb{P}(X=0) \le \frac{\mathbb{V}\mathrm{ar}[X]}{\mathbb{E}[X^2]} \le \frac{\mathbb{V}\mathrm{ar}[X]}{(\mathbb{E}[X])^2}.$$

Thus Shepp's inequality is stricter than Chebyshev's inequality in this context (although Chebyshev's inequality is more widely applicable as it does not assume the random variables are non-negative).

Example 7.8 Moment Bounds on Poisson Random Variables
As a rather trivial example of moment bounds consider $P \sim \mathrm{Poi}(\lambda)$. In this case $\mathbb{E}[P] = \mathbb{V}\mathrm{ar}[P] = \lambda$ and $\mathbb{P}(P>0) = 1-\mathrm{e}^{-\lambda}$. The first moment bound gives

$$\mathbb{P}(P>0) \le \lambda$$

while the Chebyshev (second moment) bound gives

$$\mathbb{P}(P>0) \ge 1 - \frac{\mathbb{V}\mathrm{ar}[P]}{\mathbb{E}[P]^2} = 1 - \frac{\lambda}{\lambda^2} = 1 - \frac{1}{\lambda}$$

and Shepp's inequality gives

$$\mathbb{P}(P>0) \ge 1 - \frac{\mathbb{V}\mathrm{ar}[P]}{\mathbb{E}[P^2]} = 1 - \frac{\lambda}{\lambda+\lambda^2} = \frac{\lambda}{1+\lambda}$$

These bounds are shown in Figure 7.8.

Of course, moment bounds are not generally used for simple probabilities distributions, but are generally used in complicated situations where computing the full probability distribution is intractable. We illustrated this for the first

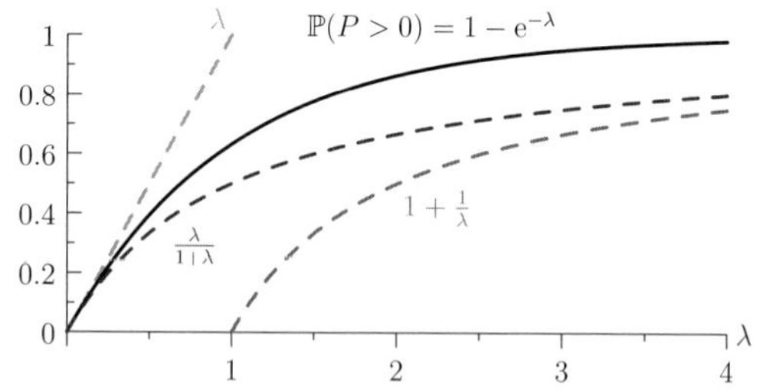

Figure 7.8 Illustration of first moment upper bound and second moment lower bound to $\mathbb{P}(P > 0)$ for $P \sim \text{Poi}(\lambda)$. The exact result is the solid curve.

moment bound on the chromatic number for random graphs. Second moment bounds are used in similar situations, although they can be a bit more intricate as they involve computing the variance for a complicated quantity (so we skip a more realistic example – for the interested reader a nice example of the use of the second moment method is given in Arclioptas et al. (2005)).

Both Markov's and Chebyshev's inequalities are often conservative since they apply to any distribution with well-defined first and second moments, respectively. As a consequence they tend not to be very tight in many practical situations. They can be viewed as worst-case bounds (we know we cannot do any worse).

7.2.4 Tail Bounds and Concentration Theorems

Often when trying to model problems it is important to show that a random variable is tightly concentrated around its mean. This is particularly true of sums of independent random variables, which (provided all the terms have a finite second moment) converge in distribution to a normal distribution. But, we often want to bound the probability of being in the tail. That is, we would like to bound the probabilities such as $\mathbb{P}(X \geq a)$. Although the Markov and Chebyshev inequalities provide such bounds they are usually too weak to be of much use. These tail bounds are sometimes referred to as *large deviation* bounds as they concern the existence of large deviations away from the mean. If there exists both upper and lower tail bounds then we have a *concentration bound* for the random variable to lie within an interval.

The main tool for obtaining tight tail bounds is a trick due to Bernstein (1924) and Chernoff (1952). For any $\lambda > 0$,

$$\mathbb{P}(X \geq a) \stackrel{(1)}{=} \mathbb{P}\left(e^{\lambda X} \geq e^{\lambda a}\right) \stackrel{(2)}{\leq} \frac{\mathbb{E}\left[e^{\lambda X}\right]}{e^{\lambda a}} \stackrel{(3)}{=} e^{-\left(\lambda a - \log\left(\mathbb{E}\left[e^{\lambda X}\right]\right)\right)}$$

(1) Where we multiply both sides of $X \geq a$ by λ and exponentiate. The inequality is unchanged provided $\lambda > 0$.
(2) Using the Markov bound $\mathbb{P}(Z \geq t) \leq \mathbb{E}[Z]/t$ with $Z = e^{\lambda X}$ and $t = e^{\lambda a}$.
(3) Writing $\mathbb{E}\left[e^{\lambda X}\right]$ as $\exp(\log(\mathbb{E}\left[e^{\lambda X}\right]))$.

This inequality is true for any $\lambda > 0$. Therefore, it is true for the value of $\lambda > 0$ that minimises the right-hand side (thus making the inequality as tight as possible). Choosing the optimum λ we have

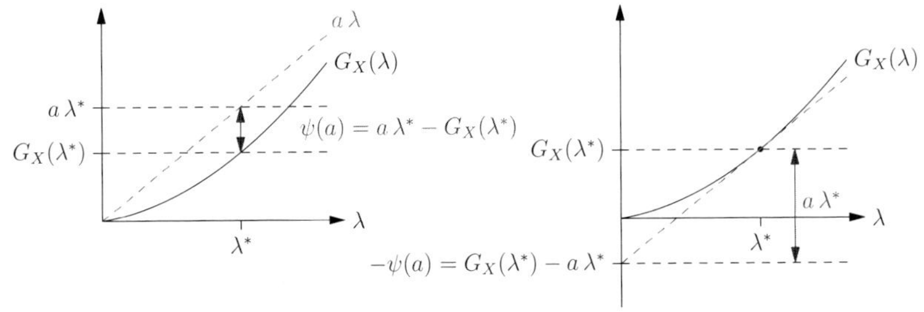

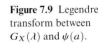

Figure 7.9 Legendre transform between $G_X(\lambda)$ and $\psi(a)$.

$$\mathbb{P}\left(X \geq a\right) \leq \mathrm{e}^{-\psi(a)} \qquad \psi(a) = \max_{\lambda>0}\left(\lambda\, a - G(\lambda)\right) \tag{7.2}$$

where $G_X(\lambda) = \log\!\left(\mathbb{E}\left[\mathrm{e}^{\lambda X}\right]\right)$ is the cumulant generating function for the random variable X.

In passing we remark that $\psi(a)$ is known as the *(Fenchel–)Legendre transform* of the function $G_X(\lambda)$. We note that $G_X(\lambda)$ is convex since its second derivative is equal to the variance which is always non-negative (see Section 7.2.6). For a differentiable function, $G_X(\lambda)$, the value of λ that maximises $a\,\lambda - G_X(\lambda)$ satisfies $a = G_X'(\lambda)$ or $\lambda^* = G_X'^{-1}(a)$ so that

More precisely, $G''(\lambda)$ is a variance with respect to a distribution proportional to $f_X(x)\,\mathrm{e}^{\lambda x}$.

$$\psi(a) = \lambda^*\, a - G(\lambda^*).$$

This is illustrated in Figure 7.9. The left-hand plot in Figure 7.9 shows that for a given a, λ^* is the value of λ that maximises the gap between the line $a\,\lambda$ and the curve $G_X(\lambda)$, where the size of the gap is $\psi(a)$. The right-hand plot show that for a given λ^*, a is the gradient $G_X'(\lambda^*)$ and $-\psi(a)$ is the intersection between the gradient through the point $(\lambda^*, G_X(\lambda^*))$ and the axis corresponding to $\lambda = 0$. Legendre transforms crop up rather frequently in inference problems and statistical mechanics.

Returning to the problem of calculating tail bounds. Often we are interested in the probability of X being at least some value greater than the mean. By a simple change of variables

$$\mathbb{P}\left(X - \mathbb{E}\left[X\right] \geq a\right) \leq \mathrm{e}^{-\bar{\psi}(a)}, \qquad \bar{\psi}(a) = \max_{\lambda>0}\left(\lambda\, a - \bar{G}(\lambda)\right) \tag{7.3}$$

where $\bar{G}(\lambda)$ is the cumulant generating function for $X - \mathbb{E}\left[X\right]$,

$$\begin{aligned}\bar{G}_X(\lambda) &= \log\!\left(\mathbb{E}\left[\mathrm{e}^{\lambda(X-\mathbb{E}[X])}\right]\right) = \log\!\left(\mathrm{e}^{-\lambda\mathbb{E}[X]}\,\mathbb{E}\left[\mathrm{e}^{\lambda X}\right]\right)\\ &= \log\!\left(\mathbb{E}\left[\mathrm{e}^{\lambda X}\right]\right) - \lambda\,\mathbb{E}\left[X\right] = G_X(\lambda) - \lambda\,\mathbb{E}\left[X\right].\end{aligned}$$

(The only difference between the cumulants of X and $X - \mathbb{E}\left[X\right]$ is in the first cumulant or mean, all other cumulants are identical.) Similarly, we can obtain a tail bound for the 'left-hand' tail

$$\begin{aligned}\mathbb{P}\left(X - \mathbb{E}\left[X\right] \leq -a\right) &\overset{(1)}{=} \mathbb{P}\left(\mathrm{e}^{-\lambda(X-\mathbb{E}[X])} \geq \mathrm{e}^{\lambda\, a}\right)\\ &\overset{(2)}{\leq} \frac{\mathbb{E}\left[\mathrm{e}^{-\lambda(X-\mathbb{E}[X])}\right]}{\mathrm{e}^{\lambda\, a}} \overset{(3)}{=} \mathrm{e}^{-\lambda\, a + \bar{G}(-\lambda)}\end{aligned}$$

(1) Multiplying through by $-\lambda$ and exponentiating. Since, by assumption $-\lambda < 0$, the inequality changes direction.
(2) Using Markov's inequality.
(3) From the definition $\bar{G}(-\lambda) = \log\big(\mathbb{E}\left[e^{-\lambda(X-\mathbb{E}[X])}\right]\big)$.

Since this is true for any $\lambda > 0$:

$$\mathbb{P}\left(X - \mathbb{E}\left[X\right] \leq -a\right) \leq e^{-\max\limits_{\lambda>0}(\lambda\, a - \bar{G}(-\lambda))}.$$

These are variational bounds in that we vary a parameter, λ, to give us as tight a bound as possible. Variational techniques are often very effective, providing surprisingly tight bounds.

Hoeffding's Inequality

For many random variables we find that their cumulant generating function satisfies an inequality of the form

$$G_X(\lambda) \leq \lambda\, \mathbb{E}\left[X\right] + \lambda^2\, c \qquad (7.4)$$

(or $\bar{G}_X(\lambda) \leq \lambda^2\, c$). Variables that satisfy inequality (7.4) are often referred to as *sub-Gaussian random variables*. We return to the problem of proving that certain random variables satisfy inequality (7.4) later on. For sub-Gaussian random variables

$$\bar{\psi}(a) = \max_{\lambda>0}\left(\lambda\, a - \bar{G}(\lambda)\right) \geq \max_{\lambda>0}\left(\lambda\, a - \lambda^2\, c\right)$$

where the right-hand side is maximised when $\lambda = a/2c$ so that $\bar{\psi}(a) \geq a^2/(4\,c)$; see also Figure 7.10. Using this bound,

$$\mathbb{P}\left(X - \mathbb{E}\left[X\right] \geq a\right) \leq e^{-a^2/(4\,c)}.$$

This inequality is particularly useful when we consider sums of random variables $S_n = \sum_{i=1}^{n} X_i$. For the sum of random variables, the cumulative generating function adds (see Section 5.3 on page 81)

$$\bar{G}_{S_n}(\lambda) = \sum_{i=1}^{n} \bar{G}_{X_i}(\lambda).$$

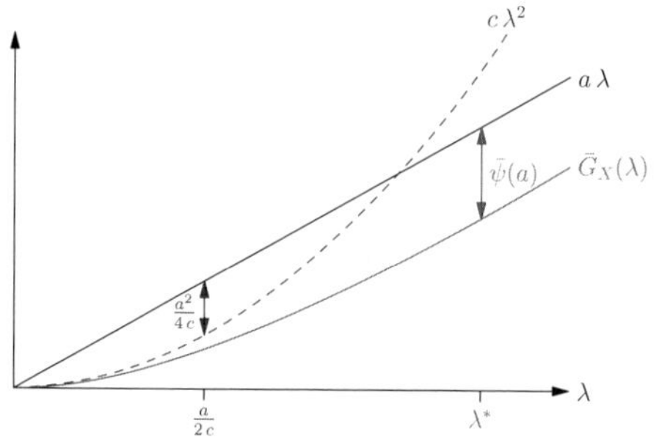

Figure 7.10 Bound on the Legendre transform $\bar{\psi}(a)$ obtained from a quadratic upper bound on $\bar{G}_X(\lambda)$.

If the variables are identically distributed (i.e. $X_i \sim X$) then $\bar{G}_{S_n}(\lambda) = n\bar{G}_X(\lambda)$, so if $\bar{G}_X(\lambda) \le c\,\lambda^2$ then $\bar{G}_{S_n}(\lambda) = n\,c\,\lambda^2$. Thus, for a sum of n *iid* variables with $\bar{G}_{X_i}(\lambda) \le c\,\lambda^2$ (for all λ) then

$$\mathbb{P}\left(S_n - \mathbb{E}\left[S_n\right] \ge a\right) \le \mathrm{e}^{-a^2/(4\,n\,c)}.$$

If $\mathbb{E}\left[X\right] = p$ so that $\mathbb{E}\left[S_n\right] = n\,p$, then (putting $a = \epsilon\left|\mathbb{E}\left[S_n\right]\right|$)

$$\mathbb{P}\left(S_n - \mathbb{E}\left[S_n\right] \ge \epsilon\mathbb{E}\left[S_n\right]\right) \le \mathrm{e}^{-\epsilon^2\,n\,p^2/(4\,c)} \tag{7.5}$$

but as p and c don't depend on n we find that the probability of being this far in the tail of a distribution is exponentially unlikely (in n). We can obtain a similar bound for the left-hand tail

$$\mathbb{P}\left(S_n \le (1-\epsilon)|\mathbb{E}\left[S_n\right]|\right) \le \mathrm{e}^{-\epsilon^2\,n\,p^2/(4\,c)}. \tag{7.6}$$

This is a typical kind of concentration result that is required in many proofs. Often you just want to show with overwhelming probability a result such as $S_n \le 2\,\mathbb{E}\left[S_n\right]$.

Of course, to pursue this line of proof you would need to show that $\bar{G}_X \le c\,\lambda^2$. This is certainly not true for all random variables and even when it is true the value of c depends on the distribution of the random variable. A particularly useful bound of this type due to Wassily Hoeffding (1963) is for random variables that lie in the range $[a, b]$, for which

$$\bar{G}_X(\lambda) \le \frac{(b-a)^2}{8}\lambda^2 \tag{7.7}$$

or

$$G_X(\lambda) = \log\left(\mathbb{E}\left[\mathrm{e}^{\lambda X}\right]\right) \le \lambda\mathbb{E}\left[X\right] + \frac{(b-a)^2}{8}\lambda^2.$$

The proof is rather elegant. We first note that, because the exponential function is convex (we discuss convexity in the next section, but for our purposes here we are just interested in the fact that the exponential function between two points lies below the line segment connecting those points), for any X between a and b

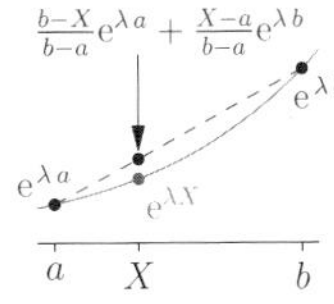

$$\mathrm{e}^{\lambda X} \le \frac{b-X}{b-a}\mathrm{e}^{\lambda a} + \frac{X-a}{b-a}\mathrm{e}^{\lambda b}.$$

Multiplying each side by $\mathrm{e}^{-\lambda\mathbb{E}[X]}$,

$$\mathrm{e}^{\lambda(X-\mathbb{E}[X])} \le \frac{b-X}{b-a}\mathrm{e}^{\lambda(a-\mathbb{E}[X])} + \frac{X-a}{b-a}\mathrm{e}^{\lambda(b-\mathbb{E}[X])}.$$

Taking the expectation of both sides

$$\mathbb{E}\left[\mathrm{e}^{\lambda(X-\mathbb{E}[X])}\right] \le \frac{b-\mathbb{E}\left[X\right]}{b-a}\mathrm{e}^{\lambda(a-\mathbb{E}[X])} + \frac{\mathbb{E}\left[X\right]-a}{b-a}\mathrm{e}^{\lambda(b-\mathbb{E}[X])}.$$

Letting $r = (\mathbb{E}\left[X\right]-a)/(b-a)$ then $1-r = (b-\mathbb{E}\left[X\right])/(b-a)$ and

$$\begin{aligned}\mathbb{E}\left[\mathrm{e}^{\lambda(X-\mathbb{E}[X])}\right] &\le (1-r)\,\mathrm{e}^{-\lambda\,r\,(b-a)} + r\,\mathrm{e}^{(1-r)\,\lambda\,(b-a)}\\ &= \left(1-r+r\,\mathrm{e}^{\lambda\,(b-a)}\right)\mathrm{e}^{-\lambda\,r\,(b-a)} = \mathrm{e}^{\phi(\lambda\,(b-a))}\end{aligned}$$

where

$$\phi(u) = \log(1 - r + r\,\mathrm{e}^u) - r\,u.$$

But, $\phi(u)$ is bounded above by $u^2/8$, which we can show by considering its Taylor expansion (see Appendix 7.A),

$$\phi(u) = \phi(0) + u\,\phi'(0) + \frac{u^2}{2}\phi''(\theta)$$

for some θ. Taking derivatives

$$\phi'(u) = r + \frac{r\,\mathrm{e}^u}{1 - r + r\,\mathrm{e}^u}, \qquad \phi''(u) = \frac{r\,(1-r)\,\mathrm{e}^u}{(1 - r + r\mathrm{e}^u)^2}$$

and noting $\phi(0) = \phi'(0) = 0$, we have

$$\phi(u) = \frac{u^2}{2}\phi''(\theta).$$

But we observe

$$\phi''(\theta) \overset{(1)}{=} \frac{r(1-r)\mathrm{e}^\theta}{(1 - r + r\mathrm{e}^\theta)^2} \overset{(2)}{=} \frac{r\,(1-r)\mathrm{e}^\theta}{(1 - r - r\mathrm{e}^\theta)^2 + 4\,r\,(1-r)\mathrm{e}^\theta}$$
$$\overset{(3)}{=} \frac{1}{4 + (1 - r - r\mathrm{e}^\theta)^2/r\,(1-r)\mathrm{e}^\theta} \overset{(4)}{\leq} \frac{1}{4}$$

(1) Writing out the second derivative in functional form.
(2) Rewriting the denominator using $(a+b)^2 = (a-b)^2 + 4\,a\,b$ with $a = 1 - r$ and $b = r\,\mathrm{e}^\theta$.
(3) Cancelling top and bottom by $r\,(1-r)\mathrm{e}^\theta$.
(4) Because $0 \leq r \leq 1$ we have that $(1 - r - e\mathrm{e}^\theta)^2/r\,(1-r)\mathrm{e}^\theta \geq 0$ so that the denominator is greater than 4.

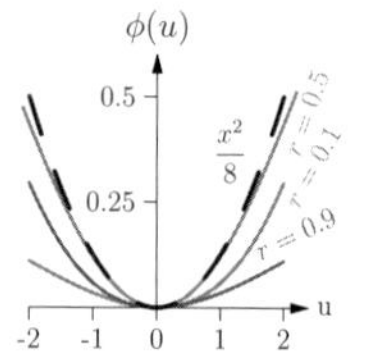

This bound does not depend on r. Using this result and the Taylor expansion

$$\phi(u) = \log(1 - r + r\,\mathrm{e}^u) - r\,u \leq \frac{u^2}{8} \tag{7.8}$$

and consequently,

$$\bar{G}_X(\lambda) \leq \phi(\lambda(b-a)) \leq \frac{(b-a)^2}{8}\lambda^2,$$

proving Hoeffding's lemma, Equation (7.7).

Putting this together with the tail bounds for sums of *iid* variables, inequalities (7.5–7.6), we have Hoeffding's inequality that if S_n is a sum of n *iid* variables each bounded in the range $a \leq X \leq b$ then

$$\mathbb{P}\left(S_n > (1+\epsilon)\,\mu\right) \leq \mathrm{e}^{-\frac{2\,\epsilon^2\,\mu^2}{n\,(b-a)^2}}, \qquad \mathbb{P}\left(S_n < (1-\epsilon)\,\mu\right) \leq \mathrm{e}^{-\frac{2\,\epsilon^2\,\mu^2}{n\,(b-a)^2}} \tag{7.9}$$

where $\mu = n\,p = \mathbb{E}\left[S_n\right]$. Note that if the X_is are *not* independent then $S_n = \sum_{i=1}^n X_i$ potentially varies between $n\,a$ and $n\,b$ so that $\bar{G}_S(\lambda) \leq n^2\,(b-a)^2\,\lambda^2/8$ and

$$\mathbb{P}\left(S_n > (1+\epsilon)\left|\mathbb{E}\left[S_n\right]\right|\right) \le \mathrm{e}^{-\frac{2\epsilon^2\mu^2}{(b-a)^2}} \qquad \text{(when } X_i\text{s are dependent).}$$

This is weaker by a factor of n in the exponent and would often be useless. Hoeffding, however, showed that in some special cases his inequality could be applied to some sums of dependent variables, provided the dependency was not too strong. For example, when sampling without replacement occurs. As this is rather specialised and technical, we relegate a treatment of this to Appendix 7.B.

Hoeffding's inequality is a very useful and powerful inequality when applied to variables whose range lives close to the middle of the interval $[a, b]$. For example, for a succession of Bernoulli trials $p \sim \text{Bern}(p)$ then $a = 0$, $b = 1$ and $\mu = n\,p$ so that

$$\mathbb{P}\left(S_n > (1+\epsilon)\mu\right) \le \mathrm{e}^{-2\epsilon^2\mu^2/n} \qquad \text{or} \qquad \mathbb{P}\left(S_n > \mu + t\right) \le \mathrm{e}^{-2t^2/n}.$$

If p is sufficiently large then this is a reasonably sharp upper tail bound. But, for $p = O(1/\sqrt{n})$, Hoeffding's inequality is rather poor. In these cases, far tighter bounds exist. The most famous of these is the Chernoff bound.

7.2.5 Chernoff Bounds for Independent Bernoulli Trials

One of the most common use of tail bounds is for independent Bernoulli trials. This was the problem considered by Chernoff. Consider bounding the probability that $S_n = \sum_{i=1}^n X_i$ is some distance from its mean value, given that $X_i \in \{0, 1\}$ are independent Bernoulli trials, $X_i \sim \text{Bern}(p_i)$. We note that

$$\mathbb{E}\left[\mathrm{e}^{\lambda X_i}\right] = p_i\,\mathrm{e}^{\lambda} + 1 - p_i.$$

So if the X_is are independent

$$\mathbb{E}\left[\mathrm{e}^{\lambda S_i}\right] = \prod_{i=1}^{n}\left(p_i\,\mathrm{e}^{\lambda} + 1 - p_i\right).$$

In the next section we prove the arithmetic mean–geometric mean (AM–GM) inequality (Equation (7.17) on page 164), which states

$$\left(\prod_{i=1}^{n} y_i\right)^{\frac{1}{n}} \le \frac{1}{n}\sum_{i=1}^{n} y_i.$$

Using this inequality

$$\mathbb{E}\left[\mathrm{e}^{\lambda S_i}\right] \overset{(1)}{\le} \left(\frac{1}{n}\sum_{i=1}^{n}\left(p_i\,\mathrm{e}^{\lambda} + 1 - p_i\right)\right)^{n} \overset{(2)}{=} \left(1 + p\left(\mathrm{e}^{\lambda} - 1\right)\right)^{n}$$

(1) Using the AM–GM inequality in the form $\prod_{i=1}^n y_i \le \left(\frac{1}{n}\sum_{i=1}^n y_i\right)^n$.
(2) Performing the sum and using $p = \frac{1}{n}\sum_i p_i$. Note that p is the mean of the Bernoulli probabilities.

Using Equation (7.2) on page 153, we find

$$\mathbb{P}\left(S_n \geq n\,a\right) \leq \exp\left(-n\,a\,\lambda + n\log(1 + p(\mathrm{e}^{\lambda} - 1))\right).$$

Choosing λ to minimise the left-hand side we find (putting $a = p + t$),

$$\mathbb{P}\left(S_n \geq (p+t)\,n\right) \leq \exp\left(-n\left((p+t)\log\!\left(\frac{p+t}{p}\right) + (q-t)\log\!\left(\frac{q-t}{q}\right)\right)\right). \tag{7.10}$$

This is the Chernoff bound. It is a remarkably good bound, but the result is not particularly easy to work with. The exponent can be interpreted as a relative entropy (see Chapter 9), but, at least speaking for myself, this doesn't shed at lot of light on Equation (7.10). A similar bound is available for the left-hand tail, which is useful for when $1 - p$ is small.

For many proofs we don't need a very tight bound, it suffices to show that large deviations are suppressed. By being less insistent on obtaining the sharpest possible bound we can show that for $\epsilon < 1$

$$\mathbb{P}\left(S_n \geq \mu\,(1+\epsilon)\right) \leq \mathrm{e}^{-\mu\epsilon^2/3} \qquad \mathbb{P}\left(S_n \leq \mu\,(1-\epsilon)\right) \leq \mathrm{e}^{-\mu\,\epsilon^2/2}$$

where $\mu = \mathbb{E}\left[S_n\right]$. If, for example, $\mu \approx \sqrt{n}$, then this tells us that S_n have tails that initially fall off like a Gaussian. Deviations greater than μ away from the mean are super-exponentially unlikely to occur in this example.

Tail bounds abound. Two useful inequalities that involve the variance of the random variables are due to Bennett and Bernstein. Bennett's inequality applies to independent variables, X_i, such that for all variable $X_i - \mathbb{E}\left[X_i\right] \leq b$ then for any $t > 0$

$$\mathbb{P}\left(S_n > \mathbb{E}\left[S_n\right] + t\right) \leq \exp\!\left(-\frac{\sigma^2}{b^2}\,h\!\left(\frac{b\,t}{\sigma}\right)\right) \tag{7.11}$$

where $h(u) = (1+u)\,\log(1+u) - u$ and $\sigma^2 = \sum_{i=1}^{n} \mathbb{V}\mathrm{ar}\left[X_i\right]$ (all random variables in the sum must have a finite variance). We leave the proof to Exercise 7.2. Bennett's inequality can be a bit awkward to work with. However, it turns out that

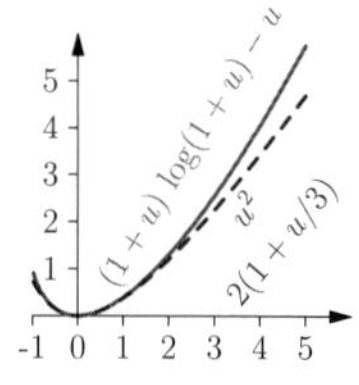

$$h(u) \geq \frac{u^2}{2(1 + u/3)}$$

(see Exercise 7.3); from this we obtain Bernstein's inequality

$$\mathbb{P}\left(S_n > \mathbb{E}\left[S_n\right] + t\right) \leq \exp\!\left(-\frac{t^2}{2\,\sigma^2\,(1 + b\,t/(3\,\sigma^2))}\right). \tag{7.12}$$

If $b\,t \ll 3\,\sigma^2$ then the tails of S_n are nearly bounded by a Gaussian. For very large t the tails are bounded by $\exp(-3\,t/(2\,b))$ (random variables that asymptotically fall off, at least exponentially fast, are sometimes referred to as *sub-Gamma variables*). For many more inequalities see the books by Boucheron et al. (2013) or Dubhashi and Panconesi (2009).

Example 7.9 Tail Bounds for Sums of Bernoulli Trials
We consider the random variable $S_{100} = \sum_{i=1}^{100} X_i$ where $X_i \sim \text{Bern}(0.05)$. Thus $S_{100} \sim \text{Bin}(0.05, 100)$ with $\mu = \mathbb{E}[S_n] = 5$. In Figure 7.11 we show the tail bounds using Hoeffding's (7.9), Bernstein's (7.12), Bennett's (7.11), and Chernoff's (7.10) inequalities. In the inset we also show the approximate Chernoff bound $\exp(-\mu\epsilon^2/3)$. Note that this is a regime where Hoeffding's inequality is poor. Although Chernoff's inequality is extremely good, its form sometimes makes it cumbersome to use in proofs.

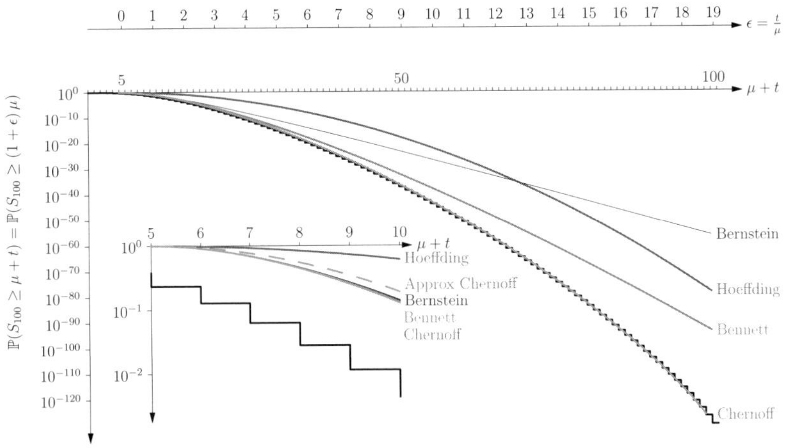

Figure 7.11 Tails and tail bounds for $S_{100} \sim \text{Bin}(100, 0.05)$. The staircase shows $\mathbb{P}(S_n > \mu + t)$. The Chernoff bound (Equation (7.10)) falls on top of this probability.

Tail bounds are used in many theoretical analyses where we need to show that large deviations are exponentially unlikely. These arise in many areas, such as proving bounds on the performance of machine learning algorithms (so-called probably approximately correct (PAC) learning) or in proofs of computational complexity of random algorithms.

7.2.6 *Jensen's Inequality*

Jensen's inequality, like the Cauchy–Schwarz inequality, is important beyond the realms of probability. It is one of the cornerstones of convexity theory. It is applicable to *convex functions*. These are functions that satisfy the inequality

$$f(t\,x + (1-t)\,y) \le t\,f(x) + (1-t)\,f(y). \tag{7.13}$$

In other words, if we choose any two points, $(x, f(x))$ and $(y, f(y))$, and we draw a straight line between them, then the curve, $f(x)$, lies on or beneath the line. This is illustrated in Figure 7.12. Notice that the region on or above the curve (sometimes known as the epigraph or supergraph) is a convex region.

Examples of convex functions include quadratic functions with positive curvature (i.e. x^2 but not $-x^2$) and exponentials. Many functions exist which satisfy the

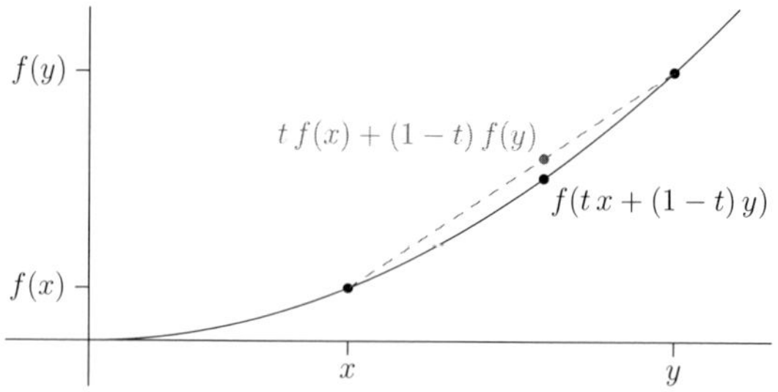

Figure 7.12 Example of a convex function.

opposite inequality (e.g. logarithms, $-x^2$, $-e^x$, etc.) which are known as concave functions or convex-down functions (the region on or below the curve is convex). Note that the inverse of a convex-up function is convex-down and vice versa, so that $\exp(x)$ is convex-up and $\log(x)$ is convex-down, similarly x^2 is convex-up and $\sqrt{x}$ is convex-down. But this is not true of reciprocals, so, for example, $\exp(x)$ and $\exp(-x)$ are both convex-up functions. Of course, most functions (e.g. $\sin(x)$, $\tanh(x)$, x^3) are neither convex-up or convex-down everywhere, although they may be locally convex which might be sufficient if you know x is confined in a region.

Jensen's inequality, in the language of probability, is that for a convex(-up) function, $f(x)$,

$$f(\mathbb{E}[X]) \leq \mathbb{E}[f(X)]. \tag{7.14}$$

Note that f here is some convex function *not* the probability distribution for X. For any convex-up function then, at any point, m, on the curve $f(x)$ we can draw a line through the point $(m, f(m))$ such that $f(x)$ is on or above that line. That is, for any m, there exists a b such that

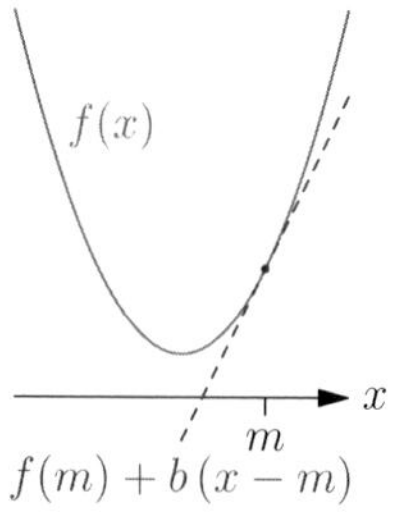

$$f(x) \geq f(m) + b\,(x - m).$$

If $f(m)$ is differentiable at m then $b = f'(x)$. If it is not differentiable there may be many values of b. Taking X to be a random variable and putting $m = \mathbb{E}[X]$ then

$$f(X) \geq f(\mathbb{E}[X]) + b\,(x - \mathbb{E}[X]).$$

Taking the expectation of both sides the last term vanishes and we are left with inequality (7.14). For a function that is concave (convex-down function $f(x)$), i.e.

$$f(t\,x + (1-t)\,y) \geq t\,f(x) + (1-t)\,f(y),$$

then by a similar argument

$$f(\mathbb{E}[X]) \geq \mathbb{E}[f(X)] \tag{7.15}$$

A function is said to be strictly convex if

$$f(t\,x + (1-t)\,y) < t\,f(x) + (1-t)\,f(y)$$

for all t in the open interval $(0, 1)$ provided $x \neq y$. In other words, there are no straight sections in the function $f(x)$. For a *strictly* convex function, Jensen's

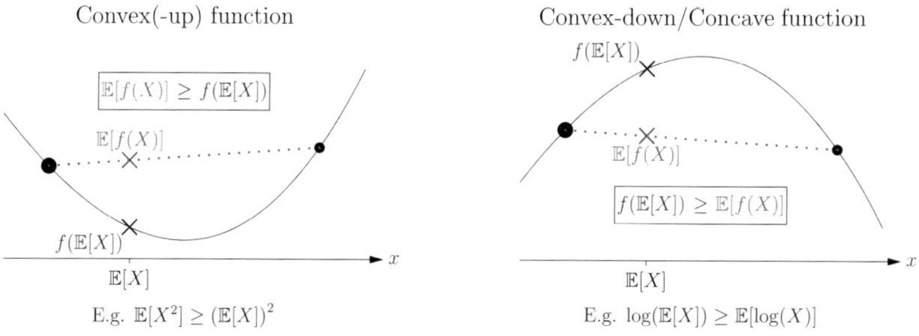

Figure 7.13 Getting Jensen's inequality the right way around. Consider the probability distribution to be two delta peaks, then ask if the average of the function values at those two peaks is greater than the function value taken at the average of the two peaks.

inequality is an equality only if X does not vary; that is, when X is a constant and not a proper random variable. Jensen's inequality does not require the function to be strictly convex. For example, the absolute-value function, $\text{abs}(x) = |x|$ is convex although not strictly convex, but Jensen's inequality still holds $\left|\mathbb{E}\left[X\right]\right| \leq \mathbb{E}\left[|X|\right]$. If the function is linear everywhere then the function is both convex and concave and Jensen's inequality becomes an equality. One of the hardest parts of using Jensen's inequality is to remember which way around the inequality works. I remember this pictorially – see Figure 7.13.

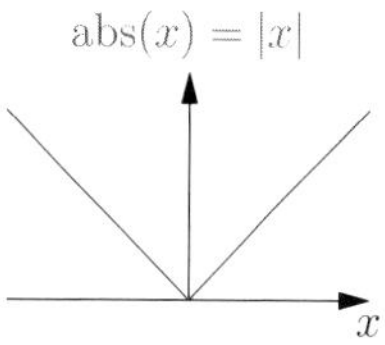

Often the challenge in using Jensen's inequality is to prove that the function is convex. Sometimes, this can be done directly from the definition, Equation (7.13). Sometimes it is easier to use the fact that for a function to be convex-up the second derivative $f''(x)$ must be non-negative. It will be strictly convex if $f''(x) > 0$ everywhere. To show that convex functions have a non-negative second derivative, we note that

$$f''(x) = \lim_{\epsilon \to 0} \frac{f(x+\epsilon) + f(x-\epsilon) - 2\,f(x)}{\epsilon^2}.$$

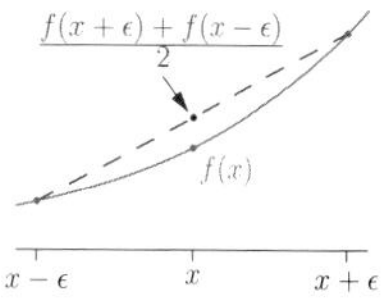

But, for a convex function,

$$f(x) \leq \frac{f(x+\epsilon) + f(x-\epsilon)}{2}$$

or $f(x+\epsilon)+f(x-\epsilon) \geq 2f(x)$, with strict inequality for a strictly convex function. From this it follows that $f''(x) \geq 0$. Proving the converse, that all functions with a non-negative second derivative are convex takes a bit more work – for full proof we refer the interested reader to the book by Steele (2004). However, when $f(x)$ is doubly differentiable then we can show that $f(x)$ lies above a tangent line at any point (we used this fact to prove Jensen's inequality). The second-order Taylor expansion of $f(x)$ around a point m is given by

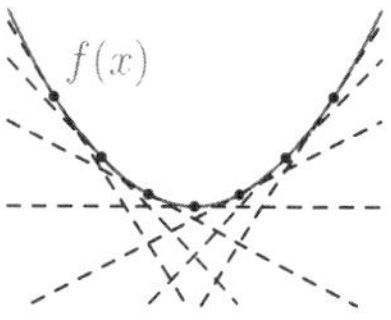

$$f(x) = f(m) + f'(m)\,(x-m) + \frac{(x-m)^2}{2} f''(\theta)$$

where θ is some value in the interval between x and m (see Appendix 7.A on page 177 for a discussion on the Taylor expansion). However, we know that for a convex function, $f''(\theta) \geq 0$ for all θ, so the last term is always non-negative, implying

$$f(x) \geq f(m) + f'(m)\,(x-m).$$

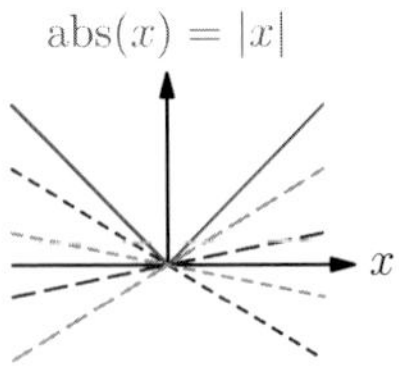

If $f(x)$ is not differentiable at a point, we can nevertheless find a constant b such that $f(x) \geq f(m) + b\,(x - m)$.In the example of the absolute-value function b takes any value in the range $[-1, 1]$. Such values are sometimes referred to as *sub-gradients*. If $f''(x) > 0$ for all x then the function will be strictly convex. To construct more complicated convex functions from simpler ones we observe that if $f(x)$ and $g(x)$ are convex and $a, b > 0$, then $a\,f(x) + b\,g(x)$ and $f(g(x))$ will also be convex.

Jensen's inequality also applies to random vectors, $\boldsymbol{X}$. That is, if $f(\boldsymbol{x})$ is a function mapping a vector to a real number such that for any two points $\boldsymbol{x}$ and $\boldsymbol{y}$

$$f(t\,\boldsymbol{x} + (1 - t)\,\boldsymbol{y}) \leq t\,f(\boldsymbol{x}) + (1 - t)\,f(\boldsymbol{y})$$

then the function is said to be convex(-up) and $\mathbb{E}\left[f(\boldsymbol{X})\right] \geq f(\mathbb{E}\left[\boldsymbol{X}\right])$. Again for a complex funtion $f(\boldsymbol{x})$, the (hyper)volume on or above $f(\boldsymbol{x})$ is convex and for all points, $(\boldsymbol{x}, f(\boldsymbol{x}))$, we can define a (hyper)plane such that all other points $(\boldsymbol{y}, f(\boldsymbol{y}))$ lies on or above that plane (we can use this to prove Jensen's inequality in higher dimensions).

Convex region

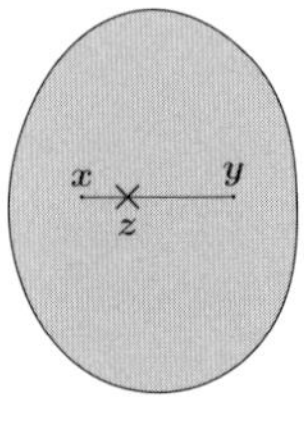

If a convex function is restricted to a convex region then the restricted function is also convex (since the function is defined for all points connecting any two points). However, if the region is not convex, then beware: there will be points $z = t\,\boldsymbol{x} + (1 - t)\,\boldsymbol{y}$ for which the function is not defined and so the inequality above is not defined.

Although Jensen's inequality is one of the primary properties of convex functions, convexity theory goes well beyond probabilities. For example, *strictly* convex functions are guaranteed to have a unique minimum, while for a general convex function the minimum will be a single connected region (very often containing only a single point). Many texts cover convexity theory, one of the classics being Boyd and Vandenberghe (2004).

Non-convex region

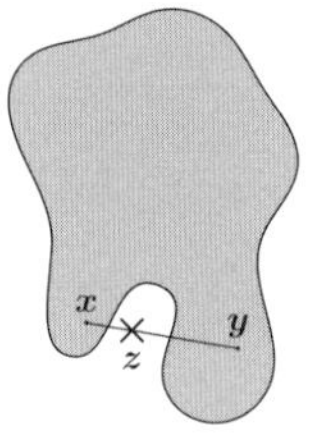

Jensen's inequality crops up in all sorts of applications. Like the Cauchy–Schwarz inequality it is a real gem and it is often just what you need. We give some examples of its use below. For people with a lot to prove the irritation is that it only gives you a one-sided bound and, frustratingly, this can be in the wrong direction. Proving inequalities in the other direction is often much harder. Hoeffding's lemma (Equation (7.7)) is an example of an inequality in the other direction. That is, provided the random variable X is bounded then

$$0 \leq \log\!\left(\mathbb{E}\left[e^{\lambda\,(X - \mathbb{E}[X])}\right]\right) \leq (b - a)\,\lambda^2 \quad \text{when } a \leq X \leq b.$$

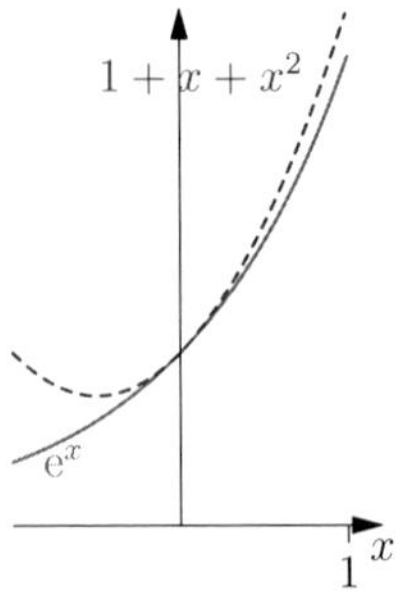

The lower bound is a consequence of Jensen's inequality while the upper bound is Hoeffding's inequality (but only applies when X is bounded). Another useful trick to get a bound in the opposite way to Jensen is to note that when $X \leq 1$

$$e^x \leq 1 + x + x^2 \quad \text{when } x \leq 1.$$

Furthermore, $\log(z) \leq z - 1$ (which can be proved from the second-order Taylor expansion around 1; see Appendix 7.A on page 177). Thus, if $X \leq 1$

$$\log(\mathbb{E}\left[e^{\lambda X}\right]) \overset{(1)}{\leq} \mathbb{E}\left[e^{\lambda X}\right] - 1 \overset{(2)}{\leq} \mathbb{E}\left[1 + \lambda\,X + \lambda\,X^2\right] - 1 \overset{(3)}{=} \lambda\,\mathbb{E}\left[X\right] + \lambda^2\,\mathbb{E}\left[X^2\right]$$

(1) Using $\log(z) \leq z - 1$ (which can be proved from the second-order Taylor expansion around 1).
(2) Using $\mathrm{e}^x \leq 1 + x + x^2$ for $x \leq 1$. We also use that if $f(X) \leq g(X)$ then $\mathbb{E}\left[f(X)\right] \leq \mathbb{E}\left[g(X)\right]$.
(3) Using the linearity of the expectation operator and cancelling the constant term.

Using this inequality we arrive at

$$\log\left(\mathrm{e}^{\lambda\,(X-\mathbb{E}[X])}\right) \leq \mathbb{E}\left[X^2\right]\,\lambda^2 \quad \text{when } X \leq 1.$$

This can, for example, be used together with Equation (7.3) to obtain tail bounds. If $X < 0$, then we can sharpen the tail bound by a factor of two (since then $\mathrm{e}^x \leq 1 + x + x^2/2$ – see Appendix 7.A).

Example 7.10 Benefits of Price Fluctuations
In the classic economic model of prices, the cost of producing a commodity is assumed to increase with the number that are produced. Clearly this does not fit commodities where there is an 'economy of scale', but it is a reasonable model for goods such as raw materials. If only a small quantity is needed then it can be obtained from the cheapest source. However, to produce greater quantities requires using more expensive sources, thus increasing the cost per unit. This is particularly marked, for example, in the energy supply industries. The cost of supply is, according to this model, a (convex) monotonically increasing curve.

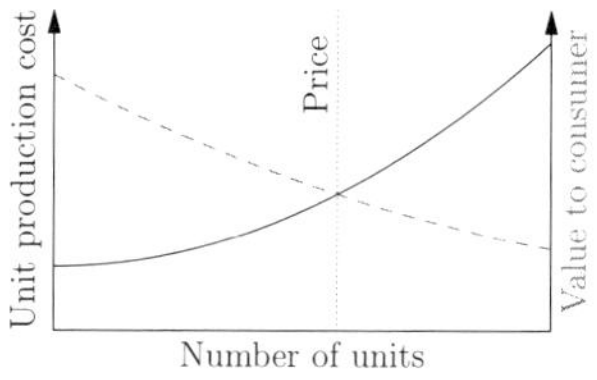

Consumers attach different values to purchasing the goods. As the price drops below a customer's valuation they would then buy the goods. The demand curve shows the minimum price the consumers are prepared to pay given a fixed number of goods. The price according to this model is determined where this demand curve crosses the supply curve.

Consider what happens when the demand fluctuates so that the number of units sold, n, changes. The price is given by $f(n)$ defined by the supply curve. As the supply curve is convex-up

$$\mathbb{E}\left[f(n)\right] \geq f(\mathbb{E}\left[n\right])$$

thus the expected price in a fluctuating market is higher than if the demand was stable. At least for the producers, demand fluctuations leads to a higher price and more profit. Producers often consider fluctuating demand as a headache they could do without, but if this model is true, perhaps it not such a bad thing for them. Of course, for the consumer the story is reversed! The price of domestic electricity is a classic example of costs often being dominated by fluctuating demands.

Example 7.11 Geometric Mean

Since the exponential is a convex-up function, Jensen's inequality implies

$$e^{\mathbb{E}[X]} \leq \mathbb{E}\left[e^X\right].$$

Writing $\mathbb{E}\left[X\right] = \sum_{i=1}^{n} p_i\, x_i$ then

$$e^{\sum_{i=1}^{n} p_i\, x_i} \leq \sum_{i=1}^{n} p_i\, e^{x_i}.$$

Making the substitution $e^{x_i} = y_i$ (so that $e^{\sum_i p_i\, x_i} = \prod_i e^{p_i\, x_i} = \prod_i (e^{x_i})^{p_i} = \prod_i y_i^{p_i}$), we find

$$\prod_{i=1}^{n} y_i^{p_i} \leq \sum_{i=1}^{n} p_i\, y_i. \tag{7.16}$$

A special case of this inequality occurs when $p_i = 1/n$ then

$$\left(\prod_{i=1}^{n} y_i\right)^{\frac{1}{n}} \leq \frac{1}{n}\sum_{i=1}^{n} y_i. \tag{7.17}$$

The left-hand side $\left(\prod_{i=1}^{n} y_i\right)^{\frac{1}{n}}$ is known as the *geometric mean*. The right-hand side is the standard or *arithmetic* mean. The geometric mean is sometimes used to compare items described by a number of factors which are measured on different scales. The arithmetic mean would be dominated by the quantity measured on the largest scale, but the geometric mean can take all scales into account. We note that the geometric mean is always smaller than the arithmetic mean (equality only holds if y_i is the same for all items).

In the UK the official measure of inflation was swapped to the consumer price index (CPI) from the retail price index (RPI). Although there are a number of differences between these measures, one difference is that the CPI uses the geometric mean to measure the increase in price while the RPI used the arithmetic mean. Although this allows price increases on small items to have a similar influence as price increases of large items, the overall cost-of-living increase would intuitively be better measured by an arithmetic mean. Some cynics have suggested that the fact that the geometric mean is always smaller than the arithmetic mean might have something to do with the choice.

Equations (7.16) and (7.17) are important inequalities in themselves, known as AM–GM inequalities. They are particularly useful for obtaining bounds involving the product of random variables.

Example 7.12 Annealed Approximation
In statistical physics of disordered systems we often have energy functions that depend on a set of variables $\boldsymbol{X}$ and a random set of couplings $\boldsymbol{J}$. The typical behaviour of these systems is described by the average free energy

$$f = -kT\,\mathbb{E}_{\boldsymbol{J}}\left[\log\bigl(Z(\boldsymbol{X},\boldsymbol{J})\bigr)\right]$$

where k is Boltzmann's constant, T is the temperature, and $Z(\boldsymbol{X},\boldsymbol{J})$ is the partition function which depends on both sets of variables. Since the logarithm is a convex-down function Jensen's inequality tells us

$$\mathbb{E}\left[\log(X)\right] \leq \log\bigl(\mathbb{E}\left[X\right]\bigr)$$

so that $-\mathbb{E}\left[\log(X)\right] \geq -\log\bigl(\mathbb{E}\left[X\right]\bigr)$. Thus,

$$f = -kT\,\mathbb{E}_{\boldsymbol{J}}\left[\log\bigl(Z(\boldsymbol{X},\boldsymbol{J})\bigr)\right] \geq -kT\,\log\bigl(\mathbb{E}_{\boldsymbol{J}}\left[Z(\boldsymbol{X},\boldsymbol{J})\right]\bigr)$$

where the right-hand side is known as the annealed free energy. The annealed free energy is sometimes used as an approximation for the true free energy, although in many cases it is a poor approximation. We see, however, that it is at least a lower bound on the true free energy.

7.2.7 Union Bound

A very useful and easy bound to use is the union bound, also known as Boole's inequality. Given a countable set of events A_1, A_2, ..., then

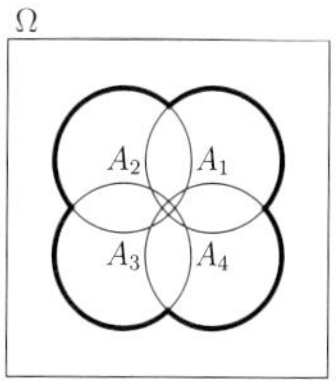

$\mathbb{P}\left(\bigcup_{i=1}^{4} A_i\right) \leq \sum_{i=1}^{4} \mathbb{P}(A_i)$

$$\mathbb{P}\left(\bigcup_i A_i\right) \leq \sum_i \mathbb{P}\left(A_i\right)$$

where $\bigcup_i A_i$ is the event in which at least one of the sub-events occurs. The importance of the bound is that we do not require the events to be mutually exclusive (obviously, when they are then the equality holds). The union bound is very easy to prove by induction. Clearly,

$$\mathbb{P}\left(A_1\right) \leq \mathbb{P}\left(A_1\right).$$

Now assume $\mathbb{P}\left(\bigcup_{i=1}^{n} A_i\right) \le \sum_{i=1}^{n} \mathbb{P}(A_i)$ then

$$\mathbb{P}\left(\bigcup_{i=1}^{n+1} A_i\right) \overset{(1)}{=} \mathbb{P}\left(\bigcup_{i=1}^{n} A_i\right) + \mathbb{P}(A_{n+1}) - \mathbb{P}\left(\left(\bigcup_{i=1}^{n} A_i\right) \cap A_{n+1}\right)$$
$$\overset{(2)}{\le} \mathbb{P}\left(\bigcup_{i=1}^{n} A_i\right) + \mathbb{P}(A_{n+1}) \overset{(3)}{\le} \sum_{i=1}^{n+1} \mathbb{P}(A_i)$$

(1) Using $\mathbb{P}(A \cup B) = \mathbb{P}(A) + \mathbb{P}(B) - \mathbb{P}(A \cap B)$ with $A = \bigcup_{i=1}^{n} A_i$ and $B = A_{n+1}$.
(2) Since $\mathbb{P}\left(\left(\bigcup_{i=1}^{n} A_i\right) \cap A_{n+1}\right) \ge 0$ (i.e. probabilities take non-negative values).
(3) Using the inductive hypothesis.

The main use of the union bound is that if you can bound the marginal probabilities for a set of events then the probability of one or more events occurring is bounded by the sum of the marginals. Note that these events don't have to be independent. This is particularly useful if these marginals are all small enough, then the union bound is often sufficient to show that with high probability none of the events will occur.

7.3 Comparing Distributions

We often want to measure the degree of similarity between probability distributions, $f(\boldsymbol{x})$ and $g(\boldsymbol{x})$, say. To compare distributions they need to have the same domain (i.e. $\boldsymbol{x}$ can take the same set of values in the two functions, although there can be times when the probability is zero for some values of $\boldsymbol{x}$). Our discussion will cover the case of both discrete and continuous random variables so f and g may be either probability mass or probability density functions. One common application will be to approximate some given probability, $f(\boldsymbol{x})$, with a simpler distribution $g(\boldsymbol{x}|\boldsymbol{\theta})$. To achieve this we choose the parameters $\boldsymbol{\theta}$ to minimise our measure of dissimilarity. We present two measures of dissimilarity: the Kullback–Leibler divergence (commonly known as the KL divergence) and the Wasserstein distance.

7.3.1 Kullback–Leibler Divergence

The KL divergence between two distribution, $f(\boldsymbol{x})$ and $g(\boldsymbol{x})$, is defined as

$$\mathrm{KL}(f \,\|\, g) = \mathbb{E}_f\left[\log\left(\frac{f}{g}\right)\right] = \begin{cases} \sum_{\boldsymbol{x}} f(\boldsymbol{x}) \log\left(\frac{f(\boldsymbol{x})}{g(\boldsymbol{x})}\right) \\ \int f(\boldsymbol{x}) \log\left(\frac{f(\boldsymbol{x})}{g(\boldsymbol{x})}\right) \mathrm{d}\boldsymbol{x}. \end{cases}$$

In this definition we take $f(\boldsymbol{x}) \log(f(\boldsymbol{x}))$ to be equal to 0 when $f(\boldsymbol{x}) = 0$. The KL divergence has two properties in common with a distance, namely,

- $\mathrm{KL}\left(f \,\|\, g\right) \geq 0$ (Gibbs' inequality);
- $\mathrm{KL}\left(f \,\|\, g\right) = 0$ if and only if $f(\boldsymbol{x}) = g(\boldsymbol{x})$ in all measurable regions.

These properties mean that the smaller the KL divergence the closer the two distributions are to each other. However, the measure is not symmetric (i.e. in general $\mathrm{KL}\left(f \,\|\, g\right) \neq \mathrm{KL}\left(g \,\|\, f\right)$). We discuss the difference between $\mathrm{KL}\left(f \,\|\, g\right)$ and $\mathrm{KL}\left(g \,\|\, f\right)$ later in this section. It also does not satisfy the triangular inequality,

$$\mathrm{KL}\left(f \,\|\, g\right) + \mathrm{KL}\left(g \,\|\, h\right) \leq \mathrm{KL}\left(f \,\|\, h\right) \qquad \text{is not generally true.}$$

As a consequence the KL divergence should be not regarded as a distance. (In mathematics, spaces with a proper distance measure – also known as a metric – have well-studied and useful properties. Alas, these don't apply the KL diverence.) In the next section we will consider the Wasserstein distance, which is a true distance.

We prove Gibbs' inequality assuming that the distributions are continuous (the generalisation to discrete variable is straightforward),

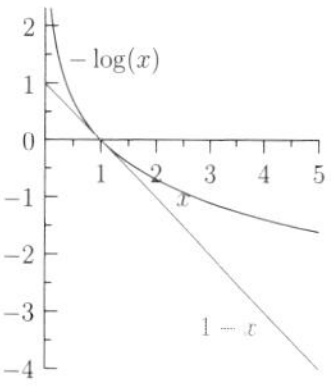

$$\begin{aligned}
\mathrm{KL}\left(f \,\|\, g\right) &\stackrel{(1)}{=} -\int \left[\!\left[f(\boldsymbol{x}) \neq 0 \right]\!\right] f(\boldsymbol{x}) \log\!\left(\frac{g(\boldsymbol{x})}{f(\boldsymbol{x})}\right) \mathrm{d}\boldsymbol{x} \\
&\stackrel{(2)}{\geq} \int \left[\!\left[f(\boldsymbol{x}) \neq 0 \right]\!\right] f(\boldsymbol{x}) \left(1 - \frac{g(\boldsymbol{x})}{f(\boldsymbol{x})}\right) \mathrm{d}\boldsymbol{x} \\
&\stackrel{(3)}{=} \int \left[\!\left[f(\boldsymbol{x}) \neq 0 \right]\!\right] \left(f(\boldsymbol{x}) - g(\boldsymbol{x})\right) \mathrm{d}\boldsymbol{x} \\
&\stackrel{(4)}{=} 1 - \int \left[\!\left[f(\boldsymbol{x}) \neq 0 \right]\!\right] g(\boldsymbol{x}) \,\mathrm{d}\boldsymbol{x} \stackrel{(5)}{=} \geq 0
\end{aligned}$$

(1) We write $\log\!\left(\frac{f}{g}\right) = -\log\!\left(\frac{g}{f}\right)$ with hindsight of the next step. As explained, we take $f(\boldsymbol{x}) \log\left(f(\boldsymbol{x})\right) = 0$ when $f(\boldsymbol{x}) = 0$. We make this explicit by putting in the indicator function $\left[\!\left[f(\boldsymbol{x}) \neq 0 \right]\!\right]$ (we will need this when we replace the logarithm with its bound).
(2) Using the inequality $-\log(z) \geq 1 - z$ with $z = \frac{g(\boldsymbol{x})}{f(\boldsymbol{x})}$.
(3) Multiply through by $f(\boldsymbol{x})$.
(4) Using $\int \left[\!\left[f(\boldsymbol{x}) \neq 0 \right]\!\right] f(\boldsymbol{x}) \,\mathrm{d}\boldsymbol{x} = \int f(\boldsymbol{x}) \,\mathrm{d}\boldsymbol{x} = 1$ (since $f(\boldsymbol{x})$ is a probability density).
(5) Since $\int \left[\!\left[f(\boldsymbol{x}) \neq 0 \right]\!\right] g(\boldsymbol{x}) \,\mathrm{d}\boldsymbol{x} \leq \int g(\boldsymbol{x}) \,\mathrm{d}\boldsymbol{x} = 1$.

When $g(\boldsymbol{x}) \neq f(\boldsymbol{x})$ then $-\log\!\left(\frac{g(\boldsymbol{x})}{f(\boldsymbol{x})}\right)$ is strictly greater than $1 - \frac{g(\boldsymbol{x})}{f(\boldsymbol{x})}$. If this happens on a measurable set (i.e. a set of points, $\mathcal{S}$, where $\int f(\boldsymbol{x}) \left[\!\left[\boldsymbol{x} \in \mathcal{S} \right]\!\right] \mathrm{d}\boldsymbol{x} > 0$) then the KL divergence will be strictly greater than 0. It is simple to see that

$$\mathrm{KL}\left(f \,\|\, f\right) = \mathbb{E}_f\!\left[\log\!\left(\frac{f}{f}\right)\right] = \mathbb{E}_f\left[\log(1)\right] = 0.$$

Note that if $g(\boldsymbol{x}) = 0$ when $f(\boldsymbol{x}) \neq 0$ then $\mathrm{KL}\left(f \,\|\, g\right)$ will diverge. This can be a serious problem when using the KL divergence in some applications.

Example 7.13 KL Divergence between Normal Distributions
The KL divergence between $f_1(\boldsymbol{x}) = \mathcal{N}(\boldsymbol{\mu}_1, \boldsymbol{\Sigma}_1)$ and $f_2(\boldsymbol{x}) = \mathcal{N}(\boldsymbol{\mu}_2, \boldsymbol{\Sigma}_2)$ is equal to

$$
\begin{aligned}
&\mathrm{KL}(f_1 \,\|\, f_2) \\
&\stackrel{(1)}{=} \int \mathcal{N}(\boldsymbol{x}|\boldsymbol{\mu}_1, \boldsymbol{\Sigma}_1) \log\left(\frac{\mathcal{N}(\boldsymbol{x}|\boldsymbol{\mu}_1, \boldsymbol{\Sigma}_1)}{\mathcal{N}(\boldsymbol{x}|\boldsymbol{\mu}_2, \boldsymbol{\Sigma}_2)}\right) \mathrm{d}\boldsymbol{x} \\
&\stackrel{(2)}{=} \int \mathcal{N}(\boldsymbol{x}|\boldsymbol{\mu}_1, \boldsymbol{\Sigma}_1) \left(-\frac{1}{2}(\boldsymbol{x}-\boldsymbol{\mu}_1)^\mathsf{T}\boldsymbol{\Sigma}_1^{-1}(\boldsymbol{x}-\boldsymbol{\mu}_1) - \frac{1}{2}\log\left((2\pi)^n |\boldsymbol{\Sigma}_1|\right)\right. \\
&\qquad \left. + \frac{1}{2}(\boldsymbol{x}-\boldsymbol{\mu}_2)^\mathsf{T}\boldsymbol{\Sigma}_2^{-1}(\boldsymbol{x}-\boldsymbol{\mu}_2) + \frac{1}{2}\log\left((2\pi)^n |\boldsymbol{\Sigma}_2|\right)\right) \mathrm{d}\boldsymbol{x} \\
&\stackrel{(3)}{=} -\frac{n}{2} + \frac{1}{2}\mathrm{Tr}\,\boldsymbol{\Sigma}_2^{-1}\boldsymbol{\Sigma}_1 + \frac{1}{2}(\boldsymbol{\mu}_1-\boldsymbol{\mu}_2)^\mathsf{T}\boldsymbol{\Sigma}_2^{-1}(\boldsymbol{\mu}_1-\boldsymbol{\mu}_2) + \frac{1}{2}\log\left(\frac{|\boldsymbol{\Sigma}_2|}{|\boldsymbol{\Sigma}_1|}\right)
\end{aligned}
$$

(1) By definition.
(2) Taking the logarithm of the normal density function.
(3) Using $\boldsymbol{a}^\mathsf{T}\mathbf{M}\,\boldsymbol{a} = \mathrm{Tr}\,\mathbf{M}\,\boldsymbol{a}\,\boldsymbol{a}^\mathsf{T}$ with $a = \boldsymbol{x} - \boldsymbol{\mu}_i$ and $\mathbf{M} = \boldsymbol{\Sigma}_i^{-1}$ together with $\int \mathcal{N}(\boldsymbol{x}|\boldsymbol{\mu}_1, \boldsymbol{\Sigma}_1)\,(\boldsymbol{x}-\boldsymbol{\mu}_1)(\boldsymbol{x}-\boldsymbol{\mu}_1)^\mathsf{T}\,\mathrm{d}\boldsymbol{x} = \boldsymbol{\Sigma}_1$ and $\int \mathcal{N}(\boldsymbol{x}|\boldsymbol{\mu}_1, \boldsymbol{\Sigma}_1)\,(\boldsymbol{x}-\boldsymbol{\mu}_2)(\boldsymbol{x}-\boldsymbol{\mu}_2)^\mathsf{T}\,\mathrm{d}\boldsymbol{x} = \boldsymbol{\Sigma}_1 + (\boldsymbol{\mu}_2-\boldsymbol{\mu}_1)(\boldsymbol{\mu}_2-\boldsymbol{\mu}_1)^\mathsf{T}$. Finally, $\mathrm{Tr}\,\boldsymbol{\Sigma}_1^{-1}\boldsymbol{\Sigma}_1 = \mathrm{Tr}\,\mathbf{I} = n$ where n is dimensional of the space.

$\boldsymbol{a}^\mathsf{T}\mathbf{M}\boldsymbol{a} = \sum_{i,j} a_i\, M_{ij}\, a_j$

$\mathrm{Tr}\,\mathbf{A} = \sum_i A_{ii}$

$\mathrm{Tr}\,\mathbf{A}\,\mathbf{B} = \sum_{i,j} A_{ij}\, B_{ji}$

$\mathrm{Tr}\,\mathbf{M}\,\boldsymbol{a}\,\boldsymbol{a}^\mathsf{T} = \sum_{i,j} M_{ij}\, a_i\, a_j$

In the case of two univariate normal distributions

$$
\mathrm{KL}\left(\mathcal{N}(\mu_1, \sigma_1^2) \,\middle\|\, \mathcal{N}(\mu_2, \sigma_2^2)\right) = \frac{\sigma_1^2 - \sigma_2^2 + (\mu_1 - \mu_2)^2}{2\,\sigma_2^2} + \log\left(\frac{\sigma_2}{\sigma_1}\right).
$$

(Note that $\boldsymbol{\Sigma}_1$ and $\boldsymbol{\Sigma}_2$ are covariance matrices while σ_1 and σ_2 are standard deviations, hence the factor of 2 in the last term.) We show some example, of KL divergences between different normal distributions in Figure 7.14. Note the lack of symmetry.

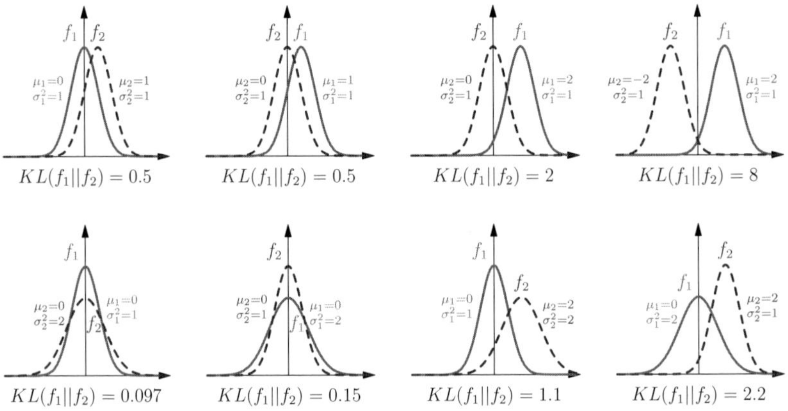

Figure 7.14 Examples of KL divergences between different normal distributions.

One of the main uses of KL divergence is to find a simple distribution $g(\boldsymbol{x}|\boldsymbol{\theta})$ that is close to some given distribution $f(\boldsymbol{x})$. To achieve this, we can choose

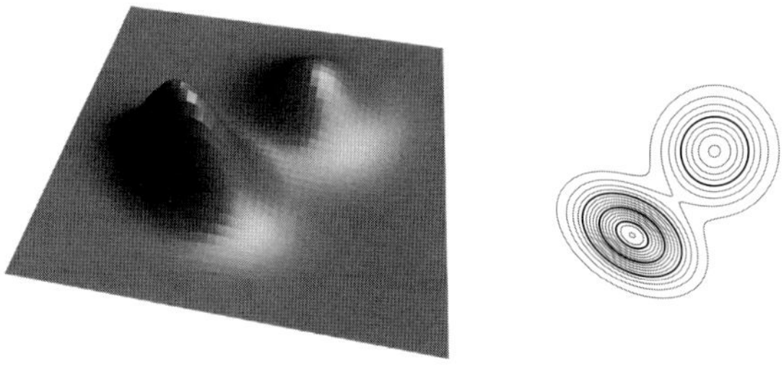

Figure 7.15 Probability density $f(\boldsymbol{x})$ that we wish to approximate. The left figure shows a three-dimensional representation where the height represents the probability density. The figure on the right shows a contour plot of the same density.

$\boldsymbol{\theta}$ to minimise the KL divergence. We can use the KL divergence either way around, but we will get different results. To illustrate this we consider the problem of fitting the distribution shown in Figure 7.15 by a two-dimensional normal distribution $g(\boldsymbol{x}|\boldsymbol{\theta}) = \mathcal{N}(\boldsymbol{x}|\boldsymbol{\mu}, \boldsymbol{\Sigma})$.

We consider first the case of minimising

$$\mathrm{KL}\left(g(\boldsymbol{x}|\boldsymbol{\theta}) \,\|\, f(\boldsymbol{x})\right) = \mathbb{E}_g\left[\log\left(g(\boldsymbol{x}|\boldsymbol{\theta})\right)\right] - \mathbb{E}_g\left[\log\left(f(\boldsymbol{x})\right)\right].$$

The first term is equal to the negative entropy of $g(\boldsymbol{x}|\boldsymbol{\theta})$ (we discuss entropy in considerable detail in Chapter 9). This term is reduced when $g(\boldsymbol{x}|\boldsymbol{\theta})$ becomes more widely distributed (i.e. the random variable $\boldsymbol{X} \sim g$ becomes more uncertain). The second term is large if $f(\boldsymbol{x})$ is small in a region where $g(\boldsymbol{x}|\boldsymbol{\theta})$ has substantial probability mass. As a consequence, if $g(\boldsymbol{x}|\boldsymbol{\theta})$ is confined to be unimodal (in our example, we enforce this by making it normally distributed), then it will tend to sit on a region where $f(\boldsymbol{x})$ has large values (so that $-\log\left(f(\boldsymbol{x})\right)$ is small). We show the best fit in Figure 7.16(a).

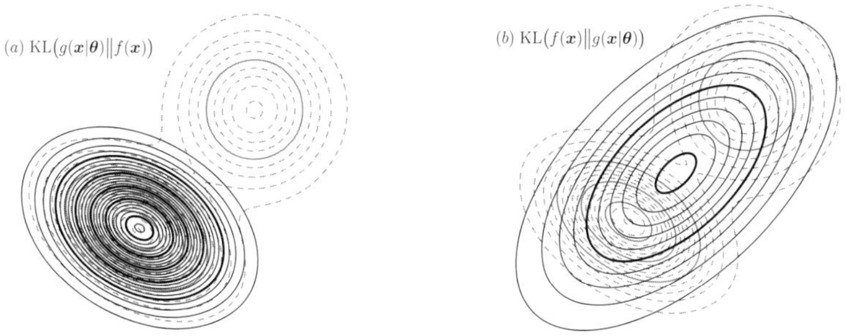

Figure 7.16 Fitting a normal distribution, $g(\boldsymbol{x}|\boldsymbol{\theta}) = \mathcal{N}(\boldsymbol{x}|\boldsymbol{\mu}, \boldsymbol{\Sigma})$, to the probability density $f(\boldsymbol{x})$ shown in Figure 7.15 by choosing the parameters to for figures (a) $\mathrm{KL}\left(g \,\|\, f\right)$ and (b) $\mathrm{KL}\left(f \,\|\, g\right)$.

The alternative is to minimise

$$\mathrm{KL}\left(f \,\|\, g\right) = \mathbb{E}_f\left[\log\left(f(\boldsymbol{x})\right)\right] - \mathbb{E}_f\left[\log\left(g(\boldsymbol{x}|\boldsymbol{\theta})\right)\right].$$

The first term is independent of $g(\boldsymbol{x}|\boldsymbol{\theta})$ so that

$$\underset{\boldsymbol{\theta}}{\operatorname{argmin}}\, \mathrm{KL}\left(f(\boldsymbol{x}) \,\|\, g(\boldsymbol{x}|\boldsymbol{\theta})\right) = \underset{\boldsymbol{\theta}}{\operatorname{argmin}} -\mathbb{E}_f\left[\log\left(g(\boldsymbol{x}|\boldsymbol{\theta})\right)\right].$$

That is, to compute the optimum parameter values we only need to consider the second term. In this case, we don't want $g(\boldsymbol{x}|\boldsymbol{\theta})$ to be small in regions where $f(\boldsymbol{x})$ has substantial probability mass. That is, we want to choose the parameter $\boldsymbol{\theta}$

so that $g(\boldsymbol{x}|\boldsymbol{\theta})$ covers all areas where $f(\boldsymbol{x})$ is relatively large. We show the best parameter fit for the density function, $f(\boldsymbol{x})$, in Figure 7.16(b).

We can symmetrise the KL divergence. This is known as the Jensen–Shannon (JS) divergence

$$JS(f\|g) = \mathrm{KL}\left(f \,\middle\|\, \frac{f+g}{2}\right) + \mathrm{KL}\left(g \,\middle\|\, \frac{f+g}{2}\right).$$

This also doesn't suffer from diverging if either $f(\boldsymbol{x}) = 0$ or $g(\boldsymbol{x}) = 0$ and so is finite everywhere. Furthermore, the square root of $JS(f\|g)$ even satisfies the triangular inequality, thus being a proper distance or metric. However, the JS divergence is rather complicated to compute in practice and is not as commonly used as the KL divergence.

7.3.2 Wasserstein Distance

Another measure of the dissimilarity between distributions that has recently found a growing number of applications is the Wasserstein distance. There is a family of such distances $W_p(f,g)$, although we will mainly concentrate on $W_1(f,g)$, which is known also known as the *earth-mover's distance*. $W_p(f,g)$ satisifes all the properties of a distance or metric:

- $W_p(f,g) \geq 0$ (non-negativity),
- $W_p(f,g) = 0 \Leftrightarrow f = g$ (identity of indiscernibles),
- $W_p(f,g) = W_p(g,f)$ (symmetry),
- $W_p(f,g) + W_p(g,h) \leq W_p(f,h)$ (triangular inequality).

The definition of the Wasserstein distance is rather subtle. We need to consider the family, Γ, of joint probability distributions, $\gamma(\boldsymbol{x},\boldsymbol{y})$, that satisfies the constraints

$$\int \gamma(\boldsymbol{x},\boldsymbol{y})\,\mathrm{d}\boldsymbol{y} = f(\boldsymbol{x}) \qquad \int \gamma(\boldsymbol{x},\boldsymbol{y})\,\mathrm{d}\boldsymbol{x} = g(\boldsymbol{y}).$$

That is, the two marginal distributions of $\gamma(\boldsymbol{x},\boldsymbol{y})$ are the two distributions whose distance from one another we are measuring. Let $d(\boldsymbol{x},\boldsymbol{y})$ be a distance measure between $\boldsymbol{x}$ and $\boldsymbol{y}$ (usually this is just taken to be the Euclidean distance), then

$$W_p(f,g) = \min_{\gamma\in\Gamma} \mathbb{E}_\gamma\left[d^p(\boldsymbol{x},\boldsymbol{y})\right].$$

That is, we find the joint probability distribution $\gamma^*(\boldsymbol{x},\boldsymbol{y})$ with the correct marginals that minimise the expected distance $d(\boldsymbol{x},\boldsymbol{y})$ to the power p.

To get an intuition about the Wasserstein distance we restrict ourselves to the case $p = 1$. In this case $\gamma(\boldsymbol{x},\boldsymbol{y})$ can be viewed as a *transport policy* that tells us how to move probability mass/density from $f(\boldsymbol{x})$ to transform it to $g(\boldsymbol{x})$. We illustrate this for two univariate probability masses in Figure 7.17.

For each pair, $(\boldsymbol{x},\boldsymbol{y})$, we can view $\gamma(\boldsymbol{x},\boldsymbol{y})$ as the amount of probability mass we move from $f(\boldsymbol{x})$ to construct $g(\boldsymbol{y})$. The work required to transform $f(\boldsymbol{x})$ into

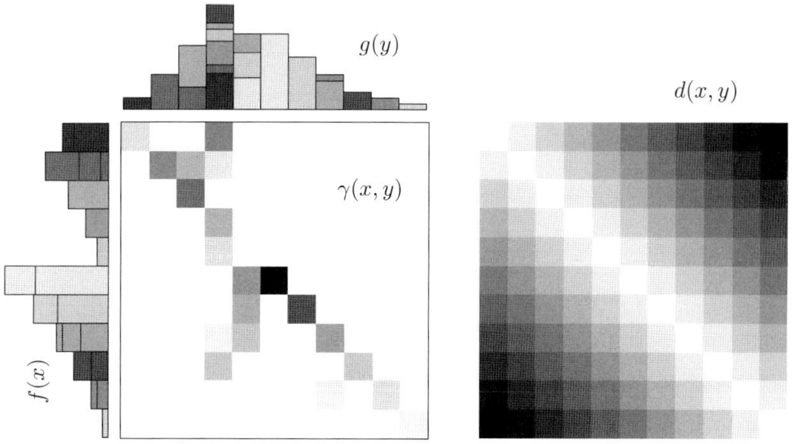

Figure 7.17 Illustration of an optimal transport policy for finding the Wasserstein distance. The joint distribution $\gamma(x, y)$ shows how to move probability mass from $f(x)$ to create $g(y)$. We also show the corresponding distance $d(x, y)$.

$g(\boldsymbol{y})$ is equal to probability mass times the distance moved. That is, we can define a transportation cost

$$c(\gamma) = \mathbb{E}_{\gamma}\left[d(\boldsymbol{x}, \boldsymbol{y})\right].$$

The Wasserstein, $W_1(f, g)$, or earth-mover's distance is equal to $c(\gamma^*)$ where the joint probability function $\gamma^*(\boldsymbol{x}, \boldsymbol{y})$ defines the optimal transport policy that minimises the cost. We note that $W_1(f, g) \geq 0$, since we are taking the expectation of a positive quantity (the distance). If $g = f$ then $\gamma(\boldsymbol{x}, \boldsymbol{y}) = f(x)\,[\![x = y]\!]$. That is, all the probability mass will be along the diagonal, but $d(\boldsymbol{x}, \boldsymbol{x}) = 0$ so $W_1(f, f) = 0$. If $f(\boldsymbol{x}) \neq g(\boldsymbol{x})$ at any point $\boldsymbol{x}$ then some mass needs to moved to transform $f(\boldsymbol{x})$ into $g(\boldsymbol{x})$, thus $W_1(f, g) = 0$ implies $f = g$. We note that the optimal transport policy for $W_1(g, f)$ is $\gamma^*(\boldsymbol{y}, \boldsymbol{x})$ (the transpose of $\gamma^*(\boldsymbol{x}, \boldsymbol{y})$) but as $d(\boldsymbol{x}, \boldsymbol{y})$ is a proper distance it is symmetric, from which it follows that $W_1(f, g) = W_1(g, f)$. Finally, we can see that if we transform $f(\boldsymbol{x})$ to $g(\boldsymbol{x})$ and then $g(\boldsymbol{x})$ into $h(\boldsymbol{x})$ this is going to take at least as much work as transforming $f(\boldsymbol{x})$ directly to $h(\boldsymbol{x})$ (since we always choose the optimal transport policy). Thus the Wasserstein distance satisfies all the conditions of a proper distance.

More importantly, the Wasserstein distance behaves linearly in the sense that

$$W_1(f(\boldsymbol{x}), f(\boldsymbol{y} - \boldsymbol{c})) = d(\boldsymbol{0}, \boldsymbol{c}).$$

In contrast, the KL depends on the relative probability. The KL divergence between two normal distributions with the same variance but different means will be equal to $(\mu_2 - \mu_1)^2/2\sigma^2$. If σ^2 is very small then the KL divergence can be very large even when the two distributions are shifted by a small amount. For sharply concentrated distributions in high dimensions the KL divergence can be very large, even when the distributions are rather close together. We illustrate such distributions in Figure 7.18.

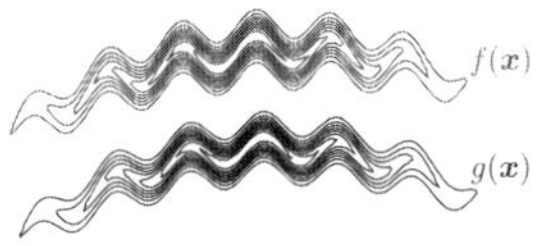

Figure 7.18 Example of two two-dimensional distributions (shown as a contour plot) with a relatively small Wasserstein distance, but a large KL divergence.

Example 7.14 Wasserstein Distance between Normal Distributions
The Wasserstein distance is usually not that easy to compute in closed form. For normal distributions we can compute the Wasserstein-2 distance, W_2. For two normals $f_1(\boldsymbol{x}) = \mathcal{N}(\boldsymbol{\mu}_1, \boldsymbol{\Sigma}_1)$ and $f_2(\boldsymbol{x}) = \mathcal{N}(\boldsymbol{\mu}_2, \boldsymbol{\Sigma}_2)$, this is equal to

$$W_2(f_1, f_2) = \|\boldsymbol{\mu}_1 - \boldsymbol{\mu}_2\|^2 + \mathrm{Tr}\left(\boldsymbol{\Sigma}_1 + \boldsymbol{\Sigma}_2 - 2\,(\boldsymbol{\Sigma}_2^{1/2}\boldsymbol{\Sigma}_1\boldsymbol{\Sigma}_2^{1/2})^{1/2}\right).$$

For univariate normal distributions

$$W_2\left(\mathcal{N}(\mu_1, \sigma_1^2), \mathcal{N}(\mu_2, \sigma_2^2)\right) = (\mu_1 - \mu_2)^2 + \sigma_1^2 + \sigma_1^2 - 2\,\sigma_1\,\sigma_2.$$

If $\sigma_1 = \sigma_2$ then the $W_1(f_1, f_2) = |\mu_2 - \mu_1|$, while if $\mu_1 = \mu_2$ then $W_1(f_1, f_2) = \sqrt{2/\pi}\,|\sigma_1 - \sigma_2|$. Otherwise, W_1 is rather complicated to compute. In Figure 7.19 we show the Wasserstein distances W_1 and W_2 between different univariate normal distributions.

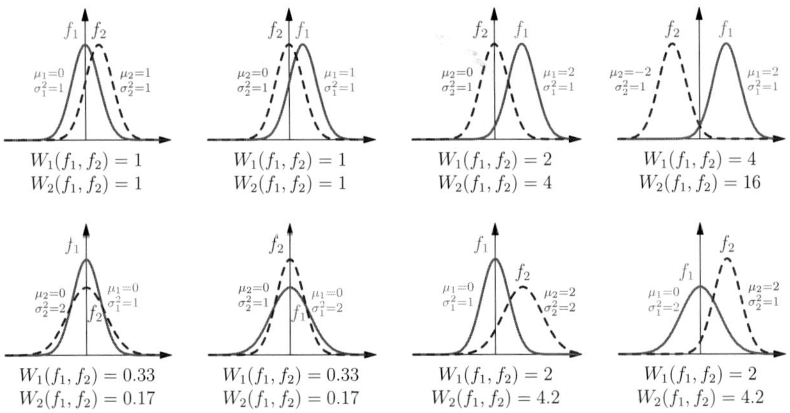

Figure 7.19 Examples of W_1 and W_2 distances between different normal distributions. Compare this with the KL distances in Example 7.13 on page 168.

Unfortunately, Wasserstein distance is not that easy to compute. For a discrete probability distribution, the problem is a classic linear programming task. That is, we need to optimise a linear objective

$$\sum_{i,j} d(\boldsymbol{x}_i, \boldsymbol{y}_i)\,\gamma(\boldsymbol{x}_i, \boldsymbol{y}_j)$$

subject to the linear constraints

$$\forall_i \sum_j \gamma(\boldsymbol{x}_i, \boldsymbol{y}_j) = f(\boldsymbol{x}_i) \qquad \forall_j \sum_i \gamma(\boldsymbol{x}_i, \boldsymbol{y}_j) = g(\boldsymbol{y}_j) \qquad \forall_{ij}\ \gamma(\boldsymbol{x}_i, \boldsymbol{y}_j) \geq 0.$$

Note that the normalisation constraint on $\gamma(\boldsymbol{x}_i, \boldsymbol{y}_j)$ is guaranteed by the marginalisation conditions. When the number of states that our random variable can take is *not* too large then we can solve this problem using a standard linear programming solver. However, this is a slow procedure and for a complex task it can be impractical.

Fortunately, linear programs can be solved in a dual form that in this case makes the problem more efficient. To obtain this dual form we write down the constrained optimisation problem as an unconstrained optimisation problem using the introduction of Lagrange multipliers (see Appendix 7.C). In our case we can impose the marginalisation constraints by introducing Lagrange multipliers $\alpha(\boldsymbol{x}_i)$ and $\beta(\boldsymbol{y}_j)$ to form a Lagrangian

$$\mathcal{L} = \sum_{i,j} d(\boldsymbol{x}_i, \boldsymbol{y}_j)\,\gamma(\boldsymbol{x}_i, \boldsymbol{y}_j) - \sum_i \alpha(\boldsymbol{x}_i)\left(\sum_j \gamma(\boldsymbol{x}_i, \boldsymbol{y}_j) - f(\boldsymbol{x}_i)\right)$$
$$- \sum_j \beta(\boldsymbol{y}_j)\left(\sum_i \gamma(\boldsymbol{x}_i, \boldsymbol{y}_j) - g(\boldsymbol{y}_j)\right)$$

where we still impose the constraint that $\gamma(\boldsymbol{x}_i, \boldsymbol{y}_j) \geq 0$. Although $\alpha(\boldsymbol{x}_i)$ and $\beta(\boldsymbol{y}_j)$ look rather unusual as Lagrange multipliers (since they look like functions), they are just numbers at the value that our random variables can take. The solution to the constrained optimisation problem is given by the extemum value problem

$$W_1(f, g) = \max_{\alpha(\boldsymbol{x}_i), \beta(\boldsymbol{y}_j)} \; \min_{\gamma(\boldsymbol{x}_i, \boldsymbol{y}_j)} \mathcal{L}.$$

To obtain the dual problem we rearrange the Lagrangian as

$$\mathcal{L} = \sum_i \alpha(\boldsymbol{x}_i)\, f(\boldsymbol{x}_i) + \sum_j \beta(\boldsymbol{y}_j)\, g(\boldsymbol{y}_j) - \sum_{i,j} \gamma(\boldsymbol{x}_i, \boldsymbol{y}_j)\, \big(\alpha(\boldsymbol{x}_i) + \beta(\boldsymbol{y}_j) - d(\boldsymbol{x}_i, \boldsymbol{y}_j)\big).$$

(Observe that there is a one-to-one correspondence between terms in this expression and terms in the expression above.) We can now interpret $\gamma(\boldsymbol{x}_i, \boldsymbol{y}_j)$ as a Lagrange multiply that imposes the constraint

$$\alpha(\boldsymbol{x}_i) + \beta(\boldsymbol{y}_j) \leq d(\boldsymbol{x}_i, \boldsymbol{y}_j).$$

(Note that because $\gamma(\boldsymbol{x}_i, \boldsymbol{y}_j) \geq 0$ this turns the constraint into an inequality constraint rather than an equality constraint.) We are left with a maximisation problem

$$W_1(f, g) = \max_{\alpha(\boldsymbol{x}_i), \beta(\boldsymbol{y}_j)} \sum_i \alpha(\boldsymbol{x}_i)\, f(\boldsymbol{x}_i) + \sum_j \beta(\boldsymbol{y}_j)\, g(\boldsymbol{y}_j)$$

subject to the constraint above. The constraint takes a very simple form when we choose $\boldsymbol{y}_j = \boldsymbol{x}_i$, then

$$\alpha(\boldsymbol{x}_i) + \beta(\boldsymbol{x}_i) \leq 0.$$

That is, $\beta(\boldsymbol{x}_i) \leq -\alpha(\boldsymbol{x}_i)$ or $\beta(\boldsymbol{x}_i) = -\alpha(\boldsymbol{x}_i) - \epsilon(\boldsymbol{x}_i)$ where $\epsilon(\boldsymbol{x}_i) \geq 0$. Putting this into the objective function

$$W_1(f, g) = \max_{\alpha(\boldsymbol{x}_i), \epsilon(\boldsymbol{y}_j)} \sum_i \alpha(\boldsymbol{x}_i) \, \big(f(\boldsymbol{x}_i) - g(\boldsymbol{x}_i)\big) - \sum_j \epsilon(\boldsymbol{y}_j) \, g(\boldsymbol{y}_j).$$

However, since both $\epsilon(\boldsymbol{y}_j) \geq 0$ and $g(\boldsymbol{x}_i) \geq 0$, the optimal value of $\epsilon(\boldsymbol{y}_j)$ is 0 and the dual problem simplifies to

$$W_1(f, g) = \max_{\alpha(\boldsymbol{x}_i)} \sum_i \alpha(\boldsymbol{x}_i) \, \big(f(\boldsymbol{x}_i) - g(\boldsymbol{x}_i)\big) = \max_{\alpha(\boldsymbol{x})} \big(\mathbb{E}_f\left[\alpha(\boldsymbol{x})\right] - \mathbb{E}_g\left[\alpha(\boldsymbol{x})\right]\big)$$

subject to the constraint

$$\alpha(\boldsymbol{x}_i) - \alpha(\boldsymbol{y}_j) \leq d(\boldsymbol{x}_i, \boldsymbol{y}_j).$$

Although we assumed that our probability distributions, $f(\boldsymbol{x}_i)$ and $g(\boldsymbol{y}_j)$, describe discrete random variables, we can consider the limit where the random variables become continuous (i.e. the gap between $\boldsymbol{x}_i$ and $\boldsymbol{x}_{i+1}$ goes to zero). In this limit the dual problem becomes

$$W_1(f, g) = \max_{\alpha(\boldsymbol{x})} \int \alpha(\boldsymbol{x}) \, \big(f(\boldsymbol{x}) - g(\boldsymbol{x})\big) \, \mathrm{d}\boldsymbol{x} = \max_{\alpha(\boldsymbol{x})} \big(\mathbb{E}_f\left[\alpha(\boldsymbol{x})\right] - \mathbb{E}_g\left[\alpha(\boldsymbol{x})\right]\big)$$

subject to $\alpha(\boldsymbol{x}) - \alpha(\boldsymbol{y}) \leq d(\boldsymbol{x}, \boldsymbol{y})$. This inequality on $\alpha(\boldsymbol{x})$ is called a Lipschitz-1 condition. (The *Lipschitz condition* on a function $f(x)$ is that

$$|f(\boldsymbol{x}) - f(\boldsymbol{y})| \leq M \, \|\boldsymbol{x} - \boldsymbol{y}\|.$$

For the function $\alpha(\boldsymbol{x})$ we require $M = 1$. This condition is often imposed as a smoothness condition on functions.)

Example 7.15 Wasserstein GAN

Generative adversarial networks (GANs) are one of the most exciting developments in deep learning (Goodfellow et al., 2014). GANs are unsupervised methods for generating images. They consist of two networks. The first is a generator that takes a random vector, z, and generates an image $\hat{\boldsymbol{x}} = g(z|\boldsymbol{\theta})$. The second network is a discriminator, $D(\boldsymbol{x}|\boldsymbol{\phi})$, that has to decide whether the image, $\boldsymbol{x}$, being generated is a fake image (from the generator) or a true image (from some data set of images) – it outputs a number between 0 and 1 where 1 denotes a true image and 0 a fake image. The $\boldsymbol{\theta}$ and $\boldsymbol{\phi}$ vectors denote parameters (weights in the neural networks) that are chosen to optimise the objective of the networks. In traditional GANs the discriminator is rewarded for distinguishing between the true images and the generated images, while the generator is rewarded for deceiving the discriminator. One reason that GANs have attracted so much attention is that they don't require the training set to be labelled, yet nevertheless the discriminator and generator learn a very high-level representation of the data set. Although this scenario provides very impressive results, GANs are tricky to train.

To solve some of the difficulties involved in training traditional GANs a new approach has been proposed by Arjovsky et al. (2017) based on minimising the Wasserstein distance between the probability distribution, $P(\boldsymbol{x})$, of images in a database and the probability distribution, $Q(g(\boldsymbol{z}|\boldsymbol{\theta}))$, for images generated by the generator. The Wasserstein distance in its dual form is

$$W_1(P,Q) = \max_{\alpha(\boldsymbol{x})} \mathbb{E}_P\big[\alpha(\boldsymbol{x})\big] - \mathbb{E}_Q\big[\alpha(\boldsymbol{x})\big]$$

where $\alpha(\boldsymbol{x})$ is a Lipschitz-1 function. The function $\alpha(\boldsymbol{x})$ acts as a discriminator – that is, it is chosen to maximise the discrepency between its expected response to true images (from distribution $P(\boldsymbol{x})$) and images produced by the generator (with a distribution $Q(g(\boldsymbol{z}|\boldsymbol{\theta}))$. We choose $\alpha(\boldsymbol{x}|\boldsymbol{\phi})$ to be a neural network with parameters $\boldsymbol{\phi}$ that are choosen to maximise $\mathbb{E}_P\big[\alpha(\boldsymbol{x})\big] - \mathbb{E}_Q\big[\alpha(\boldsymbol{x})\big]$ subject to the constraint that it is Lipschitz-1. To enforce (or, at least, encourage) this constraint we add a penalty term that punishes the absolute size of gradient in the output as a function of the input when it is greater than one (in fact, it is slightly more efficient when we force this gradient to always be one) (Gulrajani et al., 2017). We therefore simultanously try to learn $\alpha(\boldsymbol{x}|\boldsymbol{\phi})$ (to maximise $\mathbb{E}_P\big[\alpha(\boldsymbol{x})\big] - \mathbb{E}_Q\big[\alpha(\boldsymbol{x})\big]$) at the same time as choosing the parameters $\boldsymbol{\theta}$ of the generator $Q(g(\boldsymbol{z}|\boldsymbol{\theta}))$. These two sets of parameters are learnt using standard gradient descent algorithms.

Mathematical proofs involving probabilities require a dedicated tool set. Mathematicians worry a lot about convergences. However, in my experience these results are not always as useful as they at first appear. Very often the interesting phenomenon occur because $n \ll \infty$. On the other hand, inequalities have very wide-ranging uses. They are not conceptually difficult, but can lead to quick and neat proofs. By far the most common are the Cauchy–Schwarz and Jensen's inequalities, which have applications far beyond probability theory. The Markov, Chebyshev, and Chernoff inequalities are more specialised, but can be extremely powerful. In certain domains, such as proofs of algorithmic complexity and proofs of learning bounds, they are used very heavily. Another common task is to measure the dissimilarity between probability distributions. There are many such measures. One of the most widely used is the KL divergence. Another measure that has found a growing number of applications is the Wasserstein distance.

Additional Reading

A nice introduction to inequalities is by Steele (2004), although its focus is much broader than probability. Tail-bound inequalities are covered in great length in the books by Boucheron et al. (2013) and Dubhashi and Panconesi (2009).

Exercise for Chapter 7

Exercise 7.1 (answer on page 414)

Use the Cauchy–Schwarz inequality to show that Pearson's correlation

$$\rho = \frac{\mathbb{E}\left[(X - \mathbb{E}\left[X\right])(Y - \mathbb{E}\left[Y\right])\right]}{\sqrt{\mathbb{V}\mathrm{ar}\left[X\right]\ \mathbb{V}\mathrm{ar}\left[Y\right]}} = \frac{\mathbb{C}\mathrm{ov}\left[X, Y\right]}{\sqrt{\mathbb{V}\mathrm{ar}\left[X\right]\ \mathbb{V}\mathrm{ar}\left[Y\right]}}$$

satisfies $-1 \leq \rho \leq 1$.

Exercise 7.2 (answer on page 414)

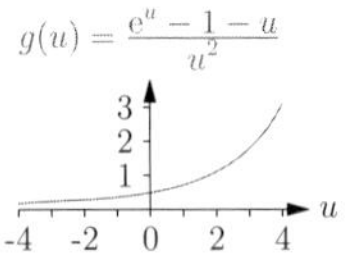

Using the fact that $g(u) = (\mathrm{e}^u - 1 - u)/u^2$ is monotonically increasing, show that if $X_i - \mathbb{E}\left[X_i\right] \leq b$ then

$$\mathbb{E}\left[\mathrm{e}^{\lambda\,(X_i - \mathbb{E}[X_i])}\right] \leq 1 + \frac{\mathbb{V}\mathrm{ar}\left[X_i\right]}{b^2}(\mathrm{e}^{b\,\lambda} - 1 - b\,\lambda).$$

Assuming X_i are independent variables and defining $S_n = \sum_{i=1}^{n} X_i$ and $\mu = \mathbb{E}\left[S_n\right]$ and using the fact that

$$\mathbb{P}\left(S_n \leq \mu + t\right) \leq \mathrm{e}^{-\psi(t)}, \qquad \psi(t) = \max_{\lambda} \lambda\,t - \log\!\left(\mathbb{E}\left[\mathrm{e}^{\lambda\,(S_n - \mu)}\right]\right)$$

derive Bennett's inequality

$$\mathbb{P}\left(S_n \leq \mu + t\right) \leq \exp\left(-\frac{\sigma^2}{b^2}\,h\left(\frac{b\,t}{\sigma^2}\right)\right)$$

where $\sigma^2 = \mathbb{V}\mathrm{ar}\left[S_n\right]$ and $h(u) = (1 + u)\,\log(1 + u) - u$.

Exercise 7.3 (answer on page 417)

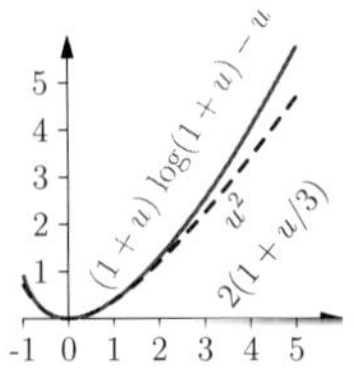

Use Taylor's expansion to second order to show that

$$g(u) = (1 + u)\,\log(1 + u) - u - \frac{u^2}{2(1 + u/3)} \geq 0,$$

hence derive Bernstein's inequality from Bennett's inequality.

Exercise 7.4 (answer on page 418)

In Example 7.3 on page 143, we showed, using the Cauchy–Schwarz inequality, that variances have to be positive. Use Jensen's inequality to obtain the same result.

Exercise 7.5 (answer on page 418)

One way of defining the median, m, of a distribution is as the value that minimises the mean absolute error

$$m = \underset{c}{\mathrm{argmin}}\,\mathbb{E}\left[|X - c|\right].$$

Use Jensen's inequality for the absolute function and for the square to show that the median satisfies the inequality

$$|\mu - m| \leq \sigma$$

where $\mu = \mathbb{E}\left[X\right]$ is the mean and σ the standard deviation of the distribution.

Exercise 7.6 (answer on page 419)

By considering the second derivative show that the generating function $G(\lambda) = \log\!\left(\mathbb{E}\left[\mathrm{e}^{\lambda X}\right]\right)$ is a convex function of λ.

Exercise 7.7 (answer on page 419)

Show that the negative logarithm of the partition function

$$G(\beta) = \log(Z), \qquad Z = \sum_{\boldsymbol{X}} \mathrm{e}^{-\beta E(\boldsymbol{X})}$$

is a convex-up function of β, where the Boltzmann probability of the random variable, $\boldsymbol{X}$, is given by

$$p(\boldsymbol{X}) = \frac{\mathrm{e}^{-\beta E(\boldsymbol{X})}}{Z}.$$

Show that for any function $G(\beta)$ (where $\beta = 1/(kT)$ – it is common to work in a systems of units where $k = 1$ so that $\beta = 1/T$, i.e. the inverse temperature)

$$\frac{\partial^2 T\, G(\frac{1}{T})}{\partial T^2} = \frac{1}{T^3} G''(\tfrac{1}{T})$$

and hence show that the free energy defined by

$$f = -kT \log\left(\sum_{\boldsymbol{X}} \mathrm{e}^{-E(\boldsymbol{X})/kT} \right)$$

is a concave (convex-down) function of T.

Exercise 7.8 (answer on page 421)

Use Jensen's inequality to show that the KL divergence

$$\mathrm{KL}\left(f \,\|\, q\right) = -\sum_i f_i \log\left(\frac{q_i}{f_i}\right)$$

is greater or equal to zero (the variables f_i and q_i are probabilities).

Appendix 7.A Taylor Expansion

The Taylor series is a great source for finding bounds on functions since for any suitably differentiable function, $f(x)$, its expansion to order n around x_0 is given by

$$f(x) = \sum_{i=0}^{n-1} \frac{(x - x_0)^i}{i!} f^{(i)}(x_0) + \frac{(x - x_0)^n}{n!} f^{(n)}(\theta) \tag{7.18}$$

where $f^{(i)}(x_0)$ is the ith derivative of $f(x)$ at the point x, and the last term is a reminder where θ takes some value in the interval between x and x_0. If we can bound the last term we can obtain a bound on $f(x)$.

For example, expanding the exponential around 0 to second order

$$\mathrm{e}^x = 1 + x + \frac{x^2}{2}\mathrm{e}^\theta$$

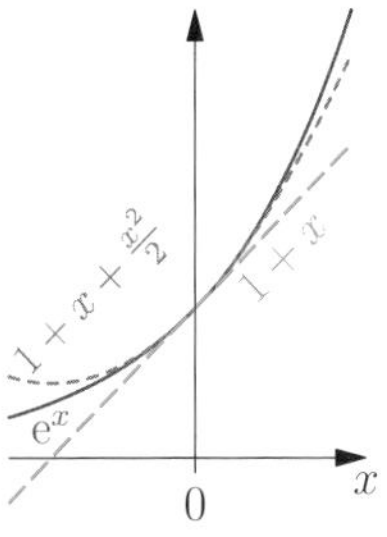

and since the last term is always positive $\mathrm{e}^x \leq 1 + x$. If we expand to third order

$$\mathrm{e}^x = 1 + x + \frac{x^2}{2} + \frac{x^3}{3!}\mathrm{e}^\theta.$$

Since $e^\theta > 0$, the last term is positive if $x > 0$ and negative if $x < 0$,

$$e^x \geq 1 + x + \frac{x^2}{2} \qquad \text{if } x \geq 0$$

$$e^x \leq 1 + x + \frac{x^2}{2} \qquad \text{if } x \leq 0.$$

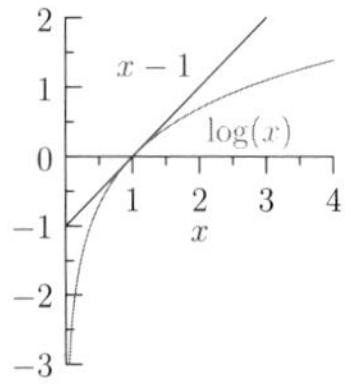

As another example consider Taylor expanding $\log(x)$ around $x = 1$

$$\log(x) = (x-1) + \frac{(x-1)^2}{2}\frac{1}{\theta^2}.$$

Again the last term is always positive (since we consider real x for which the logarithm is only defined if $x > 0$ so $\theta > 0$) then $\log(x) \leq x - 1$.

Taylor's expansion should be very familiar to anyone reading this book, but the residual term, R_n (essential for obtaining bounds) may be less familiar to readers not trained as mathematicians. To prove Equation (7.18) we integrate $f^{(n)}(t)$ n times (we use $f^{(n)}(t)$ to denote the n derivative of $f(t)$) and use the fundamental law of calculus (the integral acts as the anti-derivative). We show a few steps to see how the terms in the Taylor series emerge

$$\int_{x_0}^{t'} f^{(n)}(t)\,\mathrm{d}t = f^{(n-1)}(t') - f^{(n-1)}(x_0)$$

$$\begin{aligned}\int_{x_0}^{t''}\int_{x_0}^{t'} f^{(n)}(t)\,\mathrm{d}t\,\mathrm{d}t' &= \int_{x_0}^{t''} f^{(n-1)}(t') - f^{(n-1)}(x_0)\,\mathrm{d}t' \\ &= f^{(n-2)}(t'') - f^{(n-2)}(x_0) - (t - x_0)\,f^{(n-1)}(x_0)\end{aligned}$$

$$\begin{aligned}\int_{x_0}^{t'''}\int_{x_0}^{t''}\int_{x_0}^{t'} f^{(n)}(t)\,\mathrm{d}t\,\mathrm{d}t'\,\mathrm{d}t'' &= f^{(n-3)}(t''') - f^{(n-3)}(x_0) - (t''' - x_0)\,f^{(n-2)}(x_0) \\ &\quad - \frac{(t''' - x_0)^2}{2} f^{(n-1)}(x_0)\end{aligned}$$

$$\begin{aligned}\int_{x_0}^{x} \cdots \int_{x_0}^{t''}\int_{x_0}^{t'} f^{(n)}(t)\,\mathrm{d}t\,\mathrm{d}t' \ldots \mathrm{d}t^{n-1} &= f(x) - f(x_0) - (x - x_0)\,f'(x_0) - \cdots \\ &\quad - \frac{(x - x_0)^{n-1}}{(n-1)!} f^{(n-1)}(x_0).\end{aligned}$$

The left-hand side is R_n. On rearranging

$$f(x) = \sum_{i=0}^{n-1} \frac{(x - x_0)^i}{i!} f^{(i)}(x_0) + R_n,$$

where

$$R_n = \int_{x_0}^{x} \cdots \int_{x_0}^{t''}\int_{x_0}^{t'} f^{(n)}(t)\,\mathrm{d}t\,\mathrm{d}t' \ldots \mathrm{d}t^{n-1}.$$

The inner integral, $I = \int_{x_0}^{t'} f^{(n)}(t)\,\mathrm{d}t$, is usually quite complicated, however, we can bound it

$$f^{(n)}(t_{\min}) \int_{x_0}^{t'} \mathrm{d}t \leq \int_{x_0}^{t'} f^{(n)}(t)\,\mathrm{d}t \leq f^{(n)}(t_{\max}) \int_{x_0}^{t'} \mathrm{d}t$$

or

$$(t' - x_0)\, f^{(n)}(t_{\min}) \le I \le (t' - x_0)\, f^{(n)}(t_{\max})$$

where $t_{\max}$ and $t_{\min}$ are the values of t that maximise and minimise $f^{(n)}(t)$ in the interval $[x_0, x]$ (note that x is the upper limit that t' can take in the multiple integral R_n). Assuming that $f^{(n)}(t)$ is continuous in the interval from x to x_0 (this is a necessary condition for the Taylor expansion of this order to be a useful approximation), then there exists a θ such that $f^{(n)}(\theta)$ can take any value between $f^{(n)}(t_{\min})$ and $f^{(n)}(t_{\max})$. Thus we can bound the integral, I, as

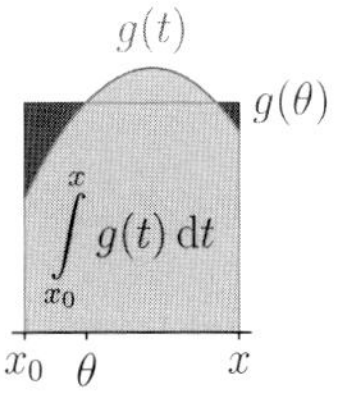

$$I = \int_{x_0}^{t'} f^{(n)}(t)\,\mathrm{d}t = (t' - x_0)\, f^{(n)}(\theta).$$

(The inequality $\int_a^b g(x)\,\mathrm{d}x \le (b-a)\,g(\theta)$, where $\theta \in [a, b]$, is known as the (integral form of the) *mean value theorem* and holds for any continuous function $g(x)$.) We can now easily compute the other integrals to find

$$R_n = \frac{(x - x_0)^n}{n!}\, f^{(n)}(\theta),$$

thereby completing the proof.

Appendix 7.B Hoeffding's Bound for Negatively Correlated Random Boolean Variables

We consider random Boolean variables, $\{X_i | i \in \mathcal{I}\}$ (where $\mathcal{I}$ is an index set), that are 'negatively correlated' in the sense that for any subset of $\mathcal{S} \subseteq \mathcal{I}$

$$\mathbb{E}\left[\prod_{i\in\mathcal{S}} X_i\right] \le \prod_{i\in\mathcal{S}} p_i$$

$$\mathbb{E}\left[\prod_{i\in\mathcal{S}} (1 - X_i)\right] \le \prod_{i\in\mathcal{S}} (1 - p_i)$$

where p_i are equal to the marginal probabilities $\mathbb{P}\left(X_i = 1\right)$. Negatively correlated variables arise in situations where we sample in a way which reduces the probability of all the events being 1 or 0. As we will show later in this subsection, this occurs when we sample without replacement. A typical result we can obtain is that for a weighted sum of negatively correlated binary random variables

$$S_n = \sum_{i\in\mathcal{I}} w_i\, X_i$$

where $w_i \ge 0$ and $n = |\mathcal{I}|$ is the number of variables, then

$$\mathbb{P}\left(|S_n - \mu_n| \ge t\right) \le \mathrm{e}^{-2t^2/\sum_{i\in\mathcal{I}} w_i^2} \qquad \text{with} \qquad \mu_n = \sum_{i\in\mathcal{I}} w_i\, p_i.$$

This is exactly the same bound as would apply to independent random Boolean variables.

The proof starts in a similar way to our derivation of the standard Hoeffding result for independent random variables. We use the standard Bernstein–Chernoff result that

$$\mathbb{P}\left(S_n - \mu_m > t\right) \le \mathrm{e}^{\hat{\psi}(t)}$$

where $\hat{\psi}(t)$ is the Legendre transform of $\psi(\lambda) = \log(\mathbb{E}\left[\mathrm{e}^{\lambda(S_n-\mu_n)}\right])$. That is,

$$\hat{\psi}(t) = \max_{\lambda>0} \quad \log\left(\mathbb{E}\left[\mathrm{e}^{\lambda(S_n-\mu_n)}\right]\right) - \lambda\, t.$$

Expanding the exponential of $\lambda\, S_n$ we find

$$\begin{aligned}
\mathrm{e}^{\lambda\, S_n} &\overset{(1)}{=} \mathrm{e}^{\lambda \sum_{i\in\mathcal{I}} w_i\, X_i} \overset{(2)}{=} \prod_{i\in\mathcal{I}} \mathrm{e}^{\lambda\, w_i\, X_i} \overset{(3)}{=} \prod_{i\in\mathcal{I}} \left(1 + X_i \left(\mathrm{e}^{\lambda\, w_i} - 1\right)\right) \\
&\overset{(4)}{=} \sum_{\mathcal{S}\in\mathcal{P}(\mathcal{I})} \prod_{i\in\mathcal{S}} X_i \left(\mathrm{e}^{\lambda\, w_i} - 1\right)
\end{aligned}$$

(1) Using the definition of S_n.
(2) Using $\mathrm{e}^{\sum_i a_i} = \prod_i \mathrm{e}^{a_i}$.
(3) As $X_i \in \{0,1\}$ we see that $\mathrm{e}^{\lambda\, w_i\, X_i} = 1$ when $X_i = 0$, and $\mathrm{e}^{\lambda\, w_i\, X_i} = \mathrm{e}^{\lambda\, w_i}$ when $X_i = 1$. Thus, $\mathrm{e}^{\lambda\, w_i\, X_i} = 1 + X_i(\mathrm{e}^{\lambda\, w_i} - 1)$. This is a neat trick when working with binary variables that occur in non-linear functions.
(4) Using the expansion $\prod_{i\in\mathcal{I}}(1+a_i) = \sum_{\mathcal{S}\in\mathcal{P}(\mathcal{I})} \prod_{i\in\mathcal{S}} a_i$, where $\mathcal{P}(\mathcal{I})$ denotes the power set (set of all subsets) of the variable labels, $\mathcal{I}$. We are assuming that $\prod_{i\in\emptyset} a_i = 1$. This identity requires some thought, but is another useful expansion to have in your back pocket.

Taking expectations and using the fact that the variables are negatively correlated

$$\begin{aligned}
\mathbb{E}\left[\mathrm{e}^{\lambda\, S_n}\right] &\overset{(1)}{=} \sum_{\mathcal{S}\in\mathcal{P}(\mathcal{I})} \mathbb{E}\left[\prod_{i\in\mathcal{S}} X_i\right] \prod_{i\in\mathcal{S}} \left(\mathrm{e}^{\lambda\, w_i} - 1\right) \\
&\overset{(2)}{\le} \sum_{\mathcal{S}\in\mathcal{P}(\mathcal{I})} \prod_{i\in\mathcal{S}} p_i \left(\mathrm{e}^{\lambda\, w_i} - 1\right) \overset{(3)}{=} \prod_{i\in\mathcal{I}} \left(1 + p_i \left(\mathrm{e}^{\lambda\, w_i} - 1\right)\right)
\end{aligned}$$

(1) Using the result above and $\prod_i a_i\, b_i = \left(\prod_i a_i\right)\left(\prod_i b_i\right)$ with $a_i = X_i$ and $b_i = \mathrm{e}^{\lambda\, w_i} - 1$.
(2) As we assume the X_is are negatively correlated, $\mathbb{E}\left[\prod_{i\in\mathcal{S}} X_i\right] \le \prod_{i\in\mathcal{S}} p_i$. For the inequality to hold we also require $\prod_{i\in\mathcal{S}}(\mathrm{e}^{\lambda\, w_i} - 1) \ge 0$, which is true when $\lambda, w_i \ge 0$.
(3) Using the expansion above in reverse $\sum_{\mathcal{S}\in\mathcal{P}(\mathcal{I})} \prod_{i\in\mathcal{S}} a_i = \prod_{i\in\mathcal{I}}(1 + a_i)$.

Taking logarithms we obtain

$$\log(\mathbb{E}\left[\mathrm{e}^{\lambda\, S_n}\right]) \le \sum_{i\in\mathcal{I}} \log(1 + p_i \left(\mathrm{e}^{\lambda\, w_i} - 1\right)).$$

Using Hoeffding's bound (Equation (7.8))

$$\log\bigl(1 + p(\mathrm{e}^u - 1)\bigr) - p\,u \le \frac{u^2}{8}$$

we find

$$\log\bigl(\mathbb{E}\left[\mathrm{e}^{\lambda S_n}\right]\bigr) \le \lambda \sum_{i\in\mathcal{I}} p_i\, w_i - \frac{\lambda^2}{8}\sum_{i\in\mathcal{I}} w_i^2 = \lambda\,\mu_n + \frac{\lambda^2}{8}\sum_{i\in\mathcal{I}} w_i^2.$$

Absorbing the first term on the right-hand side into the logarithm we find

$$\psi(\lambda) = \log\Bigl(\mathbb{E}\left[\mathrm{e}^{\lambda\,(S_n-\mu_n)}\right]\Bigr) \le \frac{\lambda^2}{8}\sum_{i\in\mathcal{I}} w_i^2.$$

We can obtain a bound on the Legendre transform

$$\begin{aligned}\hat{\psi}(t) &= \max_{\lambda>0} \quad \psi(\lambda) - \lambda\,t \\ &\le \max_{\lambda>0} \frac{\lambda^2}{8}\sum_{i\in\mathcal{I}} w_i^2 - \lambda\,t = -\frac{2\,t^2}{\sum_{i\in\mathcal{I}} w_i^2}\end{aligned}$$

leading to the bound given for the upper tail.

To compute a bound for the lower tail we need to bound

$$\mathbb{E}\left[\mathrm{e}^{-\lambda S_n}\right] = \mathbb{E}\left[\mathrm{e}^{-\lambda\sum_{i\in\mathcal{I}} w_i\,X_i}\right]$$

which we can rewrite as

$$\mathbb{E}\left[\mathrm{e}^{-\lambda S_n}\right] = \mathrm{e}^{-\lambda\sum_{i\in\mathcal{I}} w_i}\;\mathbb{E}\left[\mathrm{e}^{\lambda\sum_{i\in\mathcal{I}} w_i\,(1-X_i)}\right].$$

The bound follows from the assumption $\mathbb{E}\left[\prod_{i\in\mathcal{S}}(1 - X_i)\right] \le \prod_{i\in\mathcal{S}}(1 - p_i)$ in exactly the same way as the bound on the upper tail.

The classic example of negatively correlated random variables is when we sample without replacement. We consider a bag of N balls where m are red. Thus the probability of any randomly chosen ball being red is $p = m/N$. We choose a sample of n balls. Let $X_i = 1$ if the ith sample is a red ball and $X_i = 0$ otherwise. The first condition of negative correlation is that, for any subset of samples $\mathcal{S} \subseteq \mathcal{I}$,

$$\mathbb{E}\left[\prod_{i\in\mathcal{S}} X_i\right] = \mathbb{P}\left(\bigvee_{i\in\mathcal{S}} X_i = 1\right) \le p^{|\mathcal{S}|}$$

where $\bigvee_{i\in\mathcal{S}} A_i$ implies that all events A_i occur. The probability that of a subset of variables $\{X_i | i \in \mathcal{S}\}$ all equal 1 is given by

$$\mathbb{P}\left(\bigvee_{i\in\mathcal{S}} X_i = 1\right) = \frac{m\,(m-1)\cdots(m-|\mathcal{S}|+1)}{N\,(N-1)\cdots(N-|\mathcal{S}|+1)} \le \left(\frac{m}{N}\right)^{|\mathcal{S}|} = p^{|\mathcal{S}|}$$

(note that if $0 \le m \le N$ then for $0 \le x \le m$ we have $x/m \ge x/N$ and thus $1 - x/m \le 1 - x/N$ or equivalently $(m - x)/m \le (N - x)/N$, then on rearranging

$(m-x)/(N-x) \le m/N$). An almost identical argument shows that

$$\mathbb{E}\left[\prod_{i\in\mathcal{S}}(1-X_i)\right] = \mathbb{P}\left(\bigvee_{i\in\mathcal{S}} X_i = 0\right) \le (1-p)^{|\mathcal{S}|}.$$

That is, we have shown that

$$\mathbb{E}\left[\prod_{i\in\mathcal{S}} X_i\right] \le p^{|\mathcal{S}|} = \prod_{i\in\mathcal{S}} p$$

$$\mathbb{E}\left[\prod_{i\in\mathcal{S}}(1-X_i)\right] \le (1-p)^{|\mathcal{S}|} = \prod_{i\in\mathcal{S}}(1-p),$$

which are the conditions that X_i are negatively correlated. As a consequence, if $X \sim \text{Hyp}(N, m, n)$ then

$$\mathbb{P}\left(\left|X - \frac{m}{N}\right| > t\right) \le \mathrm{e}^{-2t^2/n}.$$

In other words, when sampling n items without replacement, deviations of order of n away from the expected value are exponentially unlikely.

Appendix 7.C Constrained Optimisation

To find a local maximum or minimum of a function $f(\boldsymbol{x})$ we look for a point, $\boldsymbol{x}^*$, where the gradient equals zero,

$$\boldsymbol{\nabla} f(\boldsymbol{x}^*) = \begin{pmatrix} \frac{\partial f(\boldsymbol{x}^*)}{\partial x_1} \\ \vdots \\ \frac{\partial f(\boldsymbol{x}^*)}{\partial x_n} \end{pmatrix} = \begin{pmatrix} 0 \\ \vdots \\ 0 \end{pmatrix}.$$

This provides a set of simultaneous equations. When $f(\boldsymbol{x})$ is sufficiently simple these can be solved in closed form. More generally they can be solved iteratively, by moving in the direction of the gradient if we are maximising $f(\boldsymbol{x})$ or in the opposite direction if we are minimising. Note that $\boldsymbol{\nabla} f(\boldsymbol{x}^*) = \boldsymbol{0}$ is also satisfied at saddle points. That is, at points where $f(\boldsymbol{x})$ is a minimum in some directions and a maximum in others. To determine the type of extremal point we are at, we would have to look at the signs of the eigenvalues of the Hessian (i.e. the matrix of second derivatives $\partial^2 f(\boldsymbol{x})/\partial x_i \partial x_j$).

In constrained optimisation we optimise a function subject to one or more constraints. We consider first the case of a single constraint, $g(\boldsymbol{x}) = 0$. The constraint requires that the solution lies on a (hyper)surface defined by the constraint. The standard way to solve this (assuming that the gradient is continuous) is to find the extremal point of a *Lagrangian*

$$\mathcal{L}(\boldsymbol{x}, \lambda) = f(\boldsymbol{x}) + \lambda\, g(\boldsymbol{x})$$

where λ is known as a *Lagrange multiplier*. The solution we seek is always at a saddle point. That is, if we are maximising with respect to $\boldsymbol{x}$ we must minimise with respect to λ and vice versa (we can prove this by looking at the eigenvalues of the Hessian for both $\boldsymbol{x}$ and λ).

To find the extremal points of the Lagrangian we seek a solution, $(\boldsymbol{x}^*, \lambda^*)$, that satisfies the constraints

$$\boldsymbol{\nabla}\mathcal{L}(\boldsymbol{x}^*, \lambda^*) = \boldsymbol{\nabla} f(\boldsymbol{x}^*) + \lambda^* \boldsymbol{\nabla} g(\boldsymbol{x}^*) = 0, \qquad \frac{\partial \mathcal{L}(\boldsymbol{x}^*, \lambda^*)}{\partial \lambda} = g(\boldsymbol{x}^*) = 0. \quad (7.19)$$

The second condition is just the constraint. But the first condition $\boldsymbol{\nabla} f(\boldsymbol{x}^*) = -\lambda^* \boldsymbol{\nabla} g(\boldsymbol{x}^*)$ looks a bit more mysterious. To understand this condition we note that the gradient, $\boldsymbol{\nabla} f(\boldsymbol{x}^*)$ at the point $\boldsymbol{x}^*$ is orthogonal to the contour $f(\boldsymbol{x}) = f(\boldsymbol{x}^*)$ (in two dimensions the contours are curves, in three dimensions surfaces and in general hypersurfaces). We can see this from the Taylor expansion of $f(\boldsymbol{x})$ around $\boldsymbol{x}^*$

$$f(\boldsymbol{x}) = f(\boldsymbol{x}^*) + (\boldsymbol{x} - \boldsymbol{x}^*)^{\mathsf{T}} \boldsymbol{\nabla} f(\boldsymbol{x}) + \frac{1}{2} (\boldsymbol{x} - \boldsymbol{x}^*)^{\mathsf{T}} \mathbf{H} (\boldsymbol{x} - \boldsymbol{x}^*) + \ldots$$

where $\mathbf{H}$ is a matrix of second derivatives (known as the Hessian). Now if we consider the set of points $\mathcal{T}$ perpendicular to the gradient

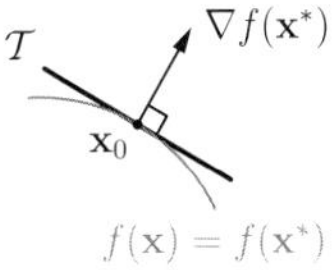

$$\mathcal{T} = \left\{ \boldsymbol{x} \middle| (\boldsymbol{x} - \boldsymbol{x}^*)^{\mathsf{T}} \boldsymbol{\nabla} f(\boldsymbol{x}^*) = 0 \right\}$$

then for $\boldsymbol{x} \in \mathcal{T}$ we have from the Taylor expansion $f(\boldsymbol{x}) = f(\boldsymbol{x}^*) + O(|\boldsymbol{x} - \boldsymbol{x}^*|^2)$, that is, the function value is approximately constant. In other words, these points are on the tangent plane to the contour surface. Now the condition $\boldsymbol{\nabla} f(\boldsymbol{x}^*) = -\lambda \boldsymbol{\nabla} g(\boldsymbol{x}^*)$ occurs at a point where a contour of $f(\boldsymbol{x}) = c_f$ touches a contour of $g(\boldsymbol{x}) = c_g$. However the second condition in Equation (7.19) ensures that $g(\boldsymbol{x}^*) = 0$, so the point where both conditions are satisfied corresponds to a point, $\boldsymbol{x}^*$, where a contour of $f(\boldsymbol{x})$ just brushes up against the surface $g(\boldsymbol{x}) = 0$.

Example 7.16 Constraint Optimisation

We consider the problem of finding the maximum value of $f(\boldsymbol{x}) = \mathcal{N}(\boldsymbol{x}|\boldsymbol{\mu}, \boldsymbol{\Sigma})$ with

$$\boldsymbol{\mu} = \begin{pmatrix} -0.4 \\ 0.1 \end{pmatrix}, \qquad \boldsymbol{\Sigma} = \begin{pmatrix} 0.6 & 0.2 \\ 0.2 & 0.6 \end{pmatrix}$$

subject to the constraint that $g(\boldsymbol{x}) = -x^2 - 4y - 2 = 0$. The constraint implies that the solution should live on the quadratic curve $y = -x^2/4 - 1/2$. To solve this problem we form the Lagrangian

$$\mathcal{L} = f(\boldsymbol{x}) + \lambda\, g(\boldsymbol{x})$$

and seek the extremum value where

$$\boldsymbol{\nabla}\mathcal{L} = 0, \qquad \frac{\partial \mathcal{L}}{\partial \lambda} = 0$$

that is, we seek the point, $(\boldsymbol{x}^*, \lambda^*)$, where

$$\boldsymbol{\nabla} f(\boldsymbol{x}^*) + \lambda^* \, \boldsymbol{\nabla} g(\boldsymbol{x}^*) = 0 \qquad g(\boldsymbol{x}^*) = 0.$$

We have thus replaced a two-dimensional constrained optimisation problem by a three-dimensional unconstrained optimisation problem (as mentioned, we are actually seeking a saddle-point solution that is a local maximum with respect to $\boldsymbol{x}$, but a local minimum with respect to λ). Note that we have three extremal equations and three unknowns. For this problem there is no closed form solution, but it is easy enough to solve these equations numerically. The extremal solution is at $\boldsymbol{x}^{\mathsf{T}} \approx (-0.50, -0.56)$ and $\lambda \approx 0.092$. Figure 7.20 shows the solution. We see that the optimal solution occurs where the contour lines of the function $f(\boldsymbol{x})$ kiss the constraint $g(\boldsymbol{x}) = 0$.

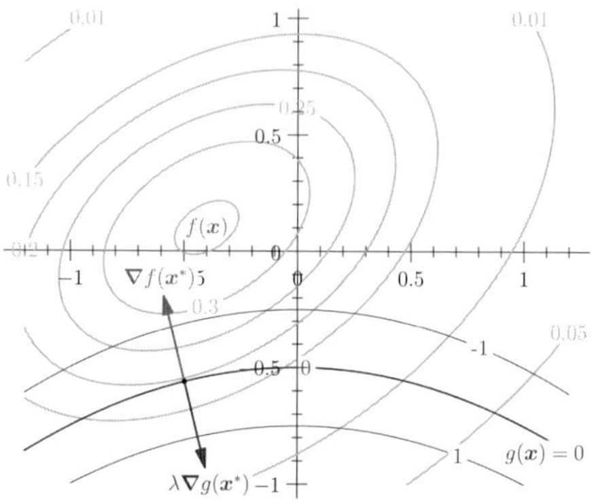

Figure 7.20 Extremum condition of the Lagrangian. We seek a maximum of $f(\boldsymbol{x})$, subject to the constraint $g(\boldsymbol{x}) = 0$. This occurs at the point, $\boldsymbol{x}^*$, where a contour line for $f(\boldsymbol{x})$ touches the contour line $g(\boldsymbol{x}) = 0$.

If we have more than one constraint then we can just add additional Lagrange multipliers: one for each constraint. One of the powers of the Lagrange method is that it works in any dimension. In some cases there can be multiple solutions to Equation (7.19) each corresponding to a different local optimum (maximum or minimum).

We can also use Lagrange multipliers to cope with inequality constraints. Although this appears significantly more complicated, it turns out that we only have to consider two outcomes; either

1. an optimum lies within the constraint, in which case we can ignore the constraint or
2. an optimum lies on the constraint, in which case the contained optimum will lie on the constraint. In this case we can use the Lagrange multiplier as normal.

We illustrate these two cases in Figure 7.21.

For an equality constraint we saw that in order for a point $\boldsymbol{x}^*$, that lies on the constraint $g(\boldsymbol{x}) = 0$, to be a (local) optimum we required that $\nabla g(\boldsymbol{x}^*)$ must be proportional to $\nabla f(\boldsymbol{x}^*)$. In the case of an inequality constraint the situation is slightly more complicated. If we are maximising $f(\boldsymbol{x})$ subject to the inequality constraint $g(\boldsymbol{x}) \geq 0$ then for a point, $\boldsymbol{x}^*$, that lies on the constraint $g(\boldsymbol{x}) = 0$ to be a (local) optimum we required that $\nabla f(\boldsymbol{x}^*)$ be in the opposite direction to $\nabla g(\boldsymbol{x}^*)$. Otherwise we could find a better solution by moving in the direction of $\nabla g(\boldsymbol{x}^*)$ (recall that the gradient points in the direction where the function

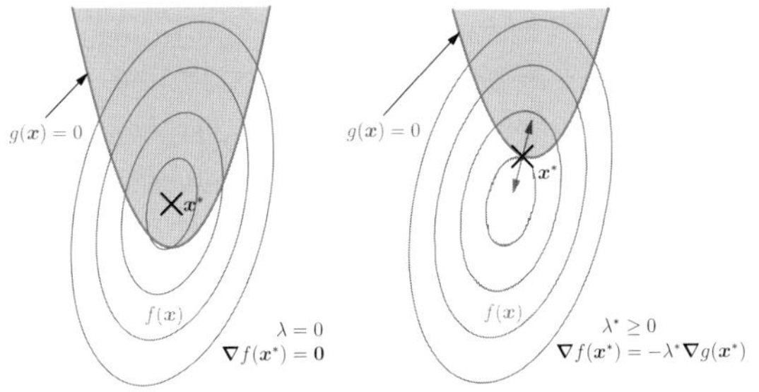

Figure 7.21 Optimisation of a function $f(\boldsymbol{x})$ subject to an inequality constraint $g(\boldsymbol{x}) \geq 0$. These illustrate two possibilities as described in the text.

increases maximally so if $\nabla f(\boldsymbol{x})$ is in the same direction as $\nabla g(\boldsymbol{x}^*)$ then we can increase $f(\boldsymbol{x})$ in a region where $g(\boldsymbol{x}) > 0$ by moving in the directions of the gradient). Thus, a point will be a local maximum of $f(\boldsymbol{x})$ subject to the inequality constraint $g(\boldsymbol{x}) \geq 0$ if it satisfies

$$\max_{\boldsymbol{x}} \min_{\lambda} f(\boldsymbol{x}) + \lambda\, g(\boldsymbol{x})$$

subject to the constraint that $\lambda \geq 0$. Note that if $\lambda = 0$ then the constraint will be ignored. Similarly, if we are seeking a minimum value of $f(\boldsymbol{x})$ subject to the constraint $g(\boldsymbol{x}) \geq 0$ then we need to find a solution to

$$\min_{\boldsymbol{x}} \max_{\lambda} f(\boldsymbol{x}) - \lambda\, g(\boldsymbol{x})$$

again subject to the constraint that $\lambda \geq 0$. These conditions are known as Karush–Kuhn–Tucker (KKT) conditions. They trivially generalise to multiple inequality constraints by using a different Lagrange multiplier for each constraint.

In general, if we wish to maximise a function, $f(\boldsymbol{x})$, subject to some constraints then we can write down a Lagrangian

$$\mathcal{L}(\boldsymbol{x}, \boldsymbol{\lambda}) = f(\boldsymbol{x}) + \sum_i \lambda_i\, g_i(\boldsymbol{x})$$

where $\lambda_i \geq 0$ if $g_i(\boldsymbol{x}) \geq 0$, or $\lambda_i \leq 0$ if $g_i(\boldsymbol{x}) \leq 0$, or λ_i is unconstrained if $g_i(\boldsymbol{x}) = 0$. A local maximum subject to these constraints satisfies

$$\max_{\boldsymbol{x}} \min_{\boldsymbol{\lambda}} \mathcal{L}(\boldsymbol{x}, \boldsymbol{\lambda}).$$

In some cases we can find a maximal solution for $\boldsymbol{x}^*(\boldsymbol{\lambda})$ for arbitrary $\boldsymbol{\lambda}$ that leaves us with the minimisation problem

$$\min_{\boldsymbol{\lambda}} \mathcal{L}(\boldsymbol{x}^*(\boldsymbol{\lambda}), \boldsymbol{\lambda}).$$

This is known as the *dual problem*. In the special case when $f(\boldsymbol{x})$ is a linear function and $g_i(\boldsymbol{x})$ are linear constraints then we can swap a maximisation problem for $\boldsymbol{x}$ with a minimisation problem for $\boldsymbol{\lambda}$, where the variables $\boldsymbol{x}$ act as Lagrange multipliers.

The extremal points of a Lagrangian correspond to local optima and potentially we could have a large number of such points. If, however, we have a convex function $f(\boldsymbol{x})$ with convex constraints then we are guaranteed to have a unique optimum, or at most a convex set of optima, whenever such a solution exists (it may be that the constraints cannot all be satisfied or the optimum exists at some point at infinity).

8

Bayes

Contents

If you are trying to infer properties about real systems where there is uncertainty in the data and you have a good model of the data-generating process, then Bayesian methods are likely to be the best you can do. This field is now vast and continues to grow. In this chapter, we give an overview of the approach and explain why it has caused both interest and controversy. We then discuss some of the technical details including conjugate distributions, uninformative priors, and model selection. We show some of the tools for approximating the posterior and finish by showing some of the Bayesian-based machine learning tools.

Bayesian inference underlies a lot of modern technology. Its contribution can range from providing a state-of-the-art machine learning tool, such as Gaussian processes, to being the cornerstone of a field such as in statistical machine translation where the translation problem is posed entirely in the language of probabilistic inference. The basis of Bayesian inference is a simple mathematical identity, but its application requires a philosophical understanding of how to use it and a body of mathematical tools required to make the inference tractable.

8.1 Bayesian Statistics

Bayes' rule provides a very simple way to perform probabilistic inference. It was proposed by the Reverend Thomas Bayes in an article published posthumously in 1763. It was invented independently, but slightly later by Laplace. Given its long history and the authority of Laplace it is perhaps surprising that it was shunned in the first half of the twentieth century as 'subjective probability'. This arose because it requires putting a prior probability on unseen events – this was seen as subjective rather than objective, which no self-respecting scientist could tolerate. During this period, being Bayesian meant being unscientific and few people defended the Bayesian approach (the economist John Maynard Keynes and the mathematical physicist Harold Jeffreys were notable exceptions). Only in the latter half of the twentieth century did the Bayesian view reassert itself, most notably through the advocacy of Ed Jaynes. Eventually though it was computers that won the day. Inference is a significant part of machine learning, and (computerised) Bayesian techniques beat many more traditional methods. The more ardent Bayesians often proclaim that the only right way of doing inference is using a Bayesian approach, although there remain grounds for being critical of many techniques that are claimed to be Bayesian. Nevertheless, the Bayesian approach can no longer be dismissed.

Historically the Bayesian approach was seen to lead to inconsistencies as you would get different results by assuming different prior distribution. One of the many contributions of Ed Jaynes was to show that by applying symmetry principles it is possible to unambiguously decide what is the correct prior given no information – see Appendix 8.A on Bertrand's paradox.

So what is Bayesian inference? Suppose we have a number of mutually exclusive hypotheses. These might be, for example, *who is the murderer in Cluedo?* or *how many fish are there in the sea?* We label these hypotheses $\mathcal{H}_i$ for $i = 1, 2, \ldots, n$. We also have some data, $\mathcal{D}$, pertaining to the hypotheses. Bayes' rule – see Equation (1.4) on page 7 – then tells us the probability of the hypotheses given the data

$$\mathbb{P}\left(\mathcal{H}_i|\mathcal{D}\right) = \frac{\mathbb{P}\left(\mathcal{D}|\mathcal{H}_i\right)\,\mathbb{P}\left(\mathcal{H}_i\right)}{\mathbb{P}\left(\mathcal{D}\right)}. \tag{8.1}$$

This is a mathematical identity so in itself it is rather uncontroversial. We call the probability on the left-hand side, $\mathbb{P}\left(\mathcal{H}_i|\mathcal{D}\right)$, the *posterior* (as it is the probability *after* seeing the data). The probability $\mathbb{P}\left(\mathcal{H}_i\right)$ is known as the *prior* (being the probability of the hypothesis *before* seeing the data). The other terms are the *likelihood*, $\mathbb{P}\left(\mathcal{D}|\mathcal{H}_i\right)$, and the probability of the data, $\mathbb{P}\left(\mathcal{D}\right)$, which is just equal to

$$\mathbb{P}\left(\mathcal{D}\right) = \sum_{i=1}^{n} \mathbb{P}\left(\mathcal{D}, \mathcal{H}_i\right) = \sum_{i=1}^{n} \mathbb{P}\left(\mathcal{D}|\mathcal{H}_i\right)\,\mathbb{P}\left(\mathcal{H}_i\right) \tag{8.2}$$

and guarantees that the posterior probability is normalised. We will see later in this subsection, however, that this term – sometimes referred to as the *evidence* – can play a very important role in model selection.

Bayes' rule also passes seamlessly to continuous variables, where instead of considering probabilities we consider probability densities. If we are trying to infer the value of a continuous parameter, x, describing the distribution of discrete data, $\boldsymbol{Y}$, we can encode our uncertainty in terms of a probability density function $f(x)$. Then we can again use Bayes' rule

$$\mathbb{P}\left(x < X < x + \delta x | \boldsymbol{Y}\right) = \frac{\mathbb{P}\left(\boldsymbol{Y}|x\right) \mathbb{P}\left(x < X < x + \delta x\right)}{\mathbb{P}\left(\boldsymbol{Y}\right)}$$

or dividing through by δx and taking the limit

$$\lim_{\delta x \to 0} \frac{\mathbb{P}\left(x < X < x + \delta x | \boldsymbol{Y}\right)}{\delta x} = \lim_{\delta x \to 0} \frac{\mathbb{P}\left(\boldsymbol{Y}|x\right) \mathbb{P}\left(x < X < x + \delta x\right)}{\mathbb{P}\left(\boldsymbol{Y}\right) \delta x}$$

$$f(x|\boldsymbol{Y}) = \frac{\mathbb{P}\left(\boldsymbol{Y}|x\right) f(x)}{\mathbb{P}\left(\boldsymbol{Y}\right)}.$$

Similarly, when our data is continuous, the probability that a data-point $\boldsymbol{Y}$ is in some ball, $B_\epsilon(\boldsymbol{y})$ of radius ϵ around the point $\boldsymbol{y}$ is

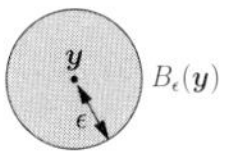

$$\mathbb{P}\left(\boldsymbol{Y} \in B_\epsilon(\boldsymbol{y})\right) \approx f(\boldsymbol{y}) \left|B_\epsilon(\boldsymbol{y})\right|$$

where $|B_\epsilon(\boldsymbol{y})|$ is the volume of the ball; the approximation becomes increasingly accurate as $\epsilon \to 0$. Putting this into Bayes' rule

$$f(x|\boldsymbol{Y} = \boldsymbol{y}) \approx \frac{\mathbb{P}\left(\boldsymbol{Y} \in B_\epsilon(\boldsymbol{y})|x\right) f(x)}{\mathbb{P}\left(\boldsymbol{Y} \in B_\epsilon(\boldsymbol{y})\right)} \approx \frac{f(\boldsymbol{y}|x) \left|B_\epsilon(\boldsymbol{y})\right| f(x)}{f(\boldsymbol{y}) \left|B_\epsilon(\boldsymbol{y})\right|} = \frac{f(\boldsymbol{y}|x) f(x)}{f(\boldsymbol{y})},$$

which becomes exact (assuming the density is sufficiently smooth) in the limit $\epsilon \to 0$. Similarly, if we were trying to infer a discrete parameter with continuous data we would get

$$\mathbb{P}\left(x|\boldsymbol{Y}\right) = \frac{f(\boldsymbol{Y}|x) \mathbb{P}\left(x\right)}{f(\boldsymbol{Y})}.$$

We see that the form of Bayes' rule does not vary whether we have probability masses or probability densities. It is easy to forget the difference between probability masses and probability densities when working in the Bayesian formalism, but remember that $f(x)$ and $f(x|\boldsymbol{Y})$ are not probabilities (although positive they can be greater than 1).

The likelihood. Before discussing the technicalities of performing a Bayesian calculation some comments are in order. Firstly, Bayes' rules allows us to reverse the conditional probabilities. That is, rather than having to compute the posterior, $\mathbb{P}\left(\mathcal{H}_i|\mathcal{D}\right)$, directly we need to compute the likelihood, $\mathbb{P}\left(\mathcal{D}|\mathcal{H}_i\right)$. But the likelihood is often much easier to compute than the posterior. To understand why this is so let us consider collecting data from some instrument. For example, we might want to deduce the structure of a crystal from some scattering experiment. The scattering pattern is usually related to the Fourier transform of the crystal structure. In principle, then we could deduce the crystal structure by taking the inverse Fourier transform. Here we are having to solve an *inverse problem*.

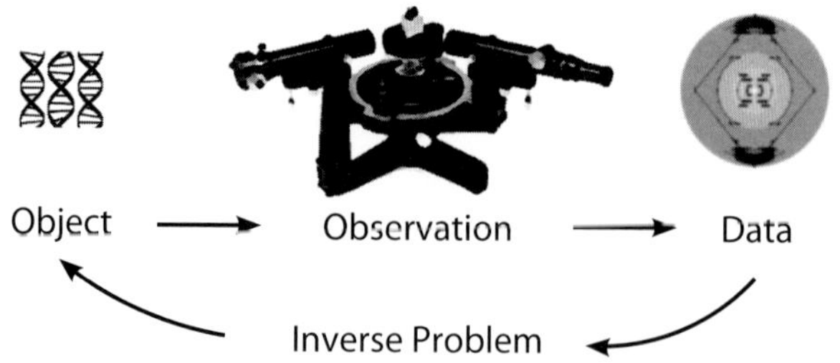

However, most instruments don't collect data at all possible angles. Thus, when you perform the inverse transform you have to invent data where there isn't any. Usually you do this by setting the intensities you have not measured to zero. This often leads to spurious *artefacts* when you perform the inverse transform. Worse happens when the instrument has inaccuracies or the data is noisy. Although you can calibrate your instruments to measure its accuracy, what you need to do is unravel all these inaccuracies in performing the inverse problem – a task that is often intractable. However, in the Bayesian formalism these problems become significantly less daunting because you are asked to calculate the probability of the data given the hypothesis. Missing data is not a problem – you only need calculate the likelihood of the data you measure – and instrument errors are also more easily modelled as you are not solving an inverse problem. Thus, the Bayesian approach offers some real advantages over trying to 'invert' the experiment. Nevertheless, constructing a likelihood function is a non-trivial exercise. It is the place where you model what is happening. That said, many 'Bayesian techniques' don't make much attempt at constructing an accurate likelihood function. Often a likelihood is chosen to keep the problem of computing the posterior tractable. Even when a poor likelihood function is used these methods sometimes work remarkably well.

It takes time to get used to the Bayesian approach as, at first, it seems backwards. That is, you tend to construct a 'generative model' which generates the data given a model, even though the data is given and actually you are trying to infer the probability of the model or hypothesis. Once you have a plausible generative model you then use Bayes' rule to infer the model given the data. When you get used to constructing models this way around it is actually straightforward.

The prior. The prior, $\mathbb{P}\left(\mathcal{H}_i\right)$, encodes our belief about the hypotheses before looking at the data. There are situations where we have good reasons for the way we assign these probabilities. For example, if the hypothesis is that someone has a particular cancer then our prior belief might be the prevalence of cancer cases in the population. Most of the time we have some prior belief about the values we are likely to obtain, after all we usually know whether we should use a microscope, ruler, or telescope to perform our measurement. However, often our prior knowledge is weak and rather than specify it precisely we prefer to say we know nothing (hopefully avoiding any bias). It might seem very natural when we have no idea which hypothesis is true to assign equal probability to all hypotheses – an approach advocated by Laplace amongst others. This leads to a posterior that has the same functional form as the likelihood used in the *maximum likelihood* approach, introduced by R. A. Fisher and accepted in the dark days when Bayesian statistics was seen as unscientific (in fact, maximum likelihood was initially criticised for being Bayesian, but its utility and the naturalness of the assumption won it grudging acceptance).

Surprisingly, assuming all hypotheses are equally likely is not always the correct assumption when we have no knowledge (we discuss so-called uninformative priors in Section 8.2.2). The true power of the Bayesian formalism comes when we have very flexible models capable of modelling very complex situations. In these cases there is a danger of 'over-fitting' the data. That is, we are apt to find a very implausible model that gives a very high likelihood for that particular data set. Priors are then used which make simpler models more probable than complicated models. How to assign probabilities to different hypotheses is often not easy to determine and, more often than not, a rather ad hoc prior is used. There are a couple of approaches to make the task of choosing a prior more principled. One is to parametrise the prior using so-called *hyperparameters*; these are then determined from the data in a Bayesian manner. A somewhat deeper approach is known as the *minimum description length* approach which we discuss briefly in Chapter 9.

The evidence. The probability of the data, $\mathbb{P}\left(\mathcal{D}\right)$, acts as a normalisation term. Unlike the likelihood and prior it has not got a universally accepted name. It is sometimes known as the *marginal likelihood* or the *model evidence* for reasons we give below. It is useful to give it a name, so I will stick with calling this the evidence. Often, to make our lives simpler, we content ourselves with finding the most probable hypothesis as the solution of our problem. This is known as the *maximum a posteriori* (or MAP) solution. When seeking the MAP solution the probability of the data is irrelevant since this term doesn't depend on the hypotheses. This can save considerable work as computing the probability of the data can be computationally expensive. However, there are good reasons for doing a little (or sometimes a lot) more work. When you compute the MAP solution you throw away the probabilistic information that the Bayesian formalism gives. This information can be useful, for example for computing error bars. Inference is often performed to make decisions. Including the full probabilistic information often allows you to make better (more accurate) decisions in terms of maximising the expectation of some utility function. There is another reason why you might want to compute the evidence that is to do with model selection. When performing inference we may be uncertain of part of the process that generates our data. That is, we might be unsure either about our prior or our likelihood function. Worse, we might have several plausible models, $\mathcal{M}_j$, which would explain the data. Within the probabilistic framework, we can encode this uncertainty by writing our probabilities as dependent on our model

$$\mathbb{P}\left(\mathcal{H}_i|\mathcal{D}, \mathcal{M}_j\right) = \frac{\mathbb{P}\left(\mathcal{D}|\mathcal{H}_i, \mathcal{M}_j\right) \mathbb{P}\left(\mathcal{H}_i|\mathcal{M}_j\right)}{\mathbb{P}\left(\mathcal{D}|\mathcal{M}_j\right)}. \tag{8.3}$$

Recall that $\mathbb{P}\left(\mathcal{H}_i|\mathcal{D}, \mathcal{M}_j\right)$ denotes the probability that the hypothesis $\mathcal{H}_i$ is true, assuming both the data $\mathcal{D}$ has been observed and that model $\mathcal{M}_j$ is correct.

The *evidence framework* is the procedure whereby we select the model that maximises the probability of the data given a particular model, $\mathbb{P}\left(\mathcal{D}|\mathcal{M}_j\right)$. This is just a maximum likelihood estimation of our models. We can make this model selection Bayesian by putting a prior probability on our models. If we have a continuum of models we can introduce a 'hyperparameter' or a set of hyperparameters that parametrise the different models. The hyperparameters can also have their own prior distribution. Given some data we can update each hyperparameter by computing a posterior distribution. We can then marginalise out the hyperparameter. An example of this would be when we believe some process is noisy, but may not know how large the noise is. We can make the magnitude of the noise a hyperparameter and put a prior on this parameter.

Example 8.1 Bayesian Carbon Dating: Example 2.3 Revisited
In Example 2.3 on page 29, we considered the problem of inferring the age of an organic specimen based on the number of carbon 14 decays observed in one hour. Assuming the number of observed radioactive decays is Poisson distributed then the probability of observing n decays is

$$\mathbb{P}\left(n|\mu\right) = \frac{\mu^n}{n!} e^{-\mu}$$

where in our problem we found that $\mu = 852.7 e^{-\lambda t}$ with $\lambda = 1.245 \times 10^{-4}$ and t is the unknown age of the sample. Assuming we observed 100 decays then the likelihood of the observation given the age of the sample, t, is shown in Figure 8.1 (this is a reproduction of Figure 2.4 on page 30).

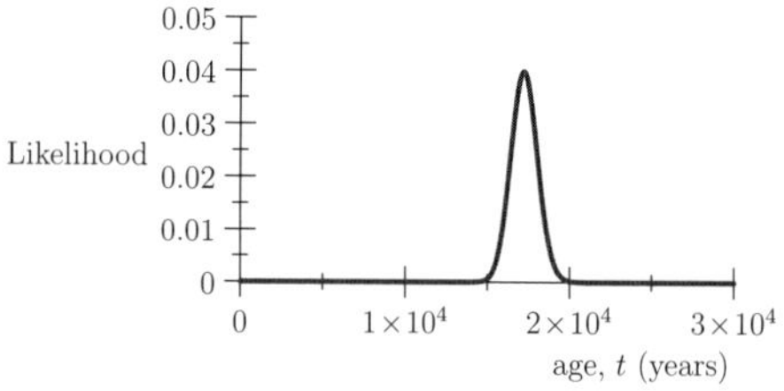

Figure 8.1 Probability of observing 100 decays in an hour as a function of the age of the sample.

There is clearly a significant likelihood that the sample is in the range from 15 000 to 20 000 years old. However, there is a natural uncertainty in our estimate due to the randomness of the radioactive process. It seems very reasonable to ask what is the probability of the sample having a particular age (or, at least, lying in any particular range of ages). There are many other uncertainties in the problem, due to natural variation in the concentration of carbon 14, how much of the sample is carbon, etc. With great care these other uncertainties can be minimised, but we are always left with a residual uncertainty due to the radioactive decay process. To turn our likelihood $\mathbb{P}\left(n|t\right)$ into a probability density for t given our observation n, we can exploit Bayes' rule

$$f(t|n) = \frac{\mathbb{P}\left(n|t\right)\, f(t)}{\mathbb{P}\left(n\right)}.$$

Now a non-Bayesian (sometimes disparagingly called a *frequentist* because of their presumed insistence that probabilities can only be interpreted as the expected frequency of many independent trials, rather than a fluffy notion of one's 'degree of belief') would complain that we cannot use Bayes' rule as there is no unambiguous way to specify the prior. Different researchers would assign different priors and obtain different posteriors! Of course, in general different researchers have different prior beliefs and as a result come to different conclusions. Those with more accurate prior knowledge come up with better conclusions (cf. the history of science – take your pick). To think of science as a purely objective enterprise demonstrates a remarkable ability to ignore all the evidence.

In a somewhat childish protest against the pretence that science is purely objective, in this book I've deliberately flouted the many conventions of scientific writing designed to further this misconception.

The frequentists used to point to examples where there seemed to be equally valid ways of assigning an uninformative prior. This arises because a uniform distribution in one representation of a problem can become non-uniform in another (see Appendix 8.A for an example of this). One of Ed Jaynes' many contributions was showing that by considering natural invariances in a problem you can uniquely determine an uninformative prior. These days there are not that many pure frequentists around (although a few still exist). This may be due to the force of the Bayesians' argument. Indeed, in the 1980s Bayesians would argue with a zeal which would put many evangelical preachers to shame. They would often win the argument, even if they did not always win friends. The demise of the frequentist school may also be in part attributed to Max Planck's pithy observation 'science advances one funeral at a time'. Of course, this remark undoubtedly refers to us all, including Planck himself, who, despite inventing quantum mechanics, never reconciled himself to its later interpretations. Perhaps we should not be too depressed by Planck's observation for surely there would be little scientific progress without embracing ideas which are in some ways wrong.

I've digressed. The challenge remains: How do we specify a prior in our example? Let us suppose the object was an ancient wooden pipe, which we are trying to date. We think that it is probably old, but perhaps it is a fake and is actually new. In that case I should assign some probability to it being quite modern. On the other hand, I guess it can't be more than a million years old because our human ancestors were not up to making pipes. But, to be honest, I don't really know much about ancient homanids so perhaps I should talk to a palaeontologist (or should that be an anthropologist?). How am I meant to transfer all my beliefs (many of which may well be wrong) into a prior probability distribution? Actually, I don't have to. For our task we have a relatively sharply concentrated likelihood. Whatever distribution I assign to the prior outside of the range 10 000–30 000 years is largely irrelevant as the posterior distribution is virtually zero outside of this range. All I care about is assigning a prior that is meaningful within this range. If I have no idea about the age of my pipe in this range, a very reasonable assumption would seem to be to assign an equal prior probability. This would have been the advice of Laplace. Rather surprisingly, Harold Jeffreys argued quite convincingly that a better prior to use would be $f(t) \propto 1/t$. In Figure 8.2 we show both posteriors – they are indistinguishable on this scale.

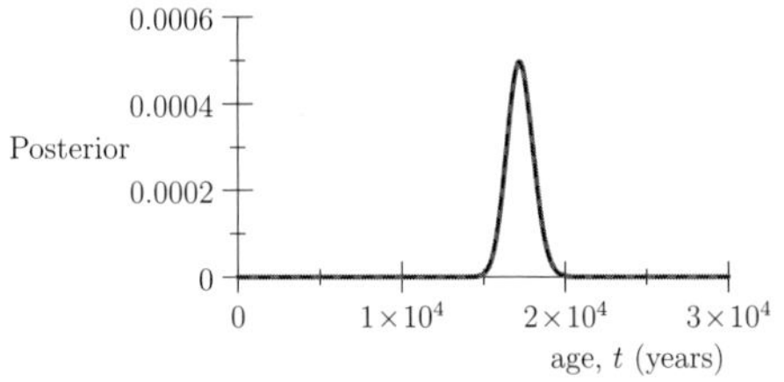

Figure 8.2 Posterior probability distribution, $f(t|n)$, for the age of the pipe assuming a Laplace prior, $f(t) \propto 1$ (solid curve), and a Jeffreys prior, $f(t) \propto 1/t$ (dashed curve). The difference between the curves is imperceptible on this scale.

We have finally arrived at our destination. We have a probabilistic description of our uncertainty. Unfortunately, we have two – maybe the frequentists are right in saying the Bayesian approach is subjective. So who is correct, Laplace or Jeffreys? In this example, the information obtained in the measurement dominates my prior beliefs so the argument of Laplace versus Jeffreys seems of little importance. However, often we work with much more flexible models (particularly in machine learning) where the addition of a prior makes a more significant difference to the result. When we have a flexible model they tend to over-fit the data so introducing a prior usually gives a dramatic improvement in performance. We therefore need a principled method for choosing a prior. We revisit this question in Section 8.2.2 when we discuss uninformative priors, and in particular we revisit this example once more in Example 8.5 where we argue for using Jeffreys' prior.

When misused, probabilities can often lead to considerable injustice. A cool-headed Bayesian can often bring light to bear on such problems.

Example 8.2 The Bayesian Lawyer

Consider a case where there is DNA evidence against the accused in a criminal case. From past experience it is known that only one person in a million will match the DNA. There are two common errors that occur in interpreting this evidence. The first is the *prosecutor fallacy*, which states that since the accused has a one in a million chance of having a DNA match then with overwhelming probability the accused must be guilty. However, in a nation like the UK, with roughly 60 million people, there would be (in expectation) 59 other people who match the DNA. If the suspect was caught by trawling a large database to find a DNA match then it is very likely that the defendant is just one of the expected 59 innocent people in the country who matches the DNA. The converse of the prosecutor fallacy is known as the *defence attorney fallacy*, which asserts that because there are 60 people in the country who match this DNA then there is only a one in 60 chance of the defendant being guilty. Thus this evidence is irrelevant and should be discounted.

To settle this dispute requires a Bayesian lawyer. Let us denote the event that the accused is guilty by G and of not being guilty by $\neg G$ (not G). The event of the accused having the DNA that matched that found at the scene of the crime we denote by E (for evidence). Bayes' rule tells us

$$\mathbb{P}\left(G|E\right) = \frac{\mathbb{P}\left(E|G\right)\,\mathbb{P}\left(G\right)}{\mathbb{P}\left(E\right)}, \qquad \mathbb{P}\left(\neg G|E\right) = \frac{\mathbb{P}\left(E|\neg G\right)\,\mathbb{P}\left(\neg G\right)}{\mathbb{P}\left(E\right)}.$$

The odds of being guilty is the ratio of the probability of being guilty to the probability of being innocent. The odds after seeing the evidence is

$$\frac{\mathbb{P}\left(G|E\right)}{\mathbb{P}\left(\neg G|E\right)} = \frac{\mathbb{P}\left(E|G\right)}{\mathbb{P}\left(E|\neg G\right)}\,\frac{\mathbb{P}\left(G\right)}{\mathbb{P}\left(\neg G\right)}.$$

Now the probability of having a DNA match given the accused is guilty is $\mathbb{P}\left(E|G\right) = 1 - \epsilon$ (where ϵ encodes the small probability that the accused is guilty but does not have a DNA match because the investigators accidentally collected the wrong DNA). The probability of the DNA match given that the defendant is not guilty is 10^{-6}. So the posterior odds of being guilty is

$$\frac{\mathbb{P}\left(G|E\right)}{\mathbb{P}\left(\neg G|E\right)} = (1-\epsilon)\,10^{6}\,\frac{\mathbb{P}\left(G\right)}{\mathbb{P}\left(\neg G\right)}.$$

In other words, the evidence increases the odds of being guilty by a factor of approximately a million. However, the posterior odds of being guilty are largely dependent on the prior odds of being guilty. If the only evidence against the defendant is the DNA evidence then the prior odds would be 1 : 60 000 000 and the posterior odds would be around 1 : 60, providing very reasonable doubt. If, on the other hand, the accused was seen behaving suspiciously close to the scene of the crime, so that the prior odds would be, say, 1 : 10, then the posterior odds would be $10^5 : 1$, and beyond reasonable doubt for most people.

The Bayesian lawyer provides the only rational way of interpreting the evidence. Interestingly, in current English law the use of probabilitics by expert witnesses in persuading juries is strictly forbidden. Although this seems rather backwards, I have some sympathy for this position, having seen rather a large number of supposed numerate people stumble over Bayes' rule. In the past, expert witnesses have made terrible mistakes using probabilistic arguments. In one notorious case an expert witness assumed the probability of two cot deaths happening to the same woman was just the square of the proportion of cot deaths in the country (thus assuming independence where it is clearly unwarranted). Judges are rightfully wary of having to adjudicate between opposing expert witnesses arguing numbers. The counter argument is that people are very poor at assessing probabilities (a fact that has been shown empirically by psychologists), therefore it is necessary to guide jurors to make the correct judgement. However, the poor understanding of probability may reflect that people have a more realistic understanding of the unpredictability of the world than most mathematical models. An example of this is when we have a number of independent pieces of evidence E_1, E_2, etc. Then

$$\frac{\mathbb{P}\left(G|E_1, E_2, \ldots, E_n\right)}{\mathbb{P}\left(\neg G|E_1, E_2, \ldots, E_n\right)} = \frac{\mathbb{P}\left(E_1|G\right)}{\mathbb{P}\left(E_1|\neg G\right)} \frac{\mathbb{P}\left(E_2|G\right)}{\mathbb{P}\left(E_2|\neg G\right)} \cdots \frac{\mathbb{P}\left(E_n|G\right)}{\mathbb{P}\left(E_n|\neg G\right)} \frac{\mathbb{P}\left(G\right)}{\mathbb{P}\left(\neg G\right)}.$$

Humans are poor at estimating how large or small numbers can get by multiplying a few numbers together. However, the expert evidence $\mathbb{P}\left(E_i|G\right)/\mathbb{P}\left(E_i|\neg G\right)$ is often uncertain. As we have seen in Example 5.3 on page 84, taking the product of random variables leads to a log-normal distribution, which can have a huge tail. As a consequence, the distribution of the log-odds is not tightly concentrated. A naive Bayesian might quote the expected log-odds and deduce that the accused is guilty beyond reasonable doubt. However, because of the thick tail in the log-normal distribution there may still be a reasonable probability that the suspect is innocent. Illogical and fallible humans might just be using their experience of the real world to come up with better answers than a naive Bayesian lawyer.

The Bayesian detective. For me, one of the great disappointments of detective fiction is that Sherlock Holmes was portrayed as a logician using the power of deduction. Alas, deductive logic is a hopelessly feeble tool in a world with uncertainty. The only rational way to reason in the world in which we find ourselves is using probabilities. Conan Doyle would have been more prescient if he had portrayed the archetypal rationalist as an exponent of Bayesian inference rather than deductive logic. Of course, Holmes would have suffered the fatal blow of being labelled 'a subjectivist' throughout the first half of the twentieth century, but in the latter half of that century he would have been exonerated and come back from the dead. Alas, this was not to be. I digress.

Bayesian inference still rouses considerable passion, with the fundamentalists asserting that the only rational approach is Bayesian. There is some element of truth in this; it is the best you can do, provided that you really do it honestly. But there's the rub. Being really Bayesian requires that you compute the real likelihood and use your true prior. However, when you do this you might end up with a very awkward posterior which might be intractable to compute. Many 'Bayesians' cheat and use a convenient likelihood and prior, but then the posterior is an approximation to the true posterior and any claims of optimality are bogus. More principled souls use Monte Carlo techniques when analytic techniques fail, but this often requires a considerable investment in time and can lose the elegance provided by an analytic but approximate solution. Philosophical dogma rarely serves science well. The persuasive arguments for or against the Bayesian approach is its performance in practical application. Undoubtedly, it has had some notable successes, but many non-Bayesian approaches also hold their own entirely on merit.

8.2 Performing Bayesian Inference

Given two independent sets of data, $\mathcal{D}_1$ and $\mathcal{D}_2$, we can either compute the posterior using the combined data or else we can compute the posterior iteratively, by using Bayes' rule on the first set of data to compute $\mathbb{P}\left(\mathcal{H}_i|\mathcal{D}_1\right)$ and then use this as our new prior to compute the posterior

$$\mathbb{P}\left(\mathcal{H}_i|\mathcal{D}_1\right) = \frac{\mathbb{P}\left(\mathcal{D}_1|\mathcal{H}_i\right)\,\mathbb{P}\left(\mathcal{H}_i\right)}{\mathbb{P}\left(\mathcal{D}_1\right)}$$

$$\mathbb{P}\left(\mathcal{H}_i|\mathcal{D}_2,\mathcal{D}_1\right) = \frac{\mathbb{P}\left(\mathcal{D}_2|\mathcal{H}_i,\mathcal{D}_1\right)\,\mathbb{P}\left(\mathcal{H}_i|\mathcal{D}_1\right)}{\mathbb{P}\left(\mathcal{D}_2|\mathcal{D}_1\right)}.$$

This is sometimes referred to as *Bayesian updating*. If the data came in one data point at a time then the posterior distribution after the last data point would become the prior distribution for the next data point. Iterative Bayesian updating can be useful numerically as the likelihood for many data points

$$\mathbb{P}\left(\mathcal{H}|\mathcal{D}\right) = \prod_{i=1}^{n} \mathbb{P}\left(\mathcal{H}|\mathcal{D}_i\right)$$

can numerically underflow. This can also be handled by working with log-probabilities, but in this case the normalisation term $\log(\mathbb{P}\left(\mathcal{D}\right))$ is slightly more awkward to compute.

One major difficulty of performing Bayesian updating is representing the posterior in a convenient form. For a finite set of hypotheses representing the posterior is relatively straightforward, although there are cases when the number of hypotheses is so great that representing all possible hypotheses is infeasible. An example of this would be if we were trying to deduce a DNA sequence given some data. Since the number of such sequences is 4^n it becomes impractical to represent the probability of all possible sequences as soon as n becomes large (i.e. much greater than 17[1]). Even the shortest DNA sequence has $n \approx 10^5$ base pairs so you will have to wait a long time before a direct Bayesian calculation of this problem is feasible. When we are trying to infer information about a continuous variable then the posterior will be a density and, in general, it might be extremely hard to represent the posterior in a convenient closed form.

8.2.1 Conjugate Priors

Fortunately, there are times when Bayesian calculations become easy. For certain likelihood functions there are so-called *conjugate prior distributions*, such that the posterior distribution will have the same form as the prior. That is, if we are interested in estimating a parameter θ and we make some measurement, X, the likelihood of the measurement is given by a distribution $f(X|\theta) = \text{Dist}(X|\theta)$ (this would be a probability mass if X is discrete or a probability density if X is continuous). If we choose the prior to be of a particular form, $f(\theta) = \text{Conj}(\theta|a_0)$, then, provided the likelihood and prior are conjugate, we obtain a posterior which is the same form as the prior

$$
\begin{aligned}
f(\theta|X) &= \frac{f(X|\theta)\, f(\theta)}{f(X)} = \frac{\text{Dist}(X|\theta)\, \text{Conj}(\theta|a_0)}{f(X)} \\
&= \text{Conj}(\theta|a_1)
\end{aligned}
$$

where a_1 depends on a_0 and X. This makes life very simple, as performing a Bayesian update just causes a change in the parameters. Alas, not all likelihood distributions have a conjugate distribution so this is only possible with certain likelihood functions. Nevertheless, many of the naturally occurring likelihood functions have a conjugate distribution.

As an example, consider estimating the success probability, $\mu \in (0, 1)$, of some process (e.g. we want to calculate the probability of a star being large enough to turn into a neutron star). The data we are given is a sequence of independent Bernoulli variables $X_i \in \{0, 1\}$ (that is, we observe many stars and set $X_i = 1$ if we believe the star is large enough to become a neutron star otherwise we set $X_i = 0$). If we choose as our prior distribution a beta distribution $\text{Bet}(\mu|a_0, b_0)$ (see Section 2.2.3 on page 34 for details of the beta distribution) then the posterior after seeing X_1 is given by

[1] If Moore's law continues, with computer speeds doubling every three years, you can add 1 to this value for every six years after 2018, when this passage was written.

$$f(\mu|X_1) = \frac{\text{Bern}(X_1|\mu) \times \text{Bet}(\mu|a_0)}{\mathbb{P}(X_i)}$$

$$= \frac{\mu^{X_1}(1-\mu)^{1-X_1} \times \mu^{a_0-1}(1-\mu)^{b_0-1}/B(a_0,b_0)}{\int_0^1 \mu'^{X_1}(1-\mu')^{1-X_1} \times \mu'^{a-1}(1-\mu')^{b-1}/B(a_0,b_0)\,\mathrm{d}\,\mu'}$$

$$= \frac{\mu^{X_1+a_0-1}(1-\mu)^{1-X_1+b_0-1}}{B(a_0+X_i,b_0+1-X_i)} = \text{Bet}(\mu|a_0+X_1,b_0+1-X_1).$$

If we make n observation with s successes and $f = n-s$ failures then the posterior, after seeing all the data, is $f(\mu|s, f) = \text{Bet}(\mu|a_0 + s, b_0 + f)$, so that $a_n = a_0 + s$ and $b_n = b_0 + f$.

If we have no knowledge beforehand, what is the appropriate prior distribution? That is, what should a_0 and b_0 be? The maximum likelihood choice is equivalent to letting $a_0 = b_0 = 1$, so that the prior is uniformly distributed between 0 and 1. However, another possible choice is to let $a_0 = b_0 = 0$, which is an example of an *uninformative prior*. It seems counter-intuitive that this prior, which is proportional to $x^{-1}(1-x)^{-1}$, should be less informative than a uniformed prior, but there are coherent arguments supporting this. We will discuss uninformative priors in the next section. It should also be pointed out that this uninformative prior is even *improper* in that it is impossible to normalise. Nevertheless, as soon as we have seen sufficient data we obtain a normalisable posterior. Assuming an uninformative prior ($a_0 = b_0 = 0$) then

$$\mathbb{E}[\mu] = \frac{a_n}{a_n+b_n} = \frac{s}{n}, \qquad \mathbb{V}\text{ar}[\mu] = \frac{a_n\,b_n}{(a_n+b_n)^2(a_n+b_n+1)} = \frac{s\,f}{n^2\,(n+1)}.$$

We note that our uncertainty in μ (i.e. the standard deviation) decreases asymptotically as $1/\sqrt{n}$ if both s and f are of order n, and decreases as $1/n$ if either s or f are of order 1.

Example 8.3 Trust Model

We illustrate Bayesian learning by considering the problem of learning the success probability, μ, of a Bernoulli trial from data. An application where this is useful is in trying to access the trustworthiness of someone you negotiate with assuming this person has a strategy of occasionally being dishonest for personal gain. You are trying to work out how often the person is dishonest by monitoring your interactions. Let $X_i = 1$ if the person is honest and 0 otherwise. We can use a Bernoulli likelihood with a beta distribution prior. Figure 8.3 shows the posterior after 10, 20, and 30 data points – in this example we chose $X_i \sim \text{Bern}(0.7)$. Notice that the Bayesian formalism provides a full description of our uncertainty. We also sketch the form of the prior distribution, $\text{Bet}(\mu|0,0)$ – the actual distribution cannot be normalised so it is not possible to put a meaningful scale on the y-axis. The prior is very strongly peaked around $\mu = 0$ and 1. This model is used in multi-agent systems and known as the beta trust model.

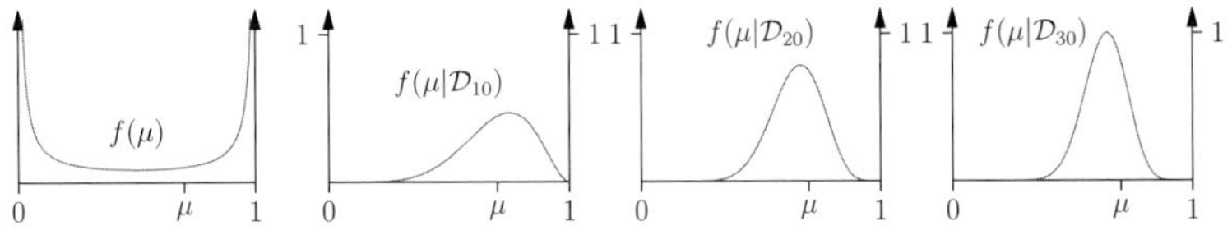

Figure 8.3 Example of Bayesian updating starting from a prior distribution $\text{Bet}(\mu|0,0)$. We show the posterior after 10, 20, and 30 data points. We are trying to learn the success probability for a Bernoulli trial. In this example, our data is generated according to $X_i \sim \text{Bern}(0.7)$. Notice that the prior is unnormalisable – we show the relative proportions of the density away from 0 and 1.

Example 8.4 Expected Traffic Rate

Suppose we want to find the rate of traffic along a road between 1:00 p.m. and 2:00 p.m. on weekdays. We assume the number of cars is given by a Poisson distribution

$$\mathbb{P}\left(N\right) = \text{Pois}(N|\mu) = \frac{\mu^N}{N!}\mathrm{e}^{-\mu}$$

where μ is the rate of traffic per hour that we want to infer from observations taken on different days. This is a reasonable model provided that the occurrence of cars are entirely independent of each other (which is often a good approximation). Let us assume a gamma-distributed prior

$$f(\mu) = \text{Gam}(\mu|a_0, b_0) = \frac{b_0^{a_0}\,\mu^{a_0-1}\mathrm{e}^{-b_0\mu}}{\Gamma(a)}.$$

The data consists of counts made on different days $\mathcal{D} = \{N_1, N_2, \ldots, N_n\}$. The likelihood of an observation is taken to be $\text{Pois}(N_i|\mu)$. The posterior after seeing the first piece of data is

$$\begin{aligned} f(\mu|N_1) &\overset{(1)}{\propto} \text{Pois}(N_1|\mu)\,\text{Gam}(\mu|a_0, b_0) \\ &\overset{(2)}{\propto} \mu^{N_1}\mathrm{e}^{-\mu}\,\mu^{a_0-1}\mathrm{e}^{-b_0\mu} \\ &\overset{(3)}{=} \mu^{N_1+a_0-1}\mathrm{e}^{-(b_0+1)\mu} \overset{(4)}{\propto} \text{Gam}(\mu|a_0+N_1, b_0+1) \end{aligned}$$

(1) Using Bayes' rule. We ignore the evidence $\mathbb{P}\left(N_1\right)$ as it is just a constant.
(2) Keeping only terms that involve the parameter μ, whose posterior we are inferring.
(3) Rearranging terms.
(4) From the functional form of $\text{Gam}(\mu|a_0 + N_1, b_0 + 1)$. In fact, as $f(\mu|N_1)$ is a probability distribution it must be equal to this gamma distribution.

Thus, the posterior is also a gamma distribution $\text{Gam}(\mu|a_1, b_1)$ with $a_1 = a_0 + N_1$ and $b_1 = b_0 + 1$. We assume $a_0 = b_0 = 0$, thus our prior is $f(\mu) \propto 1/\mu$. We show later in this section that this corresponds to a totally uninformative prior. We simulate the situation with $\mu = 5$.

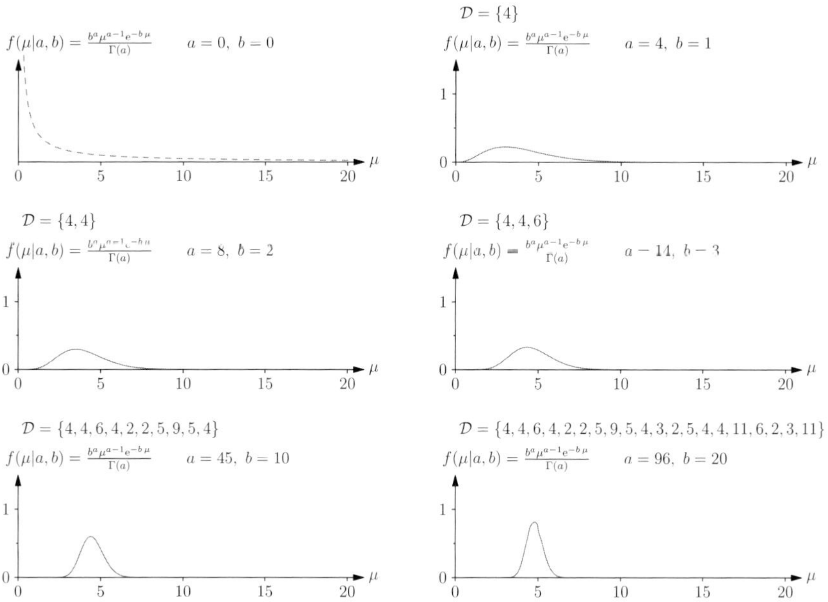

Figure 8.4 Example of Bayesian inference of the rate of a Poisson process. The (unnormalised) prior distribution and the posterior distributions after 0, 1, 2, 3, 10, and 20 data points are shown. The data was drawn from a Poisson process with a rate of $\mu = 5$.

In Figure 8.4 we show the (unnormalised) prior followed by the posterior after seeing 1, 2, 3, 10, and 20 data points. We see that the posterior becomes more concentrated around the true value of μ.

In a real situation the parameter μ may vary for different days in the week. In such a case, the posterior will not converge to a delta spike, but rather to a distribution. Worse, the distribution may not be *stationary*. That is, over time the distribution may shift slightly as the number of cars on the roads increases (or perhaps decreases). In all these cases our model does not perfectly capture reality (but then models never do). We could adapt our model to make it more accurate, but we may not be too worried about capturing everything. For a rough estimate of the average number of cars that pass in a weekday, our model will often be quite sufficient.

There are quite a few conjugate distributions. Table 8.1 lists some examples of conjugate distributions. It is just a question of algebra to prove these are conjugate and demonstrate that the update rule for the parameters are correct.

Updating normal distributions. We haven't included the normal distribution as it is slightly more complicated so the results don't easily fit into a table. When dealing with conjugate priors it is more convenient to work with the inverse variance or *precision*, $\tau = \sigma^{-2}$. If we are trying to estimate the mean, μ, of a normal distribution with known precision, τ, where the likelihood of the data is also normally distributed, starting from a prior $\mathcal{N}(\mu|\mu_0, \tau_0^{-1})$ and given data

Table 8.1 Examples of conjugate distributions. In the case of a multinomial likelihood the probabilities are confined to an k-dimensional unit simplex and the data to the discrete simplex Λ_n^k. We have omitted the normal distribution which we discuss below.

Parameter θ	**Data** x	**Prior** $f(\theta)$	**Likelihood** $f(x\|\theta)$	**Posterior** $f(\theta\|x)$
$p \in (0,1)$	$k \in \{0, 1, \ldots, n\}$	$\mathrm{Bet}(p\|a_0, b_0)$	$\mathrm{Bin}(k\|n, p)$	$\mathrm{Bet}(p\|a_0 + k,\ b_0 + n - k)$
$\mu \in (0,\infty)$	$n \in \{0, 1, \ldots\}$	$\mathrm{Gam}(\mu\|a_0, b_0)$	$\mathrm{Poi}(n\|\mu)$	$\mathrm{Gam}(\mu\|a_0 + n\ b_0 + n)$
$\boldsymbol{p} \in \Lambda^k$	$\boldsymbol{n} \in \Lambda_n^k$	$\mathrm{Dir}(\boldsymbol{p}\|\boldsymbol{\alpha})$	$\mathrm{Mult}(\boldsymbol{n}\|n, \boldsymbol{p})$	$\mathrm{Dir}(\boldsymbol{p}\|\boldsymbol{\alpha} + \boldsymbol{n})$

$X_1 \sim \mathcal{N}(\mu, \tau^{-1})$ we obtain a posterior

$$f(\mu|X_1) \propto \mathcal{N}(X|\mu, \tau^{-1})\,\mathcal{N}(\mu|\mu_0, \tau_0^{-1}) \propto \exp\Big(-\frac{\tau}{2}(X_1 - \mu)^2 - \frac{\tau_0}{2}(\mu - \mu_0)^2\Big)$$
$$\propto \exp\left(-\frac{(\tau + \tau_0)}{2}\mu^2 + \mu\,(\tau\,X_1 + \tau_0\mu_0)\right)$$
$$\propto \exp\left(-\frac{(\tau + \tau_0)}{2}\left(\mu - \frac{\tau\,X_1 + \tau_0\,\mu_0}{\tau + \tau_0}\right)^2\right) \propto \mathcal{N}(\mu|\mu_1, \tau_1^{-1})$$

where

$$\tau_1 = \tau + \tau_0, \qquad \mu_1 = \frac{\tau\,X_1 + \tau_0\,\mu_0}{\tau + \tau_0} = \frac{\tau\,X_1 + \tau_0\,\mu_0}{\tau_1}.$$

Note that in this derivation we kept all terms that depended on μ. All other terms are just constant, which can be deduced at the end because we know the posterior must be normalised. If we now repeat this exercise using $\mathcal{N}(\mu|\mu_1, \tau_1^{-1})$ as our prior and perform a Bayesian update with a second data point X_2, we find the posterior is equal to $\mathcal{N}(\mu|\mu_2, \tau_2^{-1})$, with

$$\tau_2 = \tau + \tau_1 = 2\,\tau + \tau_0, \qquad \mu_2 = \frac{\tau\,X_2 + \tau_1\,\mu_1}{\tau_2} = \frac{\tau\,(X_1 + X_2) + \tau_0\,\mu_0}{\tau_2}.$$

Given n data points the posterior will be $\mathcal{N}(\mu|\mu_n, \tau_n^{-1})$, with

$$\tau_n = n\,\tau + \tau_0, \qquad \mu_n = \frac{n\,\tau\,\hat{\mu} + \tau_0\,\mu_0}{\tau_n}$$

where

$$\hat{\mu} = \frac{1}{n}\sum_{i=1}^{n} X_i.$$

Using an uninformative prior $\tau_0 = 0$, we find $\tau_n = n\tau$ and $\mu_n = \hat{\mu}$. Thus, $\mathbb{E}\left[\mu\right] = \hat{\mu}$ and $\mathbb{V}\mathrm{ar}\left[\mu\right] = 1/\tau_n = \sigma^2/n$ where $\sigma^2 = 1/\tau$ is the variance of the normal distribution whose mean we are trying to estimate.

In the calculation above we assumed that we knew the variance (or precision) in the data even though we did not know the mean. The commoner situation is when you try to estimate a normally distributed random variable and you know neither the mean or variance (precision). For example, suppose we weigh a set of n objects (e.g. one-year-old pigs) where X_i is the weight we record. We assume that the likelihood of a data point X_i is given by $\mathcal{N}(X_i|\mu, \tau)$ – i.e. it is normally distributed with unknown mean and variance. A conjugate prior for μ and τ is the *normal-gamma* or *Gaussian-gamma* distribution defined by

$$f(\mu, \tau|\mu_0, u_0, a_0, b_0) = \mathcal{N}(\mu|\mu_0, (u_0\,\tau)^{-1})\,\mathrm{Gam}(\tau|a_0, b_0).$$

Plugging this into Bayes' rule we find the posterior is

$$\begin{aligned}
f(\mu, \tau|X_1) &\propto \mathcal{N}(X_1|\mu, \tau^{-1})\,\mathcal{N}(\mu|\mu_0, (u_0\,\tau)^{-1})\,\mathrm{Gam}(\tau|a_0, b_0)\\
&\propto \sqrt{\tau}\,\mathrm{e}^{-\frac{\tau}{2}(X_1-\mu)^2} \times \sqrt{\tau}\,\mathrm{e}^{-\frac{u_0\,\tau}{2}(\mu-\mu_0)^2} \times \tau^{a_0-1}\,\mathrm{e}^{-b_0\,\tau}\\
&\propto \tau^{a_0} \exp\Big(-\frac{\tau}{2}\Big((X_1-\mu)^2 + u_0(\mu-\mu_0)^2 + 2b_0\Big)\Big)\\
&\propto \sqrt{\tau}\,\exp\left(-\frac{(u_0+1)\tau}{2}\left(\mu - \frac{X_1 + u_0\,\mu_0}{u_0+1}\right)^2\right)\\
&\quad\times \tau^{a_0-\frac{1}{2}} \exp\left(-\left(b_0 + \frac{u_0(X_1-\mu_0)^2}{2(u_0+1)}\right)\tau\right)\\
&= \mathcal{N}(\mu|\mu_1, (u_1\,\tau)^{-1})\,\mathrm{Gam}(\tau|a_1, b_1) = f(\mu, \tau|\mu_1, u_1, a_1, b_1)
\end{aligned}$$

where

$$\mu_1 = \frac{X_1 + u_0\,\mu_0}{u_0+1}, \qquad u_1 = u_0 + 1,$$
$$a_1 = a_0 + \frac{1}{2}, \qquad b_1 = b_0 + \frac{u_0\,(X_1-\mu_0)^2}{2\,(u_0+1)}.$$

(Although this is cumbersome it is not complicated; we just multiply our likelihood (a normal) by our prior: a normal-gamma distribution. We then have to group together terms in μ and τ. The most complicated part is completing the square in the exponent.)

If we have a collection of independent data points $(X_i|i = 1, 2, \ldots, n)$ with empirical mean $\hat{\mu}$ and variance $\hat{\sigma}^2$ whose likelihoods are all normally distributed then the posterior would be a normal-gamma distribution, $f(\mu, \lambda|\mu_n, u_n, a_n, b_n)$, where

$$\mu_n = \frac{u_0\,\mu_0 + n\,\hat{\mu}}{u_0 + n}, \qquad u_n = u_0 + n,$$
$$a_n = a_0 + \frac{n}{2}, \qquad b_n = b_0 + \frac{1}{2}\sum_{i=1}^{n}(X_i - \hat{\mu})^2 + \frac{u_0\,n\,(\hat{\mu}-\mu_0)^2}{2\,(u_0+n)}.$$

This can be proved by induction using the Bayesian update formula for a single data point derived above. For an uninformative prior we require $f(\mu, \sigma) \propto 1/\sigma$ (see next section). Through the standard change of variables formula (recall that $\tau = 1/\sigma^2$)

$$f_{\mu,\tau}(\mu, \tau) = f_{\mu,\sigma}(\mu, \sigma(\tau))\left|\frac{\mathrm{d}\sigma}{\mathrm{d}\tau}\right| \propto \tau^{1/2}\left|\frac{\mathrm{d}\tau^{-1/2}}{\mathrm{d}\tau}\right| \propto \frac{1}{\tau}.$$

The uninformative prior for the normal distribution is obtained by taking $a_0 = b_0 = u_0 = 0$ (μ_0 is irrelevant). This is equivalent to a prior of $f(\mu, \sigma) \sim 1/\sigma$. Then

$$\mu_n = \hat{\mu} = \frac{1}{n}\sum_{i=1}^{n} X_i, \qquad u_n = n, \qquad a_n = \frac{n}{2}, \qquad b_n = \frac{1}{2}\sum_{i=1}^{n}(X_i - \hat{\mu})^2$$

or

$$\begin{aligned} f(\mu, \tau | X_1, X_2, \ldots, X_n) &= \mathcal{N}(\mu | \hat{\tau}, (n\,\tau)^{-1})\, \mathrm{Gam}\left(\tau \middle| \tfrac{n}{2}, \tfrac{S}{2}\right) \\ &= \frac{\sqrt{n}}{\sqrt{2\pi}\,\Gamma\left(\frac{n}{2}\right)} \left(\frac{S}{2}\right)^{\frac{n}{2}} \tau^{\frac{n-1}{2}}\, \mathrm{e}^{-\frac{n\,\tau}{2}(\mu-\hat{\mu})^2 - \frac{\tau}{2}S} \end{aligned} \tag{8.4}$$

where $S = \sum_{i=1}^{n}(X_i - \hat{\mu})^2$. We explore the marginal distributions and the expected values of the posterior in Exercise 8.4.

It seems an overly complex business to infer the mean of a normally distributed variable using Bayesian inference, but the approach follows a very standard pattern. When we don't know the precision (which is very often the case) then we are forced to infer it. We do this by assuming a distribution over the precision (in this case a Gamma distribution). This introduces a new set of parameters (often called *hyperparameters*), u_0, a_0, and b_0. We can use Bayesian update rules to infer these (hyper)parameters.

Multivariate normal likelihood. There are multidimensional generalisations of the conjugate distributions. Assume a d-dimensional normal distribution for the likelihoods $\mathcal{N}(\boldsymbol{x}|\boldsymbol{\mu}, \boldsymbol{\Lambda}^{-1})$ where $\boldsymbol{\Lambda}^{-1} = \boldsymbol{\Sigma}$ is the *precision matrix*, then the conjugate prior distribution for the mean $\boldsymbol{\mu}$ is again a normal distribution, while the conjugate distribution for the precision is the *Wishart* distribution

$$\mathcal{W}(\boldsymbol{\Lambda}|\mathbf{W}, \nu) = B\,|\boldsymbol{\Lambda}|^{(\nu-d)/2-1}\mathrm{e}^{-\mathrm{Tr}(\mathbf{W}^{-1}\boldsymbol{\Lambda})/2}$$

where $\mathrm{Tr}(\cdot)$ is the trace of the matrix and B is a normalisation constant given by

$$B = \frac{1}{|\mathbf{W}|^{\nu/2}\, 2^{\nu d/2} \pi^{d(d-1)/4} \prod_{i=1}^{d} \Gamma\left(\frac{\nu+1-i}{2}\right)}.$$

The Wishart distribution can be viewed as a multidimensional generalisation of the gamma distribution where instead of a single random variable we now have a matrix $\boldsymbol{\Lambda}$. The conjugate prior is given by

$$f(\boldsymbol{\mu}, \boldsymbol{\Lambda}|\boldsymbol{\mu}_0, u_0, \mathbf{W}_0, \nu_0) = \mathcal{N}(\boldsymbol{\mu}|\boldsymbol{\mu}_0, (u_0\,\boldsymbol{\Lambda})^{-1})\, \mathcal{W}(\boldsymbol{\Lambda}|\mathbf{W}_0, \nu_0).$$

Substituting in the likelihood and prior we find a posterior of the form

$$f(\boldsymbol{\mu}, \boldsymbol{\Lambda}|X) = \mathcal{N}(\boldsymbol{\mu}|\boldsymbol{\mu}_1, (u_1\,\boldsymbol{\Lambda})^{-1})\, \mathcal{W}(\boldsymbol{\Lambda}|\mathbf{W}_1, \nu_1)$$

where

$$\boldsymbol{\mu}_1 = \frac{u_0\,\boldsymbol{\mu}_0 + \boldsymbol{x}}{u_0 + 1}, \qquad\qquad u_1 = u_0 + 1,$$

$$\mathbf{W}_1^{-1} = \mathbf{W}_0^{-1} + \frac{1}{2(u_0+1)}\left(u_0\,\boldsymbol{X} + \boldsymbol{\mu}_0\right)\left(u_0\,\boldsymbol{X} + \boldsymbol{\mu}_0\right)^{\mathsf{T}}, \qquad \nu_1 = \nu_0 + d.$$

For n data points the posterior becomes $\mathcal{N}(\boldsymbol{\mu}|\boldsymbol{\mu}_n, (u_n\,\boldsymbol{\Lambda})^{-1})\,\mathcal{W}(\boldsymbol{\Lambda}|\mathbf{W}_n, \nu_n)$

$$\boldsymbol{\mu}_n = \frac{u_0\,\boldsymbol{\mu}_0 + n\hat{\boldsymbol{\mu}}}{u_0 + n}, \qquad u_n = u_0 + n, \qquad \nu_n = \nu_0 + n\,d,$$

with $\hat{\boldsymbol{\mu}} = \frac{1}{2}\sum_{i=1}^{n} \boldsymbol{X}_i$, and finally

$$\mathbf{W}_n^{-1} = \mathbf{W}_0^{-1} + \sum_{i=1}^{n}(\boldsymbol{X}_i - \boldsymbol{\mu}_0)\,(\boldsymbol{X}_i - \boldsymbol{\mu}_0)^{\mathsf{T}} + \frac{u_0\,n}{2(u_0+n)}(\hat{\boldsymbol{\mu}} - \boldsymbol{\mu}_0)\,(\hat{\boldsymbol{\mu}} - \boldsymbol{\mu}_0)^{\mathsf{T}}.$$

Conjugates for the exponential family. For any likelihood distribution from the exponential family

$$f(\boldsymbol{x}|\boldsymbol{\eta}) = h(\boldsymbol{x})\,g(\boldsymbol{x})\,\mathrm{e}^{\boldsymbol{\eta}^{\mathsf{T}}\boldsymbol{u}(\boldsymbol{x})}$$

there exist, a conjugate distribution

$$f(\boldsymbol{\eta}|\chi_0, \nu_0) = f(\chi, \nu_0) g(\boldsymbol{\eta})^{\nu_0} \mathrm{e}^{\nu_0\,\boldsymbol{\eta}^{\mathsf{T}}\,\chi_0}$$

where $f(\chi, \nu)$ is a normalisation constant. Given data $\mathcal{D} = (\boldsymbol{x}_i|i = 1\,\ldots, n)$ the posterior is proportional to

$$f(\boldsymbol{\eta}|\mathcal{D}, \chi_0, \nu_0) \propto g(\boldsymbol{\eta})^{\nu_0+n} \exp\left(\boldsymbol{\eta}^{\mathsf{T}}\left(\sum_{i=1}^{n} \boldsymbol{u}(\boldsymbol{x}_i) + \nu_0\,\chi_0\right)\right)$$

which follows from inspection. The posterior therefore has the form $f(\boldsymbol{\eta}|\chi_n, \nu_n)$ where

$$\nu_n = \nu_0 + n, \qquad \chi_n = \frac{1}{\nu_0 + n}\left(\nu_0\,\chi_0 + \sum_{i=1}^{n} \boldsymbol{u}(\boldsymbol{x}_i)\right).$$

Conjugate distributions clearly makes Bayesian updating very simple, but in many real applications the likelihood may not have a conjugate prior or if it does the conjugate prior might not accurately encode our prior knowledge.

8.2.2 Uninformative Priors

The Bayes formalism requires us to specify a prior that encapsulates our prior knowledge. If we have no prior knowledge then we should use an *uninformative prior* (also called a *non-informative prior*). Given a likelihood distribution $f(x|\theta)$, what is an uninformative prior for θ? It may seem that weighing every value of θ equally, so that $f(\theta) = \mathit{const}$, would be maximally uninformative. However, what we regard as a constant depends on how we parametrise our distribution. For example, if we change variables from θ to $\theta = \lambda^3$ then our uniform distribution would change according to

$$f_\lambda(\lambda) = f_\theta(\theta) \left| \frac{\mathrm{d}\,\theta}{\mathrm{d}\,\lambda} \right| = f_\theta(\lambda^3)\, 3\,\lambda^2 \propto \lambda^2$$

which is no longer a constant. If we are to set a prior distribution to a constant we must be careful what parametrisation we use (see Appendix 8.A on page 256 for an example of this).

This is a subtle issue that can lead to apparent paradoxes if you don't choose the correct parametrisation. Ed Jaynes showed that the trick to choosing an uninformative prior is to consider the desired invariance properties. We illustrate this with a number of examples. There are many likelihood distributions where

$$f_X(x|\mu) = g(x - \mu)$$

for some function $g(\cdot)$. In such cases μ is known as a *location parameter*. The obvious example of this is where μ is a mean and the likelihood is a function of the Euclidean distance from the mean (as in the normal distribution). These densities exhibit a translation invariance because if we make the shift $x' = x - c$ and $\mu' = \mu - c$ then

$$f_X(x'|\mu') = f_X(x|\mu) = g(x - \mu).$$

For an uninformative prior, we would like to choose a probability density that respects this invariance by assigning the same probability for μ to be in the interval (A, B) and in the interval $(A - c, B - c)$. That is,

$$\mathbb{P}\left(A \leq \mu \leq B\right) = \mathbb{P}\left(A - c \leq \mu \leq B - c\right)$$

$$\int_B^A f_\mu(\mu)\,\mathrm{d}\mu = \int_{B-c}^{A-c} f_\mu(\mu)\,\mathrm{d}\mu = \int_B^A f_\mu(\mu - c)\,\mathrm{d}\mu,$$

which holds for all choices of A and B if

$$f_\mu(\mu - c) = f_\mu(\mu),$$

which implies $f_\mu(\mu)$ is constant. There is a complication here in that when $\mu \in (-\infty, \infty)$ (which is necessary for translational invariance) it is not possible to normalise $f_\mu(\mu)$. This is an example of an *improper* distribution. If μ represented the unknown mean of a normally distributed random variable, then the conjugate prior would be $\mathcal{N}(\mu|\mu_0, \sigma_0^2)$ and we obtain an uninformative prior by taking the limit $\sigma_0^2 \to \infty$. Although this distribution is improper, after viewing some data we would end up with a proper posterior. Of course, in almost any conceivable real-world problem we usually have some bound on μ, so any paradoxes that are caused by improper priors are not likely to manifest themselves in the real world.

We get a more unexpected result if we consider a density of the form

$$f_X(x|\sigma) = \tfrac{1}{\sigma}\, g\left(\tfrac{x}{\sigma}\right)$$

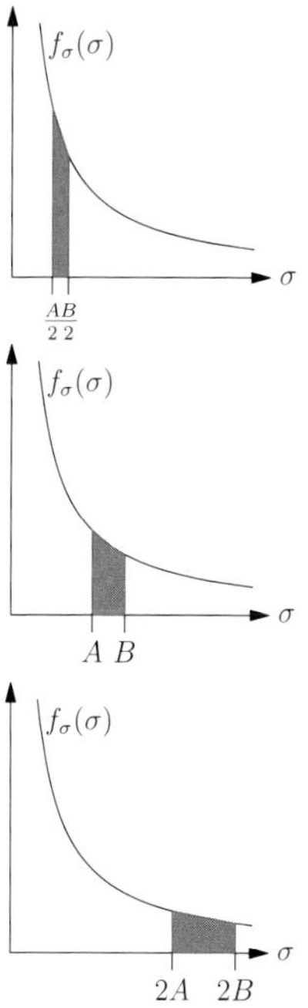

where $\sigma > 0$. We note that provided $f_X(x|1) = g(x)$ is normalised so will $f_X(x|\sigma)$ be. The parameter σ is known as a *scale parameter*. As our notation hints, the prototypical scale parameter would control the standard deviation of a distribution. The probability density has a scale invariance in so far as its value remains unaltered if we make the change of variables $x' = c\,x$ and $\sigma' = c\,\sigma$. This invariance corresponds to a change in scale, for example from kilometres to inches. If we really have no idea about the scale of a variable then we should assign the same probability of σ lying in the interval (A, B) as in the interval $(A/c, B/c)$. If this wasn't the case, and there was a higher probability of being in the kilometre range rather than the inch range, say, then we would have some prior knowledge of the scale. Thus, in the uninformative case we require

$$\mathbb{P}\left(A \le \sigma \le B\right) = \mathbb{P}\left(A/c \le \sigma \le B/c\right)$$
$$\int_A^B f_\sigma(\sigma)\,\mathrm{d}\sigma = \int_{A/c}^{B/c} f_\sigma(\sigma)\,\mathrm{d}\sigma = \int_A^B \frac{1}{c} f_\sigma\!\left(\tfrac{\sigma'}{c}\right)\,\mathrm{d}\sigma'$$

which has to hold for all A and B, implying

$$f_\sigma(\sigma) = \tfrac{1}{c} f_\sigma\!\left(\tfrac{\sigma}{c}\right).$$

Differentiating both sides of $c f_\sigma(\sigma) = f_\sigma(\sigma/c)$ with respect to c we find

$$f_\sigma(\sigma) = \tfrac{-\sigma}{c^2} f'_\sigma\!\left(\tfrac{\sigma}{c}\right),$$

setting $c = 1$ we get $f_\sigma(\sigma) = -\sigma f'_\sigma(\sigma)$. Or

$$\frac{f'_\sigma(\sigma)}{f_\sigma(\sigma)} = \frac{\mathrm{d}\log\big(f_\sigma(\sigma)\big)}{\mathrm{d}\sigma} = \frac{-1}{\sigma}$$

on integrating we find $f_\sigma(\sigma) \propto 1/\sigma$. We can also verify directly that $f_\sigma(\sigma) \propto 1/\sigma$ satisfies the relation $c f_\sigma(\sigma) = f_\sigma(\sigma/c)$. This distribution is known as the Jeffreys' prior after Harold Jeffreys who first proposed it. Again this is improper.

This may seem to be of entirely theoretical interest, but it has a startling real-world manifestation. Most naturally occurring numbers are scale invariant. This is true whether they are physical constants or they are monetary values appearing in accounts. This shouldn't be entirely surprising since we rarely choose the units we measure in to suit the quantity we are measuring. Thus the occurrence of numbers tend to be distributed, at least approximately, in inverse proportion to their value. That is, $f(x) \propto 1/x$. A consequence of this is that when we write numbers in decimal notation then the probability of the most significant figure being $n \in \{1, 2, \ldots, 9\}$ is not uniformly distributed. If we rescale x by a suitable factor of 10 so that it lies in the range $1 \le x < 10$ (which doesn't change the probability distribution due to its invariance property), then the probability of the most significant figure is

$$\mathbb{P}\left(\text{most s.f. of } x = n\right) = \frac{\int_n^{n+1} \frac{1}{x}\,\mathrm{d}x}{\int_1^{10} \frac{1}{x}\,\mathrm{d}x} = \frac{\log(n+1) - \log(n)}{\log(10)} = \log_{10}\left(\frac{n+1}{n}\right).$$

This is known as Benford's law. It provides an excellent fit to data taken from many sources. The frequency of occurrence of n as the most significant figure is

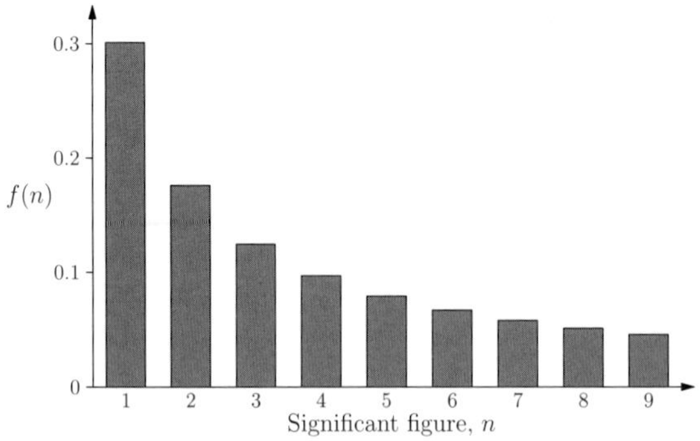

Figure 8.5 Frequency of occurrence, $f(n)$, of the most significant figure, n, as predicted by Benford's law.

shown in Figure 8.5. Note that it predicts that numbers beginning with 1 occur $\log(2)/\log(10/9) \approx 6.6$ times more often than numbers beginning with 9. Benford discovered his law by noticing the unequal use of logarithm tables – all the pages with low numbers seemed to be more worn than those for higher numbers. This strange behaviour of numbers is currently used to detect fraud in accounts.

If this is the first time you have seen this then your reaction is probably that it can't be true. Let us therefore review the argument. Consider a (non-negative) random number, X, measured in some arbitrary units. The assertion I'm making is that if the units are chosen independently of the variables then I cannot have any expectation about the scale of X. In other words I should believe that X is as likely to lie in the interval (A, B) as in the interval $(10\,A, 10\,B)$. Now that seems strange since $(10\,A, 10\,B)$ is 10 times bigger than (A, B), however that is what I am saying. It is as likely to be in the millimetre-length scales as to be in the centimetre-length scales as to be in the metre- or kilometre-length scales since the unit of distance (metres in SI units) was not chosen to make our measurement to be around 1. The only way for this scale invariance to be true is if the probability density for X is $1/x$. If you accept that the probability distribution is $1/x$ then Benford's law follows. If you still don't believe it try it out on some real-world data (see Exercise 8.5 for an example, although it doesn't matter much what data you use).

Example 8.5 Carbon Dating Priors: Example 8.1 Revisited (Again)
In Example 8.1 on page 192, we tried to infer a probability density for our belief about the age of a wooden pipe based on measurements of the number of radioactive decays of carbon 14 that occurred in one hour. We argued that there was a smallish probability that the sample was new and that it was surely less than 1 million years old. Two plausible priors are shown in Figure 8.6, with a 10% chance of the pipe being 10 years old and zero probability of it being greater than 1 million years old. We also show the uninformative prior (which we cannot properly normalise). Using an uninformative prior clearly does not substantially alter the computation of the posterior.

We saw in Figure 8.2 on page 194 that in practice it made almost no difference whether we went with Laplace or Jeffreys. But to avoid

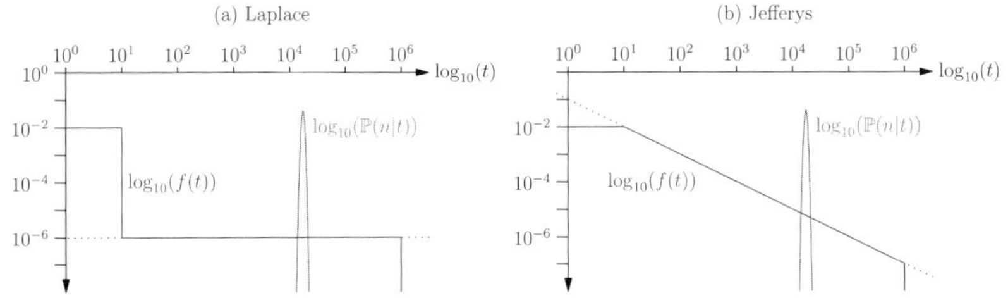

Figure 8.6 Two plausible priors together with the likelihood shown on a log-log scale. The dotted lines show a completely uninformative prior according to (a) Laplace and (b) Jeffreys.

the accusation of being subjective we must choose. Jeffreys' prior very slightly decreases our estimate of the age of the pipe. It hurts my intuition to go with Jeffreys, but I cannot see why the scale invariance argument does not apply. I really have no idea whether I was expecting an answer in the range of thousands of years or tens of thousands of years. Rather reluctantly, I have to ditch my intuition and go with Jeffreys. My nervousness in doing this is assuaged by the fact that I've tested Benford's laws on a lot of different data sets and it always seems to match the data very well despite my intuition that it shouldn't be true.

We finish this section by considering the problem of assigning an uninformative prior probability to numbers that lie in the range between 0 and 1. This might be a probability of a probability. Earlier in this chapter (Example 8.3) we gave an example of the beta trust model where we tried to predict the probability of being honest, $f(X = 1) = \mu$. A natural assumption is that the prior probability for μ should be uniformly distributed between 0 and 1 – indeed this is the assumption made by the founding fathers of the subject, Bayes and Laplace. However, there seems to be something very special about the end points. If I wanted to encode my belief about the probability that a unicorn might appear at the bottom of my garden, then after making 10 failed observations I would like to conclude the probability is overwhelmingly likely to be zero (unicorns don't exist) – using a uniform distribution, my expectation of there being a unicorn would be 1/12. Ed Jaynes has argued that the correct prior is not uniform.

The argument is a bit tricky and can safely be skipped, but it has its elegance. Jaynes asks us to imagine a population of people all with different prejudices about the probability of success. For example, Ms Q holds the belief that $f(X = 1|Q) = \mu_Q$ based on her prejudices. Now we provide to each member of the population some additional piece of useless information, $\mathcal{E}$ about X. Each member of the population will update their prior belief depending on their prejudice (that is, the new information might increase the belief that X happens, decrease it, or leave it unaltered). The posterior probability for Ms Q is

$$\mu' = f(X = 1|\mathcal{E}, Q) = \frac{f(X = 1)\, f(\mathcal{E}|X = 1, Q)}{f(X = 1)\, f(\mathcal{E}|X = 1, Q) + f(X = 0)\, f(\mathcal{E}|X = 0, Q)}.$$

Denoting $a = f(\mathcal{E}|X = 1, Q)/f(\mathcal{E}|X = 0, Q)$, this can be written

$$\mu' = \frac{a\,\mu}{1 - \mu + a\,\mu}.$$

Jaynes argues that for a totally confused (i.e. maximally uninformed) population this new piece of information should not change the distribution of the whole population, they are all just guessing based on different sets of prejudices. That is, the proportion of people, $f_{\mu'}(\theta)$, with belief $\mu' = \theta$ after seeing the additional information should be the same as the proportion of people, $f_{\mu}(\theta)$, with belief $\mu = \theta$ before seeing the evidence. Now

$$\begin{aligned} f_{\mu'}(\theta) &\overset{(1)}{=} \int_0^1 \delta\left(\theta - \frac{a\,\mu}{1 - \mu + a\,\mu}\right) f_\mu(\mu)\,\mathrm{d}\,\mu \\ &\overset{(2)}{=} \int_0^1 \left|\frac{\mathrm{d}\mu}{\mathrm{d}y}\right| \delta\left(\mu - \frac{\theta}{a + \theta - a\,\theta}\right) f_\mu(\mu)\,\mathrm{d}\,\mu \\ &\overset{(3)}{=} \left|\frac{\mathrm{d}\mu}{\mathrm{d}y}\right| f_\mu\left(\frac{\theta}{a + \theta - a\,\theta}\right) \\ &\overset{(4)}{=} \frac{a}{(a + \theta - a\,\theta)^2} f_\mu\left(\frac{\theta}{a + \theta - a\,\theta}\right) \end{aligned}$$

(1) Using the Dirac delta function to express a change of variables.
(2) Solving $\theta = \frac{a\,\mu}{1-\mu+a\,\mu}$ for μ we find $\mu = \frac{\theta}{a+\theta-a\,\theta}$. We also use the property of the Dirac delta function $\delta(y - g(x)) = |\mathrm{d}g^{-1}(y)/\mathrm{d}y|\,\delta(x - g^{-1}(y))$ (see Appendix 5.A on page 103).
(3) Performing the integral.
(4) Taking the derivative of $\mu = \frac{\theta}{a+\theta-a\,\theta}$ with respect to θ.

If we require $f_{\mu'}(\theta) = f_\mu(\theta)$ then

$$f_\mu(\theta) \propto \frac{1}{\theta\,(1 - \theta)}.$$

The easiest way to see this is to show that if $f_\mu(\theta)$ is proportional to $\frac{1}{\theta\,(1-\theta)}$ then so is $f_{\mu'}(\theta)$ (just plough through the algebra). An alternative is to differentiate with respect to a the equation

$$a f_\mu\left(\frac{\theta}{a + \theta - a\,\theta}\right) = (a + \theta - a\,\theta)^2 f_\mu(\theta),$$

then setting $a = 1$ we get

$$\theta\,(1 - \theta)\,f'_\mu(\theta) = (2\,\theta - 1)\,f_\mu(\theta).$$

Solving for $f_\mu(\theta)$ we obtain the form given above. Once again this is an improper (i.e. non-normalisable distribution). It has the nice property that for a set of n independent Bernoulli trials with s successes the mean of the posterior distribution is $\mathbb{E}\left[\mu\right] = s/n$, in accordance with common intuition. Had we used a flat distribution then the mean of the posterior would be $(s+1)/(n+2)$, which is less intuitive.

Improper Priors

Interestingly, the uninformative priors for the classic continuous distributions all appear to be improper. That is, they are not normalisable. This seems shocking as we are plugging in distributions into our formulae which aren't well defined. Yet once we have some data our once undefined priors give perfectly well-defined posteriors. All seems well. However, one needs always to be a bit cautious of things that are undefined. A better approach is to see these priors as limits of well-defined priors. Thus, we can think of the uninformative prior for probabilities to be

$$f_{\text{uninf}}(\mu) = \lim_{\epsilon \to 0} \text{Bet}(\mu|\epsilon, \epsilon) = \lim_{\epsilon \to 0} \frac{\mu^{\epsilon-1}(1-\mu)^{\epsilon-1}}{B(\epsilon, \epsilon)}$$

where $\epsilon > 0$ and $B(a, b)$ is the beta function defined in Section 2.2.3. For this particular prior, if we see n failures then the posterior remains improper

$$f(\mu|n \text{ failures}) = \lim_{\epsilon \to 0} \frac{\mu^{\epsilon-1}(1-\mu)^{n+\epsilon-1}}{B(\epsilon, \epsilon)}$$

which has mean

$$\mathbb{E}\left[\mu\right] = \lim_{\epsilon \to 0} \frac{B(1+\epsilon, n+\epsilon)}{B(\epsilon, n+\epsilon)} = \lim_{\epsilon \to 0} \frac{\epsilon}{n} = 0$$

where we have used the fact that $B(a, b) = \Gamma(a)\Gamma(b)/\Gamma(a+b)$ and $\Gamma(a+1) = a\,\Gamma(a)$. We couldn't reach this conclusion without defining the prior in terms of a limit. This conclusion accords with the intuition that I would like to say the probability of there being a unicorn in my garden is zero. It doesn't seem so logical if I wanted to conclude that football team A will always beat football team B if the only data I have is that football team A has beaten football team B once. However, in this case I have a lot of prior information (football often has a large element of luck, and football teams change over time). Indeed, in most cases if we had absolutely no prior beliefs we probably wouldn't be able to make an observation. For example, if we are measuring a position of an object that really could be anywhere then we wouldn't know where to make the observation. Similarly, if we really didn't know the scale of an object we were measuring, it would be difficult to build an instrument to measure it. Thus, improper priors are usually just helpful mathematical devices saying that our prior knowledge is so weak that it is not worth bothering to specify them – that is, our measurements will dominate the weak prior information we have.

8.2.3 Model Selection

When we first introduced Bayes' rule we said that the evidence term can be used to perform model selection. We illustrate this in the following example.

Example 8.6 Binomial or Poisson Distribution?
We present a somewhat fabricated example where we are given a set of data $\mathcal{D} = (K_1, K_2, \ldots, K_N)$ and we have to decide whether this is best modelled by Poisson likelihood, $\text{Poi}(K_i|\mu)$, or a binomial likelihood, $\text{Bin}(K_i|n,p)$. We denote these two models $\mathcal{M}_{Poi}$ and $\mathcal{M}_{Bin}$, respectively. For the Poisson model the joint probability of the data and μ, given the model, is equal to

$$\begin{aligned}\mathbb{P}\left(\mathcal{D}, \mu|\mathcal{M}_{Poi}\right) &= \text{Gam}(\mu|a_0, b_0) \prod_{K\in\mathcal{D}} \text{Poi}(K|\mu)\\ &= \frac{b_0^{a_0}\,\mu^{a_0-1}\mathrm{e}^{-b_0\mu}}{\Gamma(a_0)} \prod_{K\in\mathcal{D}} \frac{\mu^K\,\mathrm{e}^{-\mu}}{K!}.\end{aligned}$$

Denoting $S = \sum_{K\in\mathcal{D}} K$ and letting N be the number of data points, then the evidence is equal to

$$\begin{aligned}\mathbb{P}\left(\mathcal{D}|\mathcal{M}_{Poi}\right) &= \frac{b_0^{a_0}}{\Gamma(a_0)}\left(\prod_{K\in\mathcal{D}}\frac{1}{K!}\right)\int_0^\infty \mu^{S+a_0-1}\,\mathrm{e}^{-(N+b_0)\mu}\,\mathrm{d}\mu\\ &= \frac{b_0^{a_0}}{\Gamma(a_0)}\left(\prod_{K\in\mathcal{D}}\frac{1}{K!}\right)\frac{\Gamma\left(S+a_0\right)}{(N+b_0)^{S+a_0}}.\end{aligned}$$

Numerically, the evidence is likely to underflow (i.e. the numbers become so small that they are set to zero). To prevent this it is more useful to work with the log-evidence

$$\begin{aligned}\log\!\left(\mathbb{P}\left(\mathcal{D}|\mathcal{M}_{Poi}\right)\right) &= a_0\log(b_0) - \log(\Gamma(a_0)) - (a_0+S)\log(N+b_0)\\ &\quad + \log(\Gamma(S+a_0)) - \sum_{K\in\mathcal{D}}\log(\Gamma(K+1))\end{aligned}$$

where we have used the fact that $K! = \Gamma(K+1)$.

A second possible model of the data is that it comes from a binomial likelihood, $\text{Bin}(K_i|n,p)$. We suppose we know n. How we would know n without knowing that the data is binomial distributed makes little sense; nevertheless, as a very simple example of using the evidence framework let us suspend our scepticism in the rationale for doing this test. The joint probability of the data and p is given by

$$\begin{aligned}\mathbb{P}\left(\mathcal{D}, p|\mathcal{M}_{Bin}\right) &= \text{B}(p|a_0, b_0)\prod_{K\in\mathcal{D}}\text{Bin}(K|n,p)\\ &= \frac{p^{a_0-1}\,(1-p)^{b_0-1}}{\text{B}(a_0,b_0)}\prod_{K\in\mathcal{D}}\binom{n}{K}p^K\,(1-p)^{n-K}.\end{aligned}$$

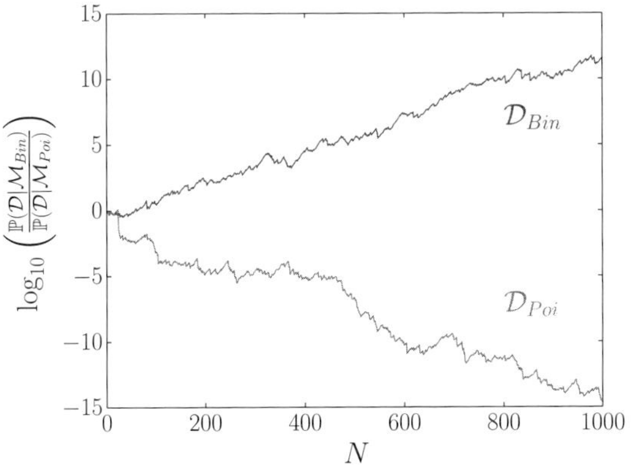

Figure 8.7 Log-odds of the evidence for models $\mathcal{M}_{Bin}$ and $\mathcal{M}_{Poi}$, and for two data sets $\mathcal{D}_{Bin}$ and $\mathcal{D}_{Poi}$.

Again denoting $S = \sum_{K \in \mathcal{D}} K$, then the evidence is equal to

$$\begin{aligned}\mathbb{P}\left(\mathcal{D}|\mathcal{M}_{Bin}\right) &= \frac{1}{\mathrm{B}(a_0, b_0)} \left(\prod_{K \in \mathcal{D}} \binom{n}{K}\right) \int_0^1 p^{S+a_0-1} \, (1-p)^{N\,n-S+b_0-1} \mathrm{d}p \\ &= \frac{1}{\mathrm{B}(a_0, b_0)} \left(\prod_{K \in \mathcal{D}} \binom{n}{K}\right) \mathrm{B}(S + a_0, \; N\,n - S + b_0).\end{aligned}$$

Again it is more convenient to work with the log-evidence

$$\begin{aligned}\log\!\big(\mathbb{P}\left(\mathcal{D}|\mathcal{M}_{Bin}\right)\big) = &- \log\!\big(\mathrm{B}(a_0, b_0)\big) + \log\!\big(\mathrm{B}(S + a_0, \; N\,n - S + b_0)\big) \\ &- N \log(n+1) - \sum_{K \in \mathcal{D}} \log\!\big(\mathrm{B}(K+1, \; n-K+1)\big)\end{aligned}$$

where we use $\binom{n}{K} = \big((n+1)\mathrm{B}(K+1, n-K+1)\big)^{-1}$.

To compare the two models we considered the log-odds of the evidence

$$\log\left(\frac{\mathbb{P}\left(\mathcal{D}|\mathcal{M}_{Bin}\right)}{\mathbb{P}\left(\mathcal{D}|\mathcal{M}_{Poi}\right)}\right).$$

Although it is usually innocuous to use an improper prior, for model comparison we can't use an improper prior because then the evidence diverges (the posterior is well defined, at least if we take the limit, because the divergence from the prior and evidence cancels). To use the evidence we choose $a_0 = b_0 = 0.001$ in both models. This choice will just shift the log-odds curve up or down. In Figure 8.7 we show the log-odds for two different data sets: $\mathcal{D}_{Bin}$ where $K_i \sim \mathrm{Bin}(30, 0.3)$ and $\mathcal{D}_{Poi}$ where $K_i \sim \mathrm{Poi}(9)$.

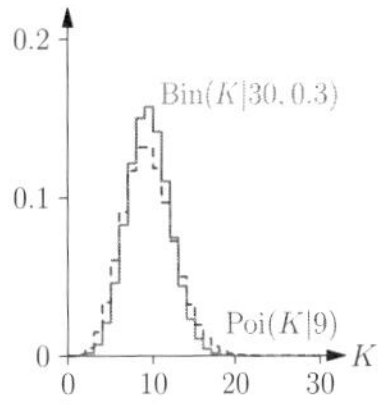

The two distributions $\mathrm{Bin}(K|30, 0.3)$ and $\mathrm{Poi}(K|9)$ are quite similar so it requires a considerable amount of data before we have firm evidence which model is preferred. However, after 1000 data points we see that the correct model is more than 10^{10} more likely than the incorrect model.

Model selection is more frequently relevant when the underlying mechanism generating the data is unclear. We might be making a measurement with some piece of equipment, but are unsure how best to model the errors. In this case we usually have a complex likelihood and computing the model evidence is non-trivial. In such cases we often have to resort to Monte Carlo methods to compute the evidence.

8.3 Bayes with Complex Likelihoods

So far we have considered classical Bayesian calculations where the likelihood has a conjugate distribution. This occurs in a number of important and natural situations. However, in many situations we gather data from rather intricate experiments where the likelihood of the data is not a simple function. In such cases, even if we start with a nice prior, our posterior is likely to be extremely cumbersome.

8.3.1 Hierarchical Models

One common way in which models become more complex is when the random variables are generated from different distributions. That is, we have a number of separate observations $\boldsymbol{X}_i$, each of which might be generated by a distribution with different parameters $\boldsymbol{X}_i \sim f_{\boldsymbol{X}_i}(\boldsymbol{\theta}_i)$, where the parameters, $\boldsymbol{\theta}_i$, come from some other distribution $\boldsymbol{\theta}_i \sim f_{\boldsymbol{\theta}}(\boldsymbol{\theta}_i|\boldsymbol{\eta})$. The parameters $\boldsymbol{\eta}$ are known as hyperparameters and they may have some hyperprior $f_{\boldsymbol{\eta}}(\boldsymbol{\eta})$.

Example 8.7 Basins of Attraction

Many optimisation problems have a huge number of local optima (this is what usually makes optimisation difficult). It is of interest to estimate the number of local optima at different cost levels. However, for even medium-sized problems the number of configurations is so large that we cannot exhaustively examine the whole search space to count the number of optima. What we can do is choose a large number, N, of starting points and run an optimisation algorithm until we have reached a local optimum. If we do this many thousands of times we can obtain a count, X_i, of the number of times we visit each local optimum, i, that we have found (of course, there may well be many local optima we have not found). Now we can model the probability of reaching a particular local optimum by a Poisson distribution $X_i \sim \text{Poi}(\lambda_i)$, where $\lambda_i = N p_i$ and p_i is the probability of landing in the basin of attraction of local optimum i. Typically, p_i will be different for different local optima. A reasonable assumption is to suppose that for a given cost the distribution of probabilities could be approximated by a beta distribution $p_i \sim \text{Bet}(a, b)$.

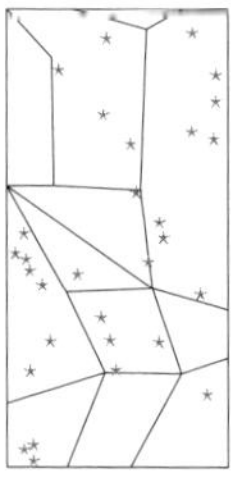

Estimating basins of attraction by random sampling

This would be a simple hierarchical model where a and b are hyperparameters. As $a, b \geq 0$, we could impose a gamma distribution hyperprior or use the uninformative hyperpriors $f(a) \propto 1/a$ and $f(b) \propto 1/b$. Using the Bayesian formulation we can obtain a posterior for $f(a, b|\{X_1, X_2, \ldots\})$ and from that estimate the expected number of local optima, many of which we have not observed.

To construct a probabilistic model it often helps to think up a mechanism that will generate the data. This often leads to a hierarchical model in which there is a series of processes that lead to the observed data. The source and intermediate stages are typically not observed and their parameters have to be deduced. This is usually done by maximising the likelihood of the observed data. We return to the problem of constructing generative models in Section 8.4 when we consider latent variable models.

When we have hierarchical models or complex likelihoods for some other reason, being Bayesian is a lot more work than when we had simple conjugate distributions to work with. There are two distinct approaches to overcoming the problem described above. The first is to use Monte Carlo techniques and in particular Markov Chain Monte Carlo (MCMC). We defer a discussion of that until later (see Section 11.2 on page 317). The second approach is to perform approximate inference. This is cheating and we lose any claim to optimality, however, it often provides a reasonably good solution with less computational effort.

8.3.2 MAP Solution

The simplest approach is to give up on a fully probabilistic treatment and instead look for the most likely solution. That is, we seek the hypothesis that maximises the posterior. This is the Bayesian equivalent of maximum likelihood and goes by the name of the *maximum a posteriori* or MAP approach. Since the evidence doesn't depend on the hypothesis we only need to consider the likelihood and prior. Furthermore, we don't attempt to get a full description of the posterior, but just the hypothesis that maximises the posterior. It is often easier to work with the log-posterior. For a discrete problem,

$$\begin{aligned}\text{MAP Hypothesis} &= \underset{H_i}{\operatorname{argmax}} \log\big(\mathbb{P}\left(H_i|\mathcal{D}\right)\big)\\ &= \underset{H_i}{\operatorname{argmax}}\Big(\log\big(\mathbb{P}\left(\mathcal{D}|H_i\right)\big) + \log\big(\mathbb{P}\left(H_i\right)\big)\Big),\end{aligned}$$

while for continuous variables

$$\text{MAP}(\boldsymbol{x}) = \underset{\boldsymbol{x}}{\operatorname{argmax}} \log\big(f(\boldsymbol{x}|\mathcal{D})\big) = \underset{\boldsymbol{x}}{\operatorname{argmax}}\Big(\log\big(f(\mathcal{D}|\boldsymbol{x})\big) + \log\big(f(\boldsymbol{x})\big)\Big).$$

(Note that $\underset{x}{\operatorname{argmax}} f(x)$ denotes the value of x that maximises the function $f(x)$.) The log-prior $\log\big(\mathbb{P}\left(H_i\right)\big)$ or $\log\big(f(\boldsymbol{x})\big)$ acts as 'regulariser', making the MAP solution more robust when there is limited data, and we are using a flexible model.

Example 8.8 A MAP Recommender

Recommendation systems have become ubiquitous on the web to help with the information overload from which many of us suffer. One common form of recommender systems are so-called collaborative filtering techniques that use ratings supplied by their users to produce personalised recommendations. The key idea is that users who rate items similarly to you can be used to recommend items which you have not seen.

We denote the set of users by $\mathcal{U}$, and the set of items by $\mathcal{I}$. The users have supplied ratings for some of the items $\mathcal{R} = (r_{ui} | (u,i) \in \mathcal{D})$ where $\mathcal{D} \subset \mathcal{U} \times \mathcal{I}$ is the set of user-item pairs which have been rated. Typically the ratings will be on some scale, e.g. $r_{ui} \in \{1, 2, 3, 4, 5\}$. In large systems it is usual for the number of ratings to be quite sparse in the sense that $|\mathcal{D}| \ll |\mathcal{U}| \times |\mathcal{I}|$, with a sparsity of less than 1% being common. A reasonable first guess at a rating is to use either the mean rating of the item $\bar{r}_i$ or the mean rating of the user $\bar{r}_u$. We can incorporate both these using a rating $\bar{r}_i + \bar{r}_u - \bar{r}$ where $\bar{r}$ is the mean rating for all the items

$$\bar{r}_i = \frac{1}{|\mathcal{U}_i|} \sum_{u \in \mathcal{U}_i} r_{ui}, \qquad \bar{r}_u = \frac{1}{|\mathcal{I}_u|} \sum_{i \in \mathcal{I}_u} r_{ui}, \qquad \bar{r} = \frac{1}{|\mathcal{D}|} \sum_{(u,i) \in \mathcal{D}} r_{ui},$$

where $\mathcal{U}_i = \{u | (u,i) \in \mathcal{D}\}$ is the set of users that have rated item i, and $\mathcal{I}_u = \{i | (u,i) \in \mathcal{D}\}$ is the set of items that have been rated by user u. These mean ratings do not, however, take into account users with similar tastes. To do that we try to infer the residue ratings

$$\mathbf{R} = \begin{pmatrix} \cdot & \cdot & \cdot & 3 & 5 & 3 \\ 2 & \cdot & \cdot & \cdot & \cdot & \cdot \\ \cdot & \cdot & \cdot & 3 & \cdot & \cdot \\ 1 & \cdot & 4 & 3 & \cdot & \cdot \\ \cdot & \cdot & \cdot & \cdot & \cdot & \cdot \\ \cdot & \cdot & \cdot & 5 & 4 & \cdot \\ \cdot & \cdot & 2 & \cdot & \cdot & 3 \\ 2 & \cdot & \cdot & 3 & 1 & 2 \\ \cdot & \cdot & \cdot & \cdot & \cdot & \cdot \\ \cdot & \cdot & 3 & \cdot & \cdot & 1 \end{pmatrix}$$

$$\tilde{r}_{ui} = r_{ui} - \bar{r}_i - \bar{r}_u + \bar{r}.$$

We can imagine the residues as being components of a residue ratings matrix $\tilde{\mathbf{R}}$. The problem of predicting the residues is thus a problem of matrix completion. That is, we know the residual ratings $\tilde{r}_{ui}$ at positions $(u,i) \in \mathcal{D}$ in the matrix $\tilde{\mathbf{R}}$. Our task is to infer the values of the residual ratings at the other positions $(u,i) \notin \mathcal{D}$.

One of the most effective means of solving the matrix completion problem is to look for a low-rank approximation. There are many different ways of accomplishing this. Here we consider approximating the residue ratings matrix by

$$\tilde{\mathbf{R}} \approx \mathbf{A}\,\mathbf{B}^{\mathsf{T}}$$

where $\mathbf{A}$ is a $|\mathcal{U}| \times K$ matrix and $\mathbf{B}$ is a $|\mathcal{I}| \times K$ matrix (we get to choose K). Thus $\tilde{\mathbf{R}}$ is approximated as a $|\mathcal{U}| \times |\mathcal{I}|$ matrix of rank K. We can determine the rank, K, empirically using a validation set. We are finally in a position to make this inference problem probabilistic. We now assume that the elements of $\mathbf{A}$ and $\mathbf{B}$ are normally distributed

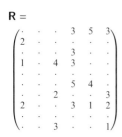

$$A_{uk} \sim \mathcal{N}(0, \sigma_U^2) \qquad\qquad B_{ik} \sim \mathcal{N}(0, \sigma_I^2).$$

This provides a prior distribution over the elements of $\mathbf{A}$ and $\mathbf{B}$. The parameters σ_U^2 and σ_I^2 encode the variation in the user's ratings and the item's ratings. These could either be estimated empirically or treated as hyperparameters to be optimised by measuring the generalisation performance on a validation set. We assume that the likelihood of the ratings are given by

$$f(\mathcal{R}|\mathbf{A},\mathbf{B}) = \prod_{r_{ui} \in \mathcal{R}} \mathcal{N}\left(r_{ui} - \sum_{k=1}^{K} A_{uk}\, B_{uk} \middle| 0, \sigma^2 \right)$$

where σ^2 measures the error in the rating (due to the inconsistency of the rater and the discretisation of the rating scale). The log-posterior is (up to a constant) equal to

$$\begin{aligned}\log\big(f(\mathbf{A},\mathbf{B}|\mathcal{R})\big) = &- \sum_{r_{ui} \in \mathcal{R}} \frac{\left(r_{ui} - \sum_{k=1}^{K} A_{uk}\, B_{ik}\right)^2}{2\,\sigma^2} \\ &- \sum_{u \in \mathcal{U}} \sum_{k=1}^{K} \frac{A_{uk}^2}{2\,\sigma_U^2} - \sum_{i \in \mathcal{I}} \sum_{k=1}^{K} \frac{B_{ik}^2}{2\,\sigma_I^2}.\end{aligned}$$

The MAP solution cannot be computed in closed form. However, we can solve this problem iteratively. We initialise the elements of $\mathbf{B}$ (for example, by drawing the components $B_{i,k}$ from the prior distribution $\mathcal{N}(0, \sigma_I^2)$). We then maximise the posterior with respect to the elements of A_{uk}. Setting the partial derivative of the log-posterior with respect to A_{uk} to zero

$$\frac{\partial \log\big(f(\mathbf{A},\mathbf{B}|\mathcal{R})\big)}{\partial A_{uk}} = \sum_{i \in \mathcal{I}_u} \frac{B_{i,k}\left(r_{ui} - \sum_{k'=1}^{K} A_{uk'}\, B_{ik'}\right)}{\sigma^2} - \frac{A_{uk}}{\sigma_U^2} = 0$$

we find

$$\sum_{k'=1}^{K} A_{uk'} \left(\frac{1}{\sigma^2} \sum_{i \in \mathcal{I}_u} B_{ik'}\, B_{ik} + \frac{1}{\sigma_U^2} [\![k' = k]\!] \right) = \sum_{i \in \mathcal{I}_u} B_{ik}\, r_{iu}.$$

Defining the matrix $\mathbf{C}^u$ and vectors $\boldsymbol{v}^u$ and $\boldsymbol{a}^u$, such that their components are given by

$$C_{kk'}^u = \sum_{i \in \mathcal{I}_u} B_{ik'}\, B_{ik} + \frac{\sigma^2}{\sigma_U^2} [\![k' = k]\!], \quad v_k^u = \sum_{i \in \mathcal{I}_u} B_{ik}\, r_{iu}, \quad a_k^u = A_{uk},$$

then $\boldsymbol{a}^u = (\mathbf{C}^u)^{-1} \boldsymbol{v}^u$. That is, for each user we find a row of $\mathbf{A}$ by inverting a $K \times K$ matrix. Similarly, once we have optimised the log-posterior with respect to $\mathbf{A}$, we can then optimise it with respect to $\mathbf{B}$. Following a similar calculation we find the ith row of $\mathbf{B}$, which we denote by $\boldsymbol{b}^i$ (with components B_{ik}), is given by $\boldsymbol{b}^i = (\mathbf{D}^i)^{-1} \boldsymbol{w}^i$, where $\mathbf{D}^i$ is a matrix and $\boldsymbol{w}^i$ a vector whose components are equal to

$$D^i_{kk'} = \sum_{u \in \mathcal{U}_i} A_{uk'} A_{uk} + \frac{\sigma^2}{\sigma_I^2} [\![k' = k]\!], \qquad w^i_k = \sum_{u \in \mathcal{U}_i} A_{uk} r_{iu}.$$

To maximise the posterior we iterate between maximising **A** and **B**. Such a procedure only guarantees that we converge to a local optimum of the posterior. In practice, this convergence is found to be relatively quick. For matrices where we only have data on a small proportion of elements the whole procedure is quite efficient as it only involves computing sums over the ratings that exist. Note that the priors ensure that the matrices **C** and **D** are positive definite and thus invertible. On a number of data sets where this approach was tested excellent performance was obtained.

The MAP solution is just a hack. That is, it often gets us an answer much more easily than by computing the full posterior, however it has no statistical relevance. When the posterior is sharply peaked (i.e. we have a reasonable amount of data or a good prior) then the MAP solution is often adequate. But when the data is ambiguous then the posterior may not be sharply peaked and the MAP solution may be poor. A better estimate in such cases may be the mean of the posterior (as opposed to the mode) – see Figure 8.8. Particularly in high dimensions the mean of the posterior distribution is very often significantly different from the MAP solution.

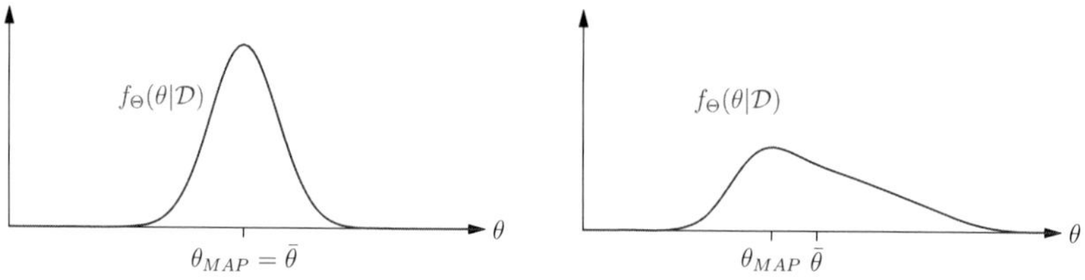

Figure 8.8 Schematic illustration of a sharply peaked posterior and a broad posterior. Also shown is both the MAP solution, θ_{MAP}, and the mean of the posterior, $\bar{\theta}$.

It is worth observing that the MAP solution depends on the representation of the variables. That is, if we make a change of variables $\theta \to \phi$ so that

$$f_\Phi(\phi) = f_\Theta(\theta(\phi)) \frac{\mathrm{d}\theta(\phi)}{\mathrm{d}\phi}$$

then the MAP solution in terms of ϕ is given by $f'_\Phi(\phi_{\text{MAP}}) = 0$, but

$$\frac{\mathrm{d} f_\Phi(\phi)}{\mathrm{d}\phi} = \frac{\mathrm{d} f_\Theta(\theta(\phi))}{\mathrm{d}\phi} + f_\Theta(\theta(\phi)) \frac{\mathrm{d}^2\theta(\phi)}{\mathrm{d}\phi^2},$$

so in general $\theta_{\text{MAP}} \neq \theta(\phi_{\text{MAP}})$.

Example 8.9 MAP Solution for Normal Deviates

We saw earlier that for data $\mathcal{D} = (X_1, X_2, \ldots, X_n)$ where $X_i \sim \mathcal{N}(\mu, \tau^{-1})$ then the posterior distribution assuming an uninformative prior is given by

$$
\begin{aligned}
f_{\mu,\tau}(\mu, \tau|\mathcal{D}) &= \mathcal{N}(\mu|\mu_n, (n\,\tau)^{-1})\ \text{Gam}\big(\tau\big|\tfrac{n}{2}, \tfrac{S}{2}\big) \\
&= \frac{\sqrt{n}}{\sqrt{2\pi}\,\Gamma\big(\frac{n}{2}\big)} \left(\frac{S}{2}\right)^{\frac{n}{2}} \tau^{\frac{n-1}{2}}\ \mathrm{e}^{-\frac{n\,\tau}{2}(\mu-\hat{\mu})^2 - \frac{\tau}{2}S}
\end{aligned}
$$

where $\hat{\mu} = \frac{1}{n}\sum_{i=1}^{n} X_i$ and $S = \sum_{i=1}^{n}(X_i - \hat{\mu})^2$. The log-posterior is

$$
\log\big(f_{\mu,\tau}(\mu, \tau|\mathcal{D})\big) = \frac{n-1}{2}\log(\tau) - \frac{n\,\tau}{2}(\mu - \hat{\mu})^2 - \tfrac{\tau}{2}S + \text{const.}
$$

To find the MAP solution we set the derivatives with respect to μ and τ equal to zero. From the derivative with respect to μ we find $\mu_{\text{MAP}} = \hat{\mu}$. Setting $\mu = \hat{\mu}$ and taking derivatives with respect to τ we find

$$
\frac{\mathrm{d}\log\big(f(\hat{\mu}, \tau|\mathcal{D})\big)}{\mathrm{d}\tau} = \frac{n-1}{2\,\tau} - \frac{S}{2}
$$

so that $\tau_{MAP}^{-1} = \frac{1}{n-1}\sum_{i=1}^{n}(X_i - \hat{\mu})^2$. Since $\tau^{-1} = \sigma^2$ we might be pleased at this answer – we have obtained the unbiased estimator of the variance.

As an aside, note that to compute the MAP solution we need not have computed the posterior. The likelihood of the data is $f(\mathcal{D}|\mu, \tau) = \prod_{i=1}^{n} f(X_i|\mu, \tau)$ where

$$
f(X_i|\mu, \tau) = \sqrt{\frac{\tau}{2\pi}}\ \mathrm{e}^{-\frac{\tau}{2}(X_i-\mu)^2};
$$

the uninformative prior is

$$
f(\mu, \tau) \lim_{\epsilon\to 0} \mathcal{N}(\mu|\mu_0, (\epsilon\,\tau)^{-1})\ \Gamma(\tau|\epsilon, \epsilon) \propto \sqrt{\tau}\,\frac{1}{\tau} = \tau^{-1/2}
$$

so that our log-posterior is (up to a constant)

$$
\log\big(f(\mu, \tau|\mathcal{D})\big) = \frac{n}{2}\log(\tau) - \frac{n\,\tau}{2}\sum_{i=1}^{n}(X_i - \mu)^2 - \frac{1}{2}\log(\tau)\,.
$$

Taking derivatives we obtain the same result for the MAP solution, but avoided the complication of writing down the posterior.

Returning to our problem of computing the MAP solution. Suppose we considered the posterior to be a function of the mean and variance σ^2, rather than the precision. Using the usual rules for make a transformation $\tau \to \sigma^2 = \frac{1}{\tau}$ we find

$$
f_{\mu,\sigma^2}(\mu, \sigma^2|\mathcal{D}) \propto \sigma^{-\frac{n+3}{2}}\ \mathrm{e}^{-\frac{n}{2\sigma^2}(\mu-\hat{\mu})^2 - \frac{S}{2\sigma^2}}
$$

which gives a map solution $\mu_{\text{MAP}} = \mu$ and $\sigma^2_{\text{MAP}} = \frac{1}{n+3}\sum_{i=1}^{n}(X_i - \hat{\mu})^2$. Thus, our previous result showing that MAP solution leads to an unbiased variance was just luck. If we write the probability distribution in terms of the mean and standard deviation

$$f_{\mu,\sigma}(\mu,\sigma|\mathcal{D}) \propto \sigma^{-\frac{n+2}{2}} \, \mathrm{e}^{-\frac{n}{2\sigma^2}(\mu-\hat{\mu})^2 - \frac{S}{2\sigma^2}}$$

we find $\mu_{\text{MAP}} = \mu$ and $\sigma_{\text{MAP}} = \sqrt{\frac{1}{n+2}\sum_{i=1}^{n}(X_i - \hat{\mu})^2}$. Note that $f_{\mu,\tau}(\mu,\tau|\mathcal{D})$, $f_{\mu,\sigma^2}(\mu,\sigma^2|\mathcal{D})$, and $f_{\mu,\sigma}(\mu,\sigma|\mathcal{D})$ all describe the same probability distribution; we are just measuring it using different variables. A consequence of using different variables is that the maximal value shifts. The posterior mean is, of course, invariant under a change of variables. It is in this sense a statistically meaningful quantity.

The maximum likelihood solution does not suffer the indignity of shifting when we perform a change of the parameters that we are trying to learn. This is because the likelihood $f(\mathcal{D}|\theta)$ is not a density in the parameters being optimised. If we make a change of variables $\theta \to \phi$ the corresponding likelihood is $f(\mathcal{D}|\phi) = f(\mathcal{D}|\theta(\phi))$. The maximum likelihood estimator is given by

$$\frac{\partial f(\mathcal{D}|\phi)}{\partial \phi} = \frac{\partial \theta}{\partial \phi}\frac{\partial f(\mathcal{D}|\theta)}{\partial \theta} = 0.$$

Thus $\theta(\phi_{ML}) = \theta_{ML}$. Of course, if we change the way we represent our data then this can change the position of the maximum likelihood solution (though it won't change the MAP solution). We should not be too surprised by these apparent inconsistencies; the mode of a probability density will often shift under a change of variables. If we do full Bayes, and work with the whole distribution rather than its maximum value, then a change of variables of either the observation or the parameters of the distribution will not change the probability of any measurable event. Of course, the MAP solution is often easier to obtain or, at least, easier to estimate than the full posterior probability distribution. However, it provides much less information than the full posterior. In particular, it provides no guidance about the uncertainty in the estimator.

MAP techniques are often labelled as 'Bayesian' as they involve a prior. But be wary, they throw away all the probabilistic information. They can often give an improvement over maximum likelihood estimators (provide the prior we use captures the true prior), but full Bayesian they are not.

The variational approximation saw its genesis in statistical physics and inherits much of its language from there. This is perhaps not too surprising as statistical physics is the application of probability to physical systems.

8.3.3 Variational Approximation

The MAP approach throws out the probabilistic information which a full Bayes calculation would give. To retain that probabilistic information but keep the calculations tractable we can attempt to find an approximation to the posterior. One approach is to replace the posterior distribution, $f(\boldsymbol{x}|\mathcal{D})$, with an approximate distribution, $q(\boldsymbol{x})$, which is easier to handle. To ensure that the approximate distribution is close to the true posterior we minimise a similarity measure between the approximate distribution and the exact posterior. The most commonly used similarity measure is the *Kullback–Leibler (or KL) divergence* (see Section 7.3.1 on page 166), defined as

$$\mathrm{KL}\left(q \,\|\, f\right) = \mathbb{E}_q\left[\log\left(\frac{q}{f}\right)\right] \tag{8.5}$$

where the expectation is with respect to the probability distribution $f(\boldsymbol{x})$. Recall that the KL divergence is a non-negative quantity with a minimum of 0 when the two distributions are identical.

In the standard variational formalism we approximate the posterior, $f(\boldsymbol{x}|\mathcal{D})$, with a distribution $q(\boldsymbol{x}|\boldsymbol{\theta})$, where $\boldsymbol{\theta}$ represents a set of parameters chosen to minimise $\mathrm{KL}\left(q(\boldsymbol{x}|\boldsymbol{\theta}) \,\|\, f(\boldsymbol{x}|\mathcal{D})\right)$. We are free to use any distribution $q(\boldsymbol{x}|\boldsymbol{\theta})$ we wish (although the quality of the approximation will depend on our choice). A common choice when inferring a high-dimensional quantity, $\boldsymbol{x} = (x_1, x_2, \ldots, x_n)$, is to choose a factorised distribution, for example,

$$q(\boldsymbol{x}|\boldsymbol{\theta}) = \prod_{i=1} q_i(x_i|\boldsymbol{\theta}_i).$$

In this case we are approximating the posterior, $f(\boldsymbol{x}|\mathcal{D})$, by a probability distribution where the variables x_i are independent of each other. To accomplish this it is useful to consider the *variational free energy* defined as

$$\Phi(\boldsymbol{\theta}) = \int q(\boldsymbol{x}|\boldsymbol{\theta}) \log\left(\frac{q(\boldsymbol{x}|\boldsymbol{\theta})}{f(\boldsymbol{x},\mathcal{D})}\right) \mathrm{d}\boldsymbol{x}$$

where $f(\boldsymbol{x},\mathcal{D}) = f(\mathcal{D}|\mathbf{x})\, f(\boldsymbol{x})$ is the joint probability of the parameters $\boldsymbol{x}$ we are trying to infer and the data (i.e. the likelihood times the prior). Note, for discrete random variables, we would just replace the integral with a sum. We can rewrite $\Phi(\boldsymbol{\theta})$ as

$$\begin{aligned}
\Phi(\boldsymbol{\theta}) &\overset{(1)}{=} \int q(\boldsymbol{x}|\boldsymbol{\theta}) \log\left(\frac{q(\boldsymbol{x}|\boldsymbol{\theta})}{\left(f(\boldsymbol{x},\mathcal{D})/f(\mathcal{D})\right) f(\mathcal{D})}\right) \mathrm{d}\boldsymbol{x} \\
&\overset{(2)}{=} \int q(\boldsymbol{x}|\boldsymbol{\theta}) \left(\log\left(\frac{q(\boldsymbol{x}|\boldsymbol{\theta})}{f(\boldsymbol{x}|\mathcal{D})}\right) - \log\left(f(\mathcal{D})\right)\right) \mathrm{d}\boldsymbol{x} \\
&\overset{(3)}{=} \mathrm{KL}\left(q(\boldsymbol{x}|\boldsymbol{\theta}) \,\|\, f(\boldsymbol{x}|\mathcal{D})\right) - \log\left(f(\mathcal{D})\right)
\end{aligned}$$

(1) By definition of $\Phi(\boldsymbol{\theta})$ and dividing and multiplying by $f(\mathcal{D})$ in the denominator.
(2) Following from Bayes' rule, $f(\boldsymbol{x}|\mathcal{D}) = f(\boldsymbol{x},\mathcal{D})/f(\mathcal{D})$, and expanding the logarithm, $\log(A/B) = \log(A) - \log(B)$ with $A = \frac{q(\boldsymbol{x}|\boldsymbol{\theta})}{f(\boldsymbol{x}|\mathcal{D})}$ and $B = f(\mathcal{D})$.
(3) Using the definition of the KL divergence and the fact that $\log\left(f(\mathcal{D})\right)$ does not depend on $\boldsymbol{x}$.

Since $\log\left(f(\mathcal{D})\right)$ is independent of the parameters $\boldsymbol{\theta}$, minimising $\Phi(\boldsymbol{\theta})$ is equivalent to minimising the KL divergence. Because $\mathrm{KL}\left(q \,\|\, f\right) \geq 0$, we see that $\Phi(\boldsymbol{\theta}) \geq -\log\left(f(\mathcal{D})\right)$. Thus, $-\Phi(\boldsymbol{\theta})$ provides a lower bound on the log-evidence

$\log(f(\mathcal{D}))$. This can be useful in the context of model selection or in selecting hyperparameters. In statistical mechanics, an almost identical framework is used to approximate the free energy, which plays the role of the negative log-evidence. By assuming a completely factorised distribution for $q(\boldsymbol{x}|\boldsymbol{\theta})$ we obtain the *mean field approximation*. Using more sophisticated factorisations gives rise to more accurate approximations, however, often this is at the cost of increased complexity.

The variational free energy can also be written as

$$\Phi(\boldsymbol{\theta}) = -W_q(\boldsymbol{\theta}) - H_q(\boldsymbol{\theta})$$

where

$$W_q(\boldsymbol{\theta}) = \int q(\boldsymbol{x}|\boldsymbol{\theta}) \log\big(f(\boldsymbol{x}, \mathcal{D})\big) \, \mathrm{d}\boldsymbol{x} \quad H_q(\boldsymbol{\theta}), = -\int q(\boldsymbol{x}|\boldsymbol{\theta}) \log\big(q(\boldsymbol{x}|\boldsymbol{\theta})\big) \, \mathrm{d}\boldsymbol{x}.$$

$W_q(\boldsymbol{\theta})$ is minus the expected log-joint probability of the data and parameters with respect to the distribution $q(\boldsymbol{x}|\boldsymbol{\theta})$ and $H_q(\boldsymbol{\theta})$ is the entropy of distribution $q(\boldsymbol{x}|\boldsymbol{\theta})$ – the entropy measures the uncertainty in a distribution (we discuss entropy in Chapter 9). If we choose a sufficiently simple form for $q(\boldsymbol{x}|\boldsymbol{\theta})$ then computing $W_q(\boldsymbol{\theta})$ and $H_q(\boldsymbol{\theta})$ is tractable. Taking derivatives with respect to the parameters $\boldsymbol{\theta}$, allows us to minimise $\Phi(\boldsymbol{\theta})$ using gradient descent. This provides a method for finding an optimal $q(\boldsymbol{x}|\boldsymbol{\theta}^*)$ which minimises the KL divergence. Note that the variational free energy can be seen as a trade-off between two terms. The first term $W_q(\boldsymbol{\theta})$ becomes smaller as $q(\boldsymbol{x}|\boldsymbol{\theta})$ is more sharply peaked around the maximum of the joint probability $f(\boldsymbol{x}, \mathcal{D})$, while $H_q(\boldsymbol{\theta})$ is maximised when the distribution $q(\boldsymbol{x}|\boldsymbol{\theta})$ is as spread out (as uncertain) as possible. It is this interplay between fitting the data and maximising our uncertainty which leads to a reliable estimate of the parameters, $\boldsymbol{\theta}$.

Example 8.10 Inferring Voting Habits

This is a rather long example. Variational approximations tend to follow a similar pattern, but they are fairly involved calculations.

Consider the problem of trying to infer how people vote from a knowledge of their network of friends. We might know how a few people in the network vote and we assume that most (though clearly not all) friends will vote in a similar way. We give an example of small network in Figure 8.9. This is an example of a semi-supervised graph-labelling problem: we have a graph (the network of friends) and we want to infer labels (the political party) of the vertices (people). It is semi-supervised as we know some labels (we might have done some polling), but we also want to use the structure of the graph (vertices sharing an edge are more likely to have the same label).

In the abstract language of graphs, we consider a weighted graph $G(\mathcal{V}, \mathcal{E}, w)$ where $w(e)$ is the weight function applied to edges $e \in \mathcal{E}$ (we assume that $w(e) = 0$ if $e \notin \mathcal{E}$). In our problem the weights might denote the amount of contact between two people. For short, we will denote the weight of edge (i, j) by $w((i, j)) = w_{ij}$. In our scenario we are given labels for a subset of the vertices, $\mathcal{L} \subset \mathcal{V}$. We denote the

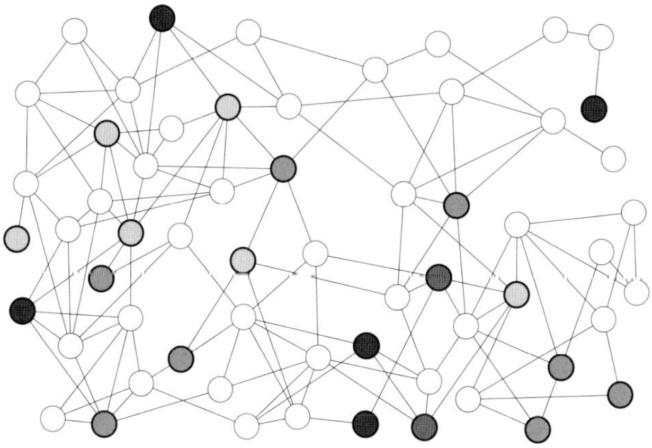

Figure 8.9 Network of friends with the voting intentions (colour coded) of a few individuals. The task is to infer likely voting preferences of the other individuals based on the assumption that there is an increased chance of friends voting similarly.

label of vertex i by S_i, which can take a value from the class set $\mathcal{C}$. Our task is to assign labels to the unobserved vertices $\mathcal{U} = \mathcal{V} \setminus \mathcal{L}$. We will denote the set of observed labels by $\boldsymbol{S}^o$ and the set of unobserved labels by $\boldsymbol{S}^u$.

We want to build a probabilistic model for the unobserved labels. We believe that vertices that share a high weight are more likely to have the same label. We model the joint probability of the observed and unobserved labels as

$$\mathbb{P}\left(\boldsymbol{S}^u, \boldsymbol{S}^o\right) = \frac{\mathrm{e}^{-\beta E(\boldsymbol{S}^u, \boldsymbol{S}^o)}}{Z(\beta)}, \qquad Z(\beta) = \sum_{\boldsymbol{S}^u, \boldsymbol{S}^o} \mathrm{e}^{-\beta E(\boldsymbol{S}^u, \boldsymbol{S}^o)} \tag{8.6}$$

where β is a parameter which reflects the degree to which neighbouring vertices have the same label and

$$E(\boldsymbol{S}^u, \boldsymbol{S}^o) = \sum_{(i,j)\in\mathcal{E}} w_{ij} \left[\!\left[S_i \neq S_j \right]\!\right]$$

(note that we use S_i to be either an observed or unobserved label). This problem has the structure of a statistical physics problem where β would be interpreted as an inverse temperature, $E(\boldsymbol{S}^u, \boldsymbol{S}^o)$ as an energy function, and $Z(\beta)$ is known as the partition function. This cost or energy function penalises edges whose vertices have different labels (in the language of combinatorial optimisation it is a min-cut cost function). We will see in Chapter 9 that this model arises naturally under a maximum entropy assumption (see Example 9.4 on page 270). The conditional probability $\mathbb{P}\left(\boldsymbol{S}^u | \boldsymbol{S}^o\right)$ (which in a Bayesian context would be the posterior probability – the observed labels $\boldsymbol{S}^o$ are our data) is given by

$$\mathbb{P}\left(\boldsymbol{S}^u | \boldsymbol{S}^o\right) = \frac{\mathrm{e}^{-\beta E(\boldsymbol{S}^u, \boldsymbol{S}^o)}}{Z'(\beta)}, \qquad Z'(\beta) = \sum_{\boldsymbol{S}^u} \mathrm{e}^{-\beta E(\boldsymbol{S}^u, \boldsymbol{S}^o)}$$

where the sum is over the labels of the unobserved vertices, i.e.

$$\sum_{S^u} \cdots = \left(\prod_{i \in \mathcal{U}} \sum_{S_i \in \mathcal{C}} \right) \cdots$$

In this model it is unclear what the parameter β should be. However, within the evidence framework one can choose β to maximise the probability of the data. This is given by

$$\mathbb{P}\left(S^o\right) = \frac{\mathbb{P}\left(S^u, S^o\right)}{\mathbb{P}\left(S^u | S^o\right)} = \frac{Z'(\beta)}{Z(\beta)}.$$

The difficulty of applying this formalism is that to compute the probabilities we have to sum over all possible labellings and there are $|\mathcal{C}|^{|\mathcal{V}|}$ labellings. This rapidly becomes intractable. The variational approximation allows us to obtain an approximation for these probabilities which is much easier to compute. In this case we minimise the variational free energy given by

$$\Phi(\boldsymbol{\theta}) = \sum_{S^u} \mathbb{Q}\left(S^u | \boldsymbol{\theta}\right) \log\left(\frac{\mathbb{Q}\left(S^u | \boldsymbol{\theta}\right)}{\mathbb{P}\left(S^u, S^o\right)} \right)$$

where $\mathbb{Q}\left(S^u | \boldsymbol{\theta}\right)$ is taken as a separable probability distribution

$$\mathbb{Q}\left(S^u | \boldsymbol{\theta}\right) = \prod_{i \in \mathcal{U}} \sum_{\mu \in \mathcal{C}} \theta_i^{\mu} \left[\!\left[S_i^u = \mu \right]\!\right]$$

with $\theta_i^{\mu} \geq 0$ and for all vertices and $\sum_{\mu \in \mathcal{C}} \theta_i^{\mu} = 1$ (note that the superscript u in S^u signifies these are unobserved labels, while the superscript $\mu \in \mathcal{C}$ in θ_i^{μ} denotes the different values that the labels can take). The parameters θ_i^{μ} can be interpreted as the marginal probability of the label for vertex i to be in class μ (in our example, this is the probability of person i voting for party μ). We have shown that the variational free energy is equal to

$$\Phi(\boldsymbol{\theta}) = \mathrm{KL}\left(\mathbb{Q}\left(S^u | \boldsymbol{\theta}\right) \| \mathbb{P}\left(S^u | S^o\right)\right) - \log(\mathbb{P}\left(S^o\right))$$

where $\mathrm{KL}\left(\mathbb{Q}\left(S^u | \boldsymbol{\theta}\right) \| \mathbb{P}\left(S^u | S^o\right)\right)$ is the KL divergence between the approximate distribution and the posterior. By minimising the variational free energy we make $\mathbb{Q}\left(S^u | \boldsymbol{\theta}\right)$ as close as possible to the posterior (as measured by the KL divergence). Furthermore, as the KL divergence is non-negative we obtain the bound that $\Phi(\boldsymbol{\theta}) \geq -\log(\mathbb{P}\left(S^o\right))$. Thus, if we choose $\boldsymbol{\theta}$ to minimise the variational free energy then we can use $-\Phi(\boldsymbol{\theta})$ as an approximation for the evidence. If we do this for different β values we can then choose the value of β which maximises $\Phi(\boldsymbol{\theta})$. That is, we can use the value of the variational free energy to select a model (in this case a β value) that maximises the probability of the data.

In statistical physics $-\beta \log(Z(\beta))$ is known as the free energy.

To compute the variational free energy we note that it can be rewritten

$$\Phi(\boldsymbol{\theta}) = -H_{\mathbb{Q}}(\boldsymbol{\theta}) - W_{\mathbb{Q}}(\boldsymbol{\theta}) = -H_{\mathbb{Q}}(\boldsymbol{\theta}) + \beta\, U_{\mathbb{Q}}(\boldsymbol{\theta}) + \log\big(Z(\beta)\big)$$

where $H(\mathbb{Q})$ is the entropy of $\mathbb{Q}\left(\boldsymbol{S}|\boldsymbol{\theta}\right)$

$$\begin{aligned}
H_{\mathbb{Q}}(\boldsymbol{\theta}) &= -\sum_{\boldsymbol{S}^u} \mathbb{Q}\left(\boldsymbol{S}^u|\boldsymbol{\theta}\right) \log\big(\mathbb{Q}\left(\boldsymbol{S}^u|\boldsymbol{\theta}\right)\big) \\
&= -\sum_{\boldsymbol{S}^u} \mathbb{Q}\left(\boldsymbol{S}^u|\boldsymbol{\theta}\right) \sum_{i\in\mathcal{U}} \log\left(\sum_{\mu\in\mathcal{C}} \theta_i^\mu \left[\!\left[S_i^u = \mu\right]\!\right]\right) \\
&= -\sum_{i\in\mathcal{U}} \sum_{\mu\in\mathcal{C}} \theta_i^\mu \log\left(\theta_i^\mu\right)
\end{aligned}$$

and

$$\begin{aligned}
W_{\mathbb{Q}}(\boldsymbol{\theta}) &= \mathbb{E}_{\mathbb{Q}}\left[\log\big(\mathbb{P}\left(\boldsymbol{S}^u, \boldsymbol{S}^o\right)\big)\right] = \mathbb{E}_{\mathbb{Q}}\left[\log\left(\frac{\mathrm{e}^{-\beta E(\boldsymbol{S}^u,\boldsymbol{S}^o)}}{Z(\beta)}\right)\right] \\
&= -\beta\, U_{\mathbb{Q}}(\boldsymbol{\theta}) - \log\big(Z(\beta)\big)
\end{aligned}$$

where $U_{\mathbb{Q}}(\boldsymbol{\theta})$ is the 'mean energy' with respect to the probability distribution $\mathbb{Q}\left(\boldsymbol{S}^u|\boldsymbol{\theta}\right)$

$$\begin{aligned}
U_{\mathbb{Q}}(\boldsymbol{\theta}) &= \sum_{\boldsymbol{S}^u} \mathbb{Q}\left(\boldsymbol{S}^u|\boldsymbol{\theta}\right) E(\boldsymbol{S}^u, \boldsymbol{S}^o) = \sum_{\boldsymbol{S}^u} \mathbb{Q}\left(\boldsymbol{S}^u|\boldsymbol{\theta}\right) \sum_{(i,j)\in\mathcal{E}} w_{ij} \left[\!\left[S_i \neq S_j\right]\!\right] \\
&= \frac{1}{2}\sum_{i\in\mathcal{U}}\sum_{j\in\mathcal{U}} w_{ij} \sum_{\mu,\nu\in\mathcal{C}} \theta_i^\mu \theta_j^\nu \left[\!\left[\mu \neq \nu\right]\!\right] + \sum_{i\in\mathcal{U}}\sum_{j\in\mathcal{L}} w_{ij} \sum_{\mu\in\mathcal{C}} \theta_i^\mu \left[\!\left[\mu \neq S_j^o\right]\!\right] \\
&\quad + \frac{1}{2}\sum_{i\in\mathcal{L}}\sum_{j\in\mathcal{L}} w_{ij} \left[\!\left[S_i^o \neq S_j^o\right]\!\right].
\end{aligned}$$

To find the minimum of the variational free energy subject to $\sum_{\mu\in\mathcal{C}} \theta_i^\mu = 1$ at each vertex we minimise the 'Lagrangian'

$$L(\boldsymbol{\theta}) = \Phi(\boldsymbol{\theta}) + \sum_{i\in\mathcal{U}} \lambda_i \left(\sum_{\mu\in\mathcal{C}} \theta_i^\mu - 1\right)$$

where the λ_is are a set of Lagrange multipliers that are chosen to ensure the constraints are satisfied (Appendix 7.C on page 182 provides a self-contained explanation of the use of Lagrange multipliers to solve constrained optimisation problems). The 'mean field equations' that are satisfied at the minima of the variational free energy are given by

$$\frac{\partial L(\boldsymbol{\theta})}{\partial \theta_i^\mu} = \log\left(\theta_i^\mu\right) + 1 + \beta \sum_{j\in\mathcal{U}} w_{ij} \sum_{\substack{\nu\in\mathcal{C}\\ \nu\neq\mu}} \theta_j^\nu + \beta \sum_{j\in\mathcal{L}} w_{ij} \left[\!\left[\mu \neq S_j^o\right]\!\right] + \lambda_i = 0.$$

Equivalently we can write this as

$$\theta_i^\mu = \mathrm{e}^{-\beta \sum_{j\in\mathcal{U}} w_{ij} \sum_{\substack{\nu\in\mathcal{C}\\ \nu\neq\mu}} \theta_j^\nu - \beta \sum_{j\in\mathcal{L}} w_{ij} \left[\!\left[\mu\neq S_j^o\right]\!\right] - \lambda_i - 1}.$$

We now choose λ_i to ensure that the constraint $\sum_{\mu\in\mathcal{C}} \theta_i^\mu = 1$ is satisfied. This gives

$$\theta_i^\mu = \frac{\mathrm{e}^{-\beta \tilde{E}_i^\mu(\boldsymbol{\theta}, \boldsymbol{S}^o)}}{\sum_{\nu\in\mathcal{C}} \mathrm{e}^{-\beta \tilde{E}_i^\nu(\boldsymbol{\theta}, \boldsymbol{S}^o)}} \tag{8.7}$$

where

$$\tilde{E}_i^\mu(\boldsymbol{\theta}, \boldsymbol{S}^o) = \sum_{j\in\mathcal{U}} w_{ij} \sum_{\substack{\nu\in\mathcal{C}\\ \nu\neq\mu}} \theta_j^\nu + \sum_{j\in\mathcal{L}} w_{ij} \left[\!\left[\mu \neq S_j^o\right]\!\right].$$

Using the identities $\sum_{\nu\in\mathcal{C}} \theta_i^\nu = 1$ and $\left[\!\left[\mu \neq S_j^o\right]\!\right] = 1 - \left[\!\left[\mu = S_j^o\right]\!\right]$, together with the observation that the equations that determine θ_i^μ are unchanged by adding a constant to the 'energy', $\tilde{E}_i^\mu(\boldsymbol{\theta}, \boldsymbol{S}^o)$ (it is cancelled by the normalisation), we can use the slightly simpler energy function

$$\tilde{E}_i^\mu(\boldsymbol{\theta}, \boldsymbol{S}^o) = -\sum_{j\in\mathcal{U}} w_{ij}\theta_j^\mu - \sum_{j\in\mathcal{L}} w_{ij} \left[\!\left[\mu = S_j^o\right]\!\right].$$

This has a very intuitive form. The energy is lower (more favourable) when the label at site i agrees with its neighbours. The right-hand side of Equation (8.7) is sometimes referred to as a soft-max function.

These equations cannot be solved in closed form. Furthermore, there can be many local solutions to the mean field equations so the quality of the solution will depend on how we solve these equations. A simple approach is to seek a solution through an iterative procedure by setting

$$\theta_i^\mu(t+1) = \frac{\mathrm{e}^{-\beta \tilde{E}_i^\mu(\boldsymbol{\theta}(t), \boldsymbol{S}^o)}}{\sum_{\nu\in\mathcal{C}} \mathrm{e}^{-\beta \tilde{E}_i^\nu(\boldsymbol{\theta}(t), \boldsymbol{S}^o)}}$$

where we start from $\theta_i^\mu(1) = 1/|\mathcal{C}|$. If we update the sites i sequentially, this algorithm can be viewed as a message-passing algorithm where the variables $(\theta_i^\mu|\mu \in \mathcal{C})$ are updated on receiving messages $(\theta_j^\mu|\mu \in \mathcal{C})$ from all of i's neighbours. To prevent ending up in a poor local optimum we can anneal the temperature (start from a low value of β and increase it to the require value) at each iteration. Although we have ended up with an iterative solution which is only an approximation to our full model, it is computable for even very large graphs. The computational complexity of each iteration is of order $|\mathcal{U}| \times |\mathcal{C}| \times N$ (where N is the average number of neighbouring vertices), which is very much smaller than the exponential complexity

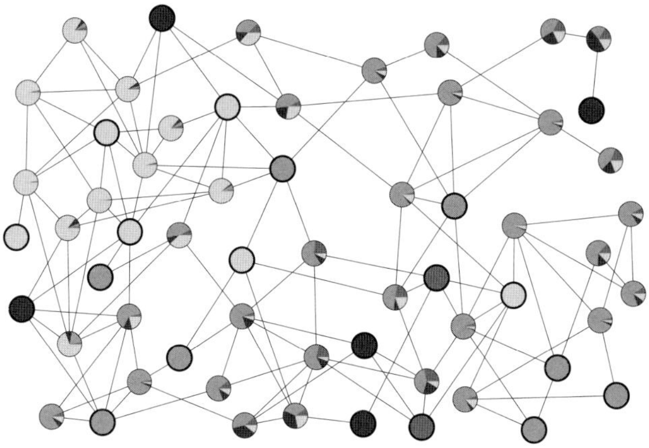

Figure 8.10 Inferred voting intentions of all members in the network. The nodes with a thick boarder are the observed labels. The other nodes show a pi-chart representation of θ_i^μ. The calculation was done at $\beta = 1$.

of the full model. We show a solution for the network introduced at the beginning of this example in Figure 8.10.

Variational approaches frequently give rise to rather natural algorithms, often with the influence of the neighbouring variables on a particular variable being replaced by their mean value (hence the name mean field approximation). Indeed, often as the number of neighbours grow these mean field theories can become more accurate, or in some cases exact. However, even in the limit where all variables are coupled so that the number of neighbours grow as n (and one might suppose the mean field to be accurate) there can be cases where the mean field approximation fails. One particular case where this happens is when some proportion of the variables are finely balanced so that if a variable changes its state it is likely to influence its neighbours, which in turn influences the original neighbour. In this case an additional reaction term is required to make the mean field equations more accurate. Interestingly, these corrections can often be derived if we minimise $\mathrm{KL}\left(f(\boldsymbol{x}|\mathcal{D}) \,\|\, q(\boldsymbol{x})\right)$ rather than minimising $\mathrm{KL}\left(q(\boldsymbol{x}) \,\|\, f(\boldsymbol{x}|\mathcal{D})\right)$ (recall from Section 7.3.1 on page 166 that reversing the KL divergence can lead to a very different solution). This goes by the name of *expectation propagation* and can be implemented using a message-passing algorithm. The scheme can be seen as a extension of *belief propagation* introduced by Pearl (1988) for computing marginal distributions in graphical models (see Section 8.5.2 on page 245). Variational approaches rapidly become rather complicated so that the interested reader is referred to the original paper of Minka (2001) and more specialised texts on machine learning such as Opper and Saad (2001); Bishop (2007), Raymond et al. (2014) and Barber (2011).

8.4 Latent Variable Models

When building a mathematical model of a process with uncertainty there are often times when the observed outcome depends on some unobserved states that we are uncertain about. We are often uninterested in the value of these unobserved states. This occurs in many hierarchical models where we introduce additional layers between the underlying model of interest and the data we actually observe (these additional layers more faithfully capture how the data would be generated). To build our model we can assign a random variable to these unobserved states. These variables are known as *latent variables* or sometimes *nuisance variables*. To compute the quantity of interest, such as a likelihood, we have to average over (marginalise out) the latent variable.

Example 8.11 Mixture of Gaussians
Consider the situation when we observe samples $(X_i | i = 1, 2, \ldots n)$ that we believe might come from two different mechanisms. For example, we might be measuring the half-lives of two short-lived species of particles. The problem is that our detector cannot distinguish the particles. Our data might look something like that shown in Figure 8.11. We would like to know the lengths of the half-lives of our two species. Note that we are in the opposite position to where we would want to apply the analysis of variance (see page 122). In that case, we know the features associated with the data, but wish to find which features explain the observed variance. Here, we believe we know the underlying cause of the variation, but we have no way of telling which data point is generated by which mechanism.

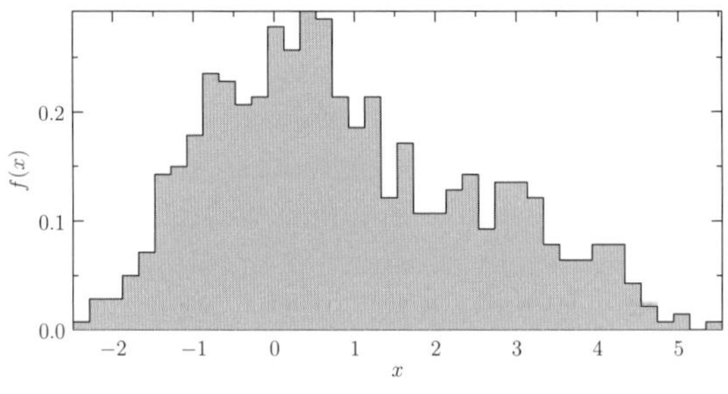

Figure 8.11 Example of data that are generated from two different mechanisms. In fact, this is a histogram of 500 deviates drawn from $\mathcal{N}(0, 1)$ and 200 deviates drawn from $\mathcal{N}(3, 1)$.

We assume that for both particles the data is normally distributed (but with different parameters). We can therefore model this process by including a latent variable Y_i equal to 1 if X_i is generated by mechanism 1 and 0 otherwise. The likelihood of a data point is

$$f(X_i|Y_i, \mu_1, \sigma_1^2, \mu_2, \sigma_2^2) = Y_i\, \mathcal{N}(X_i|\mu_1, \sigma_1^2) + (1 - Y_i)\, \mathcal{N}(X_i|\mu_2, \sigma_2^2)$$

while the joint likelihood is

$$f(X_i, Y_i) = \mathbb{P}\left(Y_i\right)\, f(X_i|Y_i, \mu_1, \sigma_1^2, \mu_2, \sigma_2^2).$$

Let $p_1 = \mathbb{P}\left(Y_i = 1\right)$ be the prior probability that the data point comes from the first distribution. Our task is to infer p_1, μ_1, σ_1^2, μ_2, and σ_2^2. To do this in a Bayesian framework we would assign a prior to our unknown parameters. Alternatively, we could assume a uniform distribution on all the priors and use maximum likelihood estimation instead (if we have sufficient data, then the exact prior we use is not that important). In both cases we have to average over the latent variables Y_i. We return to this problem in Example 8.12 on page 232.

Although latent variables add accuracy to a model they also add complexity. The resulting posteriors or likelihoods turns out to be a highly non-linear function of the parameters we are try to estimate. We can proceed by using a standard multidimensional optimisation algorithm. These algorithms are most efficient when we compute the gradient, then we can apply a standard algorithm such as the Broyden–Fletcher–Goldfarb–Shanno (BFGS) algorithm, which will rapidly converge to an optimum. However, there is an alternative route which often provides an elegant solution which is easier to program. This is the *expectation maximisation* or the EM algorithm. In general it can be used to maximise some probability (or probability density)

$$f_{\Theta}(\boldsymbol{\theta}) = \begin{cases} \sum_{\boldsymbol{y}} f_{\boldsymbol{Y},\Theta}(\boldsymbol{y},\boldsymbol{\theta}) \\ \int f_{\boldsymbol{Y},\Theta}(\boldsymbol{y},\boldsymbol{\theta})\,\mathrm{d}\boldsymbol{y} \end{cases}$$

where $\boldsymbol{Y}$ are unseen nuisance parameters (latent variables) and $\boldsymbol{\theta}$ are the parameters of the model we are trying to infer. It is used both for finding MAP and maximum likelihood solutions. It is an iterative algorithm proposed by Dempster et al. (1977), which provides a way of finding a sequence of parameter values $\boldsymbol{\theta}^{(1)}, \boldsymbol{\theta}^{(2)}, \ldots, \boldsymbol{\theta}^{(t)}$, each of which increases the probability $f_{\Theta}(\boldsymbol{\theta})$. The algorithm consists of iteratively *maximising* a function, $Q(\boldsymbol{\theta}|\boldsymbol{\theta}^{(t)})$, which acts as a surrogate for the logarithm of the true function, $\log\bigl(f_{\Theta}(\boldsymbol{\theta})\bigr)$. That is, at each cycle we compute

$$\boldsymbol{\theta}^{(t+1)} = \underset{\boldsymbol{\theta}}{\operatorname{argmax}}\, Q(\boldsymbol{\theta}|\boldsymbol{\theta}^{(t)})$$

where $Q(\boldsymbol{\theta}|\boldsymbol{\theta}^{(t)})$ is an *expectation* of the logarithm of the joint probability with respect to the old parameters

$$Q(\boldsymbol{\theta}|\boldsymbol{\theta}^{(t)}) = \mathbb{E}_{\boldsymbol{Y}|\boldsymbol{\theta}^{(t)}}\bigl[\log\bigl(f(\boldsymbol{Y},\boldsymbol{\theta})\bigr)\bigr] = \int \log\bigl(f(\boldsymbol{y},\boldsymbol{\theta})\bigr)\, f(\boldsymbol{y}|\boldsymbol{\theta}^{(t)})\mathrm{d}\boldsymbol{y}.$$

The reason this is useful is that we can often solve the maximisation in closed form, resulting in a very natural series of update equations for the parameters. (Somewhat confusingly the EM algorithm is sometimes presented as an iterative two-step algorithm. In the first (expectation) step the function $Q(\boldsymbol{\theta}|\boldsymbol{\theta}^{(t)})$ is defined, while in the second (maximisation) step we optimise it. This, however, does not correlate with the actual algorithm, which just involves applying a series of updated equations.)

To understand why the EM algorithm works we first note that

$$f(\boldsymbol{Y},\boldsymbol{\theta}) = f(\boldsymbol{Y}|\boldsymbol{\theta})\, f(\boldsymbol{\theta})$$

so that

$$\log(f(\boldsymbol{\theta})) = \log(f(\boldsymbol{Y},\boldsymbol{\theta})) - \log(f(\boldsymbol{Y}|\boldsymbol{\theta})).$$

Taking expectations of both sides with respect to $\boldsymbol{Y} \sim f(\boldsymbol{Y}|\boldsymbol{\theta}^{(t)})$ we get

$$\begin{aligned}\log(f(\boldsymbol{\theta})) &= \mathbb{E}_{\boldsymbol{Y}|\boldsymbol{\theta}^{(t)}}\left[\log(f(\boldsymbol{Y},\boldsymbol{\theta}))\right] - \mathbb{E}_{\boldsymbol{Y}|\boldsymbol{\theta}^{(t)}}\left[\log(f(\boldsymbol{Y}|\boldsymbol{\theta}))\right]\\ &= Q(\boldsymbol{\theta}|\boldsymbol{\theta}^{(t)}) + S(\boldsymbol{\theta}|\boldsymbol{\theta}^{(t)})\end{aligned}$$

(note that the left-hand side does not depend on Y so is unaffected by the expectation) where

$$S(\boldsymbol{\theta}|\boldsymbol{\theta}^{(t)}) = -\mathbb{E}_{\boldsymbol{Y}|\boldsymbol{\theta}^{(t)}}\left[\log(f(\boldsymbol{Y}|\boldsymbol{\theta}))\right] = -\int \log(f(\boldsymbol{y}|\boldsymbol{\theta}))\, f(\boldsymbol{y}|\boldsymbol{\theta}^{(t)})\mathrm{d}\boldsymbol{y}.$$

$Q(\boldsymbol{\theta}|\boldsymbol{\theta}^{(t)})$ is the expectation that we maximised above. We are now in a position to show that $f(\boldsymbol{\theta}^{(t+1)}) \geq f(\boldsymbol{\theta}^{(t)})$. That is, our updated parameters are at least as good as the current values. To see why this is, we consider

$$\log\left(f(\boldsymbol{\theta}^{(t+1)})\right) - \log\left(f(\boldsymbol{\theta}^{(t)})\right) = Q(\boldsymbol{\theta}^{(t+1)}|\boldsymbol{\theta}^{(t)}) - Q(\boldsymbol{\theta}^{(t)}|\boldsymbol{\theta}^{(t)}) + \mathrm{KL}\left(\boldsymbol{\theta}^{(t)} \,\middle\|\, \boldsymbol{\theta}^{(t+1)}\right)$$

where

$$\begin{aligned}\mathrm{KL}\left(\boldsymbol{\theta}^{(t)} \,\middle\|\, \boldsymbol{\theta}^{(t+1)}\right) &= S(\boldsymbol{\theta}^{(t+1)}|\boldsymbol{\theta}^{(t)}) - S(\boldsymbol{\theta}^{(t)}|\boldsymbol{\theta}^{(t)})\\ &= -\int f(\boldsymbol{\theta}^{(t)}) \log\left(\frac{f(\boldsymbol{\theta}^{(t+1)})}{f(\boldsymbol{\theta}^{(t)})}\right) \geq 0\end{aligned}$$

since $\mathrm{KL}\left(\boldsymbol{\theta}^{(t)} \,\middle\|\, \boldsymbol{\theta}^{(t+1)}\right)$ is the KL divergence between $f(\boldsymbol{\theta}^{(t)})$ and $f(\boldsymbol{\theta}^{(t+1)})$, and, as we have seen (Section 7.3.1 on page 166), KL divergences are non-negative. But, as

$$\boldsymbol{\theta}^{(t+1)} = \underset{\boldsymbol{\theta}}{\operatorname{argmax}}\, Q(\boldsymbol{\theta}|\boldsymbol{\theta}^{(t)})$$

we also have that

$$Q(\boldsymbol{\theta}^{(t+1)}|\boldsymbol{\theta}^{(t)}) \geq Q(\boldsymbol{\theta}^{(t)}|\boldsymbol{\theta}^{(t)});$$

consequently,

$$\log\left(f(\boldsymbol{\theta}^{(t+1)})\right) - \log\left(f(\boldsymbol{\theta}^{(t)})\right) \geq 0.$$

Thus, the series $f(\boldsymbol{\theta}^{(1)})$, $f(\boldsymbol{\theta}^{(2)})$, …, $f(\boldsymbol{\theta}^{(t)})$ is non-decreasing.

The previous discussion shows that the EM algorithm produces improving solutions, but doesn't give any intuition of why it is a reasonable method for optimising $f(\boldsymbol{\theta})$. To understand that we note that for all $\boldsymbol{\theta}$

$$\log(f(\boldsymbol{\theta})) = Q(\boldsymbol{\theta}|\boldsymbol{\theta}^{(t)}) + S(\boldsymbol{\theta}^{(t)}|\boldsymbol{\theta}^{(t)}) + \mathrm{KL}\left(\boldsymbol{\theta} \,\middle\|\, \boldsymbol{\theta}^{(t)}\right) \geq Q(\boldsymbol{\theta}|\boldsymbol{\theta}^{(t)}) + S(\boldsymbol{\theta}^{(t)}|\boldsymbol{\theta}^{(t)}).$$

We define $\hat{Q}(\boldsymbol{\theta}|\boldsymbol{\theta}^{(t)}) = Q(\boldsymbol{\theta}|\boldsymbol{\theta}^{(t)}) + S(\boldsymbol{\theta}^{(t)}|\boldsymbol{\theta}^{(t)})$, so that $\hat{Q}(\boldsymbol{\theta}|\boldsymbol{\theta}^{(t)})$ is a lower bound on $\log(f(\boldsymbol{\theta}))$. Note that $\hat{Q}(\boldsymbol{\theta}|\boldsymbol{\theta}^{(t)})$ differs from $Q(\boldsymbol{\theta}|\boldsymbol{\theta}^{(t)})$ by the constant $S(\boldsymbol{\theta}^{(t)}|\boldsymbol{\theta}^{(t)})$ so they share the same optimum value. At the point $\boldsymbol{\theta} = \boldsymbol{\theta}^{(t)}$

$$\log\left(f(\boldsymbol{\theta}^{(t)})\right) = \hat{Q}(\boldsymbol{\theta}^{(t)}|\boldsymbol{\theta}^{(t)}) + \mathrm{KL}\left(\boldsymbol{\theta}^{(t)} \,\middle\|\, \boldsymbol{\theta}^{(t)}\right) = \hat{Q}(\boldsymbol{\theta}^{(t)}|\boldsymbol{\theta}^{(t)})$$

so that $\hat{Q}(\boldsymbol{\theta}|\boldsymbol{\theta}^{(t)})$ equals $\log(f(\boldsymbol{\theta}))$ at $\boldsymbol{\theta} = \boldsymbol{\theta}^{(t)}$. We also observe that

$$\nabla_{\boldsymbol{\theta}} \log\left(f(\boldsymbol{\theta}^{(t)})\right) = \frac{1}{f(\boldsymbol{\theta})} \nabla_{\boldsymbol{\theta}} f(\boldsymbol{\theta})$$

while

$$\begin{aligned}\nabla_{\boldsymbol{\theta}}\hat{Q}(\boldsymbol{\theta}|\boldsymbol{\theta}^{(t)}) &\overset{(1)}{=} \nabla_{\boldsymbol{\theta}} \int \log(f(\boldsymbol{y},\boldsymbol{\theta}))\; f(\boldsymbol{y}|\boldsymbol{\theta}^{(t)})\,\mathrm{d}\boldsymbol{y} \\ &\overset{(2)}{=} \int \frac{f(\boldsymbol{y}|\boldsymbol{\theta}^{(t)})}{f(\boldsymbol{y},\boldsymbol{\theta})} \nabla_{\boldsymbol{\theta}} f(\boldsymbol{y},\boldsymbol{\theta})\,\mathrm{d}\boldsymbol{y} \overset{(3)}{=} \frac{1}{f(\boldsymbol{\theta})} \int \frac{f(\boldsymbol{y}|\boldsymbol{\theta}^{(t)})}{f(\boldsymbol{y}|\boldsymbol{\theta})} \nabla_{\boldsymbol{\theta}} f(\boldsymbol{y},\boldsymbol{\theta})\,\mathrm{d}\boldsymbol{y}\end{aligned}$$

(1) From the definition of $\hat{Q}(\boldsymbol{\theta}|\boldsymbol{\theta}^{(t)})$.
(2) Taking the gradient inside the integral and using the chain rule $\nabla_{\boldsymbol{\theta}} \log(f(\boldsymbol{y},\boldsymbol{\theta})) = \frac{\nabla_{\boldsymbol{\theta}} f(\boldsymbol{y},\boldsymbol{\theta})}{f(\boldsymbol{y},\boldsymbol{\theta})}$.
(3) Using Bayes' rule $f(\boldsymbol{y},\boldsymbol{\theta}) = f(\boldsymbol{\theta})\, f(\boldsymbol{y}|\boldsymbol{\theta})$.

At the point $\boldsymbol{\theta} = \boldsymbol{\theta}^{(t)}$

$$\begin{aligned}\nabla_{\boldsymbol{\theta}}\hat{Q}(\boldsymbol{\theta}|\boldsymbol{\theta}^{(t)})\Big|_{\boldsymbol{\theta}=\boldsymbol{\theta}^{(t)}} &\overset{(1)}{=} \frac{1}{f(\boldsymbol{\theta})} \int \frac{f(\boldsymbol{y}|\boldsymbol{\theta}^{(t)})}{f(\boldsymbol{y}|\boldsymbol{\theta})} \nabla_{\boldsymbol{\theta}} f(\boldsymbol{y},\boldsymbol{\theta})\,\mathrm{d}\boldsymbol{y}\Bigg|_{\boldsymbol{\theta}=\boldsymbol{\theta}^{(t)}} \\ &\overset{(2)}{=} \frac{1}{f(\boldsymbol{\theta}^{(t)})} \nabla_{\boldsymbol{\theta}^{(t)}} \int f(\boldsymbol{y},\boldsymbol{\theta}^{(t)})\,\mathrm{d}\boldsymbol{y} \\ &\overset{(3)}{=} \frac{1}{f(\boldsymbol{\theta}^{(t)})} \nabla_{\boldsymbol{\theta}^{(t)}} f(\boldsymbol{\theta}^{(t)}) \overset{(4)}{=} \nabla_{\boldsymbol{\theta}} \log(f(\boldsymbol{\theta}))\,\big|_{\boldsymbol{\theta}=\boldsymbol{\theta}^{(t)}}\end{aligned}$$

(1) Using the result obtained above.
(2) Replacing $\boldsymbol{\theta}$ by $\boldsymbol{\theta}^{(t)}$, cancelling terms, and taking the derivative outside the integral.
(3) Using $\int f(\boldsymbol{y},\boldsymbol{\theta}^{(t)})\,\mathrm{d}\boldsymbol{y} = f(\boldsymbol{\theta}^{(t)})$.
(4) From the chain rule $\nabla \log(f(\boldsymbol{\theta})) = \frac{\nabla f(\boldsymbol{\theta})}{f(\boldsymbol{\theta})}$.

Thus at the point $\boldsymbol{\theta} = \boldsymbol{\theta}^{(t)}$ the functions $\hat{Q}(\boldsymbol{\theta}|\boldsymbol{\theta}^{(t)})$ and $\log(f(\boldsymbol{\theta}))$ not only share the same function value, but also the same gradient. We leave it as an exercise for the interested reader to compute the second-order derivative for $\log(f(\boldsymbol{\theta}))$ and $\hat{Q}(\boldsymbol{\theta}|\boldsymbol{\theta}^{(t)})$. At $\boldsymbol{\theta} \neq \boldsymbol{\theta}^{(t)}$ the two quantities differ, but have a similar structure – this is important to ensure that the step size of the improving move is approximately correct. Thus, $\hat{Q}(\boldsymbol{\theta}|\boldsymbol{\theta}^{(t)})$ plays a surrogate role for $\log(f(\boldsymbol{\theta}))$ and finding the maximum of $\hat{Q}(\boldsymbol{\theta}|\boldsymbol{\theta}^{(t)})$, or equivalently $Q(\boldsymbol{\theta}|\boldsymbol{\theta}^{(t)})$, provides a reasonable approximation for the maximum of $\log(f(\boldsymbol{\theta}))$. In Figure 8.12 we schematically illustrate what $\log(f(\boldsymbol{\theta}))$ and $\hat{Q}(\boldsymbol{\theta}|\boldsymbol{\theta}^{(t)})$ might look in a one-dimensional problem.

One final note of caution though, the EM algorithm will converge towards a local optimum of $f(\boldsymbol{\theta})$, but this is not guaranteed to be the global optimum

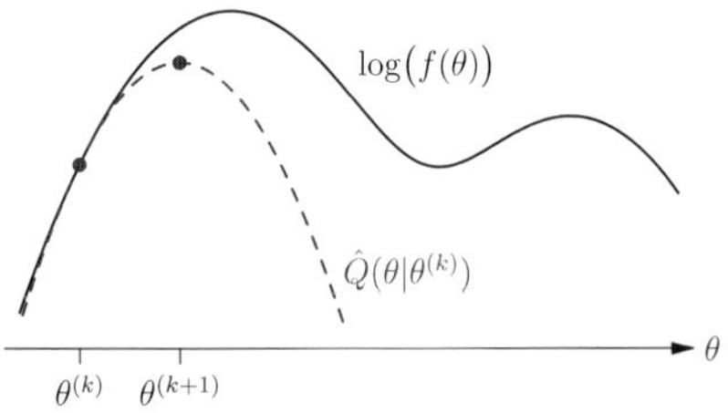

Figure 8.12 Schematic showing a one-dimensional example of the true log-probability $\log\big(f(\theta)\big)$ and the surrogate function $\hat{Q}(\theta|\theta^{(t)})$ used by the EM algorithm.

in the case when $f(\boldsymbol{\theta})$ is multimodal. This, of course, is also true of a gradient-based search method such as BFGS. To find a good solution it is important to use sensible initial parameters, $\boldsymbol{\theta}^{(1)}$.

Example 8.12 Mixture of Gaussians Revisited
We continue our discussion of Example 8.11 on page 228. To briefly recap we have some observations $\mathcal{D} = (X_i|i = 1, 2, \dots n)$, which we believe comes from two different normal distributions. We introduce latent variables Y_i equal to 1 if the data is generated by the first distribution and 0 if it is generated by the second distribution. The likelihood of a data point and the latent variables is thus

$$f(X_i, Y_i|p_1, \mu_1, \sigma_1^2, \mu_2, \sigma_2^2) = Y_i\, p_1\, \mathcal{N}(X_i|\mu_1, \sigma_1^2) + (1 - Y_i)\, p_2\, \mathcal{N}(X_i|\mu_2, \sigma_2^2)$$

where $p_2 = 1 - p_1$. Our task is to infer the unknown variables

$$\boldsymbol{\theta} = (p_1, \mu_1, \sigma_1^2, \mu_2, \sigma_2^2).$$

We could do this through maximising the posterior by putting priors on the parameters or just by maximising the likelihoods. To prevent this chapter expanding even further, we consider the slightly easier maximum likelihood calculation. In this case, we want to maximise the marginalised likelihood (or equivalently log-likelihood)

$$\log\big(f(\mathcal{D}|\boldsymbol{\theta})\big) = \sum_{i=1}^{n} \log\big(f(X_i|\boldsymbol{\theta})\big) = \sum_{i=1}^{n} \log\left(\sum_{Y_i \in \{0,1\}} f(X_i, Y_i|\boldsymbol{\theta}) \right)$$

where the marginalised likelihood is

$$\begin{aligned} f(X_i|\boldsymbol{\theta}) &= \sum_{Y_i \in \{0,1\}} f(X_i, Y_i = 1|\boldsymbol{\theta}) \\ &= p_1\, \mathcal{N}(X_i|\mu_1, \sigma_1^2) + p_2\, \mathcal{N}(X_i|\mu_2, \sigma_2^2). \end{aligned}$$

This is a non-linear equation in unknown parameters, so that it is awkward to solve. This is where the EM algorithm comes to the rescue. Instead of maximising the log-likelihood we maximise

$$
\begin{aligned}
Q(\boldsymbol{\theta}|\boldsymbol{\theta}^{(t)}) &\overset{(1)}{=} \mathbb{E}_{\boldsymbol{Y}|\boldsymbol{\theta}^{(t)}}\left[\log\left(\prod_{i=1}^{n} f(X_i, Y_i|\boldsymbol{\theta})\right)\right] \overset{(2)}{=} \sum_{i=1}^{n} \mathbb{E}_{Y_i|\boldsymbol{\theta}^{(t)}}\left[\log\big(f(X_i, Y_i|\boldsymbol{\theta})\big)\right] \\
&\overset{(3)}{=} \sum_{i=1}^{n} \sum_{Y_i \in \{0,1\}} \mathbb{P}\left(Y_i \middle| X_i, \boldsymbol{\theta}^{(t)}\right) \log\big(f(X_i, Y_i|\boldsymbol{\theta})\big) \\
&\overset{(4)}{=} \sum_{i=1}^{n} \sum_{k=1}^{2} \mathbb{P}\left(\llbracket k=1 \rrbracket \,|X_i, \boldsymbol{\theta}^{(t)}\right) \log\left(p_k \, \mathcal{N}(X_i|\mu_k, \sigma_k^2)\right) \\
&\overset{(5)}{=} \sum_{i=1}^{n} \sum_{k=1}^{2} \mathbb{P}\left(\llbracket k=1 \rrbracket \,|X_i, \boldsymbol{\theta}^{(t)}\right) \\
&\qquad \left(\log(p_k) - \frac{(X_i - \mu_k)^2}{2\sigma_k^2} - \log\left(\sqrt{2\pi}\,\sigma_k\right)\right)
\end{aligned}
$$

(1) From the definition $Q(\boldsymbol{\theta}|\boldsymbol{\theta}^{(t)}) = \mathbb{E}_{\boldsymbol{Y}|\boldsymbol{\theta}^{(t)}}\left[\log\big(f(\boldsymbol{X}, \boldsymbol{Y}, \boldsymbol{\theta})\big)\right]$ and assuming the data points are independent.
(2) Using $\log\left(\prod_i a_i\right) = \sum_i \log(a_i)$ and the linearity of expectations.
(3) Writing out the expectation explicitly.
(4) When $Y_i = 1$ then X_i comes from the first distribution while if $Y_i = 0$ they come from the second distribution.
(5) Putting in the explicit form for the normal distribution and expanding the logarithm,

where

$$
\begin{aligned}
\mathbb{P}\left(\llbracket k=1 \rrbracket \,|X_i, \boldsymbol{\theta}^{(t)}\right) &\overset{(1)}{=} \frac{\mathbb{P}\left(X_i \middle| \llbracket k=1 \rrbracket, \boldsymbol{\theta}^{(t)}\right) p_k^{(t)}}{\mathbb{P}\left(X_i|\boldsymbol{\theta}^{(t)}\right)} \\
&\overset{(2)}{=} \frac{\mathcal{N}(X_i|\mu_k^{(t)}, (\sigma_k^{(t)})^2)\, p_k^{(t)}}{\mathcal{N}(X_i|\mu_1^{(t)}, (\sigma_1^{(t)})^2)\, p_1^{(t)} + \mathcal{N}(X_i|\mu_2^{(t)}, (\sigma_2^{(t)})^2)\, p_2^{(t)}}
\end{aligned}
$$

(1) Using Bayes' rule where the prior probability before seeing data X_i is $\mathbb{P}\left(\llbracket k=1 \rrbracket \,|\boldsymbol{\theta}^{(t)}\right) = p_k^{(t)}$.
(2) Using $\mathbb{P}\left(X_i \middle| \llbracket k=1 \rrbracket, \boldsymbol{\theta}^{(t)}\right) = \mathcal{N}(X_i|\mu_k^{(t)}, (\sigma_k^{(t)})^2)$ and the fact that the denominator is a normalisation condition.

To maximise $Q(\boldsymbol{\theta}|\boldsymbol{\theta}^{(t)})$ with respect to μ_k we set the partial derivative to zero

$$
\frac{\partial Q(\boldsymbol{\theta}|\boldsymbol{\theta}^{(t)})}{\partial \mu_k} = \sum_{i=1}^{n} \mathbb{P}\left(\llbracket k=1 \rrbracket \,|X_i, \boldsymbol{\theta}^{(t)}\right) \left(\frac{\mu_k - X_i}{\sigma_k^2}\right) = 0.
$$

Solving for μ_k we find the updated means are

$$
\mu_k^{(t+1)} = \frac{\sum_{i=1}^{n} \mathbb{P}\left(\llbracket k=1 \rrbracket \,|X_i, \boldsymbol{\theta}^{(t)}\right) X_i}{\sum_{i=1}^{n} \mathbb{P}\left(\llbracket k=1 \rrbracket \,|X_i, \boldsymbol{\theta}^{(t)}\right)}. \tag{8.8}
$$

To update the variances we again set the partial derivatives to zero

$$\frac{\partial Q(\boldsymbol{\theta}|\boldsymbol{\theta}^{(t)})}{\partial \sigma_k} = \sum_{i=1}^{n} \mathbb{P}\left(\left[\!\left[k = 1 \right]\!\right] | X_i, \boldsymbol{\theta}^{(t)}\right) \left(\frac{(X_i - \mu_k)^2}{\sigma_k^3} - \frac{1}{\sigma_k} \right) = 0$$

Solving for the variance (and plugging in the updated mean) we find the updated variance is

$$(\sigma_k^{(t+1)})^2 = \frac{\sum_{i=1}^{n} \mathbb{P}\left(\left[\!\left[k = 1 \right]\!\right] | X_i, \boldsymbol{\theta}^{(t)}\right) (X_i - \mu_k^{(t+1)})^2}{\sum_{i=1}^{n} \mathbb{P}\left(\left[\!\left[k = 1 \right]\!\right] | X_i, \boldsymbol{\theta}^{(t)}\right)}. \tag{8.9}$$

Finally, to learn the priors p_k, we first note that $p_2 = 1 - p_1$. Taking derivatives with respect to p_1 we get

$$\frac{\partial Q(\boldsymbol{\theta}|\boldsymbol{\theta}^{(t)})}{\partial p_1} = \sum_{i=1}^{n} \left(\frac{\mathbb{P}\left(\left[\!\left[k = 1 \right]\!\right] | X_i, \boldsymbol{\theta}^{(t)}\right)}{p_1} - \frac{\mathbb{P}\left(\left[\!\left[k = 2 \right]\!\right] | X_i, \boldsymbol{\theta}^{(t)}\right)}{1 - p_1} \right) = 0.$$

Multiplying through by $p_1(1 - p_1)$, and using the fact that $\mathbb{P}\left(\left[\!\left[k = 1 \right]\!\right] | X_i, \boldsymbol{\theta}^{(t)}\right) + \mathbb{P}\left(\left[\!\left[k = 2 \right]\!\right] | X_i, \boldsymbol{\theta}^{(t)}\right) = 1$, we find

$$p_1^{(t+1)} = \frac{1}{n} \sum_{i=1}^{n} \mathbb{P}\left(\left[\!\left[k = 1 \right]\!\right] | X_i, \boldsymbol{\theta}^{(t)}\right). \tag{8.10}$$

Observe that Equations (8.8)–(8.10) are very natural update rules. We use them on the data in Figure 8.11, starting from $p_1^{(1)} = p_2^{(1)} = 1/2$, $\mu_1^{(1)} = -1$, $(\sigma_1^{(1)})^2 = 1.2$, $\mu_1^{(1)} = 2$, and $(\sigma_1^{(1)})^2 = 0.8$. The trajectories of the EM algorithm are shown in Figure 8.13.

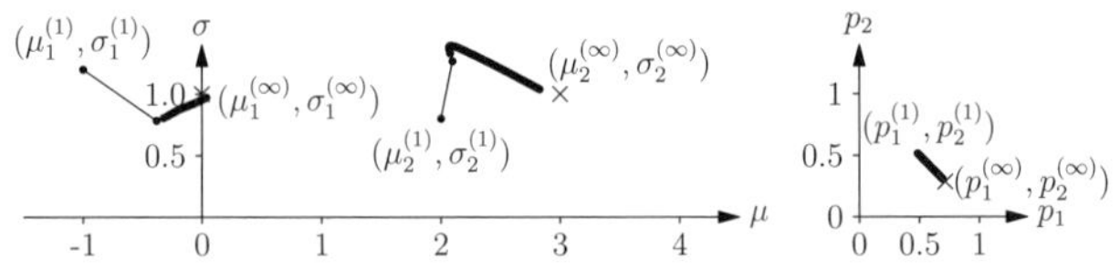

Figure 8.13 Trajectory of the maximum likelihood parameters found by the EM algorithm. Grey crosses mark the parameters used to generate the data.

As long as we start close enough to an optimum value we converge towards the same point. However, the convergence can be quite slow, as in this example. This is one of drawbacks of using the EM algorithm – the convergence can be significantly slower than a gradient-based optimiser. The final fit from the mixture of Gaussians is shown in Figure 8.14.

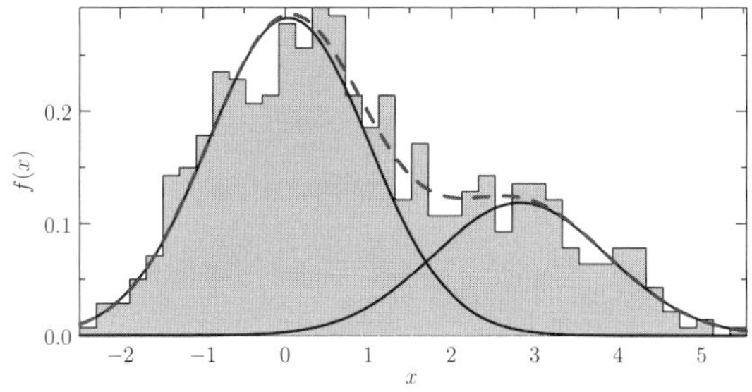

Figure 8.14 Histogram of the data together with the maximum likelihood estimates for the two-Gaussian model. The dashed line shows the sum of the two models.

One of the nice features of the EM algorithm is that it is easily extended to multiple Gaussians and to Gaussians in higher dimensions.

Latent variable models are often used in cases where we do not explicitly know the underlying mechanism generating the data, but wish to have a more flexible model. For example, if we want to have a parameterised model for a distribution (possible in high dimensions), then we could use a mixture of Gaussian models with some moderate number of Gaussians (depending on the amount of data). The hope would be to capture some unknown structure in the probability distribution. Latent variable models come in many different forms. We next consider one of the most commonly used such models: hidden Markov models (HMMs).

8.4.1 Hidden Markov Models

An HMM is a popular probabilistic model for describing sequence data. It is heavily used in speech analysis and biological sequence analysis (e.g. for DNA and proteins). The model consists of a set of states which can emit symbols. There are emission probabilities associated with each state. Usually there is a special initial state which emits no symbol. There are also transition probabilities between states. The model generates a sequence by first transitioning from the initial state to one of the emitting states according to the transition probabilities. Then it emits a symbol according to the emission probability of the state that it is in. It then moves to a new state according to the transition probabilities of the current state (there is often a transition probability back to itself so the HMM can stay in the same state for several cycles). The HMM continues to emit a symbol and then make a transition. This either happens until a sequence of a certain length has been generated or until a stop state is reached (not all HMMs will include a stop state). The HMM is acting as a probabilistic finite state machine.

Example 8.13 Switching Coins
Consider someone with two coins. One honest and one dishonest with a probability of 0.8 of coming up heads. The person choose one coin at random and then after every throw he either uses the same coin with a probability 0.7 or changes coins with a probability 0.3. We

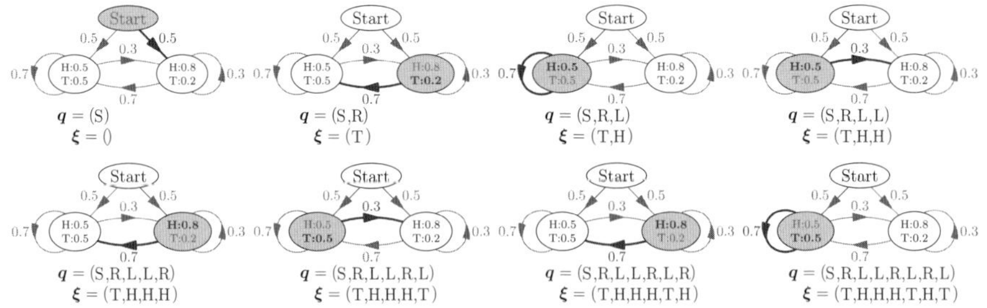

Figure 8.15 A sample run of an HMM for flipping two different coins. The vector $\boldsymbol{q}$ represents states (start, left, or right), while $\boldsymbol{\xi}$ represents the set of observed events (head or tail).

can build an HMM that models this scenario. We show a sequence generated by the HMM in Figure 8.15.

The HMM can generate many different sequences, but it can also be used to assign probabilities to sequences. That is, the probability of a sequence is the probability that it will be generated by an HMM. Given a sequence, it is possible to calculate the probability of it being generated even when we don't know which states generated the sequence (the states are hidden, hence the name). This requires us to average over all possible paths through the states. Calculating this efficiently is non-trivial as there are an exponential number of paths through the states which could generate the sequence. However, because of the Markovian property of the model we can use dynamic programming to efficiently sum over all possible paths. The model can be viewed as a latent variable model where the states correspond to latent variables.

In Example 8.13 we knew the parameters of the model (transition and emission probabilities) as well as the structure of the model (number of states, which states are connected, etc.). Often HMMs are used where we don't know the exact model. In this case, if we have enough sequences we can try to learn the model. Learning the structure of the model is very difficult (we discuss this problem at the end of this section). However, we can learn the parameters of the HMM by maximising the likelihood of the data. This can be done efficiently using an EM algorithm (which is known in this context as Baum–Welch). This is doubly clever as it uses both the idea of the EM algorithm and dynamic programming.

We briefly describe the details of making inferences and training a standard HMM. This involves a lot of details that you might find overwhelming or of little interest, in which case just skip to the next section. We want to assign a probability to an observed sequence of symbols, $\boldsymbol{\xi} = (\xi_1, \cdots, \xi_T)$. Each symbol comes from a finite alphabet $\mathcal{A}$ (in the example above this would be $\{H, T\}$, in protein analysis this is the set of 20 amino acids). We denote the set of states that make up the HMM by $\mathcal{Q}$. At each step, the HMM will emit a symbol and make a transition between states. The state of the model at time step t is denoted q_t. The transition probabilities for moving from state i to state j is denoted by $a_{ij} = \mathbb{P}\left(q_t = j | q_{t-1} = i\right)$; while emission probabilities for 'emitting' a symbol ζ, given that we are in a state i, is denoted by $e_i(\zeta) = \mathbb{P}\left(\xi_t = \zeta | q_t = i\right)$. We denote the set of transition and emission probabilities by $\boldsymbol{\theta}$ – these are the parameters of the model which we wish to learn from the data. For any given state sequence

$\mathbf{q} = (q_1, \cdots, q_T)$, the joint probability for the HMM visiting those states and emitting a symbol sequence $\boldsymbol{\xi}$ is given by

$$\mathbb{P}\left(\boldsymbol{\xi}, \boldsymbol{q}|\boldsymbol{\theta}\right) = \prod_{t=1}^{T} a_{q_{t-1}q_t}\, e_{q_t}(\xi_t) = a_{q_0 q_1}\, e_{q_1}(\xi_1)\, a_{q_1 q_2}\, e_{q_2}(\xi_2) \cdots a_{q_{T-1} q_T}\, e_{q_T}(\xi_T).$$

The initial state (q_0) is assumed not to emit a symbol. To compute the likelihood of a sequence $\boldsymbol{\xi}$ we sum over all possible paths through the hidden states

$$\mathbb{P}\left(\boldsymbol{\xi}|\boldsymbol{\theta}\right) = \sum_{\boldsymbol{q} \in \mathcal{Q}^T} \mathbb{P}\left(\boldsymbol{\xi}, \boldsymbol{q}|\boldsymbol{\theta}\right). \tag{8.11}$$

Note that the actual state of the system at any time step is hidden (since the states are marginalised over).

Naively summing over all possible state paths appears computationally expensive since there are an exponential number of paths. However, all quantities of interest can be efficiently computed from the forward and backward variables $(\alpha_t(i), \beta_t(i))$, which are defined as

$$\alpha_t(i) = \mathbb{P}\left(\xi_1, \cdots, \xi_t, q_t = i|\boldsymbol{\theta}\right)$$
$$\beta_t(i) = \mathbb{P}\left(\xi_{t+1}, \cdots, \xi_T|q_t = i,\, \boldsymbol{\theta}\right).$$

The forward variables, $\alpha_t(i)$, can be calculated using the forward algorithm

$$\alpha_t(i) = \sum_{j \in \mathcal{Q}} \mathbb{P}\left(\xi_t, q_t = i|q_{t-1} = j,\, \boldsymbol{\theta}\right)\, \mathbb{P}\left(\xi_1, \cdots, \xi_{t-1}, q_{t-1} = j|\boldsymbol{\theta}\right)$$
$$= \sum_{j \in \mathcal{Q}} e_i(\xi_t)\, a_{ji}\, \alpha_{t-1}(j) = e_i(\xi_t) \sum_{j \in \mathcal{Q}} a_{ji}\, \alpha_{t-1}(j)$$

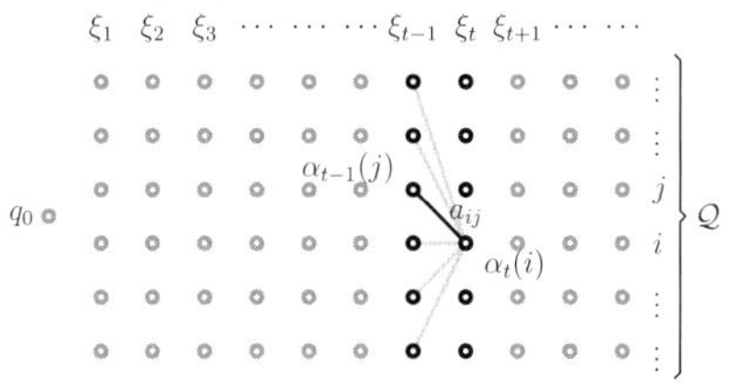

starting with $\alpha_0(i) = [\![i = s]\!]$ (where state s is taken to be initial state).

Note that the computation of $\alpha_t(i)$ is efficient because of the Markovian property of the HMM. That is, the probability of the next emission and transition depends only on the current state of the HMM and is conditionally independent of all previous history. The probability of a sequence of length T is given by

$$\mathbb{P}\left(\boldsymbol{\xi}|\boldsymbol{\theta}\right) = \sum_{i \in \mathcal{Q}} \alpha_T(i)\, a_{ie}$$

where e is the end state (often we don't have end states when we replace a_{ie} by 1). The backward variables, $\beta_t(i)$, are computed from the backward algorithm

$$\beta_{t-1}(i) = \sum_{j \in \mathcal{Q}} \mathbb{P}\left(\xi_t, q_t = j|q_{t-1} = i\right)\, \mathbb{P}\left(\xi_{t+1}, \cdots, \xi_T|q_t = j\right) = \sum_{j \in \mathcal{Q}} a_{ij}\, e_j(\xi_t)\, \beta_t(j)$$

with $\beta_T(i) = 1$. In practice some care is necessary in computing these quantities as it is easy for the variables to underflow. An example of the use of the forward and backward variables is to determine the probability of being in a particular state at step t given the observed sequence

$$\mathbb{P}\left(q_t = i|\boldsymbol{\xi}, \boldsymbol{\theta}\right) = \frac{\mathbb{P}\left(\boldsymbol{\xi}, q_t = i\right)}{\mathbb{P}\left(\boldsymbol{\xi}|\boldsymbol{\theta}\right)} = \frac{\alpha_t(i)\,\beta_t(i)}{\mathbb{P}\left(\boldsymbol{\xi}|\boldsymbol{\theta}\right)}. \tag{8.12}$$

We can train an HMM (i.e. find a good set of values of the transition and emission probabilities) if we are given a set of training sequences $\mathcal{D} = \{\boldsymbol{\xi}^\mu|\mu = 1, 2, \ldots, n\}$. If for each training sequence we know the path through the set of states which generated the sequence then we can count the number of transitions, A_{ij}, from state i to state j and, similarly, the number of times, $E_i(\zeta)$, we emit a symbol ζ in state i (we return to the problem posed by this assumption later). To maximise the likelihood of the data we set the emission and transition probability according to

$$e_i(\zeta) = \frac{E_i(\zeta)}{\sum\limits_{\xi \in \mathcal{A}} E_i(\xi)}, \qquad a_{ij} = \frac{A_{ij}}{\sum\limits_{k \in \mathcal{Q}} A_{ik}}. \tag{8.13}$$

These maximum likelihood values are somewhat dangerous as we can easily end up with transition and emission probabilities equal to zero simply because our training data doesn't have any examples of such transition or emission even though they may well occur. To overcome this we can add pseudo-counts (see Exercise 8.3), $r_i(\zeta)$ and r_{ij}, such that

$$e_i(\zeta) = \frac{E_i(\zeta) + r_i(\zeta)}{\sum\limits_{\xi \in \mathcal{A}} \left(E_i(\xi) + r_i(\xi)\right)}, \qquad a_{ij} = \frac{A_{ij} + r_{ij}}{\sum\limits_{k \in \mathcal{Q}} (A_{ik} + r_{ik})}, \tag{8.14}$$

where r_{ij} and $r_i(\zeta)$ encode our prior belief in the existence of a transition from state i to j and of emitting a symbol ζ from state i, respectively. These would often all be set to a small constant.

The problem of applying the above algorithm to calculate the emission and transition probabilities is that we don't know which state the HMM is in when it emits a symbol. What state we are in is probabilistic (hidden), but, worse, it depends on the parameters, which is what we are trying to learn. We are in a chicken and egg situation. To overcome this we use an iterative algorithm starting from some random emission and transition variables. We can then infer the probability of being in a state using Equation (8.12). We then use the update Equations (8.13) or (8.14) to increase the likelihood of the training sequences. By iterating many times we can nudge the parameters towards a local optimum.

To be more explicit, for the training sequences the expected number of times that symbol ζ is emitted while in state i is given by

$$E_i(\zeta) = \sum_{\boldsymbol{\xi}^\mu \in \mathcal{D}} \sum_{t=1}^{T^\mu} \mathbb{P}\left(q_t = i|\boldsymbol{\xi}^\mu, \boldsymbol{\theta}\right) \left[\!\left[\xi_t^\mu = \zeta\right]\!\right]$$

where T^μ is the length of sequence $\boldsymbol{\xi}^\mu$. This is just the probability that, for sequence $\boldsymbol{\xi}^\mu$, the HMM is in state i given that the observed symbol is ξ_t, summed over the set of sequences and over all elements of the sequence t. The probability of being in state i at time t for sequence $\boldsymbol{\xi}^\mu$ is

$$\mathbb{P}\left(q_t = i|\boldsymbol{\xi}^\mu, \boldsymbol{\theta}\right) = \frac{\mathbb{P}\left(q_t = i, \boldsymbol{\xi}^\mu|\boldsymbol{\theta}\right)}{\mathbb{P}\left(\boldsymbol{\xi}^\mu|\boldsymbol{\theta}\right)} = \frac{\alpha_t^\mu(i)\, \beta_t^\mu(i)}{\mathbb{P}\left(\boldsymbol{\xi}^\mu|\boldsymbol{\theta}\right)}$$

where $\alpha_t^\mu(i)$ and $\beta_t^\mu(i)$ are the forward and backward variables for sequence $\boldsymbol{\xi}^\mu$. Thus,

$$E_i(\zeta) = \sum_{\boldsymbol{\xi}^\mu \in \mathcal{D}} \frac{1}{\mathbb{P}\left(\boldsymbol{\xi}^\mu|\boldsymbol{\theta}\right)} \sum_{t=1}^{T^\mu} \alpha_t^\mu(i)\, \beta_t^\mu(i) \left[\!\left[\xi_t^\mu = \zeta \right]\!\right].$$

Similarly, the expected number of transitions from state i to state j that occur in our training sequences is

$$A_{ij} = \sum_{\boldsymbol{\xi}^\mu \in \mathcal{D}} \sum_{t=1}^{T^\mu - 1} \mathbb{P}\left(q_t = i,\, q_{t+1} = j|\boldsymbol{\xi}^\mu, \boldsymbol{\theta}\right).$$

From Bayes' rule

$$\mathbb{P}\left(q_t = i, q_{t+1} = j|\boldsymbol{\xi}^\mu,\, \boldsymbol{\theta}\right) = \frac{\mathbb{P}\left(q_t = i,\, q_{t+1} = j,\, \boldsymbol{\xi}^\mu|\boldsymbol{\theta}\right)}{\mathbb{P}\left(\boldsymbol{\xi}^\mu|\boldsymbol{\theta}\right)}$$

but (using the Markov properties of the HMM)

$$\begin{aligned}
&\mathbb{P}\left(q_t = i,\, q_{t+1} = j,\, \boldsymbol{\xi}^\mu|\boldsymbol{\theta}\right) \\
&\quad = \mathbb{P}\left(q_t = i,\, \xi_1,\, \ldots,\, \xi_t|\boldsymbol{\theta}\right) \mathbb{P}\left(q_{t+1} = j,\, \xi_{t+1}|q_t = i, \boldsymbol{\theta}\right) \\
&\qquad \mathbb{P}\left(\xi_{t+2},\, \ldots,\, \xi_T|q_{t+1} = j, \boldsymbol{\theta}\right) \\
&\quad = \alpha_t^\mu(i)\, a_{ij}\, e_j(\xi_{t+1})\, \beta_{t+1}^\mu(j).
\end{aligned}$$

Thus,

$$A_{ij} = \sum_{\boldsymbol{\xi}^\mu \in \mathcal{D}} \frac{1}{\mathbb{P}\left(\boldsymbol{\xi}^\mu|\boldsymbol{\theta}\right)} \sum_{t=1}^{T^\mu - 1} \alpha_t^\mu(i)\, a_{ij}\, e_j(\xi_{t+1})\, \beta_{t+1}^\mu(j).$$

This algorithm is known as the Baum–Welch algorithm after its two inventors (Leonard E. Baum being the driving force). The Baum–Welch algorithm can be viewed as an EM type algorithm. As we have seen this is guaranteed to converge to a local maximum of the likelihood (although it is not guaranteed to find the global maximum). More details can be found in Rabiner (1989) and Durbin et al. (1998). Baum–Welch is by far the most commonly used algorithm for training HMMs, but HMMs can also be trained by using a gradient-based method on the parameters. Although more complicated, this often has faster convergence.

HMMs are used in many application areas involving sequences. For example, they are commonly used in biological data analysis. In particular, we might want to detect which parts of a protein sequence will wrap up into an alpha helix or form part of a beta sheet. For this problem there exists a large data set of training examples. We can use this together with the Baum–Welch algorithm to train an HMM so as to maximise the likelihood of these sequences corresponding to an alpha helix, say. We can then use the HMM on new sequences to see which parts are likely to form an alpha helix. The hard part is determining the structure of the HMM. Often this is done by a domain expert, trying to represent what they believe is important. However, for complex problems like protein structure prediction, the domain expert is usually beaten by an automated search algorithm (if done sensibly). In many applications, such as speech recognition, little attempt is made to rationalise the meaning of the hidden states. The HMM is just used as a trainable network with enough flexibility to capture information about the sequences. The final rationale for using some number of states is that it works – at least for many applications. However, care has to be made in choosing the topology of the HMM (number of states and non-zero transitions between states). This is because we are maximising the likelihood of the data which is apt to over-fit the data and give poor generalisation on unseen data. To make HMMs Bayesian we would need to put prior probabilities on the topology of the HMM and the parameters. There is no easy way of doing this and it is not often done in practice.

8.4.2 Variational Auto-Encoders

In many inference models the latent variables are of little interest, however, there are models where the latent space (the values taken by the latent variables) take centre stage. One such model is the variational auto-encoder (VAE) proposed by Kingma and Welling (2013). Here we try to find an embedding from the space of objects (typically images) to a continuous latent space. Images can be viewed as vectors in some very high-dimensional spaces (each pixel or even each colour channel of each pixel is a separate dimension). The set of all images that we are interested in would correspond to a manifold (i.e. a low-dimensional hypersurface) in this very high-dimensional space. The shape of this manifold will be very non-trivial. For example, the mean of two image typically won't be an images so the manifold is not convex. The aim of VAEs is to find an embedding of this manifold to a continuous space where most points correspond to an image. Furthermore, we try to find an inverse mapping from the points in the latent space back to an image. In this sense we can think of the VAE as a generative model. We can generate a random point in the latent space and use our inverse mapping to generate an image.

The VAE uses powerful deep networks to both perform the embedding (or encoding) of an image into the latent space and the decoding from a point in the latent space back to an image. To add some additional regularisation, the encoder actually maps each image to a probability distribution in the latent space. This ensures that if we make a small move in latent space we will end up with a similar image. Figure 8.16 shows a schematic diagram of a VAE.

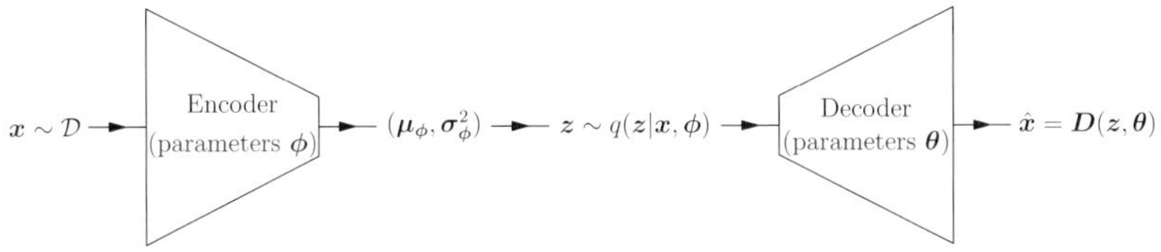

Figure 8.16 Schematic diagram of a VAE. This is described in the text.

We briefly outline how VAEs are implemented. We start with a data set, $\mathcal{D}$, of objects (typically images). This data set can be regarded as a subset of typical examples of a much large set of objects that we are interesting in modelling. From the data set we draw a particular object $\boldsymbol{x}$. This is given to a feed-forward deep neural network (the encoder) with parameters $\boldsymbol{\phi}$, which outputs a set of means, $\boldsymbol{\mu}_{\boldsymbol{\phi}}$, and a set of variances, $\boldsymbol{\sigma}^2_{\boldsymbol{\phi}}$. These outputs from the encoder network specify a probability distribution in latent space,

$$q(z|\boldsymbol{x}, \boldsymbol{\phi}) = \mathcal{N}\left(z \middle| \boldsymbol{\mu}_{\boldsymbol{\phi}}, \mathrm{diag}(\boldsymbol{\sigma}^2_{\boldsymbol{\phi}})\right),$$

where $\mathrm{diag}(\boldsymbol{\sigma}^2_{\boldsymbol{\phi}})$ denotes a diagonal matrix, $\boldsymbol{\Sigma}$, with elements $\Sigma_{ii} = \sigma_i^2$. That is, each image is mapped to an axis-aligned normal distribution in the latent space. To train the network we next introduce a decoder. We sample a vector z from $q(z|\boldsymbol{x}, \boldsymbol{\phi})$ and give this as an input to a second deep neural network that generates an image $\hat{\boldsymbol{x}}$. This new image depends on z and the parameters of the decoder network, $\boldsymbol{\theta}$. To train a VAE we try to minimise some distance measure between the initial image $\boldsymbol{x}$ and the image $\hat{\boldsymbol{x}}$ generated by the decoder, at the same time as minimising the KL divergence between the distribution $q(z|\boldsymbol{x}, \boldsymbol{\phi})$ and some 'prior distribution', $p(z)$, usually taken to be a multivariate normal $\mathcal{N}(\mathbf{0}, \mathbf{I})$. This KL divergence ensures that the distribution $q(z|\boldsymbol{x}, \boldsymbol{\phi})$ does not collapse to a point, losing the regularisation properties provided by having a probability distribution. If we consider generating images from $\hat{\boldsymbol{x}}$ with a probability $p(\boldsymbol{x}|\hat{\boldsymbol{x}})$ (where, for example, $p(\boldsymbol{x}|\hat{\boldsymbol{x}}) = \mathcal{N}(\boldsymbol{x}|\hat{\boldsymbol{x}}, \sigma^2\,\mathbf{I})$) then we can view the objective function as

$$\mathbb{E}_{\boldsymbol{x}\sim\mathcal{D}}\Big[\mathbb{E}_{z\sim q}\big[\log\big(p(\boldsymbol{x}|\hat{\boldsymbol{x}})\big)\big] - \mathrm{KL}\left(q(z|\boldsymbol{x}, \boldsymbol{\phi}) \,\|\, p(z)\right)\Big].$$

That is, we are maximising the log-probability of generating an image $\boldsymbol{x}$ as the output, at the same time as minimising the KL divergence. We note that if $p(\boldsymbol{x}|\hat{\boldsymbol{x}}) = \mathcal{N}(\boldsymbol{x}|\hat{\boldsymbol{x}}, \sigma^2\,\mathbf{I})$ then maximising $\log\big(p(\boldsymbol{x}|\hat{\boldsymbol{x}})\big)$ is equivalent to minimising the squared error between $\boldsymbol{x}$ and $\hat{\boldsymbol{x}}$. We can perform gradient descent on the parameters $\boldsymbol{\phi}$ and $\boldsymbol{\theta}$ to maximise this objective function. In the original paper the authors show that this objective function can be viewed as a variational bound on maximising the log-likelihood of generating images from the data set (Kingma and Welling, 2013). We don't give this derivation here, however, in Example 9.8 on page 283 we discuss another interpretation of the objective function in terms of the minimum description length.

In Figure 8.17 we schematically illustrate the encoding and decoding carried out by the VAE. The left figures show images (represented as points in a vector space) from the data set. These images lie close to a low-dimensional manifold representing the full set of images that we are interested in modelling. Clearly,

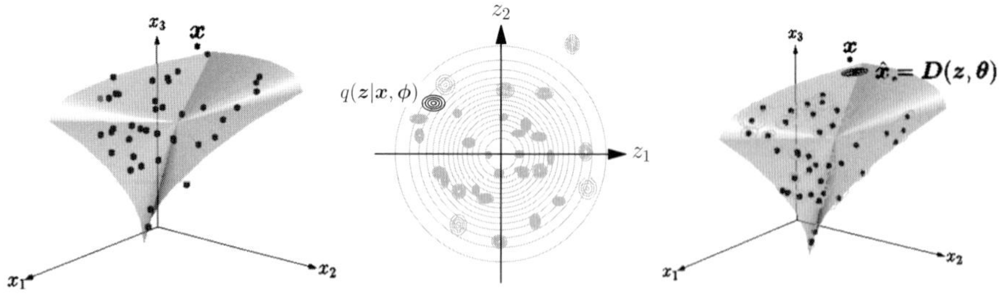

Figure 8.17 Schematic illustration of the encoding and decoding of a VAE. The left figure represents the images from the data set $\mathcal{D}$ in a vector space (each axis representing a pixel value). Also shown is a low-dimensional manifold around which most images lie. The middle figure shows the projection of the images to normal probability distributions in latent space. The right image shows the projection back from the latent space to a manifold of the same dimensions as the latent space that is embedded in image space. We highlight a single image $\boldsymbol{x}$ that is projected to a probability distribution $q(\boldsymbol{z}|\boldsymbol{x},\boldsymbol{\phi})$ and the projection of that probability distribution back to image space. The generated images $\hat{\boldsymbol{x}}$ should be close to the original image $\boldsymbol{x}$.

real images will lie in a very large dimensional space and the 'low-dimensional' manifold may involve many dimensions. The centre diagram illustrates the latent space (here shown as a two-dimensional space, but in practice this is usually considerable larger). Each image in the database is mapped to a normal probability distribution in latent space (shown as a series of contour lines). We also show contours of the 'prior' distribution $p(z) = \mathcal{N}(\mathbf{0}, \mathbf{I})$, which corresponds to concentric circles around the origin. By minimising the KL divergence between $q(\boldsymbol{z}|\boldsymbol{x},\boldsymbol{\phi})$ and this prior we prevent this distribution from collapsing to a point. The right-hand figure shows the inverse mapping generated by the decoder network. Each point, $\boldsymbol{z}$, in the latent space is mapped to a point $\hat{\boldsymbol{x}} = D(\boldsymbol{z},\boldsymbol{\theta})$ in this image point. If we consider putting a set of basis vectors around a point, $\boldsymbol{z}$, in the latent space these basis vectors will get mapped to a local coordinate system on a low-dimensional manifold in the new image space. The dimensionality of the manifold will be no greater and typically the same size as the dimensionality of the latent space. All vectors $\boldsymbol{z}$ will be mapped to a point on this low-dimensional manifold. Note that typically the space of images might be of the order of 10^4–10^6 dimensions, the latent space 10–100 dimensions, and the number of images in the data set of order 10^5–10^6.

Part of the excitement about the development of VAEs is that they provide an unsupervised method for training deep neural networks (the encoder and decoder) in so far as the images don't need to be labelled. This is similar to generative adversarial networks (GANs) discussed in Example 7.15 on page 174. This is a rapidly developing field of research, but at the time of writing GANs are seen as generating more realistic images, while VAEs are seen as easier to train. VAEs are also exciting in that they allow very non-linear mappings to latent spaces. This contrasts, for example, with principle component analysis, which similarly finds a latent variable representation, but does so by projecting all vectors in the original space to a low-dimensional subspace. That is, it models data that live on 'flat' manifolds.

8.5 Machine Learning

Machine learning evolved separately from classical statistical inference. Its motivation was to copy how humans solved problems. Unhindered by the requirements for mathematical rigour, it developed as a series of heuristic tricks. It has always been a major part of artificial intelligence. Early attempts at building intelligent machines concentrated on building rule-based systems. Although these seem to show early promise they suffered from a complexity explosion as the systems became more developed. An alternative approach was to learn simple rules from data. One of the earliest attempts at this was the perceptron developed by Frank Rosenblatt in the late 1950s. The perceptron performed binary classification by dividing the space of input patterns by a simple hyperplane. This limited the perceptron to learn only very simple rules. The field stalled until the development of the multilayer perceptron and related neural networks in the mid-1980s. This renaissance was in part due to the development of smart algorithms, but also because of the availability of computers powerful enough and cheap enough to make it worthwhile. At this stage, the discipline rebadged itself, discarding the dated name *machine learning* (or sometimes cybernetics) and labelled itself *neural networks*. A host of other neural networks were developed. After this rapid burst of creativity, a slow process of re-examination took place where it became clear machine learning was just statistical inference dressed up. Indeed, it was found that by moving from the language of neurons to that of random variables and inference, many techniques could be improved. Yet this historical digression through neural networks had brought some real progress in the field. Most important was the scale of the ambition. Whereas classical statistical inference had concentrated on interpreting experimental data using techniques that could be carried out with pen and paper, neural networks had built systems to undertake speech and handwriting recognition.

As the field of neural networks developed in the 1990s and beyond, it sought to throw off the Wild West image of the 1980s and seek mathematical respectability. New techniques were developed that were no longer inspired by the human brain. The field rebranded itself as *machine learning* (a name which a few decades earlier carried the connotation of outmoded rule-based techniques). The field has recently witnessed the resurgence of neural networks with the remarkable success of deep networks. Ironically, the desire for mathematical respectability had almost extinguished all research in traditional neural networks, which was virtually banned from major conferences and journals. Only through the tenacity of a few key researchers – most notably Geoff Hinton, who had previously been central to the resurgence of neural networks in the 1980s, and Yann LeCun, who had consistently developed convolutional neural networks – was the flame of neural networks kept alight long enough for deep learning to become competitive. Machine learning provides a case study in how rigorous mathematics can both significantly push and strengthen a field, but when allowed free reign can also stifle innovation. Currently we are back in the Wild West where the best results are produced by cowboys who play.

Although techniques such as support vector machines are not overtly Bayesian, they nevertheless claim a rigorous probabilistic foundation in terms of statistical learning theory.

Present-day machine learning splits between generally applicable black-box techniques such as multilayer perceptrons and support vector machines (a kind of perceptron on steroids) and true Bayesian techniques. In terms of ease of use, the Bayesian techniques suffer from the need to carefully model the underlying process. They therefore tend to be bespoke to the problem being solved. They can be very effective, but they are typically time-consuming to create and are often analytically intractable, so require either expensive Monte Carlo simulation or sophisticated approximation methods. There are attempts to make Bayesian methods more deployable. Three prominent examples are naive Bayes classification, graphical models, and Gaussian processes. We give a quick overview of the first two of these techniques – we defer a discussion of Gaussian processes until Section 12.1.2.

Another dichotomy in machine learning is between discriminative and generative models. In machine learning we are often given features $\boldsymbol{x}$ and wish to infer labels $\boldsymbol{y}$. Probabilistic discriminative models try to learn the conditional probability $\mathbb{P}\left(\boldsymbol{y}|\boldsymbol{x}\right)$, while generative models try to learn the full joint distribution $\mathbb{P}\left(\boldsymbol{y},\boldsymbol{x}\right)$. That is, they also model the process of generating the features. Generative models are often much more complicated to implement and are much more problem specific as they involve modelling the whole process, but, when the model accurately captures reality, it can give more accurate predictions. Generative models are also used as flexible discriminative models. For example, HMMs might be used effectively to classify genres of music, but they are usually woefully disappointing when used for composing (they will capture short-range correlations, but typically ignore high-order correlations such as repeated motifs). So even this dichotomy is often blurred.

8.5.1 Naive Bayes Classifier

In many real-world problems doing Bayes right is just too hard. An example of this is text classification. To correctly model the likelihood of some text being in a particular class or to encode all possible prior knowledge would be almost impossible with present-day algorithms (for example, we would have to build semantic analysis into our classification algorithm). In the naive Bayes classifier we make a ridiculous assumption that the order of words doesn't matter – sometimes known as a bag-of-words representation. We should therefore not expect our classifier to perform very well, but rather surprisingly, this classifier performs extraordinarily well on many text classification problems. It is, for example, the method of choice for many spam filters.

Example 8.14 Spam Filter
Let us consider the problem of classifying whether an email is spam or not. This is an important problem which a large number of people care a lot about. We assume that we have a lot of examples consisting of a list of words $\langle w_1, w_2, w_3, \ldots w_n \rangle$. Bayes' rule tells us

$$\mathbb{P}(\text{Spam}|\langle w_1, \ldots w_n \rangle) = \frac{\mathbb{P}(\langle w_1, \ldots w_n \rangle|\text{Spam})\,\mathbb{P}(\text{Spam})}{\mathbb{P}(\langle w_1, \ldots w_n \rangle)}$$

where

$$\mathbb{P}(\langle w_1, \ldots w_n\rangle) = \mathbb{P}(\langle w_1, \ldots w_n\rangle|\text{Spam})\,\mathbb{P}(\text{Spam}) + \mathbb{P}(\langle w_1, \ldots w_n\rangle|\neg\text{Spam})\,\mathbb{P}(\neg\text{Spam}).$$

If we were being honest, to compute the likelihood $\mathbb{P}(\langle w_1, \ldots w_n\rangle|$ Spam) would require knowing (or estimating) the likelihood of each message given that it is spam. In naive Bayes we make a very strong (and clearly wrong) assumption of conditional independence

$$\mathbb{P}(\langle w_1, \ldots w_n\rangle|\text{Spam}) = \prod_{i=1}^{n} \mathbb{P}(w_i|\text{Spam}).$$

Our task is now much easier. We need only estimate two types of probabilities:

- the prior $\mathbb{P}(\text{Spam})$ and
- the likelihoods $\mathbb{P}(w_i|\text{Spam})$ and $\mathbb{P}(w_i|\neg\text{Spam})$ for all words w_i.

The prior is easy to estimate empirically. Probably 80–90% of my emails are spam. The likelihoods $\mathbb{P}(w_i|\text{Spam})$ are also easy to estimate from data

$$\mathbb{P}(w_i|\text{Spam}) = \frac{\#(\text{occurrence of } w_i \text{ in spam messages})}{\#(\text{words in spam messages})}$$

($\#(\cdots)$ denotes the number of occurrences of $\cdots$). This would be unreliable for rare words, but we can 'fix' this by including some 'pseudo-counts'

$$\mathbb{P}(w_i|\text{Spam}) = \frac{\#(\text{occurrence of } w_i \text{ in spam messages}) + m}{\#(\text{words in spam messages}) + m\,n}$$

where m is a parameter (pseudo-count) encoding our prior belief in that a word might be part of a spam message and n is the total number of words ($m = 1$ would be a typical choice for the pseudo-count). Similarly, we can compute $\mathbb{P}\left(w_i|\neg\text{Spam}\right)$.

Pseudo-counts are a kludge. If we want to do a better job we could introduce a prior distribution that a word is part of a spam message. If we choose a Dirichlet distribution for this prior distribution and assume that the likelihood of the words appearing in the data was given by a multinomial distribution (an assumption compatible with our naive Bayes assumption), then we would end up with a quantity that looks just like a pseudo-count. The advantage of doing this properly is that we would get justifiable pseudo-counts rather than numbers that appear to come from nowhere. See Exercise 8.2 to see this explicitly.

8.5.2 Graphical Models

We finish our discussion of Bayesian methods by discussing one of the most explicitly Bayesian techniques for performing inferencing, namely graphical models. One of the challenges of using Bayesian methods in a complex problem is managing the model. Graphical models aid this process by building a pictorial representation of the causal dependencies between the random variables that make up a probabilistic model. There are different ways of using graphs to represent probabilities. The mainstream graphical models, known as *Bayesian belief networks*, use directed graphs to show causal relationships between variables, while a second strand, often known as *Markov random fields*, use undirected graphs. Graphical models don't solve the problem of performing inference, rather they provide a framework for representing problems. A lot of the current research is aimed at developing efficient algorithms for performing (approximate) inference for graphical models.

Bayesian Belief Networks

In Bayesian belief networks (or *Bayesian networks* for short) we represent random variables as nodes or vertices in a graph and conditional dependencies as arrows between nodes. Thus if A and C are random variables then we would show a casual dependency between A and C as shown below.

This indicates that event A causes event C. As an example, the random variable A might be 1 if Abi has an anniversary today and 0 otherwise, while C might be 1 if there are cakes in the coffee room and zero otherwise. We then have to associate probabilities with each event, e.g. $\mathbb{P}\left(C=1|A=1\right)=0.8$, $\mathbb{P}\left(C=0|A=1\right)=0.2$, $\mathbb{P}\left(C=1|A=0\right)=0.1$ and $\mathbb{P}\left(C=0|A=0\right)=0.9$, while $\mathbb{P}\left(A=1\right)=1/365$ and $\mathbb{P}\left(A=0\right)=364/365$. This requires some clarification, clearly having an anniversary is not a random event (we don't roll a dice with 365 sides each day to decide whether it is our anniversary), however, I personally am useless at dates so if we take these probabilities to encode my belief about the state of the world then this is a reasonably accurate model. Note that because we are capturing a causal relationship, then it is relatively straightforward to assign conditional probabilities. E.g. if it is Abi's anniversary then there is an 80% chance she will put cakes in the coffee room. Of course, the world is more complicated; for example, there might be cakes because Ben has a birthday, which we can represent using the graphical model below.

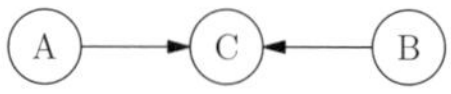

To perform probabilistic inference it is useful to rewrite a joint probability in terms of conditional probabilities. Bayes' rule allows us to do this in many different ways, e.g.

$$\mathbb{P}\left(A,B,C\right)=\mathbb{P}\left(A|B,C\right)\,\mathbb{P}\left(B|C\right)\,\mathbb{P}\left(C\right)=\mathbb{P}\left(C|A,B\right)\,\mathbb{P}\left(A|B\right)\,\mathbb{P}\left(B\right),$$

etc. However, the Bayesian belief network points to a particularly useful way of disentangling the probabilities. In our example, we can write

$$\mathbb{P}\left(A,B,C\right)=\mathbb{P}\left(C|A,B\right)\,\mathbb{P}\left(A\right)\,\mathbb{P}\left(B\right)$$

which not only involves the conditional probabilities which are easy to specify, but simplifies the relationship in that we know that $\mathbb{P}(A|B) = \mathbb{P}(A)$. That is, in our simplistic model of the world Abi's anniversary is not directly related to Ben's birthday. Although random variables A and B are not casually connected they are dependent on each other through C. Thus, if we know there are cakes and it is not Abi's anniversary then there is a higher probability of it being Ben's birthday than if it were Abi's anniversary.

In general, to infer questions such as what is the probability there are cakes, or what is the probability that it's Ben's birthday, we have to sum over all possible events, e.g.

$$\mathbb{P}(\text{It's Ben's birthday}) = \mathbb{E}[B] = \sum_{A \in \{0,1\}} \sum_{B \in \{0,1\}} \sum_{C \in \{0,1\}} B\, \mathbb{P}(A, B, C).$$

For a simple model this sum is straightforward, but as the number of random variables grow this sum quickly becomes impractical to compute. That is, if we denote our random variables by X_i and their states by $\mathcal{S}_i$ then the total number of distinct combinations of states is

$$\prod_{i=1}^{n} |\mathcal{S}_i| = e^{\sum_{i=1}^{n} \log(|\mathcal{S}_i|)}$$

where $|\mathcal{S}_i|$ denotes the cardinality (number of elements) of the set $\mathcal{S}_i$. The sum over states grow 'exponentially' in the number of random variables. There are situations when we can use conditional independence to reduce the number of states we need to sum over. For example, if we don't know whether there are cakes in the coffee room then $\sum_{C=\pm 1} \mathbb{P}(C|A, B) = 1$ and consequently

$$\mathbb{P}(\text{It's Ben's birthday}) = \sum_{B \in \{0,1\}} B\, \mathbb{P}(B) = \mathbb{P}(B = 1).$$

If we extend the model and consider two colleagues Dave and Eli who like cake, then we could let D be a binary random variable set to 1 if we see Dave eating some cake and similarly E is a random variable telling us whether we see Eli eating cake. Our model is now as shown below.

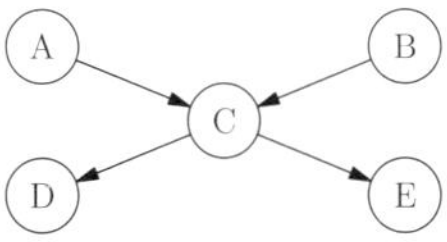

which we decompose as

$$\mathbb{P}(A, B, C, D, E) = \mathbb{P}(D|C)\, \mathbb{P}(E|C)\, \mathbb{P}(C|A, B)\, \mathbb{P}(A)\, \mathbb{P}(B).$$

In this case, if we see Dave eating cake, then we can use the model to infer the probability that it is Ben's birthday or whether we can expect to see Eli eating cake. Again the inference involves summing over all unknown states. Note that our variables D and E are dependent on each other through C. However, if we go to the meeting room and observe that there is some cake (or there is no cake), then the probabilities that Dave and Eli eat cake become independent of each other. In graphical models, if some of our random variables are observed it is traditional to shade the nodes, for example if we observe that there is cake in the coffee room we can show this using the following graphical representation.

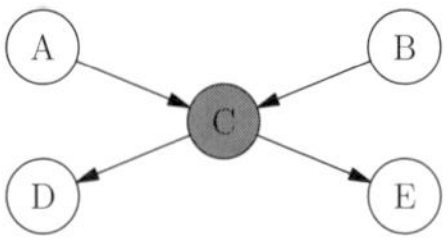

Observing C makes A and B dependent, however, D and E are conditionally independent given C. Thus, in this model, having observed there are cakes in the coffee room, C, then the probability that Eli will eat cake is just $\mathbb{P}\left(E = 1|C\right)$ (we don't care about A, B, and D). Graphical models allow us to formalise the conditions where random variables become conditionally independent – the interested reader can consult any text on graphical models. One of the major benefits of using graphical models is that they can make the dependency relationship between random variables very explicit, which can often speed up inference by many orders of magnitude.

Bayesian belief networks come into their own when we have to make inferences about complex systems. Below we consider a simplified example of trying to build an automatic fault diagnostic system for a printer.

Example 8.15 Printer Failure

Suppose the printer isn't printing and I want to know the most likely causes. There are a lot of unknowns which can be modelled as random variables: Is the printer switched on? Is there a paper jam?, etc. To determine if the printer has, for example, run out of ink, we first identify what could go wrong and their causal relations, e.g.

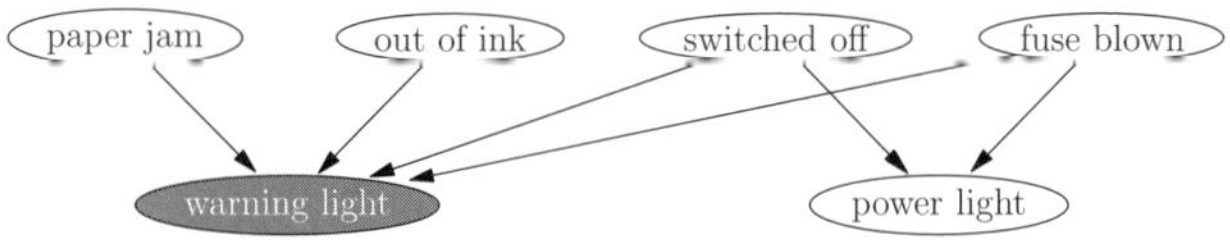

In this case we assume we have observed the warning light. In our simple model we will assume that all the random variables take binary values (in general we might have more complex random variables, for example, we might receive a warning message rather than have a warning light). For our model we can then assign prior probabilities to the top-level (parent) random variable, e.g. $\mathbb{P}\left(\text{paper jam}\right) = 0.1$, $\mathbb{P}\left(\text{switched off}\right) = 0.4$, etc. At the lower level (children) we can assign conditional probabilities

$$\mathbb{P}\left(\text{warning light} = \text{True}|\text{paper jam}, \neg\text{out of ink}, \neg\text{switched off}, \neg\text{blow fuse}\right) = 0.9$$
$$\mathbb{P}\left(\text{warning light} = \text{True}|\text{paper jam}, \neg\text{out of ink}, \text{switched off}, \neg\text{blow fuse}\right) = 0$$
$$\vdots$$

and similarly for the other pairs of random variables. If we have additional information (e.g. we have observed the warning light), or we know that the printer is switched on, then we can set these random variables to their observed state. To find the most likely cause of the fault we can sum over all possible states of the system that have not been observed. Using the causal structure of the model we can often reduce the computation considerably.

Often graphical models are just used to help visualise the dependency structure in complex probabilistic models. Many probabilistic models can be represented as a graphical model. As an example of this we consider one of the most prominent probabilistic models of recent years: *latent Dirichlet allocation* (LDA).

Example 8.16 Latent Dirichlet Allocation

LDA is also an acronym for linear discriminant analysis—a quite different algorithm.

LDA was proposed in Blei et al. (2003). Although it can and has been used in many different application areas, its prototypical application is in topic models. That is, in grouping a collection of documents into topics, where each document belongs predominantly to one or a few topics. LDA is a means for automatically finding the topics.

More specifically, we consider a set of documents or corpus $\mathcal{C} = \{d_i | i = 1, 2, \ldots |\mathcal{C}|\}$. Each document is composed of a list of words

$$d = \left(w_1^{(d)}, w_2^{(d)}, \ldots, w_{N_d}^{(d)}\right).$$

We make the assumption that the ordering of words is irrelevant. Clearly, this is not true of real documents. It is a useful simplifying assumption as it allows us to ignore grammar, which is difficult to understand. For identifying topics, however, the word ordering is perhaps less important. Our aim is to infer a set of topics $\mathcal{T} = \{t_1, t_2, \ldots, t_{|\mathcal{T}|}\}$ such that the words in each document are 'explained' by only a few topics (see Figure 8.18).

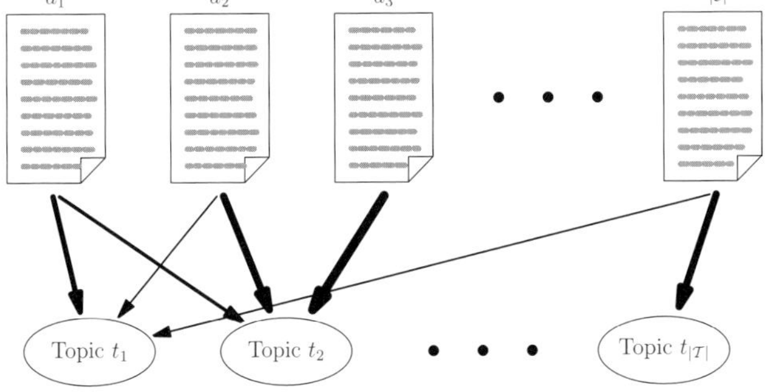

Figure 8.18 Each document has a few topics assoicated with it which are responsible for generating the words.

For each document, d, we associate a probability vector, $\boldsymbol{\theta}^{(d)} = (\theta_t^{(d)} | t \in \mathcal{T})$, which tells us the probability that a word in document d is generated by topic $t \in \mathcal{T}$. We want this to be sparse, in the sense that most of the probability mass is associated with only a few topics. In addition we associate a relatively small proportion of words in the vocabulary, $\mathcal{V}$, to each topic (see Figure 8.19).

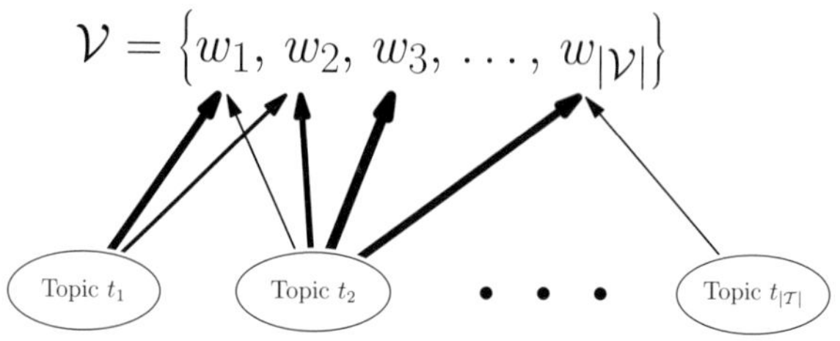

Figure 8.19 For each topic we associate a relatively few words from the vocabulary.

Again we associate a probability vector, $\boldsymbol{\phi}^{(t)} = (\phi_w^{(t)} | w \in \mathcal{V})$, with each topic, where $\phi_w^{(t)}$ is the probability of word, $w \in \mathcal{V}$, being generated by that topic, t. Again we want to choose a sparse probability vector. Here we describe a version of LDA known as smoothed LDA as the probability vector, $\boldsymbol{\phi}^{(t)}$, are Dirichlet distributed; although sparse in the sense that $\phi_w^{(t)}$ is large for only a relatively few words, nevertheless, it is non-zero for all words.

To obtain sparse vectors, $\boldsymbol{\theta}^{(d)}$ and $\boldsymbol{\phi}^{(t)}$, we draw them from a Dirichlet prior distribution

$$\boldsymbol{\theta}^{(d)} \sim \text{Dir}(\alpha\, \mathbf{1}) \qquad\qquad \boldsymbol{\phi}^{(t)} \sim \text{Dir}(\beta\, \mathbf{1})$$

where $\mathbf{1}$ is the vector of all ones (note that for $\boldsymbol{\theta}^{(d)}$ the vector is of length $|\mathcal{T}|$, while for $\boldsymbol{\phi}^{(t)}$ the vector is of length $|\mathcal{V}|$). By choosing $\alpha, \beta \ll 1$ we put a strong bias towards sparse vectors (see Figure 8.20).

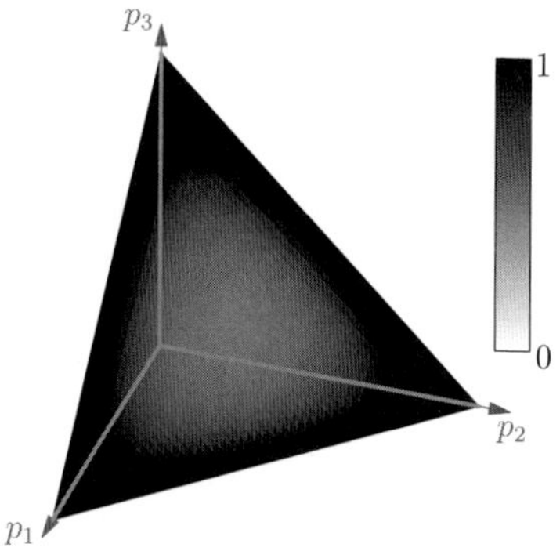

Figure 8.20 A three-dimensional Dirichlet distribution, $\text{Dir}\left(\boldsymbol{p} | \boldsymbol{\alpha} = \frac{1}{3}(1,1,1)\right)$.

LDA can be viewed as a generative model for documents. That is, for each topic we can generate a probability vector $\boldsymbol{\phi}^{(t)}$ from $\text{Dir}(\beta\, \mathbf{1})$. Then for each document we draw a probability vector $\boldsymbol{\theta}^{(d)}$ from $\text{Dir}(\alpha\, \mathbf{1})$. Finally, to generate the document for each word position, i, we first draw a topic, $\tau_i^{(d)}$, from $\boldsymbol{\theta}^{(d)}$ and then draw a word, $w_i^{(d)}$, from $\boldsymbol{\phi}^{(\tau_i^{(d)})}$.

In general, if we have a set of items $\mathcal{I}$ where $\boldsymbol{p} = (p_i | i \in \mathcal{I})$ is a probability vector such that p_i is the probability of drawing item i, then we often refer to the probability distribution for the items being drawn as the categorical distribution

$$\mathrm{Cat}(i|\boldsymbol{p}_i) = p_i = \mathrm{Mult}(\boldsymbol{\delta}_i|1, \boldsymbol{p})$$

where $\boldsymbol{\delta}_i$ is a vector with a 1 at position i and 0 elsewhere (this is often referred to as a *one-hot vector*). Note, that this is a multivariate generalisation of the Bernoulli distribution discussed in Section 4.3.3 on page 68. It is a very simple distribution (we are just choosing a single sample with probabilities p_i) so even though the categorical distribution is often used it does not always get given a name. It is, however, useful in clarifying the model underlying LDA. Thus, in LDA we choose

$$\begin{aligned}
\forall d \in \mathcal{C} \quad & \boldsymbol{\theta}^{(d)} \sim \mathrm{Dir}(\alpha\, \mathbf{1}) \\
\forall t \in \mathcal{T} \quad & \boldsymbol{\phi}^{(t)} \sim \mathrm{Dir}(\beta\, \mathbf{1}) \\
\forall d \in \mathcal{C} \wedge \forall i \in \{1, 2, \ldots, N_d\} \quad & \tau_i^{(d)} \sim \mathrm{Cat}(\boldsymbol{\theta}^{(d)}), \quad w_i^{(d)} \sim \mathrm{Cat}(\boldsymbol{\phi}^{(\tau_i^{(d)})}).
\end{aligned}$$

Note that this way of viewing LDA is as a generative model for documents. That is, we view each document as being primarily about a few topics, which we encode in $\boldsymbol{\theta}^{(d)}$. Each topic has a relatively few words associated with it, which we encode in $\boldsymbol{\phi}^{(t)}$. To generate a document for each word we select a topic, $\tau_i^{(d)}$, according to $\boldsymbol{\theta}^{(d)}$, and then we choose a word according to $\boldsymbol{\phi}^{(\tau_i^{(d)})}$. Although, this is clearly not how documents are generated, nevertheless, we have an intuitive feeling that this might capture some aspect of the distribution of words in a real set of documents. Thinking in terms of generative models is often a very useful way of coming up with natural models. Of course, the last thing we want to do with our model is to actually generate documents. In most applications we are given the documents and want to infer topics. It is typical of many *generative models* (e.g. HMMs) that they are rarely used as generative models. Thinking of them as generative models is just a useful intellectual crutch to come up with a plausible probabilistic model.

As we are not going to use LDA as a generative model, we want a description of the model in terms of a probability distribution rather than as an algorithm for drawing samples from that distribution. To succinctly express LDA as a probability distribution it is useful to denote the set of random variables in the model by

$$\begin{aligned}
&\boldsymbol{W} = (\boldsymbol{w}^{(d)} | d \in \mathcal{C}) \quad \text{with} \quad \boldsymbol{w}^{(d)} = (w_1^{(d)}, w_2^{(d)}, \ldots, w_{N_d}^{(d)}), \quad \text{and} \quad w_i^{(d)} \in \mathcal{V} \\
&\boldsymbol{T} = (\tau_i^{(d)} | d \in \mathcal{C} \wedge i \in \{1, 2, \ldots, N_d\}) \quad \text{with} \quad \tau_i^{(d)} \in \mathcal{T} \\
&\boldsymbol{\Theta} = (\boldsymbol{\theta}^{(d)} | d \in \mathcal{C}) \quad \text{with} \quad \boldsymbol{\theta}^{(d)} = (\theta_t^{(d)} | t \in \mathcal{T}) \in \Lambda^{|\mathcal{T}|} \\
&\boldsymbol{\Phi} = (\boldsymbol{\phi}^{(t)} | t \in \mathcal{T}) \quad \text{with} \quad \boldsymbol{\phi}^{(t)} = (\phi_w^{(t)} | w \in \mathcal{V}) \in \Lambda^{|\mathcal{V}|}
\end{aligned}$$

where Λ^k is the $(k-1)$-dimensional unit simplex. The joint probability distribution for the random variables is then given by

$$\mathbb{P}\left(\boldsymbol{W},\boldsymbol{T},\boldsymbol{\Theta},\boldsymbol{\Phi}\middle|\alpha,\beta\right) = \left(\prod_{t\in\mathcal{T}} \text{Dir}\left(\boldsymbol{\phi}^{(t)}\middle|\beta\mathbf{1}\right)\right) \times \left(\prod_{d\in\mathcal{C}} \text{Dir}\left(\boldsymbol{\theta}^{(d)}\middle|\alpha\mathbf{1}\right)\prod_{i=1}^{N_d} \text{Cat}\left(\tau_i^{(d)}\middle|\boldsymbol{\theta}^{(d)}\right) \text{Cat}\left(w_i^{(d)}\middle|\boldsymbol{\phi}^{(\tau_i^{(d)})}\right)\right).$$

Typically we are given the set of words $\boldsymbol{W}$ in each topic and wish to infer plausible topics. To do this it is convenient to marginalise out the Dirichlet variables $\boldsymbol{\theta}^{(d)}$ and $\phi_w^{(t)}$. As the Dirichlet distribution is conjugate to the multinomial (and categorical) distribution this marginalisation can be computed in closed form. It is still necessary to sum over all possible allocations of topics $\tau_i^{(d)}$, which can be done using a variational approximation or using MCMC. There is a large literature discussing different strategies for performing this inference.

We now return to the graphical representation of LDA. Figure 8.21 shows a traditional graphical representation of LDA showing all the dependencies between random variables. This figure is rather large and inelegant.

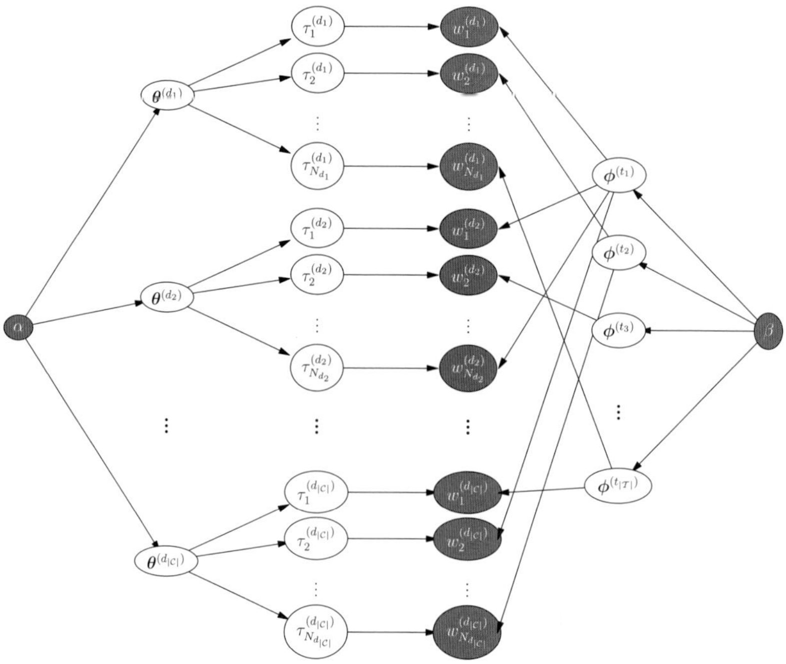

Figure 8.21 Graphical representation of smoothed LDA showing all the random variables and their dependencies. The words depends on both the topic $\tau_i^{(d)}$ and on the word vector $\boldsymbol{\phi}^{(\tau_i^{(d)})}$ for that particular topic.

Graphical models have invented a notation known as the plate representation where sets of causally independent variables are shown in boxes. Thus, to show that all the vectors $\boldsymbol{\theta}^{(d)}$ are independently generated given α we would use the diagram below.

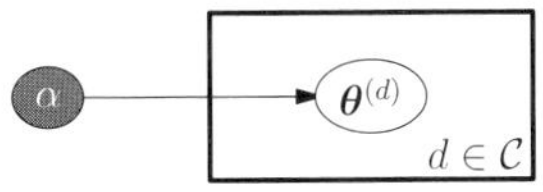

Some authors would write Dir above the arrow to show that $\boldsymbol{\theta}^{(d)}$ was generated by a distribution $\mathrm{Dir}(\boldsymbol{\theta}^{(d)}|\alpha\mathbf{1})$. LDA involves several levels of random variables so needs to be modelled as a more complex plate diagram as shown in Figure 8.22. I have to confess, I find such plate diagrams hard to digest. They compact a lot of information, but require a lot of staring at to comprehend. Clearly, they are only visual aids. Although in LDA, $w_i^{(d)}$ depends on both $\tau_i^{(d)}$ and $\boldsymbol{\phi}^{(t)}$, the graphical model fails to convey the type of dependency.

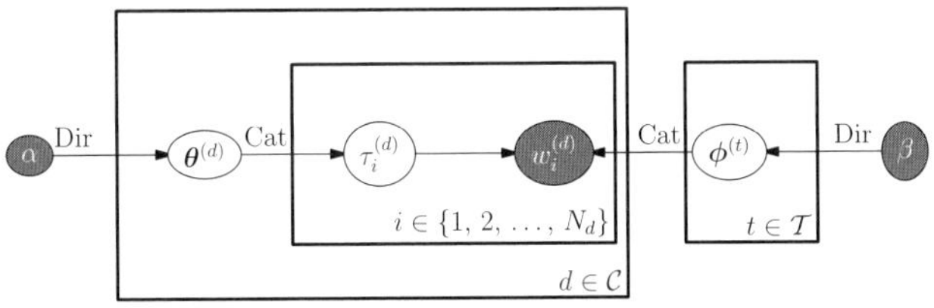

Figure 8.22 Plate diagram showing smoothed LDA.

I like visual representations. It fits in with the way I think. As such, I should be a great fan of Bayesian nets. However, sometimes I find the mechanics of setting up a graphical model and the rules for applying them take centre stage and for my taste they can sometimes obscure what the model really means. They certainly deserve study and have a significant number of advocates, although it is always important to step back and ask what the model really means and is the graphical model really useful.

Algorithms have also been developed to learn the structure of the Bayesian network (i.e. to determine which pairs of variables are conditionally independent of each other and what are the conditional probabilities). We leave the interested reader to study the relevant literature (see Additional Reading section).

It is worth pointing out that many probabilistic models can be represented as graphical models. This includes classic models such as HMMs (see Section 8.4.1) and Kalman filters (see Section 11.3.1). However, these models tend to have a rather simple structure, allowing faster algorithms to be used than those used for general Bayesian networks.

Markov Random Fields

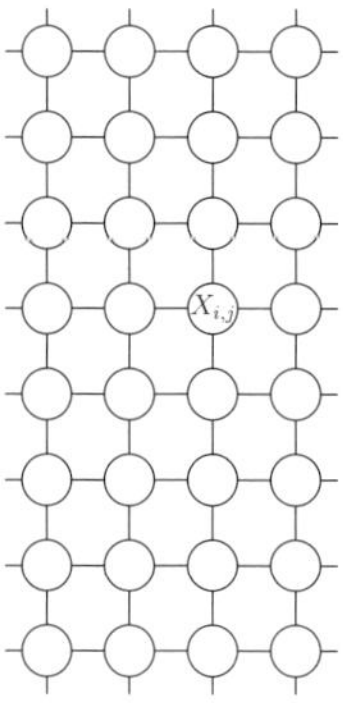

The Bayesian belief networks imposed an idea of causality with parent nodes influencing child nodes. For many situations this is perfectly appropriate, but not always. Probability is blind to causality and we can define the conditional probability $\mathbb{P}(A|B)$ as easily as the conditional probability $\mathbb{P}(B|A)$ whatever the causal relationship between A and B. In some situations we would want to represent the dependency in a symmetric manner. To do this we can use an undirected graph. A classic application of this type of graphical model is in image processing. For example, we may believe that neighbouring pixels are likely to be correlated with each other, while two non-neighbouring pixels are conditionally independent given the pixels in between (at least, to a first approximation). We can represent the dependency structure by a lattice.

Suppose we have a noisy image and wish to smooth it to remove as much noise as possible. To do this we can build a probabilistic model for the pixel values of the true image. The likelihood function depends on the neighbours (as represented by our graph) of the current pixel and on how close its value is to that of the corrupted image. We can then find the image that maximises our likelihood (we can even put a prior over the pixel values and find the MAP solutions). The problem is often set up in a statistical physics language such that maximising the likelihood or posterior is equivalent to minimising an energy.

Bayesian techniques have become some of the most important methods in analysis of, and learning from, uncertain data. In complex applications such techniques nearly always outperform traditional methods such as maximum likelihood. The reason for this is that maximum likelihood is liable to over-fit the data, while Bayesian methods, done right, won't. In a few cases, applying Bayesian methods is straightforward, however, it becomes technically challenging very quickly. A few frameworks have been developed to help cope with the technical challenges, for example graphical models and Gaussian processes. Compared with other learning methods such as neural networks and support vector machines, probabilistic techniques tend to require considerable work (modelling) before they can be applied. However, they are well grounded so that when executed properly they often give extremely good results.

Additional Reading

There are a huge number of books on Bayesian methods. A good introduction to Bayesian methods in statistics is by Lee (2003). A more advanced text is Gelman et al. (2003). If you are interested in the historical development and philosophical arguments then Jaynes (2003) is a good read. There are an enormous number of books on Bayesian approaches to machine learning. Up-to-date books in this category (that also cover graphical models) include Bishop (2007) and Barber

(2011). The modern development of graphical models can be found in the book by Pearl (1988). Markov random fields were introduced in the classic paper by Geman and Geman (1984). A nice text on HMMs is Durbin et al. (1998).

Exercise for Chapter 8

Exercise 8.1 (answer on page 421)

Harry the Axe is one of ten suspects found at the murder scene. On searching Harry a rare watch was found on him identical to that owned by the victim. The victim's watch is missing. Harry claimed it was his, but only 0.1% of the population has a similar watch. What is the probability that Harry is the murderer given the new evidence?

Exercise 8.2 (answer on page 421)

Show that the Dirichlet distribution is a conjugate prior for a multinomial likelihood and derive the update equations.

Exercise 8.3 (answer on page 424)

Suppose we have an experiment with k possible outcomes. Consider performing n independent experiments. What is the likelihood of the n experiments having outcomes $\boldsymbol{N} = (N_1, N_2, \cdots, N_k)$ if the probability of a single outcome is given by $\boldsymbol{p} = (p_1, p_2, \cdots, p_k)$?

Suppose we want to estimate $\boldsymbol{p}$ from our observations $\boldsymbol{N}$. What are the maximum likelihood probabilities? Show how using a conjugate prior to encode our prior beliefs about $\boldsymbol{p}$ will modify the expected value of $\boldsymbol{p}$ given our observation (the answer to Question 8.2 will be of use).

Exercise 8.4 (answer on page 425)

Equation (8.4) gives the posterior probability for the mean and variance given a set of data $\mathcal{D} = (X_1, X_2, \ldots, X_n)$ and assuming an uninformative prior. Show that the marginal created by averaging over the variance is distributed according to Student's t-distribution

$$f(\mu|\mathcal{D}) = \int_0^\infty f(\mu, \tau|\mathcal{D})\, \mathrm{d}\tau = \mathrm{T}\left(\frac{n\,(\mu - \hat{\mu})}{\sqrt{S}} \middle| n \right)$$

with

$$\hat{\mu} = \frac{1}{n}\sum_{i=1}^{n} X_i, \quad S = \sum_{i=1}^{n} (X_i - \hat{\mu})^2, \quad \mathrm{T}(t|\nu) = \frac{1}{\sqrt{\nu}\, B\left(\frac{1}{2}, \frac{\nu}{2}\right)} \left(1 + \frac{t^2}{\nu}\right)^{-\frac{\nu+1}{2}}.$$

Exercise 8.5 (answer on page 426)

Obtain some data, e.g. the area of different countries, their population, physical constants, etc., and plot the frequency of the most significant figure. Compare this with Benford's law.

Exercise 8.6 (answer on page 426)

Show that a scale parameter, σ, of the form $f_X(x|\sigma) = \frac{1}{\sigma} g_X(\frac{x}{\sigma})$ turns into a location parameter under the change of variables $X = e^Y$.

Exercise 8.7 (answer on page 427)

Show that the uninformative prior for a probability, P, transforms to a uniform prior for the log-odds (logit) parameter $Y = \log\left(\frac{P}{1-P}\right)$.

Appendix 8.A Bertrand's Paradox

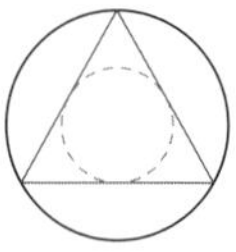

In 1889 Bertrand published a probabilistic problem which caused confusion for almost a century. The problem can be viewed as follows. We throw long (knitting) needles into an area with a circle drawn on the floor. We then ask, of the needles that cross the circle, what is the probability that the chord formed by the portion of the needle within the circle is longer than the sides of an inscribed equilateral triangle? The problem takes a bit of time to absorb. But an equivalent formulation is to ask whether a needle that crosses the circle also crosses a concentric circle of half the width (see dashed circle in margin figure).

Bertrand presented three possible solutions to the problem.

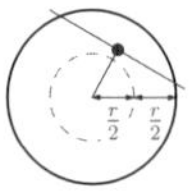

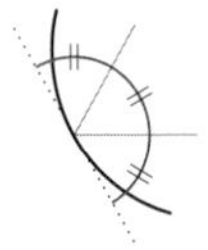

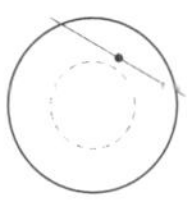

(a) If we assume that the distance from the centre of the circle to the midpoint of the line within the circle is uniformly distributed then we come up with an answer of $p = 1/2$, since the inner circle is half the radius of the outer circle.
(b) If we assume the angle between needle and the circle edge is uniformly distributed we come up with an answer of $p = 1/3$. To see this consider a needle that lies on the vertex of the equilateral triangle. It will have a chord longer than the edge of the equilateral triangle if it lies between the two edges of the triangle. That is, for 60 of a possible 180 degrees it will have a chord greater than the length of the equilateral triangle.
(c) Finally, we could assume the midpoint of the chord is uniformly distributed throughout the whole circle. As the inner circle has a quarter of the area of the outer circle we would conclude $p = 1/4$.

What is the correct answer? There seems to be an ambiguity in assigning a uninformative prior for this problem. Is assigning priors really subjective? Of course, in one sense the answer is all solutions are correct. Needles can be placed at random assuming any one of the three priors – see Figure 8.23.

You could conclude that the problem is not well posed. I haven't told you enough about the problem for you to decide. But that seems a bit of a cop out. Surely I can imagine doing the experiment where I rather inaccurately throw long knitting needle at a circle. What would be the result? Ed Jaynes, in a typically well-written and combative paper, argued the problem is sufficiently well posed to have a unique solution (Jaynes, 1973). The basic argument is that we should expect the same probability distribution if we where to rotate the room, change the size of the circle, and shift the circle slightly. That is, we require the probability distribution to be rotation, scale, and translation invariant. In fact, it is the

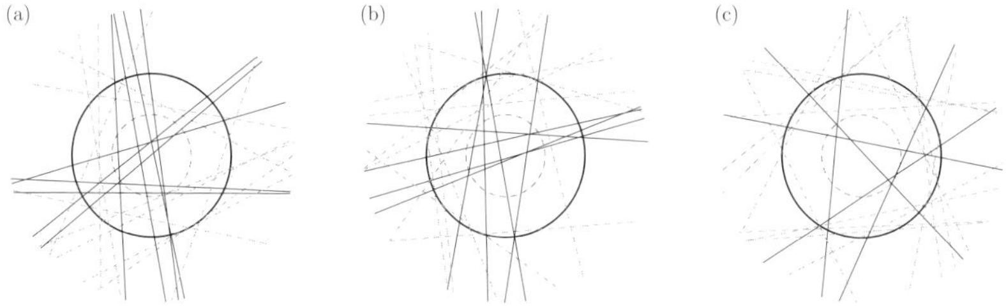

Figure 8.23 Example of samples of 20 needles for Bertrand's problem where we assume a uniform distribution with respect to (a) the distance from the centre of circle to the midpoint of the chord, (b) the angle between the chord and the circumference of the circle, and (c) the position of the midpoint of the chord. The needles represented by solid lines have a chord longer than the equilateral triangle while the needles represented by dashed lines have a shorter chord.

translation invariance that is most important. Here is a very informal argument. Consider standing east of the circle and rolling the needles so they lie in a north–south direction (obviously to achieve rotational invariance we would have to choose the starting angle at random). See Figure 8.24.

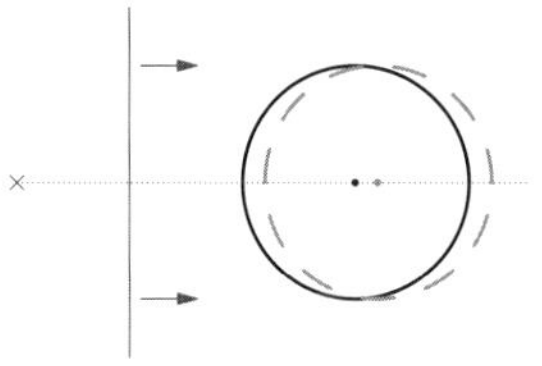

Figure 8.24 Throwing a needle from point × towards a circle (and a translated circle).

We assume we are rather unskilled so we apply very different forces. Now suppose that we translated the circle slightly, then we would expect to have the same distribution of needles. This is only true if the horizontal positions of the needles are evenly distributed. As a consequence, we would expect the distance from the midpoint of the chord made by the needles to the centre of the circle to be evenly distributed. This would be true whichever angle we started from. This is case (a), which corresponds to $p = 1/2$. Those who prefer a rigorous and mathematical proof can read Ed Jaynes' paper. Note that in any real situation we would only expect this translation symmetry to be approximately correct (the universe is finite and we can't throw infinitely hard), nevertheless, these symmetry arguments are very effective at describing what is observed in the real world to a high degree of accuracy. Of course, you can argue that by assuming these symmetries Jaynes is providing the extra information to make the problem well posed. We can set up physical situations where these symmetries don't hold and then we get a different outcome. What is worthwhile to take from this though is that there are always grounds for selecting between priors provided that the problem is properly posed. Bayesian inference is not as subjective as, at first, it may appear.

9

Entropy

Contents

Entropy is a means for quantifying the amount of uncertainty in a system. The ideas can be formalised within information theory. Information theory has many applications and is an entire discipline in itself. We give a brief overview of information theory and discuss a few of the properties of entropy before giving a glimpse of some of its applications. We finish by considering other measures of complexity, namely *Kolmogorov complexity* and *Fisher information*.

9.1 Shannon Entropy

Probability is used in situations of uncertainty, but how uncertain is an event? Is the outcome of the Grand National more uncertain than the score of a football match? These seem difficult to compare. In the first instance we have the name of winning horse (or possibly the race was cancelled), while in the second case we have the number of goals scored. Interestingly, however, we can measure uncertainty for these events in quite a precise and useful way (at least if we can attach probabilities to every possible outcome). The key idea allowing us to do this comes from *information theory* – that is, the theory of communicating information as efficiently as possible. More precisely we associate the uncertainty of an event with the expected minimum amount of information that has to be communicated for person A to tell person B about the event. It is important here to recognise that it is the *minimum possible* amount of information that needs to be communicated. The uncertainty of a discrete random variable X that takes values x_i is given by the Shannon entropy defined as

$$H_X = -\sum_{i=1}^{n} \mathbb{P}\left(X = x_i\right) \log_2\left(\mathbb{P}\left(X = x_i\right)\right) \tag{9.1}$$

which is measured in *bits*. Note that H_X is not a function of X, but rather a function (or functional) of the distribution $\mathbb{P}\left(X\right)$. If X can take two values, 0 or 1 say, with equal probability then $H_X = 1$ bit. Working with logarithms to the base 2 ($\log_2$) is often awkward. We can instead use the natural logarithm, in which case entropy is measured in *nats* with 1 bit $= \log(2)$nats.

Entropy has a much older pedigree than information theory, going back to Rudolf Clausius, Ludwig Boltzmann, Willard Gibbs, and James Clerk Maxwell, but it was Claude Shannon in 1948 who made the important connection to information. Ed Jaynes, whom we met in Chapter 8, argued that entropy in statistical mechanics can be viewed as an application of Shannon entropy – that is, statistical physicists are actually undertaking statistical inference.

9.1.1 Information Theory

What then is the connection between entropy and communication? A simple way to see this connection is to consider the problem of communicating information about a set of mutually exclusive events. Let p_i be the probability of event i. This might be the racing results from Chepstow or the text of the Queen's speech. Our goal is to send this information using the least number of bits possible. In other words, we want to minimise the *expected length* of the message – note that we only consider communicating information related to uncertain events (at least for the receiver), otherwise our message carries no information content. The way to do this is to devise a code so that the most likely outcome has a very short message. Of course, there is a limited amount of information you can encode using only short words so you will have to use longer codes for less probable events. How short can you make the expected message length? Interestingly there is a lower bound that is known to be tight (we can get arbitrarily close to the bound). *That lower bound is the entropy.*

When coding a text in English the problem arises as to what should be considered independent. For example, if we treat letters independently we could devise a code which gives very short bit strings to common letters such as 'e', 't', 'a', and 'o', while letters such as 'j', 'q', 'x', and 'z' should be coded with longer bit strings. This wouldn't lead to a very compact code as letters are far from independent. We would do better to use a coding which works on pairs of letters or some larger collection of letters. We might do even better by coding words rather than combinations of letters, or even whole sentences. Shannon made the first estimate of the entropy of English, which is estimated to be around 1 to 1.5 bits per letter (i.e. it can be compressed by a factor of between 3 to 5). Clearly, the entropy of English will depends on the speaker so an exact figure doesn't exist.

Information theory is about theoretical bounds on the length of codes (as opposed to coding theory which is about finding practical coding schemes). Shannon considered the problem of communicating through a noisy channel. This clearly has practical applications for telephone communications or communicating with space probes or even storing information on disk drives. In all these applications one wants to retrieve information despite the fact that some information might be corrupted by errors. We can correct these errors by building redundancy into the code. An interesting question is, given some error rate, how much redundancy is required to guarantee that you can correct your code up to an error rate ϵ. The common assumption before Shannon's work was that the amount of redundancy would have to grow infinitely as $\epsilon \to 0$. That is, you would need infinite redundancy to correct the code perfectly. In fact, Shannon showed this wasn't the case and you only need finite redundancy. Shannon's result provided a tight lower bound on the amount of information you would need to send, but it does not provide a mechanism for constructing such codes. The problem of devising optimally efficient error-correcting codes is unsolved and most practical codes don't get near to the bound, although a few more recently developed codes are getting closer to the Shannon bound.

Shannon formulated the problem of communication in terms of sending a message represented by a random variable, X. The message received at the end of the channel is another random variable Y. The probability of the two messages is represented by a joint probability $\mathbb{P}\left(X, Y\right)$. We can define the *joint entropy*

$$H_{X,Y} = -\sum_{x,y} P_{X,Y}(x, y) \log\big(P_{X,Y}(x, y)\big) \tag{9.2}$$

where $P_{X,Y}(x, y) = \mathbb{P}\left(X = x, Y = y\right)$ and the sum is over the set of values taken by X and Y. If X and Y are independent so that $\mathbb{P}\left(X, Y\right) = \mathbb{P}\left(X\right) \mathbb{P}\left(Y\right)$ then $H_{X,Y} = H_X + H_Y$ (the so-called *additive property of entropy*, which states the uncertainty of a set of independent events is equal to the sum of the uncertainties of the events on their own).

For communication, we are interested in the case when the signal received, Y, isn't independent of the message sent, X. In this case it is useful to consider the conditional entropy of X given $Y = y$ (i.e. Y takes a particular value) defined as

$$H_{X|Y=y} = -\sum_x P_{X|Y}(x|y)\log\bigl(P_{X|Y}(x|y)\bigr) \tag{9.3}$$

and the expected conditional entropy or uncertainty in the message sent, X, given the signal received, Y,

$$H_{X|Y} = \sum_y P_Y(y)\, H_{X|Y=y} = \sum_{x,y} P_{X,Y}(x,y)\log\bigl(P_{X|Y}(x|y)\bigr). \tag{9.4}$$

Substituting the expressions

$$\mathbb{P}\left(X,Y\right) = \mathbb{P}\left(Y|X\right)\,\mathbb{P}\left(X\right) = \mathbb{P}\left(X|Y\right)\,\mathbb{P}\left(Y\right).$$

into Equation (9.2) and using the property of logarithms, $\log(A\,B) = \log(A) + \log(B)$, we obtain the identity

$$H_{X,Y} = H_X + H_{Y|X} = H_Y + H_{X|Y}. \tag{9.5}$$

Shannon defined the *mutual information* between X and Y as

$$I_{X;Y} = H_X - H_{X|Y} \tag{9.6}$$

which measures the expected information gained (or uncertainty lost) about X by observing Y. If we had a noise-free channel we could deduce X from Y then $H_{X|Y} = 0$, so that the gain in information would be H_X (the uncertainty in X). On the other hand, if Y was independent of X (i.e. the channel completely corrupted the signal) then $P(X|Y) = P(X)$ so that $H_{X|Y} = H_X$ and $I_{X;Y} = 0$. In other words, Y would convey no information about X.

We note that

$$\begin{aligned} I_{X;Y} &= H_X - H_{X|Y} \\ &= H_X - (H_{X,Y} - H_Y) \\ &= H_Y - (H_{X,Y} - H_X) \\ &= H_Y - H_{Y|X} = I_{Y;X}. \end{aligned}$$

We can visualise the relation between the mutual information and various entropies as shown in Figure 9.1, which is worth contemplating. $H_{X,Y}$ is the total (expected) uncertainty about random variables X and Y. H_X is the uncertainty we have about X if we don't know Y, while H_Y is the uncertainty we have about Y if we don't know X. The mutual information $I_{X;Y}$ is the information we get about X by knowing Y or the reduction in uncertainty in X once we have observed Y. We note that since $I_{X;Y} = I_{Y;X}$, we gain the same amount of information about X by knowing Y as we do about Y from knowing X.

Now we return to Shannon's formulation of information theory. He characterised a noisy channel by $\mathbb{P}\left(Y|X\right)$. That is, given an input message X the probability of an output message Y is $\mathbb{P}\left(Y|X\right)$. To maximise the information rate we want to choose the input distribution $\mathbb{P}\left(X=x\right) = P_X(x)$ to maximise the

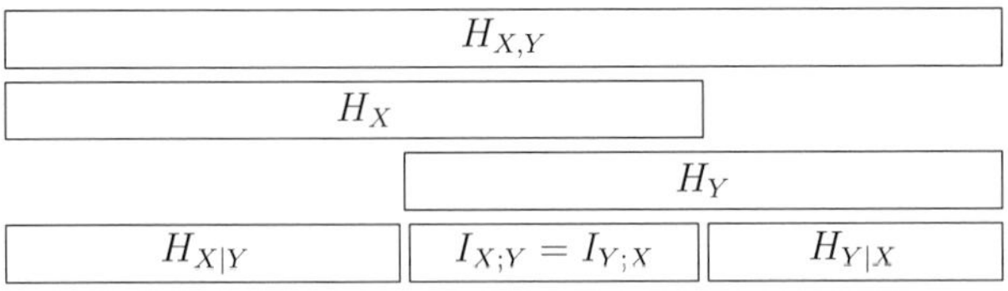

Figure 9.1 Relationship between mutual information and entropy.

mutual information. Shannon defined this maximised mutual information as the *capacity of the channel*

$$C = \max_{P_X} I_{X;Y}.$$

Shannon's remarkable theorem is that provided we send information down the channel at a rate below the capacity, then, if we use sufficiently long code words, there exists an error-correcting code whose expected error is less than ϵ for any $\epsilon > 0$. In other words, if we have a sufficiently large amount of information to transmit, there exists a code which will allow us to reach an information flow rate arbitrarily close to C. The proof of this is non-trivial, but can be found in most books on information theory. We give an outline of the key ideas in the proof after the following example showing Shannon's theorem in action.

Example 9.1 Communicating Down a Noisy Channel
To illustrate the meaning of this theorem we consider a binary channel (that is, the channel sending messages consisting of strings of 1 and 0) with an error rate of $f = 0.1$, where the error is independent on each bit. That is, $\mathbb{P}\left(X = 1|Y = 0\right) = \mathbb{P}\left(X = 0|Y = 1\right) = 0.1$. Assuming $\mathbb{P}\left(Y = 1\right) = p$, the conditional entropy of the signal received is

$$\begin{aligned}
H_{X|Y} &\overset{(1)}{=} -\sum_{X,Y\in\{0,1\}} \mathbb{P}\left(X,Y\right) \log(\mathbb{P}\left(X|Y\right)) \\
&\overset{(2)}{=} -P(1,1)\log\big(P(1|1)\big) - P(0,0)\log\big(P(0|0)\big) \\
&\quad - P(1,0)\log\big(P(1|0)\big) - P(0,1)\log\big(P(0|1)\big) \\
&\overset{(3)}{=} -(1-f)\,p\log(1-f) - (1-f)(1-p)\log(1-f) \\
&\quad - f(1-p)\log(f) - f\,p\log(f) \\
&\overset{(4)}{=} -(1-f)\log(1-f) - f\log(f) = 0.325\,\text{nats} = 0.469\,\text{bits}
\end{aligned}$$

(1) By the definition of conditional entropy.
(2) Expanding out the summation with $\mathbb{P}\left(X = x, Y = y\right) = P(x,y)$ and $\mathbb{P}\left(X = x|Y = y\right) = P(x|y)$.
(3) Using $P(x,y) = P(x|y)\,P(y)$ with $P(1|1) = P(0|0) = 1 - f$, $P(1|0) = P(0|1) = f$, $P(1) = p$, and $P(0) = 1 - p$.
(4) Combining terms.

In this particular example, because the error is the same for both $X = 0$ and 1, the conditional entropy does not depend on p. Recall the mutual information is defined as

$$I_{X;Y} = H_X - H_{X|Y}.$$

The capacity of the channel is found by maximising the mutual information with respect to p. Since $H_{X|Y}$ is independent of p, we need only choose $\mathbb{P}(X)$ to maximise H_X. This is maximised when $\mathbb{P}(X = 0) = \mathbb{P}(X = 1) = 1/2$, giving $H_X = 1$ (the amount of uncertainty in a single bit with equal probability is 1 bit). Thus the capacity of such a channel is

$$\max_{P_X} H_X - H_{X|Y} = 1 - 0.469 \approx 0.53\text{bits}.$$

Therefore, Shannon's noisy channel theorem tells us that even with a 10% error rate we should be able to construct an error-correcting code so that, if we send sufficiently large words down the channel, we can send information at a rate slightly above half the maximum sending rate and still lose no information. Alternatively, if we wanted to store information on a hard drive with an appalling 10% error rate then there exists an error-correcting code which allows us to retrieve the original information with arbitrarily small error probability that uses only twice the memory.

In information theory one commonly meets the entropy for a Bernoulli variable $X \sim \text{Bern}(f)$ given by $H(f) = -f \log_2(f) - (1 - f) \log_2(1 - f)$.

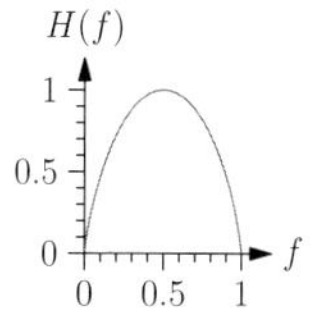

To understand Shannon's theorem we consider a binary channel with error rate f where we use code words of length n. Due to the channel noise, the typical code word will be corrupted at $f\,n$ bits. Some code words will be corrupted by more than this, but, by making n large enough, we can make the probability of more than $(f + \epsilon)n$ bits being corrupted arbitrarily small. (Recall in Section 7.2.4 we showed, using a Chernoff bound, that the probability of an error being greater than $(f + \epsilon)n$ is strictly less than $\mathrm{e}^{-f\,n\,\epsilon^2/3}$, for $\epsilon < f$). Thus, to construct an error-correcting code, each word of our code has to differ from any other by at least $2\,(f + \epsilon)\,n$ bits. We can think of our code words as points in the space of binary strings of length n surrounded by a ball of radius $(f+\epsilon)\,n$. To prevent any confusion these balls must not intersect. We show this schematically in Figure 9.2 – the actual space of message is not, of course, two-dimensional.

The number of strings which differ from some string $\boldsymbol{x}$ by at most $r = (f+\epsilon)\,n$ bits is

$$N(f) = \sum_{i=0}^{r} \binom{n}{i} = \binom{n}{r} + \binom{n}{r-1} + \binom{n}{r-2} + \cdots + \binom{n}{0}.$$

However, $\binom{n}{r-1} = \frac{r}{n-r+1}\binom{n}{r}$. If $r \approx fn$ and, $n \gg 1$ then $\binom{n}{r-1} \approx \frac{f}{1-f}\binom{n}{r}$ and, by a similar argument, $\binom{n}{r-2} \approx \left(\frac{f}{1-f}\right)^2 \binom{n}{r}$. (We cheat here slightly by setting ϵ to zero. But we can make ϵ as small as we like and still have an

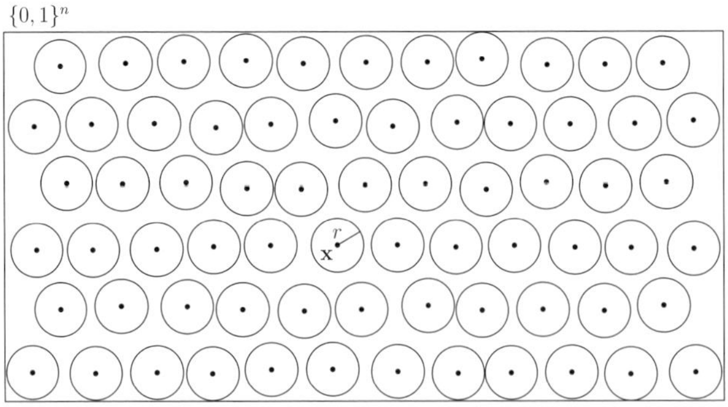

Figure 9.2 Schematic diagram showing error-correcting code words in the space of binary strings of length n. Each code word $\boldsymbol{x}$ is surrounded by a ball of radius $r = (f + \epsilon)\, n$. After transmission of the message down a noisy channel all error messages will land up in this ball with a probability $e^{-f\, n\, \epsilon^2/3}$ (assuming $\epsilon < f$).

arbitrary small probability of having more than $n\,(f + \epsilon)$ errors provided we ensure n is sufficiently large. Thus, setting $\epsilon = 0$ gives a lower bound.) Using this approximation we have

$$N(f) \approx \binom{n}{r} \left(1 + \frac{f}{1-f} + \left(\frac{f}{1-f} \right)^2 + \cdots \right) \approx \binom{n}{r} \frac{1-f}{1-2f}$$

where we used the fact that the sum of a geometric progression $1 + x + x^2 + \cdots = 1/(1-x)$. Finally, if we use Stirling's approximation for factorials

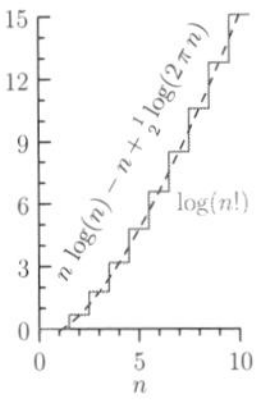

Stirling's approximation (dashed curve) plotted against $\log(n!)$.

$$n! \approx \sqrt{2\,\pi\, n}\, n^n\, e^{-n} \qquad \Longleftrightarrow \qquad \log(n!) = n \log(n) - n + \frac{1}{2} \log(2\,\pi\, n)$$

with $r = f n$ we get, after a little rearrangement,

$$N(f) \approx \frac{1-f}{1-2f} \frac{1}{\sqrt{2\,\pi\, n\, f(1-f)}} e^{-n(f \log(f) + (1-f) \log(1-f))}.$$

This is the total number of strings that are within a Hamming distance of $f\, n$ of any code word (the Hamming distance between two strings $\boldsymbol{x}$ and $\boldsymbol{y}$ is equal to the number of elements in the strings where $x_i \neq y_i$). Now the total number of strings of length n is 2^n so the maximum number of balls of radius $r = f n$ that could fit in the space of all strings of length n is

$$\frac{2^n}{N(f)} \approx \sqrt{\frac{2\,\pi\, n\, f(1-2f)}{1-f}}\; e^{n \log(2) + n(f \log(f) + (1-f) \log(1-f))}$$

where we have written $2^n = \exp(n \log(2))$ and $g(f)$ is a polynomial in f. Using the identity $\log(x) = \log(2) \times \log_2(x)$ we note that

$$\begin{aligned} \frac{2^n}{N(f)} &\approx \sqrt{\frac{2\,\pi\, n\, f(1-2f)}{1-f}}\; e^{n \log(2) (1 + f \log_2(f) + (1-f) \log_2(1-f))} \\ &= c\,\sqrt{n}\, e^{n \log(2)\, I_{X;Y}} = c\,\sqrt{n}\, 2^{n\, I_{X;Y}} \end{aligned}$$

where c is a constant depending on f and $I_{X;Y}$ is our channel capacity. In other words, the maximum number of error-correctable code words we can send in a string of length n is approximately $2^{n\, I_{X;Y}}$. That is, we cannot transmit information at a rate faster than the channel capacity. Figure 9.2 is somewhat misleading in that our space of binary codes is actually a n-dimensional hypercube (not a nice two-dimensional space), so that partitioning the space is non-trivial. Nevertheless, Shannon showed that there exists a code that achieves the channel capacity. That is, it packs these 'balls' of radius $r = (f + \epsilon)\, n$ into the space of strings of length n tightly. In fact, he showed that most codes where the code words are random chosen strings would be close to satisfying the bound. The proof does not, however, tell us how to construct the code. Shannon's theorem is more general than our discussion as it does not assume that the signals are binary or that the errors are simple.

There are a few caveats that one needs to keep in mind. Firstly, Shannon's theorem doesn't tell us what these error-correcting codes look like. Most error-correcting codes are far less efficient than Shannon's bound. A very efficient code may be exceedingly computationally costly to decode. For example, if we were to choose random code words, then to decode a message we may have to compute the distance between the signal we receive and all the code words. As there are an exponential number of code words this strategy is too slow to be practical. Secondly, there is a relationship between the size of block you need to code and the error rate you can tolerate. A code that requires you to transfer a terabyte of information in a single block might not be the solution you are looking for. You might prefer to have a slightly less efficient code or accept a higher error after decoding in order to have practical packet sizes. Finally, in the example, I made the assumption that the noise was independent on each byte. In most real systems noise is often much more correlated, so that several consecutive bits may all be corrupted. This can be overcome to some extent by shuffling information in the coded message and unshuffling the information before decoding. However, this again adds to the complexity of the coder.

9.1.2 Properties of Entropy

Entropy is a strictly convex-down (concave) function of p_i which is easily verified since the second derivative of $-p_i \log(p_i)$ is $-1/p_i$, which is always negative. The normalisation constraint for probabilities restricts the entropy to a convex region. One consequence of this is that the entropy can have only one maximum value.

The entropy of a discrete random variable is greater than or equal to zero (which follows from the fact that $\log(P_X(x)) \leq 0$) and is maximised when all probabilities are equally probable. That is, if p_i is the probability that $X = x_i$ for $i = 1, 2 \ldots, m$, then

$$H_X = -\sum_{i=1}^{n} p_i \log_2(p_i) \leq -\sum_{i=1}^{n} \frac{1}{n} \log_2\left(\frac{1}{n}\right) = \log_2(n).$$

To prove this we can maximise H_X with respect to the p_i's subject to the constraint that the probabilities sum to 1. This can be achieved by maximising the Lagrangian (see Appendix 7.C)

$$\mathcal{L} = -\sum_{i=1}^{n} p_i \log_2(p_i) + \lambda \left(1 - \sum_{i=1}^{n} p_i\right)$$

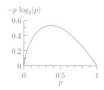

to obtain the conditions

$$\frac{\partial \mathcal{L}}{\partial p_i} = -\frac{1}{\log(2)} - \log_2(p_i) + \lambda = 0 \qquad \frac{\partial \mathcal{L}}{\partial \lambda} = 1 - \sum_{i=1}^{n} p_i = 0.$$

(The $1/\log(2)$ arises because we are differentiating $\log_2(p_i) = \log(p_i)/\log(2)$.) The first equation implies $p_i = e^{\lambda - 1/\log(2)}$, where we have to choose the Lagrange multiplier λ so that $\sum_{i=1}^{n} p_i = 1$. This gives us $p_i = 1/n$, as advertised.

The entropy associated with two *independent* random variables $H_{X,Y}$ is equal to the sum of the entropies of the two variables

$$H_{X,Y} = H_X + H_Y \quad \text{(true if } X \text{ and } Y \text{ are independent).}$$

This additive property clearly makes intuitive sense if we think of the entropy as a measure of uncertainty. A well-known argument is that if we require the entropy to have this additive property then it must have the form given in Equation (9.1). Alas, although well known, it isn't true. We will see that the Fisher information also satisfies this additive condition.

Entropy for continuous random variables. Entropy makes perfect sense when discussing discrete random variables. We know after all how to communicate information about a discrete set of events. When working with continuous random variables things get more complicated. An obvious definition of entropy would be

$$H_X = -\int f_X(x) \log\big(f_X(x)\big)\, \mathrm{d}x$$

but caution is necessary. For one thing, $f_X(x)$ is not guaranteed to be less than 1, so you can have a negative entropy (see Exercise 9.3). Also this entropy is not invariant to a change in variables. That is, if we make the change in variables $X \to Y$, such that $X = g(Y)$, then we get

$$H_X = -\int f_X(g(y)) \log\big(f_X(g(y))\big)\, g'(y)\, \mathrm{d}y$$

but $f_X(g(y))\, g'(y) = f_Y(y)$ by the normal transformation law for probability densities, so

$$H_X = -\int f_Y(y) \log\left(\frac{f_Y(y)}{g'(y)}\right) \mathrm{d}y \neq H_Y$$

unless $g'(y) = 1$. So the degree of uncertainty changes depending on how we measure the random variable! This isn't nice. We can use entropy for continuous random variables, but we have to be careful.

The awkwardness of using continuous random variables in information theory should not be so surprising. It takes an infinite amount of information to communicate a continuous variable to arbitrary accuracy. To develop an information theory for continuous variables we have to introduce some discretisation or coarse graining of the message space (sometimes this is known as regularisation). If we change variables our discretisation differs and our entropy will also change.

If we transform both the variable and the discretisation in the same way then the information will remain unaltered.

9.2 Applications

Entropy or measures of information have numerous practical applications. Obvious applications are in data compression and coding, although we do not discuss these topics here – the interested readers should consult a more specialist book. Instead we consider an application of direct use in performing probabilistic inference.

9.2.1 Mutual Information

Recall the mutual information between two random variables X and Y is given by

$$I_{X;Y} = H_X - H_{X|Y} = H_Y - H_{Y|X}.$$

This measures the amount of information you can infer about one random variable given the second one. Sometimes, we consider the information ratio

$$\frac{I_{X;Y}}{H_X} = \frac{H_X - H_{X|Y}}{H_X}$$

which is equal to 1 if you can infer X from Y perfectly and 0 if they are independent of each other. Mutual information or the information ratio is often used as a measure to compare different signals, as illustrated in the following example.

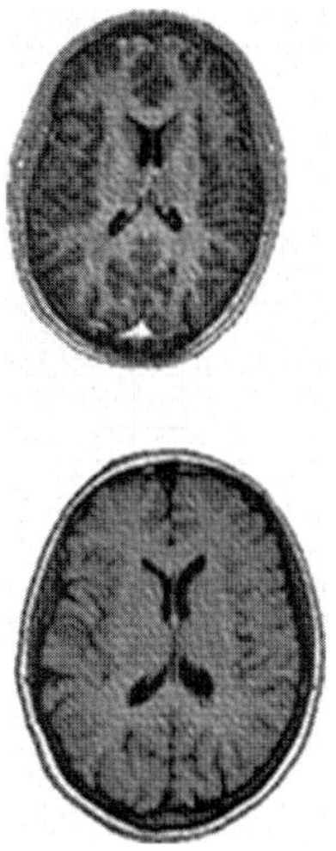

Example 9.2 Image Registration
Suppose we have two images (e.g. MRI brain scans) and we want to align them. That is, we want to transform one (e.g. by shifting, rotating, scaling, etc.) so that it is as well aligned as possible with the second, reference, image. A difficulty is that the images may not have been taken under the same conditions, so what is grey in the first image may be black in the reference image. Furthermore, this discrepancy may not be linear due to saturation, automatic contrast enhancement, etc. Thus it is difficult to find a mapping between the intensities in the two images. One way around this is to consider each pixel as a bit of information and seek the transformation that maximises the mutual information between the two images. This way you can learn the mapping between images.

9.2.2 Maximum Entropy

One of the problems faced by the Bayesian approach to inferencing is to encode our prior knowledge, or in many cases our complete lack of knowledge. As entropy is a measure of the uncertainty, by maximising the entropy we maximise the uncertainty and in so doing remove any prejudices we might have. This is a commonly used approach for defining priors. It is applicable in a very broad setting. Imagine that we have a system depending on a discrete random variable X (this may be multidimensional). We assume that we measured some average properties of our system. We impose these measurements as constraints of the form

$$\mathbb{E}\left[f_l(X)\right] = \sum_i p_i f_l(x_i) = \hat{f}_l$$

where p_i is the probability that $X = x_i$, and $l = 1, 2, \ldots, m$ labels the constraints. We suppose we can measure $\hat{f}_l$, but we don't know p_i. *Is there some way to infer the 'most likely' values of p_i?* Maximum entropy provides an answer. That is, we can maximise the entropy subject to the constraints by writing down a Lagrangian (see Appendix 7.C)

$$\mathcal{L} = -\sum_i p_i \log(p_i) + \lambda_0 \left(1 - \sum_i p_i\right) + \sum_{l=1}^{m} \lambda_l \left(\sum_i p_i\, f_l(x_i) - \hat{f}_l\right).$$

This has three terms. The first is the entropy (we use $\log(p_i)$ rather than $\log_2(p_i)$ because it is more convenient, however, it only introduces a multiplicative constant which can be absorbed into the Lagrange multipliers). The λ_0 term ensures that $\sum_i p_i = 1$, while the last term ensures all the observed constraints are satisfied. Since the entropy is concave and the constraints linear in p_i the Lagrangian will have a unique maximum which we can find by setting the partial derivatives to zero

$$\frac{\partial \mathcal{L}}{\partial p_i} = -1 - \log(p_i) - \lambda_0 + \sum_{l=1}^{m} \lambda_l f_l(x_i) = 0.$$

Solving for p_i we find

$$p_i = \frac{1}{Z} e^{\lambda_1 f_1(x_i) + \lambda_2 f_2(x_i) + \cdots + \lambda_m f_m(x_i)} \tag{9.7}$$

where the Lagrangian multipliers, λ_l, must be chosen so that all the constraints are satisfied and

$$Z = \sum_i e^{\lambda_1 f_1(x_i) + \lambda_2 f_2(x_i) + \cdots + \lambda_m f_m(x_i)} \tag{9.8}$$

is a normalisation constant (related to λ_0). This normalisation constant is known as the *partition function* and has the property that

$$\frac{\partial \log(Z)}{\partial \lambda_l} = \frac{1}{Z}\frac{\partial Z}{\partial \lambda_l} = \sum_i f_l(x_i)\, \frac{1}{Z} e^{\lambda_1 f_1(x_i) + \lambda_2 f_2(x_i) + \cdots + \lambda_m f_m(x_i)} = \sum_i p_i\, f_l(x) = \hat{f}_l.$$

This formulation lies at the heart of equilibrium statistical physics (see Exercise 9.4). Equation 9.7 can be seen as a generalised Boltzmann distribution. However, maximum entropy can be used in many other areas.

Example 9.3 Normal Distribution
Consider a continuous random variable, X, with a known mean and second moment

$$\mathbb{E}\left[X\right] = \mu, \qquad \mathbb{E}\left[X^2\right] = \mu_2 = \mu^2 + \sigma^2.$$

To compute its maximum entropy distribution we maximise the entropy subject to the above constraints. This can be achieved by maximising the Lagrangian

$$\mathcal{L}(f) = -\int f_X(x)\,\log\big(f_X(x)\big)\,\mathrm{d}x + \lambda_0\left(\int f_X(x)\,\mathrm{d}x - 1\right)$$
$$+\,\lambda_1\left(\int f_X(x)\,x\,\mathrm{d}x - \mu\right) + \lambda_2\left(\int f_X(x)\,x^2\,\mathrm{d}x - \mu_2\right).$$

Note that the term proportional to λ_0 ensures that the probability density is properly normalised. The Lagrangian is a functional (i.e. a function of a function) of f. We can solve the maximisation problem by setting the functional derivative to 0 (see Appendix 9.A)

$$\frac{\delta \mathcal{L}(f)}{\delta f_X(x)} = -1 - \log\big(f_X(x)\big) + \lambda_0 + \lambda_1\,x + \lambda_2\,x^2 = 0$$

or

$$f_X(x) = \mathrm{e}^{-1+\lambda_0+\lambda_1\,x+\lambda_2\,x^2}$$

where we have to choose λ_0, λ_1 and λ_2 so that

$$\int \mathrm{e}^{-1+\lambda_0+\lambda_1\,x+\lambda_2\,x^2}\,\mathrm{d}x = 1$$
$$\int \mathrm{e}^{-1+\lambda_0+\lambda_1\,x+\lambda_2\,x^2}\,x\,\mathrm{d}x = \mu$$
$$\int \mathrm{e}^{-1+\lambda_0+\lambda_1\,x+\lambda_2\,x^2}\,x^2\,\mathrm{d}x = \mu_2 = \mu^2 + \sigma^2.$$

It is a rather tedious exercise to compute all the integrals and solve for the Lagrange multipliers but if you do so you find

$$\lambda_0 = 1 - \frac{\mu^2}{2\,\sigma^2} - \frac{1}{2}\log\left(2\,\pi\,\sigma^2\right) \qquad \lambda_1 = \frac{\mu}{\sigma^2} \qquad \lambda_2 = \frac{1}{2\,\sigma^2}$$

so that

$$f_X(x) = \frac{1}{\sqrt{2\,\pi}\,\sigma}\mathrm{e}^{-(x-\mu)^2/(2\sigma^2)}.$$

That is, the normal distribution is the maximum entropy distribution given knowledge of the mean and variance.

Returning to an example from Chapter 8, we can use maximum entropy to provide a likelihood and prior function.

Example 9.4 Inferring Voting Habits: Example 8.10 Revisited
In Example 8.10 on page 222 we considered a set of voters who interact with each other. We encode the degree of interactions by w_{ij}. These could be a set of your Facebook friends where $w_{ij} = 1$ if i and j are friends with each other and $w_{ij} = 0$ otherwise. The voting intentions of individual i we denote by S_i. In this case, we can define an 'energy' function

$$E(\boldsymbol{S}) = \sum_{(i,j)\in\mathcal{E}} w_{ij} \left[\!\left[S_i \neq S_j \right]\!\right]$$

that just counts the number of friends (i.e. pair of individuals with $w_{ij} > 0$) who would vote differently. If we expect that the probability of two friends voting differently is p, then in expectation

$$\mathbb{E}\left[E(\boldsymbol{S})\right] = \sum_{(i,j)\in\mathcal{E}} p = p\,|\mathcal{E}|\,. \tag{9.9}$$

Under a maximum entropy assumption we would then have a probability distribution

$$\mathbb{P}\left(\boldsymbol{S}\right) = \frac{\mathrm{e}^{-\beta E(\boldsymbol{S})}}{Z(\beta)}, \qquad Z(\beta) = \sum_{\boldsymbol{S}} \mathrm{e}^{-\beta E(\boldsymbol{S})}$$

where β is to be chosen so that Equation (9.9) is satisfied. This is in accordance with Equation (8.6), which appeared from nowhere in Example 8.10 – in that example it was helpful to distinguish between the observed labels (voting intentions), $\boldsymbol{S}^o$, and unobserved labels, $\boldsymbol{S}^u$. We might not know the probability, p, of two friends voting differently, in which case we may have to choose β through model selection as suggested in the original example.

Maximum entropy is used in many different areas of inference. For example, it has been used for image enhancement and in estimating the distribution of bubble sizes.

Is the maximum entropy distribution correct? Recall the rationale behind the maximum entropy method is that by maximising our uncertainty subject to the known constraints, we are choosing the least biased distribution, which in some sense is the 'most likely'. To make this more concrete, suppose we have a large set of bins lined up on the x-axis and we put N balls in the bins at random. We illustrate this in Figure 9.3. We will treat the balls as probability masses and the position of the bin as the value of the random variable. If we consider all arrangements of the balls in bins that have the right statistical properties (e.g. if we know the mean and variance then we consider only those arrangements of

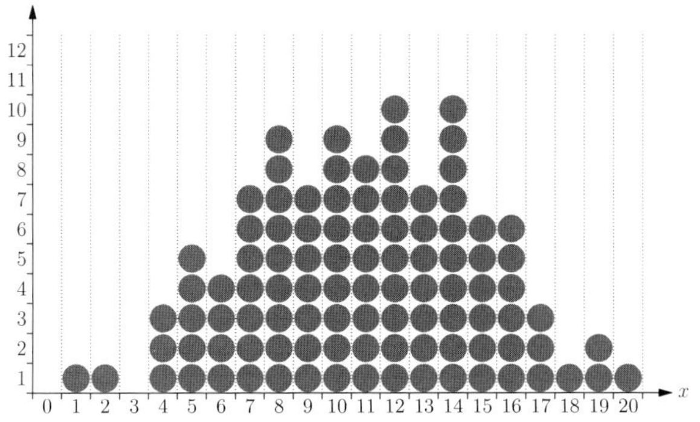

Figure 9.3 An example of putting balls into bins.

the balls that have the correct statistics) then the probability of an arrangement of the balls is given by

$$\begin{aligned}\mathbb{P}(\boldsymbol{n}) &\overset{(1)}{\propto} \text{Mult}(\boldsymbol{n}|N, \mathbf{1}/N) \prod_l \left[\!\left[\sum_i \frac{n_i}{N} f_l(x_i) = \hat{f}_l \right]\!\right] \\ &\overset{(2)}{\propto} N! \prod_i \frac{1}{n_i!} \left[\!\left[\sum_i n_i = N \right]\!\right] \prod_l \left[\!\left[\sum_i \frac{n_i}{N} f_l(x_i) = \hat{f}_l \right]\!\right]\end{aligned}$$

(1) Here we consider placing N balls in K boxes where $\boldsymbol{n} = (n_1, n_2, \ldots, n_K)$ denotes the placement of the balls in the box so that there are n_i balls are in box i. The probability of such an assignment is given by the multinomial distribution $\text{Mult}(\boldsymbol{n}|N, \mathbf{1}/N)$ (each box is assumed to be equally likely). We only consider placements that are consistent with our observations $\hat{f}_l$ (hence the indicator functions).

(2) We write out the multinomial distribution and explicitly put in the constraint that $\sum_i n_i = N$ (i.e. there are N balls in total).

Now, using Stirling's approximation,

$$n! \approx \sqrt{2\pi n}\, n^n \,\mathrm{e}^{-n} \quad \Longleftrightarrow \quad \log(n!) = n\log(n) - n + \frac{1}{2}\log(2\pi n),$$

we obtain

$$\mathbb{P}(\boldsymbol{n}) \approx C\, \mathrm{e}^{-N \sum_i \frac{n_i}{N} \log\left(\frac{n_i}{N}\right)} \left[\!\left[\sum_i \frac{n_i}{N} = 1 \right]\!\right] \prod_l \left[\!\left[\sum_i \frac{n_i}{N} f_l(x_i) = \hat{f}_l \right]\!\right].$$

In the limit of large N, where we treat n_i/N as the probability of being in state x_i, the exponent is proportional to the entropy. Thus, the most likely configuration of balls in bins is that which maximises the entropy subject to the known constraints. As $N \to \infty$, this distribution becomes overwhelmingly more probable than any other distributions.

The argument above suggests that the maximum entropy distribution is overwhelmingly more likely than any other distributions. However, be wary of these kinds of arguments. The uncertainty for most random variables doesn't arise from putting balls into boxes. So there are times when maximum entropy distributions provide poor approximations. Often if we are more careful we might discover some other statistical quantity which, for some particular reason, has a specific value, and when we add in this quantity the whole story changes. At times maximum entropy can give excellent results for very little work, but it can also let you down.

9.2.3 The Second Law of Thermodynamics

This is a digression into physics for the curious. It can be skipped by those in a hurry.

The laws of thermodynamics read more like a grim morality tale than a series of scientific principles. Thus, the first law states that energy is conserved so that you cannot build a perpetual motion machine that extracts any energy. The second law tells us that you cannot extract internal energy from a system without a temperature gradient that itself requires energy to maintain. Furthermore, even when there is a temperature gradient the efficiency with which you can extract useful work depends on the temperature you are working at. An engine would only be 100% efficient when it is run at absolute zero temperature (0° Kelvin). The third law tells us that we cannot ever reach absolute zero! Another statement of the second law is that entropy must always increase. Thus, the universe slowly slides to every greater uncertainty and chaos.

Statistical mechanics (i.e. the description of physical systems that obey thermodynamics) can be seen as an application of the maximum entropy principle. It involves a set of macroscopic observables (e.g. pressure, volume, temperature, etc.), which we have learnt through careful observation are sufficient to describe the subsequent macroscopic behaviour of the system. That is, if we run repeated experiments using the same set of macroscopic variables we obtain the same results up to the accuracy of the experiment. Statistical mechanics describes what happens to an ensemble of systems that have the same macroscopic properties. There are an astronomically huge number of *microstates* (atomic configurations) with the same *macrostate* (macroscopic variables). Different microstates can give you totally different future behaviours. In statistical mechanics we assume that the microstate that we are actually in is one that maximises the entropy of the system. The rationale for this is that such microstates are overwhelmingly more probable than other microstates. Thus, statistical mechanics is derived by maximising the entropy of the microstates given the observed macroscopic variables. It provides a statistical prediction of the evolution of the system in the sense that it tells us what will happen with overwhelming probability; it does not tell us what necessarily will happen (we could be in a very weird microstate which although compatible with the macroscopic variables leads to an entirely different future to that predicted for the maximum entropy microstates – although this is overwhelmingly unlikely).

There are many definitions of entropy used in physics (some of which are inconsistent). The most useful definition of entropy is due to Josiah Willard Gibbs, who defined the entropy in terms of the probability distribution in the phase space of the microstates. Classically the phase space, Γ, describes the set of all possible positions and momenta of the atoms making up the system. Letting z represent a point in phase space (note that for a system with N particles this would be a $6N$-dimensional vector describing the positions and momenta of all the particles) and defining $\rho(z)$ to be the probability distribution for the system being in a microstate state z then the Gibbs entropy is given by

$$S_G = -\int_{z \in \Gamma} \rho(z) \log(\rho(z)) \, dz.$$

(Although this is an entirely classical description, using a quantum mechanical definition does not substantially change this discussion so we will stay in the world of classical physics.) We will denote the set of microstates that are consistent with the macrostate, Ξ, by $\mathcal{W}(\Xi)$. Assuming that we are equally likely to be in any of the microstates, then

$$\rho(z) = \frac{[\![z \in \mathcal{W}(\Xi)]\!]}{|\mathcal{W}(\Xi)|}$$

(where $|\mathcal{S}|$ denotes the size of the set $\mathcal{S}$). In this case the Gibbs' entropy is equal to $S_G = \log(|\mathcal{W}(\Xi)|)$, where $|\mathcal{W}(\Xi)|$ is the volume of phase space consistent with the macrostate Ξ. Interestingly, under classical laws of physics (i.e. Newton's law) the volume of phase space does not change under the time evolution of the particles. This is a famous result known as Liouville's theorem, which follows because of the reversibility of the microscopic laws of physics. The same holds for quantum mechanics. Thus, Gibbs' entropy never changes!

What about the second law of thermodynamics? This is a subtle point. Let us consider what happens when we remove a partition from a box with gas molecules on one side (Figure 9.4). We denote the macroscopic variables before removing the partition by Ξ, and the variables after removing the partition (and having let the system settle down) by Ξ'. We do not know the exact microstate of the system, however, we know that we started in one of the microstates in $\mathcal{W}(\Xi)$ (shown as a circle inside the set of all possible microstates Γ in Figure 9.4). After

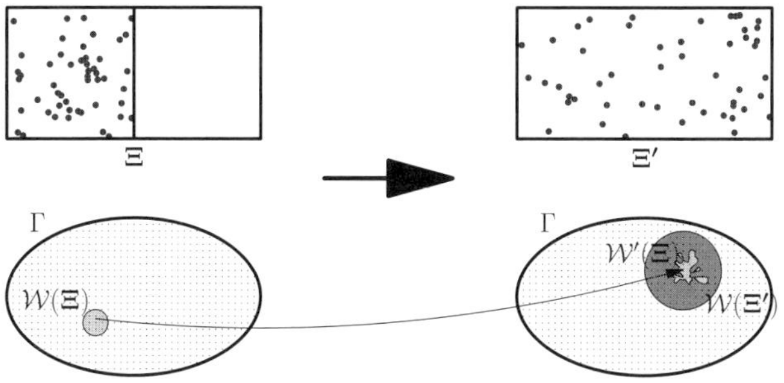

Figure 9.4 Removing a partition from a box with gas molecules on one side. The lower diagram shows a schematic of the phase space. The set $\mathcal{W}(\Xi)$ represents the set of microstates compatible with the observed macroscopic variables Ξ (e.g. temperature and pressure). When the partition is removed those states will equilibrate to a new set of microstates $\mathcal{W}'(\Xi)$ with a different set of macroscopic variables Ξ'. These will be experimentally indistinguishable from the complete set of microstates $\mathcal{W}(\Xi')$ compatible with Ξ'.

removing the partition we are in one of the microstates $\mathcal{W}'(\Xi)$ which comes from one of the initial possible microstates in $\mathcal{W}(\Xi)$. However, there are many other microstates that are also consistent with the final macrostate Ξ'. That is, it is not necessary to have started with the partitioned system. We could have just adjusted the temperature and pressure and gas density to reach the same macrostate, Ξ', without introducing a partition. Now we note that $\mathcal{W}'(\Xi) \subset \mathcal{W}(\Xi')$, since one way of achieving the macrostate Ξ' is by starting from a microstate $\mathcal{W}(\Xi)$ and removing the partition. Because phase-space volume is conserved this implies that $|\mathcal{W}(\Xi')| \geq |\mathcal{W}'(\Xi)| = |\mathcal{W}(\Xi)|$. As the entropy is a monotonically increasing function of the phase-space volume the experimental entropy increases! That is, experimental entropy increases because it considers only the current macroscopic variables and forgets the past history of the particles. The atoms actually always retain a 'memory' of their past history. Those microstates $\mathcal{W}'(\Xi)$ will never coincide with those other microstates in $\mathcal{W}(\Xi')$ which don't originate from $\mathcal{W}(\Xi)$. To see this we just recall that the laws of physics are time reversible. Consequently, if we could somehow reverse all the momenta of the atoms at time t after removing the partition, those microstates in $\mathcal{W}'(\Xi)$ would evolve in such a way that the particles at time $2t$ would all end up in the left half of the box, while this would not be true for all the other microstates in $\mathcal{W}(\Xi')$.

We have seen that experimental entropy always increases, which is a consequence of the fact that the phase-space volume is invariant (so that the Gibbs' entropy does not change). The argument requires only that there exist observable macroscopic variables which describe the behaviour of the system as accurately as we can measure it. We are not forced to assume microscopic chaos or what physicists have called ergodicity, meaning that taking an average over time is equivalent to taking an average over the set of microstates consistent with the observed macroscopic variables. However, notions such as microscopic chaos may be necessary to explain why a small number of macroscopic variables are sufficient to describe the evolution of a system. In our example, illustrated in Figure 9.4, the microstates $\mathcal{W}'(\Xi)$ will not form a tight ball in the phase space, but will be thoroughly mixed up with the other microstates in $\mathcal{W}(\Xi')$ so that they are effectively indistinguishable from these other microstates. If this wasn't the case then we could experimentally distinguish these microstates; this would tell us that there is another macroscopic variable which controls the expected evolution. The fact that we know of no such macroscopic variable shows that at a macroscopic level the past history of the microstates is lost. This is a consequence of the fact that a system of many interacting particles is in a state of microscopic chaos. That is, two microstates that differ minutely will diverge exponentially. It is easy to imagine that if we took two boxes of gas where the atoms were in identical positions and had identical momenta except for one atom being slightly displaced, the discrepancy would alter the first collision involving the perturbed particle so that now two particles would have slightly different trajectories. These two particles could then interact with two more particles, causing an exponential divergence of the two systems. A consequence of this is that $\mathcal{W}'(\Xi)$ would be dispersed throughout $\mathcal{W}(\Xi')$.

The great success of statistical mechanics is a consequence of the fact that there are a small number of macroscopic variables which are sufficient to determine the future evolution of the system with overwhelming probability. The reason why this situation arises is a consequence of microscopic chaos effectively erasing the past history by mixing microstates coming from very different histories. Maximum entropy can perfectly logically be applied to other systems that do not experience microscopic chaos. However, in these cases there are rarely a small number of macroscopic variables which sufficiently describe the system. We can still apply maximum entropy to some sets of observables that we have measured, however, in such cases we are likely to find that the predicted maximum entropy state differs from the true state. The maximum entropy prediction is then often incorrect although it frequently gives a reasonable approximation.

Maxwell's Demon

The great physicist James Clerk Maxwell proposed the following thought experiment to break the second law of thermodynamics. He considered a box containing gas with a partition down the middle (see Figure 9.5). The box has a small trapdoor in the partition controlled by a demon. The demon looks at the atoms bouncing around and opens the door if a fast particle is moving from left to right or if a slow-moving particle is moving from right to left. This way, eventually the right container will contain fast high-energy particles while the left container will contain slow low-energy particles. The demon has created a temperature gradient which can then be used to extract thermal energy. The demon has therefore broken the second law of thermodynamics.

Maxwell's demon caused a great deal of confusion until it was fully resolved by Charles Bennett, who showed that, alas, Maxwell's demon cannot break the second law of thermodynamics. The problem for the demon is that he needs energy to store information – information he needs to decide when to open the trapdoor. In fact, to store a bit of information (or more correctly to erase a bit of information) requires an energy of kT where k is Boltzmann's constant and T is the thermal temperature. The argument is rather intricate, but to store one bit of information requires reducing the degree of freedom of a state by two, which requires work. In the best case, this is the same energy that Maxwell's demon could extract by opening the trapdoor. You might think that Maxwell's demon

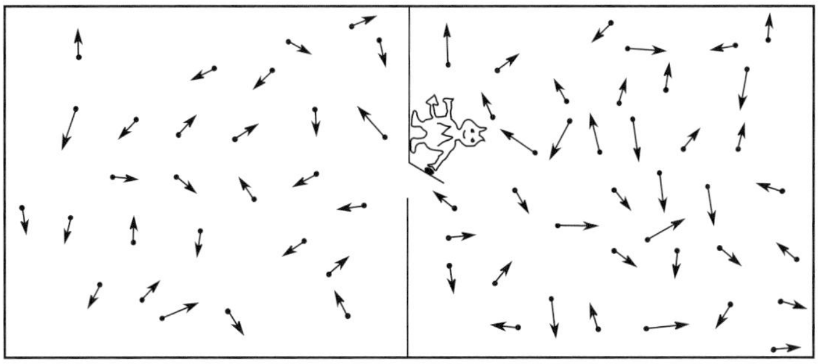

Figure 9.5 Maxwell's demon at work.

could win by living at a lower temperature than the gas, but in this case he would be just like a Carnot engine, extracting useful work from a temperature gradient (the temperature difference between himself and the gas). However, maintaining that temperature gradient would use up any energy gained by the demon. Maxwell's demon illustrates a deep connection between information entropy and thermodynamic entropy. A connection which you would not easily anticipate.

9.3 Beyond Information Theory

Shannon entropy is not the only measure of uncertainty. We briefly introduce two others which have some prominence: namely Kolmogorov complexity and Fisher information.

9.3.1 Kolmogorov Complexity

Information theory doesn't address the question of *how much information is there in a string?* That is something external to the theory. Kolmogorov complexity addresses precisely that question. It defines the complexity of a string to be the size of the shortest program which can generate that string. This sounds straightforward, but there are some complications. Firstly, this will depend slightly on the language we are using to describe the program. This, however, is a superficial difficulty because the difference in the lengths of strings when we change our computer language will always be small (a consequence of the universal nature of computation). The harder problem is that the length of the shortest program to compute a string is incomputable – that is, it is one of those questions which is impossible to answer in general. This may seem to make Kolmogorov complexity rather uninteresting, but despite this drawback there are some interesting inferences that can be made.

Although commonly referred to as Kolmogorov complexity it was developed independently by R. J. Solomonoff (1962), A. N. Kolmogorov (1963), and G. K. Chaitin (1965). It is sometimes referred to as Kolmogorov–Chaitin complexity.

The most important idea is that there exists strings for which the shortest program to generate the string will be the same length as the string. Clearly the program `print` ⟨`string`⟩ can print any string, but it is not shorter than the string itself. These strings are said to be incompressible and are truly random strings. Now a simple counting argument between possible programs and strings shows that nearly all string are incompressible. Unfortunately, determining whether a string is incompressible is incomputable. The proof of this is extremely easy. Consider the question: What is

> the least natural number that cannot be described in less than fourteen words?

If we could compute this number then we could describe it in less than 14 words (i.e. the 13 words in the sentence above). This produces a contradiction showing that compression is incomputable. However, we know from Gödel that in any powerful formal system there will be true statements that cannot be proved and this is one such question. We can, of course, compress many strings that we encounter in the real world and there is a host of algorithms for doing this (jpeg, zip, gzip, ...). However, these only work by exploiting obvious redundancies. Give a compression algorithm a sequence generated from a pseudo-random number generator, it is unlikely to give you any compression, even though pseudo-random number generators tend to be short programs so the strings generated by them have low Kolmogorov complexity. Although we cannot determine which strings are compressible we can ask questions about the compressibility of sets of strings.

Kolmogorov complexity may appear rather abstract and theoretical, but it has inspired some practical applications. One of these is a similarity measure computed by concatenating two strings (or documents) and then compressing them using a standard compression algorithm (e.g. gzip). To measure the similarity, we compare the length after compressing the two strings concatenated together with the lengths when compressing the strings separately. If the strings are similar then a good compression routine will exploit their similarity to highly compress the concatenated string.

This similarity measure has the nice feature that it picks up similarities which might be quite complex. However, compression algorithms only provide a crude approximation to the true Kolmogorov complexity, so this algorithm really has to be viewed as a neat idea inspired by Kolmogorov complexity. Had this algorithm been thought of by someone not in the Kolmogorov complexity field, I'm not sure that it would be viewed as an application of Kolmogorov complexity at all.

On a more philosophical note, the scientific method is often portrayed purely in terms of refutation. That is, in science we build theories that can be refuted and then undertake experiments to test the predictions of the theory to their limits. Although this captures one aspect of science, it misses the question of what makes a good theory and, if we cannot immediately test a theory, how plausible should we consider it? Kolmogorov complexity fills this gap. If we have a theory that drastically compresses data obtained from a huge amount of observations, then with overwhelming probability the theory must capture something about reality. This is a judgement we can make without attempting to falsify the theory. Of course, the theory might not quite be correct: Newton's laws summarise a huge number of observations, but isn't quite right (although it is sufficiently useful that we still teach it to children and use it in innumerable applications). By attempting to refute the theory we can check its domain of applicability, but it is the ability to compress information that makes Newton's law worthy of consideration. This compression might be sheer coincidence. Given any short sequence (the sunny days last year) it is reasonably likely one can find a formula to compress the data (in fact there is likely to be considerable structure so we would be able to compress the data using a compression algorithm). However, the vast majority of strings are incompressible, so unless there is real structure it is astronomically improbable that we are going to find a way of substantially compressing large amounts of data by chance.

9.3.2 Minimum Description Length

Another inspiration from Kolmogorov complexity is the *minimum description length* (MDL) method for performing inference (Rissanen, 1983). Its adherents claim that it is more consistent than Bayesian inference. The idea is that the best model for a collection of data would be the one requiring the shortest program that generates the data. We should therefore seek a model that allows us to communicate the data as efficiently as possible. The rationale behind this is that compressible data is much more likely (can be generated by far more mechanisms) than less compressible data. For example, if we could generate the data with a program 50 bits long then there are likely to be billions of programs 100 bits long that will also generate the data. Turning this idea into a workable method requires some effort.

Example 9.5 Curve Fitting

Given a set of data $\mathcal{D} = \{(\boldsymbol{x}_\alpha, y_\alpha) | \alpha = 1, \ldots, n\}$, where each data point corresponds to a set of feature $\boldsymbol{x}_\alpha$ and a target y_α, then a very common problem in data analysis is to find a function $f(\boldsymbol{x})$ such that $f(\boldsymbol{x}_\alpha) \approx y_\alpha$. This is useful if we want to make some prediction where we know the features $\boldsymbol{x}$ but not the target value. Determining such a function is known as *regression* and is a major task in machine learning. In the case when the function we are fitting takes a one-dimensional input this problem is known as *curve fitting*. Curve fitting is nice to illustrate as we can easily visualise it, however, it is only a small part of regression. We concentrate here on curve fitting, although much of what we say is applicable to more general forms of regression. A typical problem would be to fit a polynomial to all the data points (x_α, y_α). That is, we fit an m order polynomial

$$f(x|\boldsymbol{w}) = \sum_{i=0}^{m} w_i x^i = w_m\, x^m + w_{m-1}\, x^{m-1} + \cdots w_1\, x + w_0$$

as close as possible to the data points. The classical solution (going back to the great mathematical genius Carl Friedrich Gauß) is to choose the set of weights, $\boldsymbol{w} = (w_0, w_1, \ldots, w_m)$, to minimise the squared error

$$E(\boldsymbol{w}) = \sum_{\alpha=1}^{n} \left(f(x_\alpha|\boldsymbol{w}) - y_\alpha\right)^2.$$

This is known as the *least squares problem*. For polynomials, $f(x|\boldsymbol{w})$ is a linear function of the weights so this is known as *linear regression*, whose solution is obtained by solving a set of linear equations. Linear regression appears very natural, but raises the question what degree polynomial should we fit? The higher the degree of the polynomial we use the more closely we can fit the data points. If we use as many free parameters as we have data points then we can fit all the data points perfectly. See Figure 9.6 for an example. But, what is the best fit?

In the example shown in Figure 9.6 the third-order polynomial appears, at least to me, the more *reasonable* fit, but can we justify this? Here MDL has an elegant answer. Suppose we want to communicate the data to some degree of accuracy, ϵ. For concreteness, let us

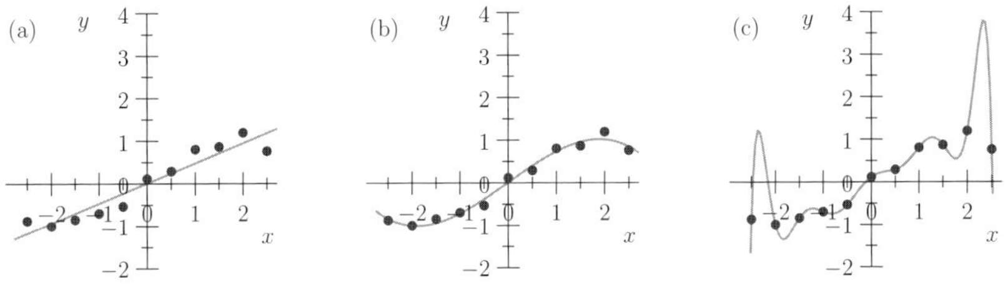

Figure 9.6 Fitting a curve to 11 data points using (a) a line, (b) a third-order polynomial, and (c) a tenth-order polynomial.

assume we want to communicate the data to two decimal places where the y values range from -1 to 1. We could just send each data point as it is (i.e. unencoded). This would take $2\log_2(10) + 1 \approx 7.6$ bits of information per data point (the extra bit is to determine the sign) or around 84 bits of information for the 11 data points in our example. However, we could compress the message by sending the parameters of our fitting models and the residual errors between the model and our data points. The message would then consists of two pieces: the message to describe the model (basically the parameters of the model), and a message describing the residual errors. The preferred model is the one with the shortest code.

How does transmitting a model help? In Figure 9.7 we show the residual errors between the data points and the third-order polynomial fit. As the model is a reasonably good fit the standard deviation in the residual error is $\sigma \approx 0.1$. We also show a number of bins of size $\epsilon = 0.01$. To communicate a data-point value y_i given the model and the x_i value we just have to communicate which bin the residual error falls into. We can compress this information using the fact that some bins are more likely to be occupied than others. If we assume that the probability of the residual error, Δ_i, falling into a bin is approximately given by $\epsilon\,\mathcal{N}(\Delta_i|0,\sigma^2)$, then Shannon's bound tells us that we need at least $-\log\big(\epsilon\,\mathcal{N}(\Delta_i|0,\sigma^2)\big)$ bits to communicate this information. Using this bound it would take approximately 41 bits of information to transmit the residual errors for this data set. As the curve we are fitting is approximately odd, the only significantly non-zero coefficients are w_1 and w_3, which would take around 15 bits of information to transmit. We would also have to transmit the

We are being slightly cavalier about assigning probabilities to bins, but doing this properly will only result in a negligible correction.

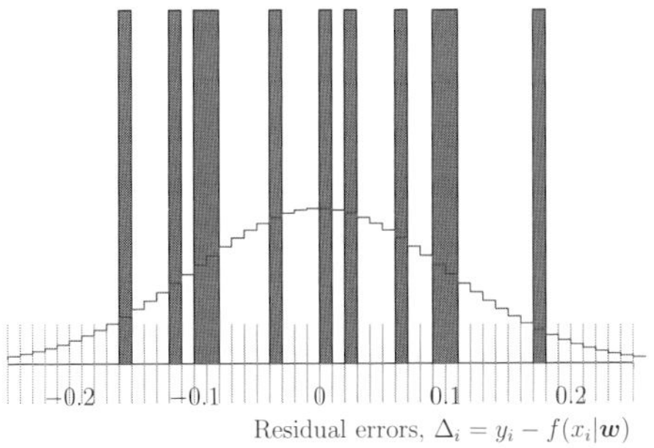

Figure 9.7 Histogram of residual errors between the data points and the third-order polynomial fit shown in Figure 9.6(b). Also shown are bins of size $\epsilon = 0.01$ and a (discretised) normal distribution with standard deviation of $\sigma = 0.1$.

variance in the residual errors, which would take another 6.6 bits, and finally we would have to transmit a 4-bit mask showing which weights are non-zero. In total we would need at least 67 bits of information (saving 17 bits compared with communicating the raw data). We can repeat this analysis for the linear fit. As this is not so accurate it would take 57 bits of information to communicate the residual errors. (We don't show the residual errors. The histogram is not too dissimilar to that shown in Figure 9.7 except now the standard deviation is close to 0.5 rather than 0.1.) We only have to transmit a single parameter, a 2-bit mask showing which parameters are non-zero, and the variance, which would take a total of 73 bits. This is an improvement over sending the raw data, but is worse than the third-order polynomial.

To communicate the exact model (i.e. the tenth-order polynomial) we have to transmit all 11 parameters, which gives us no reduction is code length. Indeed, the situation is worse in that the fit is extremely sensitive to the parameters of the model. We would therefore have to transmit the leading weights, w_{10}, w_9, etc. to many more decimal places, so that the description length for this model is worse than that if we transmitted the raw data. (We were possibly being rather generous to the third-order fit in that we may have had to transmit w_3 to higher accuracy or pay a price in having slightly larger residual errors.) If we decided to consider other regression functions such as Fourier series then MDL provides a means for deciding whether these are better or worse than polynomial-fitting functions.

In practice MDL usually leaves us with two terms, the information to transmit the parameters of our model and the information to transmit the data given the model. This is rather similar to the maximum *a posteriori* (MAP) formalism where we balance the log-prior and the log-likelihood. The information to transmit the parameters acts something like a log-prior while the information to transmit the residual errors acts like a log-likelihood.

Shannon's bounds give asymptotic results, but they are often very good estimates to the code length achieved in practice, as the following example shows.

Example 9.6 Entropy of Normally Distributed Random Variables
Although the entropy for the distribution of continuous variables can lead to bizarre results such as negative entropy (see Exercise 9.3) we can always make sense of them by discretising space. If we assume our data, X, is normally distributed, i.e. $X \sim \mathcal{N}(\mu, \sigma^2)$, then information theory tells us the length of message required to transmit the information that X lies in the interval x to $x + \epsilon$ is bounded by

$$-\log_2\left(\int_x^{x+\delta x} \mathcal{N}(y|\mu,\sigma^2)\,\mathrm{d}y\right) \approx -\log_2\left(\epsilon\,\mathcal{N}(x|\mu,\sigma^2)\right)$$

bits.

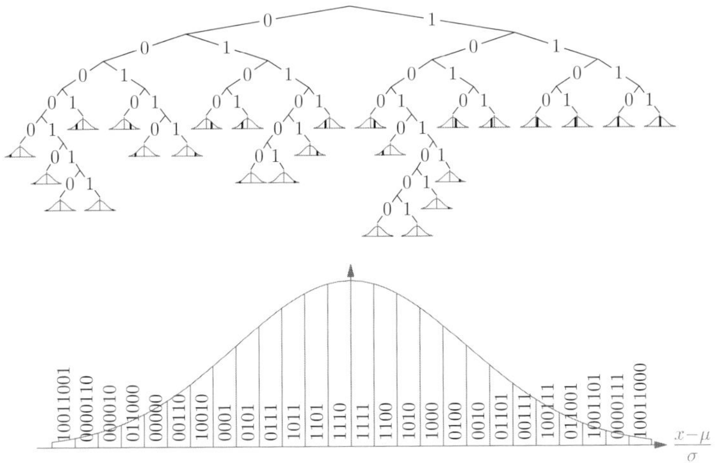

Figure 9.8 Example of Huffman tree where the 'alphabet' corresponds to different bins and the probability of each bin is used to determine its place in the tree.

Now, to encode an alphabet of symbols with different probabilities of occurrence an optimally efficient code is the Huffman code, which constructs a binary tree where the frequently occurring characters in the alphabet live at the top of the tree and the less frequently occurring characters live lower down the tree. The details of constructing Huffman trees goes beyond the scope of this book. However, in Figure 9.8 we show a Huffman tree for different bins from a normal distribution.

Note that the corresponding code is obtained by following the links (0 for a left link, 1 for a right link) until the leaf node is reached. Not only are the codes all unique, but the Huffman tree provides a means for uniquely decoding any sequence of letters in the alphabet. Notice that the most commonly occurring bins have much shorter codes. In Figure 9.9 we show the length of Huffman code for different bins. We see that the bins corresponding to rare events have significantly larger code lengths. We also see that the Shannon bound provides a remarkably good approximation to the length of the Huffman code.

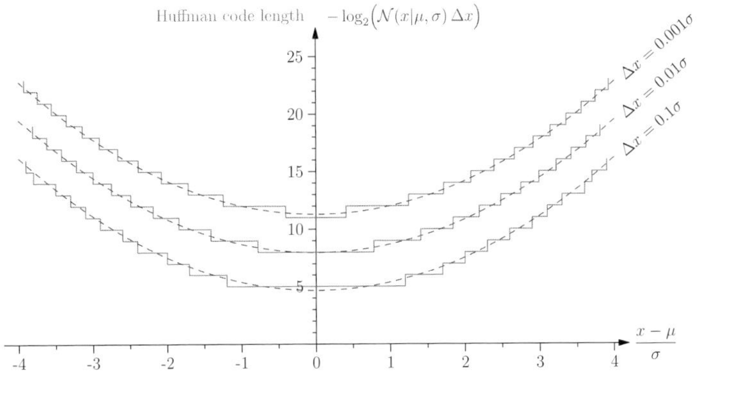

Figure 9.9 Length of Huffman code versus the bin position (measured in deviation from the mean divided by standard deviation) for different bin sizes. Also shown is the Shannon bound as the dashed line.

Note that although Huffman's code achieves the Shannon entropy bound, it is used in the case where we have no noise. The challenge for coding theory is to find efficient error-correcting codes, close to their Shannon bound.

In machine learning it is reasonably common to add regularisation terms (regularisers) to the squared error term. The squared error is interpreted as a log-likelihood (so the likelihood of the residual errors are normally distributed) and the regulariser acts like a log-prior. However, the priors often seems rather ad hoc. For example, in curve fitting a typical regulariser term added to the squared error might be $-\nu\|\boldsymbol{w}\|^2$ (often called a weight decay term) where ν is a coefficient that is chosen to get a good generalisation behaviour (which we would have to determine using additional data). Such terms would penalise large weights in the model. We could interpret the regulariser as a log-prior, but does it really encode our prior knowledge? However, by penalising large weights we make the weights lie in a tighter interval, which reduces their description length (the model is also less sensitive to the exact value of the parameters so they can be communicated to a lower accuracy). More recently, sparsity regularisers have become fashionable, where we try to find a fit using as few terms as possible. In practice, we often use a regularisation term $-\nu\|\boldsymbol{w}\|_1$ which penalises small terms as well as large terms. A whole new field of *compressed sensing* is based around finding sparse representations of data. MDL provides a coherent rationale behind finding sparse representations – they allow us to communicate the model more efficiently. Other regularisation terms directly punish the number of parameters in a model, examples of these are the *Akaike information criterion* (AIC) and the *Bayesian information criterion* (BIC). These are usually used as rules of thumb to favour simple models, but they find a natural justification in terms of MDL.

Example 9.7 Clustering

MDL can be used in contexts other than regression. One classic problem when handling data is to cluster the data. That is, we are given data $\mathcal{D} = \{\boldsymbol{x}_\alpha | \alpha = 1, \ldots, n\}$, where we treat the feature sets, $\boldsymbol{x}_\alpha$, as vectors in an m-dimensional vector space. In clustering we try to find some centres, $\boldsymbol{\mu}_k$, which capture some natural clustering in the data set. In Figure 9.10 we show an example of three clusters found by the classic K-means clustering algorithm for some two-dimensional data.

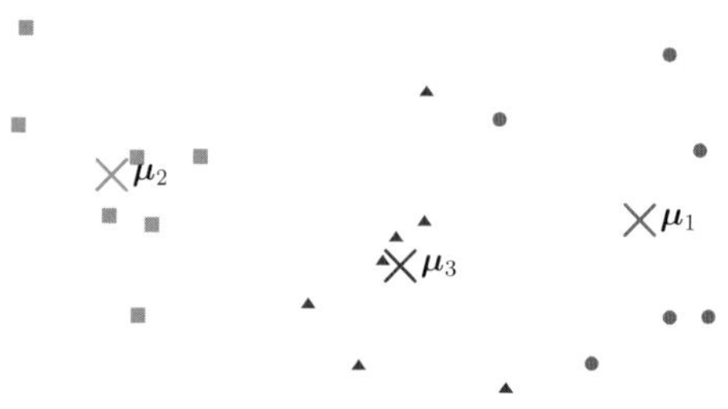

Figure 9.10 Three clusters found using the K-means clustering algorithm. The data points are shown by dots and the centres by $\times$. The points are assigned to the closest centre.

The difficulty with clustering algorithms is that they tend to find clusters whether or not the data is actually clustered. This is particularly true in high dimensions. Even determining how many clusters there should be is difficult. For many applications this does not really matter, but in some cases we really want to know if the clustering we have found represents a true clustering (e.g. can we really split some disease into separate categories). One principled way of answering this is to measure the minimum message length required to communicate the data. If data is truly clustered we would expect that we could compress the data by communicating the centre of clusters and then, for each data point, transmitting its closest centre and its position relative to that centre. If the data points are clustered then once the centre is known it should require less information to communicate a data point. In practice though, it's often too much work to even think about using MDL and most people fall back on simple heuristics for choosing the number of centres.

The MDL formalism often provides a new and useful perspective on a machine learning technique that was initially derived using a different set of arguments. We provide an example of this below.

Example 9.8 Variational Auto-Encoder

We introduced variational auto-encoders (VAEs) in Section 8.4.2 on page 240. Recall that a VAE takes an input, $\boldsymbol{x}$, (usually an image) that is passed to an encoder network that outputs the parameters $(\boldsymbol{\mu}_{\boldsymbol{\phi}}, \boldsymbol{\sigma}^2_{\boldsymbol{\phi}})$ of a probability distribution

$$q(\boldsymbol{z}|\boldsymbol{x}, \boldsymbol{\phi}) = \mathcal{N}\Big(\boldsymbol{z}\,\Big|\,\boldsymbol{\mu}_{\boldsymbol{\phi}}, \operatorname{diag}(\boldsymbol{\sigma}^2_{\boldsymbol{\phi}})\Big)\,.$$

Vectors, $\boldsymbol{z}$, can be viewed as points in a latent space. We can then generate an image close to the input $\boldsymbol{x}$ by sampling a vector in latent space, $\boldsymbol{z} \sim q(\boldsymbol{z}|\boldsymbol{x}, \boldsymbol{\phi})$, and feeding this to a decoder network with output $\hat{\boldsymbol{x}} = D(\boldsymbol{z}, \boldsymbol{\theta})$. Finally, from $\hat{\boldsymbol{x}}$ we can generate another image, $\boldsymbol{x}'$, with a probability $p(\boldsymbol{x}'|\hat{\boldsymbol{x}})$. In their original paper, Kingma and Welling (2013) obtained a variational bound on the log-likelihood of generating an output $\boldsymbol{x}$, given $\boldsymbol{x}$ as an input. This variational bound is equal to

$$\mathbb{E}_{\boldsymbol{z}\sim q}\big[\log\big(p(\boldsymbol{x}|\hat{\boldsymbol{x}})\big)\big] - \mathrm{KL}\left(q(\boldsymbol{z}|\boldsymbol{x}, \boldsymbol{\phi})\,\|\,p(\boldsymbol{z})\right). \qquad (9.10)$$

This is maximised for images coming from a particular data set, $\mathcal{D}$, which is representative of the images we want to generate.

It was quickly realised that this objective function has a very natural MDL interpretation. In this interpretation, $\boldsymbol{z} \sim q(\boldsymbol{z}|\boldsymbol{x}, \boldsymbol{\phi})$ is viewed as a message for efficiently communicating the input $\boldsymbol{x}$. The first term in the objective function can be seen as the information (in nats) needed to communicate the error, $\boldsymbol{\epsilon} = \boldsymbol{x} - \hat{\boldsymbol{x}}$, where $\hat{\boldsymbol{x}}$ is obtained by decoding a message $\boldsymbol{z}$. (Interestingly, empirically this probability function often appears to be close to a two-sided or double exponential distribution, i.e.

$$p(\boldsymbol{x}|\hat{\boldsymbol{x}}) = \frac{1}{2c}\mathrm{e}^{-c\,\|\boldsymbol{x}-\hat{\boldsymbol{x}}\|}.$$

In this case, the first term in the objective function is proportional to the absolute error $\|\boldsymbol{x} - \hat{\boldsymbol{x}}\|$.) The more intriguing term in Equation (9.10) is the Kullback–Leibler (KL) divergence. This term can be interpreted as the amount of information required to transmit a message with probability distribution $q(\boldsymbol{z}|\boldsymbol{x}, \boldsymbol{\phi})$ using the distribution $p(\boldsymbol{z})$. The KL divergence is also known as the *relative entropy* between distributions q and p. Note that if the entropy, H_q, of $q(\boldsymbol{z}|\boldsymbol{x}, \boldsymbol{\phi})$ is large then we can transmit $\boldsymbol{z}$ with less accuracy (requiring a small code word) than if H_q is small (i.e. $q(\boldsymbol{z}|\boldsymbol{x}, \boldsymbol{\phi})$ is highly concentrated around its mean). In choosing parameters, $\boldsymbol{\phi}$ and $\boldsymbol{\theta}$, for the encoder and decoder that minimise this objective function for all images from the data set, $\mathcal{D}$, we are minimising the expected message length for communicating a randomly sampled image by transmitting a code message, $\boldsymbol{z}$, and a correction term $\boldsymbol{\epsilon} = \boldsymbol{x} - \hat{\boldsymbol{x}}$.

Generally in the MDL formalism we have to determine the accuracy to which we transmit the error term $\boldsymbol{\epsilon} = \boldsymbol{x} - \hat{\boldsymbol{x}}$ and the accuracy of the code message $\boldsymbol{z}$. To a first approximation the accuracy of the error term isn't important in model selection. To see this, note that the message length for transmitting the error term to an accuracy $\delta\boldsymbol{\epsilon}$ is

$$\log\big(p(\boldsymbol{\epsilon})\,\|\delta\boldsymbol{\epsilon}\|\big) = \log\big(p(\boldsymbol{\epsilon})\big) + \log\big(\|\delta\boldsymbol{\epsilon}\|\big)$$

where $p(\boldsymbol{\epsilon})$ is the probability density of the errors. However, the last term, $\log\big(\|\delta\boldsymbol{\epsilon}\|\big)$, is common to all models so is not important for deciding between models. On the other hand, the accuracy to which we transmit $\boldsymbol{z}$ can be highly model dependent. To use the MDL criteria properly we would typically have to explore different accuracy levels for transmitting the model (in many models each different parameter of the model may have a different optimal accuracy level). An elegant property of Equation (9.10) is that it incorporates the accuracy to which each code message, z_i, needs to be communicated through the variance $\sigma^2_{\boldsymbol{\phi}_i}$ that is outputted by the encoder. Thus, by minimising Equation (9.10), the VAE learns how accurately it needs to transmit the code, $\boldsymbol{z}$. By choosing a distributions $q(\boldsymbol{z}|\boldsymbol{x}, \boldsymbol{\phi})$ with a large entropy it allows us to communicate the message with less bits (if $q(\boldsymbol{z}|\boldsymbol{x}, \boldsymbol{\phi}) = p(\boldsymbol{z})$ then we can communicate the message for free), however, when we choose H_q too large then the expected reconstruction error, $\mathbb{E}_{\boldsymbol{z}\sim q}\big[\log\big(p(\boldsymbol{x}|\hat{\boldsymbol{x}})\big)\big]$, is also likely to be very large. The optimisation procedure finds a trade-off that ensures the encoder extracts the salient information that allows us to maximally compress the data set $\mathcal{D}$. Without the KL divergence term in the objective function we could find distributions that reduced the errors, but at the cost of requiring more information to transmit the latent variables than the gain we get from reducing the error. That is, we would be apt to over-fit the training data.

The advantage of the MDL formalism is that it provides a principled way of choosing between different models. In Bayesian inference we can use the evidence to choose between models. However, to do this in a fully Bayesian way we would want to put priors on our models. Doing so goes beyond the usual Bayesian framework, but MDL provides a rational method for doing this. Furthermore, if we are only using a MAP method then we cannot compute the evidence and the MAP formalism provides no way of doing model selection. Here MDL has a significant advantage over MAP. In fact, if we do MDL carefully, then we have to consider the sensitivity of the model to determine how accurately we need to communicate its parameters. This begins to look a bit more like a full Bayesian approach. As a general philosophy, MDL has applications, for example to clustering, which are harder to motivate from a purely Bayesian perspective. Of course, the true MDL of the data would require finding the shortest program which generates the data and this task is incomputable. In practice, this difficulty is finessed by using some model and assuming there is no additional hidden structure in the data. To obtain a code length we can either use Shannon's bound or some explicit coding scheme. This can be technically difficult to accomplish and is not for the faint-hearted. If you have a philosophical bent you can spend many hours arguing the case for or against MDL. However, for the rest of us we can just take the pragmatic view *does it work?* The jury is out. There seems to be some cases where MDL helps, but it has failed to take the world by storm so far. On the other hand, a lot of apparently ad hoc methods that work well in practice, such as sparsity regularisation, AIC, and BIC, etc. find a natural explanation using MDL.

9.3.3 Fisher Information

Shannon entropy plays a pivotal role in information theory, but it is not the only measure of uncertainty and information. Another measure is *Fisher information* proposed by Ronald A. Fisher. Fisher information arises when we consider estimating a quantity from some measurements.

Consider a situation where we want to estimate a parameter θ (this might be a success probability or the mean of some quantity). The estimate will depend on the data collected. We denote the data collected by a random variable X (if we collect multiple measurements this would be a multidimensional vector). Now let us assume that we use an unbiased estimator $\hat{\theta}(X)$ to estimate the quantity of interest. The unbiased estimator has the property that

$$\mathbb{E}\left[\hat{\theta}(X)\right] - \mathbb{E}\left[\theta\right] = \int f_X(x|\theta)\left(\hat{\theta}(x) - \theta\right) \mathrm{d}\,x = 0$$

where $f_X(x|\theta)$ is a probability density function describing the likelihood of the data given the parameter θ. Differentiating this equation with respect to θ we find

$$\int \left(\left(\hat{\theta}(X) - \theta\right) \frac{\partial f_X(x|\theta)}{\partial \theta} - f_X(x|\theta) \right) \mathrm{d}\,x = 0.$$

Using the identities

$$\frac{\partial f_X(x|\theta)}{\partial \theta} = f_X(x|\theta)\,\frac{\partial \log(f_X(x|\theta))}{\partial \theta}, \qquad \text{and} \qquad \int f_X(x|\theta)\,\mathrm{d}\,x = 1$$

we find

$$\int \left(\hat{\theta}(X) - \theta\right) \frac{\partial \log(f_X(x|\theta))}{\partial \theta} f_X(x|\theta)\,\mathrm{d}\,x = 1.$$

Cunningly rewriting this as

$$\int \left(\left(\hat{\theta}(X) - \theta\right) \sqrt{f_X(x|\theta)} \right) \left(\frac{\partial \log(f_X(x|\theta))}{\partial \theta} \sqrt{f_X(x|\theta)} \right) \mathrm{d}\,x = 1.$$

and using the Cauchy–Schwarz inequality

$$\int f(x)\,g(x)\,\mathrm{d}\,x \le \sqrt{\left(\int f^2(x)\mathrm{d}x\right)\left(\int g^2(x)\,\mathrm{d}\,x\right)}$$

we find

$$\left(\int \left(\frac{\partial \log(f_X(x|\theta))}{\partial \theta} \right)^2 f_X(x|\theta)\,\mathrm{d}\,x \right) \left(\int \left(\hat{\theta}(X) - \theta\right)^2 f_X(x|\theta)\,\mathrm{d}\,x \right) \ge 1$$

or

$$\mathbb{E}\left[\left(\frac{\partial \log(f_X(x|\theta))}{\partial \theta} \right)^2 \right] \times \mathbb{E}\left[\left(\hat{\theta}(X) - \theta\right)^2 \right] \ge 1.$$

The first term,

$$I_\theta = \mathbb{E}\left[\left(\frac{\partial \log(f_X(x|\theta))}{\partial \theta} \right)^2 \right], \tag{9.11}$$

is known as the *Fisher information*. The second term

$$e^2 = \mathbb{E}\left[\left(\hat{\theta}(X) - \theta\right)^2 \right]$$

measures the *mean squared error* in the estimator. The inequality

$$e^2 \ge \frac{1}{I_\theta}$$

is known as the *Cramér–Rao bound.* It provides an absolute limit to the ability to measure a quantity. That is, the expected mean error is, at best, inversely proportional to the Fisher information, which itself measures the expected sensitivity in the likelihood. If the likelihood of a measurement, $f_X(x|\theta)$, is very sensitive to the parameter, θ, we are trying to infer then the mean squared error will be relatively low. Conversely, if the probability of a measurement is little affected by the parameter then we are likely to have a large error in our estimate of the parameter. If the equality holds in the Cramér–Rao bound so that $e^2 = 1/I(\theta)$ then the estimator is known as an *efficient estimator*.

Example 9.9 Fisher Information for a Normally Distributed Variable
If $f_X(x|\theta)$ is a normal distribution and θ the mean then

$$\log\big(f_X(x|\theta)\big) = -\frac{(x-\theta)^2}{2\sigma^2} - \frac{1}{2}\log\Big(2\pi\sigma^2\Big)$$

so that

$$\frac{\mathrm{d}\log\big(f_X(x|\theta)\big)}{\mathrm{d}\theta} = \frac{x-\theta}{\sigma^2}$$

and

$$I_\theta = \int\left(\frac{x-\theta}{\sigma^2}\right)^2 f_X(x|\theta)\,\mathrm{d}x = \frac{1}{\sigma^2}.$$

Thus, the mean squared error for any unbiased estimator of the mean of a normally distributed variable given a single measurement will be at best σ^2. If we make n independent measurements with normally distributed errors then the mean squared error would be at least σ^2/n. Note that the estimator

$$\hat{\mu} = \frac{1}{n}\sum_{i=1}^{n} X_i$$

has a mean square error of σ^2/n and is thus an efficient estimator. A consequence of this is that there is no better estimator for the mean of normally distributed variables (in the sense of giving a smaller mean squared error).

We briefly outline some of the properties of Fisher information. It can also be written

$$I_\theta = \int \frac{1}{f_X(x|\theta)}\left(\frac{\partial f_X(x|\theta)}{\partial\theta}\right)^2 \mathrm{d}x. \tag{9.12}$$

This form is occasionally more convenient. Consider two independent measurements X and Y. The joint likelihood is given by

$$f_{X,Y}(x,y|\theta) = f_X(x|\theta)\,f_Y(y|\theta).$$

Taking the logarithm of both sides and differentiating

$$\frac{\partial \log\big(f_{X,Y}(x,y|\theta)\big)}{\partial\theta} = \frac{1}{f_X(x|\theta)}\frac{\partial f_X(x|\theta)}{\partial\theta} + \frac{1}{f_Y(y|\theta)}\frac{\partial f_Y(y|\theta)}{\partial\theta}.$$

Squaring we get

$$\left(\frac{\partial \log\big(f_{X,Y}(x,y|\theta)\big)}{\partial\theta}\right)^2 = \frac{1}{f_X^2(x|\theta)}\left(\frac{\partial f_X(x|\theta)}{\partial\theta}\right)^2 + \frac{1}{f_Y^2(y|\theta)}\left(\frac{\partial f_Y(y|\theta)}{\partial\theta}\right)^2 + 2\frac{1}{f_X(x|\theta)}\frac{1}{f_Y(y|\theta)}\frac{\partial f_X(x|\theta)}{\partial\theta}\frac{\partial f_Y(y|\theta)}{\partial\theta}.$$

Using the definition of Fisher information, Equation (9.11),

$$\begin{aligned} I_\theta(x,y) &= \int f_{X,Y}(x,y|\theta)\left(\frac{\partial \log\big(f_{X,Y}(x,y|\theta)\big)}{\partial\theta}\right)^2 \mathrm{d}x\,\mathrm{d}y \\ &= \int \frac{1}{f_X(x|\theta)}\left(\frac{\partial f_X(x|\theta)}{\partial\theta}\right)^2 \mathrm{d}x + \int \frac{1}{f_Y(y|\theta)}\left(\frac{\partial f_Y(y|\theta)}{\partial\theta}\right)^2 \mathrm{d}y \\ &\quad + 2\int \frac{\partial f_X(x|\theta)}{\partial\theta}\,\mathrm{d}x \int \frac{\partial f_Y(y|\theta)}{\partial\theta}\,\mathrm{d}y. \end{aligned}$$

But

$$\int \frac{\partial f_X(x|\theta)}{\partial\theta}\,\mathrm{d}x = \frac{\partial}{\partial\theta}\int f_X(x|\theta)\,\mathrm{d}x = \frac{\partial 1}{\partial\theta} = 0$$

and using Equation (9.12) we find

$$I_\theta(x,y) = I_\theta(x) + I_\theta(y).$$

That is, just like Shannon entropy, the Fisher information for two independent variables is the sum of the Fisher information for each variable separately.

When a distribution has multiple parameters (e.g. the normal distribution $\mathcal{N}(x|\mu,\sigma^2)$ or the multivariate normal distribution $\mathcal{N}(\boldsymbol{x}|\boldsymbol{\mu},\boldsymbol{\Sigma})$), then the Fisher information will be a matrix $\boldsymbol{I}_{\boldsymbol{\theta}}$ with components

$$I_{\theta_i,\theta_j} = \mathbb{E}_X\left[\frac{\partial \log\big(f_X(X|\boldsymbol{\theta})\big)}{\partial\theta_i}\frac{\partial \log\big(f_X(X|\boldsymbol{\theta})\big)}{\partial\theta_j}\right].$$

Information Geometry

A very abstract way of thinking about families of probability distributions, $f(\boldsymbol{x}|\boldsymbol{\theta})$, is as a manifold parameterised by $\boldsymbol{\theta}$ (e.g. the space of normal distributions $\mathcal{N}(x|\mu,\sigma^2)$ is a two-dimensional manifold parameterised by the mean and variance). If we make a change of variables in the space of parameters (e.g. from variance, σ^2, to precision $\tau = \sigma^{-2}$) then the geometry of the manifold will change. However, we would hope that probability theory, if applied properly, should not care about what parameterisation we used. This raises the question of whether there exists a parameter-free representation of probability distributions. The mathematical language for exploring such questions is tensor algebra and differential geometry. Interestingly, the Fisher information matrix $\boldsymbol{I}_\theta$ plays the role of being an invariant metric (that is, it provides a means for measuring a distance between distributions that is invariant under reparameterisation). Reasoning about probability distribution in this way is known as *information geometry*. It is highly technical. Many smart theorists have worked in this area and discovered interesting connections. For example, distributions belonging to the exponential family have a nice property of being 'flat,' which has interesting consequences when considering their divergence properties. For those of us with a more practical disposition, the relevant question is does this buy you anything? There are practitioners such as Shun'ichi Amari who have demonstrated that it is possible to speed up learning in some machine learning applications by using a natural gradient which respects the invariance properties of a manifold – whether this could have been deduced without using information geometry I'm not in a position to judge. However, it's a field which hasn't quite taken off. Whether this is because we are not smart enough to know how to fully exploit these ideas or whether it's just some quirky mathematics which don't really cut the mustard I don't think we quite know.

Fisher information is much less exploited than Shannon entropy. It is occasionally used in constructing kernels in machine learning (although even there it is often set to a constant). It has been used by Frieden (1998) to explain almost the whole of physics, but this remains controversial. Although Fisher information looks very different to Shannon's entropy it shares some of its properties, such as the additive property for independent random variables. Fisher's information contains a derivative of the likelihood density and so cares about the local properties of the distribution. In contrast, Shannon's entropy doesn't measure local properties and can be seen as a more global measure of the uncertainty in a distribution. Shannon entropy and Fisher information give two different views of uncertainty, each with its own uses.

Shannon's entropy provides a consistent measure of uncertainty which has found many applications in diverse disciplines. Understanding its information theory roots is invaluable when trying to apply it to different applications. One of its most prominent applications is the so-called maximum entropy method, where we try to infer information about unobserved states given measurements of some average properties. Other measures of uncertainty exist such as Kolmogorov complexity and Fisher information. Although these are not nearly as commonly used as Shannon's entropy, they nevertheless have some strong proponents.

Additional Reading

A classic text on information theory is *Elements of Information Theory* by Cover and Thomas (2006). A book covering many different topics all related to information theory is *Information Theory, Inference, and Learning Algorithms* by

MacKay (2003). A standard text on Kolmogorov complexity is *An Introduction to Kolmogorov Complexity and Its Applications* by Li and Vitányi (1997).

Exercise for Chapter 9

Exercise 9.1 (answer on page 427)

Which has the most uncertainty: rolling an honest dice or tossing three independent fair coins assuming

i. you care about the order of the results;
ii. you only care about the number of heads.

Exercise 9.2 (answer on page 428)

Use Stirling's approximation, $n! \approx \left(\frac{n}{e}\right)^n \sqrt{2\pi n}$, to show that

$$\binom{n}{f\,n} \approx \frac{2^{n\,H(f)}}{\sqrt{2\pi\,n\,f(1-f)}}$$

where $H(f) = -f\log_2(f) - (1-f)\log_2(1-f)$ is the entropy of Bernoulli trial with success probability f.

Exercise 9.3 (answer on page 429)

Compute the entropy for a normally distributed random variable $X \sim \mathcal{N}(\mu, \sigma)$. Plot the entropy versus σ and show that it can be negative.

Exercise 9.4 (answer on page 429)

Consider a system whose state, X, we represent by a discrete random variable. We label the possible states by x_i where $i \in \mathcal{I}$ ($\mathcal{I}$ is an index set). The system is in thermal contact with a large system so that it can exchange energy. We observe the average energy of the system is

$$\mathbb{E}\left[E(X)\right] = \sum_{i\in\mathcal{I}} p_i\,E(x_i) = U$$

where $p_i = \mathbb{P}\left(X = x_i\right)$. Find the maximum entropy probability, p_i, of being in a state x_i. This is the *Boltzmann distribution*.

Appendix 9.A Functionals

A functional, $F(f)$, is a function of a function. That is, given a function $f(x)$ it returns a number. The entropy of a continuous random variable, $X \sim f_X$, is an example of a functional

$$F(f_X) = H_X = -\int f_X(x)\,\log\big(f_X(x)\big)\,\mathrm{d}x$$

where the integral is over the domain of X. Finding a function which maximises the entropy (usually subject to some constraints) is a common task which sits in the domain of the *calculus of variations*. To do this properly, we consider adding a small perturbation $\epsilon\,\nu(x)$ to our function and expanding in ϵ

$$F(f+\epsilon\,\nu) = F(f) + \epsilon\,F'(f,\nu) + O(\epsilon^2)$$

where $F'(f,\nu)$ is the first-order correction. When $F(f_X)$ is our entropy function

$$\begin{aligned}
F(f+\epsilon\,\nu) &= -\int (f_X(x) + \epsilon\,\nu(x))\,\log\big(f_X(x) + \epsilon\nu(x)\big)\,\mathrm{d}x \\
&= -\int (f_X(x) + \epsilon\,\nu(x))\,\left(\log\big(f_X(x)\big) + \log\left(1 + \epsilon\frac{\nu(x)}{f(x)}\right)\right)\,\mathrm{d}x \\
&= F(f) - \epsilon \int \nu(x)\,\big(\log\big(f_X(x)\big) + 1\big)\,\mathrm{d}x + O(\epsilon^2)
\end{aligned}$$

so that

$$F'(f,\nu) = -\int \nu(x)\,\big(\log\big(f_X(x)\big) + 1\big)\,\mathrm{d}x.$$

Now the extremum condition is that

$$F'(f,\nu) = \lim_{\epsilon\to 0} \frac{F(f) - F(f+\epsilon\nu)}{\epsilon} = 0.$$

To maximise the entropy subject to constraints on the mean of some quantities $g_i(x)$, we extremise the Lagrangian

$$\mathcal{L}(f_X) = H_X + \sum_{i=1}^{m} \lambda_i \int f_X(x)\,g_i(x)\mathrm{d}x \tag{9.13}$$

which is equivalent to

$$\mathcal{L}'(f_X,\nu) = \int \nu(x)\left(-\log\big(f_X(x)\big) - 1 + \sum_{i=1}^{m} \lambda_i\,g_i(x)\right)\mathrm{d}x = 0.$$

The perturbation function $\nu(x)$ was arbitrary (although it should vanish at the boundaries), so the only way for $\mathcal{L}'(f_X,\nu) = 0$ is that

$$-\log\big(f_X(x)\big) - 1 + \sum_{i=1}^{m} \lambda_i\,g_i(x) = 0 \tag{9.14}$$

at all points x.

An equivalent way of obtaining this equation is to define the functional derivative, such that

$$\frac{\delta f(x)}{\delta f(y)} = \delta(x-y)$$

where $\delta(x-y)$ is the Dirac delta (see Appendix 5.A). Any other function of x we treat as a constant. The functional derivative obeys the usual rules of differentiation. If we take the functional derivative of the entropy $F(f_X) = H_X$ where $X \sim f_X$ we find

$$\frac{\delta F(f_X)}{\delta f_X(y)} = -\int\left(\delta(x-y)\,\log\big(f_X(x)\big) + f_X(x)\frac{\delta(x-y)}{f_X(x)}\right)\mathrm{d}x = -\log\big(f_X(y)\big) - 1.$$

If we apply this to the Lagrangian defined in Equation (9.13) we get

$$\frac{\delta \mathcal{L}(f_X)}{\delta f_X(y)} = -\log\big(f_X(y)\big) - 1 + \sum_{i=1}^{m} \lambda_i\, g_i(y).$$

The maximum entropy condition becomes

$$\frac{\delta \mathcal{L}(f_X)}{\delta f_X(y)} = 0$$

which gives the same result as Equation (9.14).

10

Collective Behaviour

Contents

We have looked at models involving a few random variables. As we increase the number of such variables the models quickly become intractable. However, when the interactions are sufficiently simple, the models can become tractable again in the limit of a large number of variables, where laws of large numbers are working in our favour. We will see that sometimes rather surprising behaviour can emerge. To give a flavour of these models we consider a few classic models, each increasing in complexity. We start by considering one of the simplest systems, the random walk – essentially just a sum of random variables. The random walk acts as a prototypical stochastic system; we will have much more to say about this model in later chapters. We briefly discuss the branching process, which provides a playground for generating function techniques. We then consider percolation, which provides a very simple example of a model with a phase transition. We consider a simple model of a magnet, the Ising model, and discuss some of its more striking behaviours; briefly mentioning self-organised criticality on the way. Finally, we discuss disordered systems.

10.1 Random Walk

One of the most important probabilistic models is the random walk. The simplest situations is that of the one-dimension walk where the walker can move either forwards or backwards. Examples of two unbiased one-dimensional random walks are shown in Figure 10.1. At each step the height x is incremented or decremented by 1 with equal probability.

The random walk can be described by a sequence of random variables $(X_1, X_2, \ldots)$. A key feature of a random walk is the *Markov property* that X_t depends just on X_{t-1}, or in math-speak

$$\mathbb{P}\left(X_t | X_{t-1}, X_{t-2}, \ldots, X_1\right) = \mathbb{P}\left(X_t | X_{t-1}\right).$$

In the example shown in Figure 10.1,

$$\mathbb{P}\left(X_t | X_{t-1}\right) = \tfrac{1}{2}\left[\!\left[X_t = X_{t-1} + 1\right]\!\right] + \tfrac{1}{2}\left[\!\left[X_t = X_{t-1} - 1\right]\!\right].$$

We will have a lot more to say about the Markov property in Chapters 11 and 12. In a simple random walk with no bias, the expected position at any time is the position at the previous time step (they are an example of a Martingale – see Section 7.1.2). The variance is proportional to the number of steps. This is a consequence of the random variables at each instance being independent. The root mean squared (RMS) distance thus grows in proportion to $\sqrt{t}$. This is a very important property. It is true whether the random steps are on a fixed lattice or if the steps come from a continuous distribution. Since the final position is equal to a sum of random variables we know from the central limit theorem that the position of the walker after t steps will converge to a normal distribution with mean zero and variance $t\,\sigma^2$ where σ^2 is the variance of a single step.

The property for the one-dimensional random walk that the expected distance from the starting point grows as $\sqrt{t}$ also holds in higher dimensions. Figure 10.2 illustrates a random walk in two dimensions. If the walker is confined to stepping on a lattice (as in Figure 10.1) then in one and two dimensions the random walker will (with probability arbitrarily close to 1 as t becomes large) return to his or her origin. The walk is said to be recurrent. For high dimensions this is no longer true and the walk is said to be transient. Random walks are rather finely balanced so that even in one dimension, although the walk is recurrent, nevertheless the expected return time is infinite (see Section 7.1.2 and Exercise 10.1).

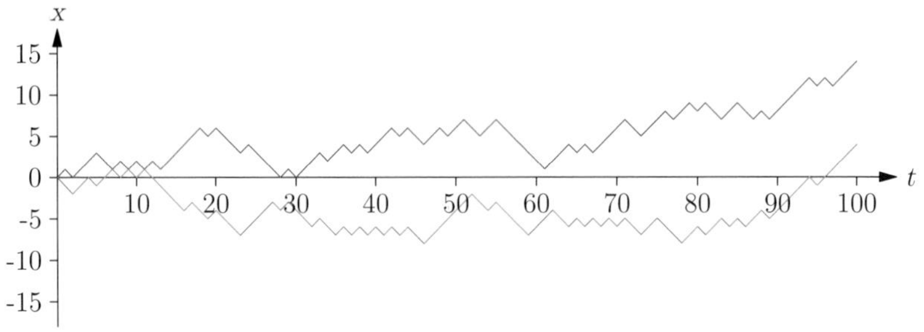

Figure 10.1 Illustration of two one-dimensional random walks. The walker takes a step up or down by one unit at each time interval.

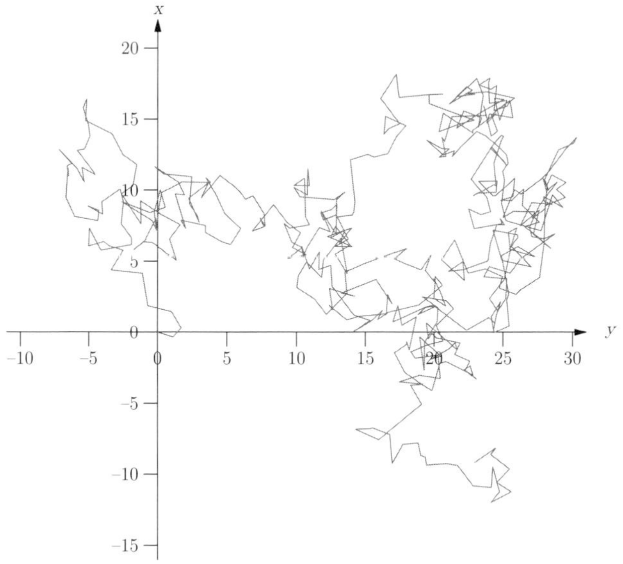

Figure 10.2 Random walk in two-dimensions. A normally distributed step in each dimension is made for 500 time steps.

Random walks have a long history, in part because they model a diffusion process which has many applications throughout science. The early work in this area was carried out to understand Brownian motion – the random dance of small particles in a liquid. Some of the fundamental work was done in 1905 by a newly qualified PhD student, A. Einstein – these works have now been republished (see Einstein 1998). Random walks and their generalisations have attracted considerable recent interest because of their role in financial prediction. There are, however, other applications which, although maybe not so lucrative, are more interesting. We will revisit random walks in Chapters 11 and 12 when we consider Markov chains and stochastic processes.

10.2 Branching Processes

Branching processes describe how a population grows over time. This may be a population of bacteria, neutrons in a nuclear chain reaction, or the spread or demise of surnames. In the simplest models we consider a population of k individuals at some time t. At the next time step, $t + 1$, each individual, i, creates K_i new individuals, where $K_i \in \{0, 1, 2, \ldots\}$ is a random variable from some distribution. In the simplest example the distribution is assumed to be the same across the population and over time. Figure 10.3 illustrates a branching processing. Once again this is an example of a Markov process, although the behaviour is richer than the random walk.

These processes are conveniently studied through their generating functions. Denote the probability of an individual generating k children by p_k (assumed to be constant for all individuals and all times), and let the size of the population at time t be $K(t)$, then the cumulant generating function is

$$G_t(\lambda) = \log\left(\mathbb{E}\left[e^{\lambda K(t)}\right]\right).$$

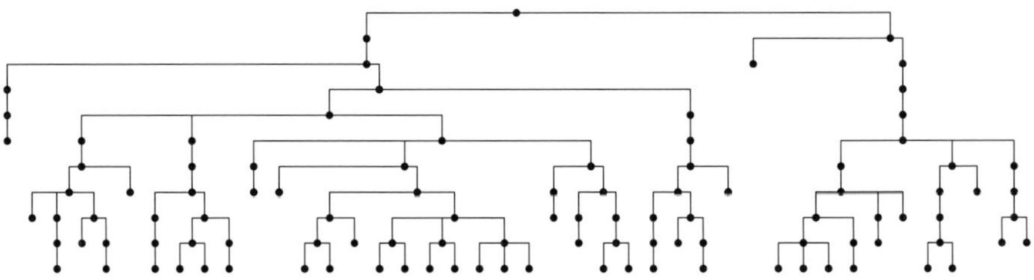

Figure 10.3 Illustration of an instance of a branching process with $p_0 = p_1 = p_2 = 2/7$ and $p_3 = 1/7$ (the numbers chosen for aesthetic reasons).

Starting from one individual at time 0, then

$$G_1(\lambda) = G(\lambda) = \log\left(\sum_{k=1}^{\infty} p_k e^{\lambda k}\right)$$

where $G(\lambda)$ is the cumulant generating function for the distribution of children at each individual. At the second generation

$$G_2(\lambda) = \log\left(\sum_{k=1}^{\infty} p_k \left(\prod_{i=1}^{k}\sum_{i=1}^{\infty} p_i\, e^{\lambda i}\right)\right) = \log\left(\sum_{k=1}^{\infty} p_k e^{k\,G(\lambda)}\right) = G(G(\lambda))$$

and by a similar argument $G_t(\lambda)$ is equal to the function G applied t times to λ – this is a consequence of the independence of the number of children produced by each individual. The expected number of individuals at generation t is given by

$$\left.\frac{dG_t(\lambda)}{d\lambda}\right|_{\lambda=0} \overset{(1)}{=} \left.\frac{dG(G_{t-1}(\lambda))}{d\lambda}\right|_{\lambda=0} \overset{(2)}{=} G'(G_{t-1}(\lambda))\left.\frac{dG_{t-1}(\lambda)}{d\lambda}\right|_{\lambda=0}$$
$$\overset{(3)}{=} G'(G_{t-1}(\lambda))\, G'(G_{t-2}(\lambda)) \cdots G'(G(\lambda))\, G'(\lambda)\Big|_{\lambda=0} \overset{(4)}{=} (\kappa_1)^t$$

(1) Using $G_t(\lambda) = G(G_{t-1}(\lambda))$.
(2) Applying the chain rule.
(3) Iteratively repeating the process of applying the chain rule.
(4) Noting that, from the definition of $G(\lambda)$, we have $G(0) = 0$ and $G'(0) = \kappa_1$ where κ_1 is the first cumulant (mean) of the distribution p_k.

The variance in the number of individuals is given by

$$\left.\frac{d^2G_t(\lambda)}{d\lambda^2}\right|_{\lambda=0} = \kappa_2\, \kappa_1^{t-1}\left(\kappa_1^{t-1} + \kappa_1^{t-2} + \cdots \kappa_1 + 1\right)$$

where $\kappa_2 = G''(0)$ is the second cumulant (variance) of the distribution p_k. This comes from a rather awkward application of the chain rule for differentiation to the expression in the previous equation. If $\kappa_1 \neq 1$ then we can sum the geometric progression to get

$$\frac{\kappa_2\, \kappa_1^{t-1}\left(\kappa_1^t - 1\right)}{\kappa_1 - 1}$$

while if $\kappa_1 = 1$ this sum equals $t\,\kappa_2$.

Obtaining a closed-form solution for $G_t(\lambda)$ is more challenging, but can be done for some distributions, such as the geometric distribution. Branching processes can be generalised in many ways, for instance, they can be defined in continuous time, made age dependent, etc. For a fuller discussion of branching processes see, for example, Grimmett and Stirzaker (2001a). Similar models have been constructed to create graphs with different attachment preferences. Many of the properties can again be studied using either the moment or cumulant generating functions (in my experience working with the cumulant generating function often leads to simpler expressions). This is an area that has received a lot of attention from the late 1990s onwards, with small-world and scale-free graphs attracting particular attention. Again I leave the interested reader to pursue this independently. There are many volumes on the subject, including the popular book by Watts (2003) and the more comprehensive book by Newman (2010).

10.3 Percolation

Consider modelling a porous rock. The rock consists of pockets that let water through and solid parts that block the flow of the liquid. We want to know under what conditions will water percolate through a thick layer of rock. To model this we consider the rock as a lattice of sites, where each site can either let water through or block the flow. At each site we choose to make the site either a pocket where water flows (with probability p) or else solid rock where the water can't flow (with probability $1 - p$). The water will percolate through the rock if there is a connected chain of pockets which reaches from the top to the bottom. This is an example of *percolation*. It has many applications besides the flow of fluids, such as water, oil, gas, or even coffee. Similar models are used to understand when a disease becomes a pandemic, when a forest fire spreads out of control, when a disordered material loses conductivity, or even when a revolution takes place. A nice introduction to percolation theory is by Stauffer and Aharony (1997).

Figure 10.4 shows two examples of *site percolation* in two dimensions. Here we 'occupy' each site independently with a probability p. In the left-hand figure we have $p = 0.5$ while in the right-hand figure we have $p = 0.6$. We highlight the largest connected cluster.

(a) $p = 0.5$ (b) $p = 0.6$

Figure 10.4 Example of site percolation for (a) $p = 0.5$ and (b) $p = 0.6$. We highlight the largest cluster.

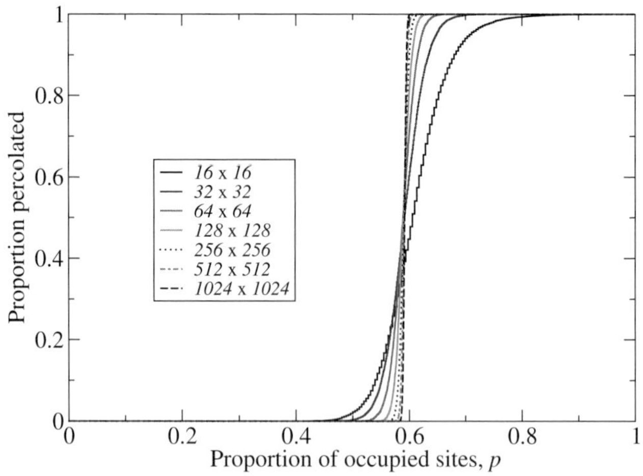

Figure 10.5 Graph of the proportion of 10 000 lattices where there is a cluster from the top to the bottom of the lattice versus the proportion of filled sites in the lattice.

As we increase p, the probability of there being a connected chain of sites that percolates from the bottom to the top of the sample increases. We can define a phase transition in p, below which most configurations don't percolate and above which they do. We illustrate this in Figure 10.5, where we show the proportion of lattices that have a cluster that reaches from the bottom of the lattice to the top, plotted against the proportion of filled sites. In the limit of very large networks this percolation transition would happen with overwhelming probability at a particular value of p – the *percolation threshold*. This is an example of a *phase transition*. The critical threshold for this problem is around $p = 0.5928$. The existence of the sharp change in the macroscopic behaviour at a particular critical value occurs frequently and is a consequence of having large numbers. There are, of course, a lot of lattice configurations which percolate with p much less than the percolation threshold (all you need is a line of sites connecting the top to the bottom) and also many lattices which do not percolate even when p is much larger than the percolation threshold (all you need is a line of unoccupied sites going across the lattice). However, for a lattice with N sites of which n are occupied there are $\binom{N}{n} \approx \exp(N\,H(n/N))$ lattices (where $H(n/N)$ is the entropy of a Bernoulli variable with success probability n/N). Of these astronomical number of graphs, the probability of generating a graph (assuming that they are generated at random) with an unusual cluster configuration is exponentially small.

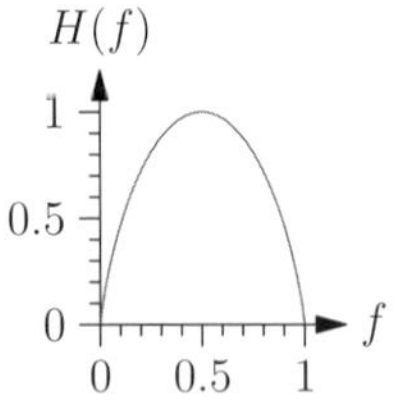

Phase transitions occur in many situations. In constraint satisfaction problems there is often a phase transition between the case where the optimisation problem can be solved perfectly and the case where it can't. An example of this is graph colouring, where the task is to shade the graph with a given set of colours so that no two vertices that share an edge have the same colour. For random graphs, as we increase the number of edges in the graph there is a transition from graphs being colourable to uncolourable, which, with overwhelming probability, occurs when a particular proportion of edges exist in the graph. Interestingly, around the phase transition, determining whether all the constraints can be satisfied is often found to be computationally extremely challenging.

10.4 Ising Model

One of the most famous and well-studied models in statistical physics is the Ising model. It has influenced many generations of physicists and statisticians. It provides a slightly more complex model than we have seen so far, in that, although the components are identical, they interact. Here, we get quite surprising *emergent behaviour* at the macroscopic scale that is difficult to see at the component level.

The Ising model can be viewed as a simple model of a magnet consisting of spins confined to a lattice. It is a special magnet in which the magnetic moments (usually called spins) can point either up or down. It is assumed to be energetically favourable for neighbouring spins to point in the same direction.

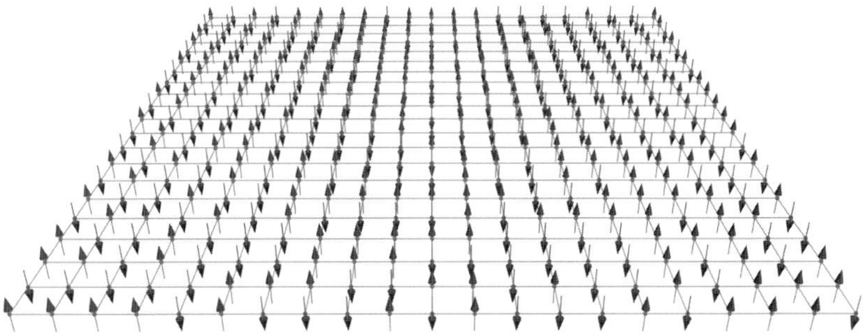

In real magnetic systems the magnetic spins would usually point in any direction in three-dimensional space, however, there are some systems where the spins will point only in two directions. It is easy to think of the spins aligning because of the magnetic field of one spin encouraging its neighbour to align with it. This, however, is not the case for real magnets at normal temperatures as the magnetic fields are simply too weak. The reason for the alignment of spins is a consequence of Pauli's exclusion principle, which says that no two particles can be in the same state at the same time. Spins tend to align or sometimes to anti-align so that they can live in lower-energy states. This coupling is known as an exchange interaction.

After this brief interlude of reality let us return to our abstract world. The energy for the Ising model is assumed to have the form

$$E(\boldsymbol{S}) = -J \sum_{(i,j)\in\mathcal{E}} S_i\, S_j \tag{10.1}$$

where $S_i = \pm 1$, J is a coupling constant and the sum is over all neighbours (if we think of the spins as living on a graph then $\mathcal{E}$ would be the set of edges – usually the graph is taken to be a regular lattice). The minus sign is a convention which simply means that the energetically favoured solutions are those with the lowest energies. Assuming our system is in thermal contact with a 'heat bath' (i.e. a large system such as the outside world) then we know from Section 9.2.2

that the maximum entropy probability distribution is given by the Boltzmann distribution (see Exercise 9.4)

$$\mathbb{P}(\boldsymbol{S}) = \frac{1}{Z} \exp\left(\frac{J}{kT} \sum_{(i,j)\in\mathcal{E}} S_i S_j \right) \tag{10.2}$$

where Z is a normalisation constant, called the *partition function*, which is equal to

$$Z = \sum_{\boldsymbol{S}\in\{-1,1\}^n} \exp\left(\frac{J}{kT} \sum_{(i,j)\in\mathcal{E}} S_i S_j \right)$$

and n is the total number of spins. Very frequently physicists will write $\beta = J/(kT)$, which they call the inverse temperature (ignoring the constants J and k).

It might look like we know everything we want to about the system; after all we know the probability of every possible configuration. However, there is a twist. Physicists are interested in the case when the system becomes large, $n \to \infty$ – this is known as the *thermodynamic limit*. For small n we can compute all the properties exactly, but this becomes intractable pretty rapidly (even for 10×10 two-dimensional lattice with 100 variables we are unable to average over all the $2^{100} \approx 10^{30}$ spin configurations).

Although the model is defined through a maximum entropy distribution, we can get the same distribution through a dynamic process. This dynamics is an example of Markov Chain Monte Carlo (MCMC), which we will study in Section 11.2. This allows us to study the properties of this model for larger systems. However, great care is needed in extrapolating results to the case of very large systems. In two dimensions an exact solution to the Ising model at the critical point was found by Lars Onsager in 1944. The thermodynamic properties of the two-dimensional Ising model are shown in Figure 10.6. This was one of the outstanding achievements in theoretical physics. The solution to the three-dimensional Ising model is one of the most sought-after open problems in statistical mechanics.

Wolfgang Pauli said of the Second World War that 'The only important thing that happened is that Onsager solved the two dimensional Ising model.'

The most startling property of the Ising model (true in two and more dimensions) is that at low temperatures it experiences a *spontaneous magnetisation* (see Figure 10.6(d)). That is, below a critical temperature the system prefers to be in a state with the majority of its spins either pointing up ($S_i = 1$) or with the majority pointing down ($S_i = -1$). In many ways this should not be so surprising as spin alignment is energetically favoured. At high temperatures the system will maximise its entropy so that all the spins act almost independently of each other. As the temperature is lowered, the spins tend to cluster in groups that either predominantly point up or predominantly point down. As the temperature is lowered further the clusters grow until the critical temperature is reached when a cluster has spread through the system. At lower temperatures the system enters either a phase where the spins all tend to point up or a phase where they all tend to point down. This is an example of a *spontaneous symmetry breaking* –

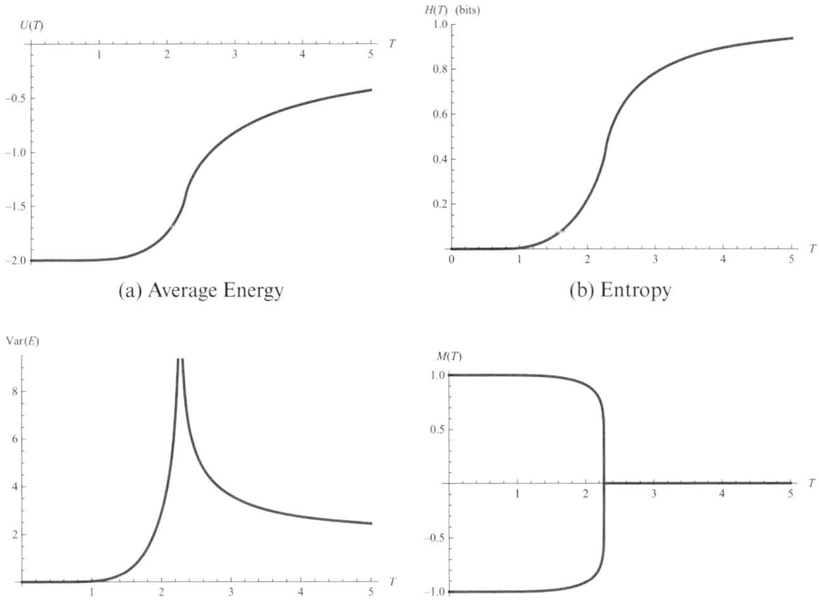

(a) Average Energy (b) Entropy

(c) Variance in Energy (d) Magnetisation

Figure 10.6 The thermodynamics of the two-dimensional Ising model. We show (a) the average energy, $U(T) = \mathbb{E}[E]$, (b) the entropy (measured in bits), $H(T)$, (c) the variance in the energy, $\mathbb{V}\mathrm{ar}[E]$, and (d) the magnetisation, $M(T)$ per spin.

it breaks the symmetry of the energy function, Equation (10.1). The maximum entropy distribution incorrectly describes the dynamics because it doesn't predict this spontaneous symmetry breaking. Actually, in finite systems the symmetry is retained because with a very small but finite probability the system will change from most of the spins pointing up to most of the spins pointing down. However, this probability is exponentially small in the system size, so that in the thermodynamic limit, $n \to \infty$, the probability of this switch around is negligible. Figure 10.7 shows typical configurations of the two-dimensional Ising model as we lower the temperature (increase the inverse temperature).

The second remarkable feature of the model is that right at the temperature where the spontaneous magnetisation occurs the system is in a *critical state*. This happens at a temperature $T_c \approx 2.269$ (in units where $J/k = 1$). At this temperature the gradient in the internal energy becomes infinite, and as a consequence its variance diverges. The specific heat is the amount of energy required to change the temperature. It is equal to the variance in the energy divided by kT^2 and so also diverges at T_c. Similarly, the fluctuations in the magnetisation also diverge at the critical temperature. In the critical state, there are clusters of up spins within clusters of down spins within clusters of up spins and so on. A consequence of this is that there is effectively an infinite correlation length. That is, the correlation between spins falls off as a power law of the distance between them (rather than falling off exponentially fast, as happens at temperatures away from the critical point). The system has a statistical scale invariance in the sense that it looks very similar at different length scales. What makes this remarkable is that at the critical point the fine details of the system

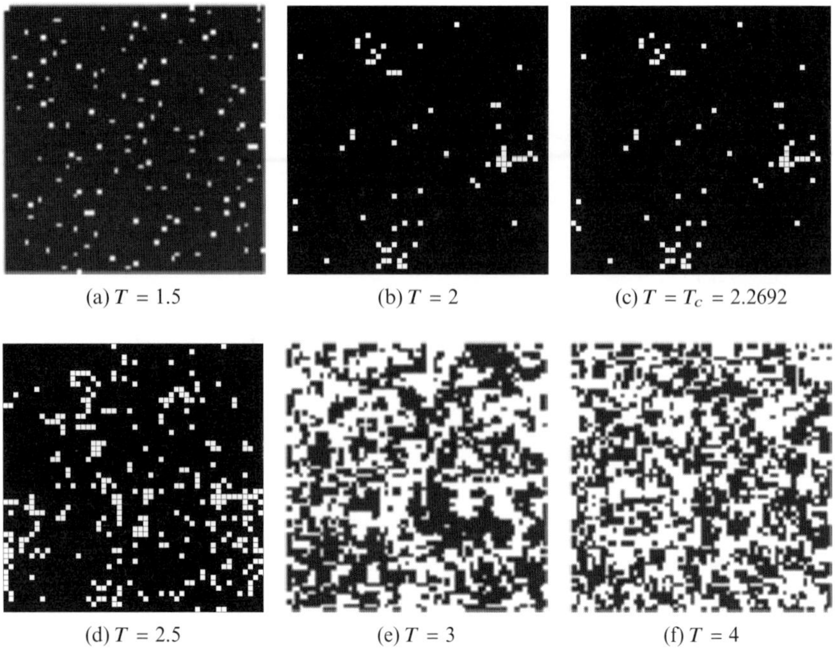

Figure 10.7 Example of instances obtained by simulations of the two-dimensional Ising model at different temperatures. (a) $T = 1.5$, (b) $T = 2$, (c) $T = T_c = 2.2692$, (d) $T = 2.5$, (e) $T = 3$, (f) $T = 4$.

become irrelevant and many very different types of system display the same properties. Thus, for a typical compound there is a liquid phase, which at the right pressure will go through a phase transition and become a gas. However, as we increase the pressure we reach a state where there is no longer a phase transition between gas and liquid (the two fluid phases are indistinguishable) – see Figure 10.8. At this point the liquid is at its critical point and many of the macroscopic features obey exactly the same power-law scaling behaviour as that found in the three-dimensional Ising model. This surprising and profound result is known as *universality*. One of the features of this is that at its critical point the density of a fluid varies on all scales so the fluid scatters all wavelengths of light and becomes opaque.

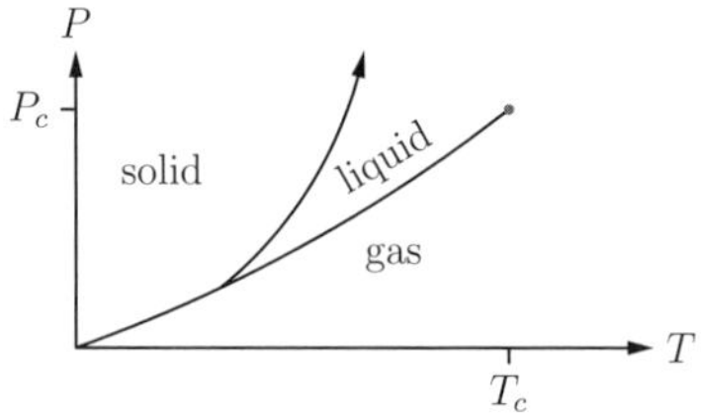

Figure 10.8 Schematic phase diagram for a compound with a liquid phase. The critical point occurs where there is no longer a transition from liquid to gas and is marked by a dot.

10.5 Self-Organised Criticality

In the systems we have discussed so far criticality only occurs at a carefully tuned point in the parameter space. There are, however, systems where it is believed they evolve toward a critical state and then show power-law behaviour thereafter.

Example 10.1 Forms Office

An amusing if fanciful example of self-organised criticality, due to Peter Grassberger, is an office of workers tasked with processing forms. The workers sit in rows of desks. Every so often the post boy arrives and places some new forms rather randomly on people's desks. When a worker feels overwhelmed, he or she will turn to one of their neighbours at random and, if the neighbour has fewer forms, they will deposit some of their forms on their neighbour's desk. To prevent the piles of forms growing for ever, the people at the edge of the room, rather then pass on the forms to a neighbour, might throw them out of the window – see Figure 10.9.

Figure 10.9 The form sorting office.

Figure 10.10 shows a simulation of this situtation with 64×64 desks where at each time step the postman simultaneously deposits Poi(1) (i.e. a Poisson deviate) new forms on each desk. The figure shows a histogram of the number of forms thrown out the window at each time step. Note that this is plotted on a log-log scale. For most time steps a passer-by will observe perhaps a couple of forms being thrown out of each window. But, there are times when tens, hundreds, and even many thousands of forms will be thrown out of each window in a single time step. The straight-line behaviour for large avalanches of forms is characteristic of a power-law behaviour, where if n is the number of forms thrown out of the window, then, for large n, the probability distribution behaves as $P(n) \propto n^{-\gamma}$. The constant γ is the exponent that measures the decay of the tail; for this problem empirically $\gamma \approx 1$. What makes this interesting is that despite the process being driven by a large number of random actions (those of the workers), $P(n)$ is very far from being normally distributed. Being a passer-by of this office could be rather dangerous. Fortunately, real bureaucrats would never work like this.

There are a lot of phenomena where the distributions of events have power-law tails. Perhaps the most prominent of these are earthquakes, which occur with a power-law frequency. This is reflected in the Richter scale, which is a logarithmic scale, such that the amplitude of the waves measured by a seismograph for a

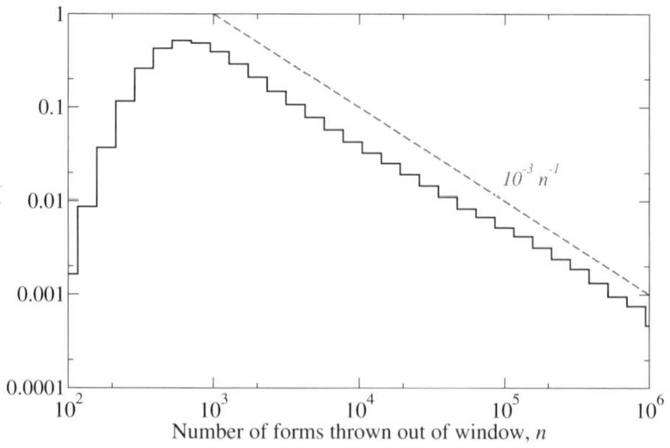

Figure 10.10 Histogram (plotted on a log-log scale) of number of forms thrown out of the window each round. Also shown as a dashed line is $10^{-3} \times n^{-1}$.

magnitude 8.0 earthquake are 10 times larger than those for a magnitude 7.0 (the actual energy is about 30 times greater). It is important to use a logarithmic scale as, although large earthquakes are much rarer than medium-sized earthquakes, they are nevertheless of great magnitude.

10.6 Disordered Systems

We finish this survey of collective phenomena by briefly discussing disordered systems. The prototypical disordered system is the simple spin glass, which is similar to the Ising model except the interactions between spins are taken to be random. The energy is given by

$$E(\boldsymbol{S}, \boldsymbol{J}) = - \sum_{(i,j)\in\mathcal{E}} J_{ij}\, S_i\, S_j$$

where, for example, $J_{ij} \sim \mathcal{N}(0, 1)$. $\mathcal{E}$ is the set of edges connecting the spins, often taken to be a regular lattice. If we have four edges which are connected in a loop and one or three of the couplings, J_{ij}, are negative then we cannot choose the spin variables around the loop to satisfy all the edges – a situation that physicists term *frustration*. As a result of this there are typically very many local energy minima. For these problems, understanding the low-energy behaviour of the system is extremely challenging.

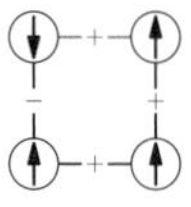

A few real alloys behave as spin glasses, but disorder systems abound in physical systems, for example, when there are impurities in some conductors. Disordered systems also appear in very different settings such as in combinatorial optimisation problems, models of the brain, and even in sociology. They differ from classical models that appear in statistical physics in that they also have some fixed (quenched) noise. Thus, one instance of a disordered system is very different at a microscopic level from another. Despite this, for large systems there are many properties which seem to be shared by almost all instances. These quantities physicists sometimes call self-averaging, which means that the distribution of these quantities are sufficiently sharply concentrated that the average behaviour is equal to the typical behaviour. In the case of the spin glass the partition function

$$Z(\beta, \boldsymbol{J}) = \sum_{\boldsymbol{S} \in \{-1,1\}^N} e^{-\beta E(\boldsymbol{S}, \boldsymbol{J})}$$

is non-self-averaging, while the free energy $F = -\beta \log\big(Z(\beta, \boldsymbol{J})\big)$ is self-averaging.

Example 10.2 Non-Self-Averaging Quantities

Non-self-averaging quantities typically occur with long- (thick-) tailed distributions. The classical example is the log-normal distribution, which we have considered previously in Example 5.3 on page 84. We construct a slightly simpler example where we focus on the non-self-averaging behaviour.

We consider a random variable

$$\Pi_n = \prod_{i=1}^{n} e^{X_i}$$

where $\boldsymbol{X} \sim \mathcal{N}(\boldsymbol{0}, \mathbf{I})$. The random variable $L_n = \log(\Pi_n)$ is just the sum of n normally distributed random variables so that $L_n = \mathcal{N}(0, n)$ and $\Pi_n \sim \text{LogNorm}(0, n)$. We show this distribution in Figure 10.11. The mean, $\mathbb{E}\left[\Pi_n\right] = e^{n/2}$, is very far from the mode of the distribution (which is at e^{-n}), and the median at 1. In contrast, $\exp(\mathbb{E}\left[\log(\Pi_n)\right]) = 1$ is at the median of the distribution. Note that the probability of Π_n being greater than or equal to its mean is given by

$$\mathbb{P}\left(\Pi_n \geq e^{n/2}\right) = \mathbb{P}\left(\log(\Pi_n) \geq \tfrac{n}{2}\right) = \int_{n/2}^{\infty} \mathrm{N}(\ell|0, n)\, d\ell$$

$$= \Phi\left(-\frac{\sqrt{n}}{2}\right) \approx \frac{e^{-n/2}}{\sqrt{\pi\, n/2}}$$

which is exponentially small in n. Thus, for large n we are exceedingly unlikely to ever see a samples even close to its mean value. As $L_n = \log(\Pi_n) \sim \mathcal{N}(0, n)$, most sample will be in the interval $-3\sqrt{n} \leq L_n \leq 3\sqrt{n}$ and a good approximation to the typical value of Π_n is around its median of $\exp(\mathbb{E}\left[L_n\right]) = 1$.

(Note that the mode is in a very narrow peak and so the probability close to the mode is also exponentially small. The median

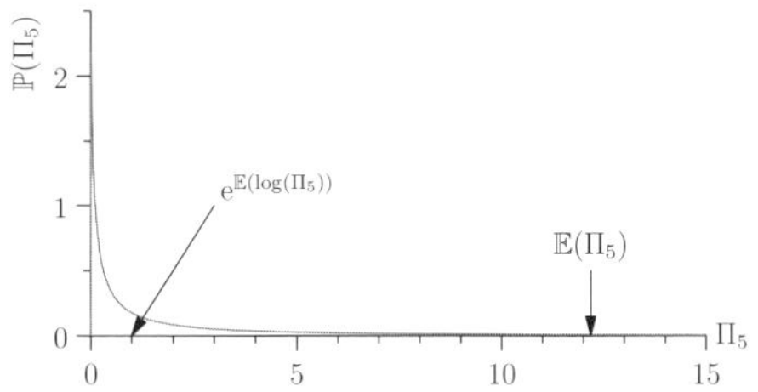

Figure 10.11 Distribution of $\Pi_n \sim \text{LogNorm}(0, n)$. We show both the median and mean. The mode is very close to 0.

gives a much better estimate of the typical behaviour than both the mean or mode. In Example 5.4 on page 87 we showed the importance of maximising the expectation of our log-gain rather than the expectation of our gain if we want to maximise our typical long-term winnings when performing sequential bets. This is because our log-gain is a self-averaging quantity while our gain is not.)

Returning to our spin-glass example the free energy is a self-averaging quantity so its typical value is close to its mean value. We can therefore obtain an understanding of the typical behaviour of a spin-glass system by considering the behaviour of the expected free energy

$$F_{\text{typical}}(\beta) = -\beta\, \mathbb{E}_{\boldsymbol{J}}\left[\log\bigl(Z(\beta, \boldsymbol{J})\bigr)\right]$$

where we take the expectation with respect to the distribution of $\boldsymbol{J}$. The advantage of taking this expectation is that the behaviour we compute does not depend on the particular instances (i.e. the particular coupling, $\boldsymbol{J}$). Technically averaging a logarithm is challenging and a huge amount of effort has gone into developing tricks for accomplishing this. This, however, goes beyond the scope of this book. The interested reader is referred to the specialised literature, e.g. Ziman (1979); Mézard et al. (1987); Fischer and Hertz (1991).

There are many probabilistic models where the random variables are extended throughout space. These include most of the classical models from statistical mechanics. Solving these models is technically challenging and often requires the development of a set of tools well beyond those covered in this book. These models often reveal some emergent or collective behaviour caused by the interactions of the variables. An example of this is the phase transition which occurs, for example, in models of percolation and the Ising model. Some of the most interesting developments since the late 1990s have come in our understanding of disordered systems.

A good way to gain an understanding of the collective phenomena is to simulate the systems. If you want to find the percolation point then a useful data structure to know about goes by the name of disjoint or union-find sets – in this context it is a very efficient algorithm for finding clusters. The Ising model is best simulated using MCMC techniques, which we cover in Chapter 11.

Exercise for Chapter 10

Exercise 10.1 (answer on page 430)

Compute a histogram of the time for a random walk, with the walker taking steps of ± 1 to return to his/her starting position. If T is the return time show that $\mathbb{P}\left(T = t/2\right) \approx t^{-3/2}/\zeta(3/2)$, where the normalisation constant $\zeta(3/2) \approx 2.61$ is the Riemann zeta function. Use this fact to show that a random walker will return to his/her starting position with probability arbitrarily close to 1, but the expected return time is infinite.

Exercise 10.2 (answer on page 431)

Simulate the branching process for some distribution and measure the mean number of particles and the variance after five iterations. Compare this with the theoretical results κ_1^5 and $\kappa_2\kappa_1^4(\kappa_1^5 - 1)/(\kappa_1 - 1)$, where κ_1 and κ_2 are the mean and variance of the distribution of children for a single individual.

11

Markov Chains

Contents

Markov chains are probabilistic systems that evolve in discrete time steps where the probability of the next step depends only on the current state of the system. This is such a commonly occurring situation that it pays handsomely to study and understand the properties of these systems. Markov chains also provide another means of sampling (Monte Carlo). Markov Chain Monte Carlo (MCMC) is so useful that it probably accounts for burning up more central processing unit (CPU) time than any other form of scientific computing.

We finish the chapter by considering the problem of inferring the states of a Markov chain from noisy observations. This is known as *filtering*. We describe the Kalman filter and particle filtering (sequential MCMC) as methods for solving this problem.

11.1 Markov Chains

A Markov chain describes a probabilistic system evolving in time, $\left(X(t)|t \in \mathbb{N}\right)$ such that

$$\mathbb{P}\left(X(t)|X(t-1), X(t-2), \ldots, X(1)\right) = \mathbb{P}\left(X(t)|X(t-1)\right).$$

In other words, the state of the system at time t only depends on the state of the system at time $t-1$ and is *conditionally* independent of the state at earlier times. This applies to a huge number of systems. For example, the random walk described in Chapter 10 is an example of a Markov chain. Here $X(t)$ would represent the position of the particle at time t. To determine the probability of the particle at the next time step we only need to know the position at the current time step.

Example 11.1 Population Genetics
A slightly more complex system which can be modelled by a Markov chain would be a population of genes evolving under selection and mutation (admittedly this is a rather crude abstraction of a real biological system). We let $X_g(t)$ denote the number of individuals in the population with genotype g. If $\mathcal{G}$ denotes the set of all possible genotypes, then we can describe the whole population at time t by a vector $\boldsymbol{X}(t) = (X_g(t)|g \in \mathcal{G})$. The population at the next generation, $\boldsymbol{X}(t+1)$ depends on the current population $\boldsymbol{X}(t)$ as well as the mutation rate and the fitness of the genotypes (i.e. the probability of them having a given number of offspring). These kinds of models, although crude, have been much studied both by population geneticists and by researchers studying the properties of genetic algorithms.

It is possible to generalise the notion of Markov chains to n^{th}-order Markov chains where the probability of being in a state $X(t)$ at time t depends on the state at the n previous time steps

$$\begin{aligned}&\mathbb{P}\left(X(t)|X(t-1), X(t-2), \ldots, X(1)\right) \\ &\quad = \mathbb{P}\left(X(t)|X(t-1), X(t-2), \ldots, X(t-n)\right).\end{aligned}$$

In this generalised notion, normal Markov chains are first-order Markov chains. Although this generalisation can be useful, by far the most frequently encountered Markov chains are those of first order (higher-order Markov chains can be modelled as first-order Markov chains acting on vectors). First-order Markov chains are also the models where the tools of analysis are most developed. In the rest of this chapter we will focus exclusively on first-order Markov chains.

In the definition of Markov chains we saw that $X(t)$ was *conditionally independent* of $X(t-\tau)$ for $\tau > 1$ given $X(t-1)$. This does not mean that $X(t)$ is uncorrelated or independent of these states. Usually $X(t)$ will be highly correlated with $X(t-2)$. What conditional independence tells us is simply that if we know $X(t-1)$ then the states at previous times are irrelevant to determining the probability that $X(t)$ takes a particular value. Markov chains describe systems with no 'memory' of their past.

11.1.1 Markov Chains and Stochastic Matrices

The dynamics of (first-order) Markov chains can be described by a simple matrix equation. Denoting the probability of a discrete Markov chain being in state i at time t by $p_i(t)$, then we can describe the probability distribution of being in a particular state by a vector $\boldsymbol{p}(t) = (p_i(t)|i \in \mathcal{S})$, where $\mathcal{S}$ is the set of possible states. Here we assume that the set of states is discrete. We can write update equations for a Markov chain as

$$\begin{aligned} p_i(t+1) &= \mathbb{P}\left(X(t+1)=i\right) = \sum_{j\in\mathcal{S}} \mathbb{P}\left(X(t+1)=i, X(t)=j\right) \\ &= \sum_{j\in\mathcal{S}} \mathbb{P}\left(X(t+1)=i|X(t)=j\right)\mathbb{P}\left(X(t)=j\right) = \sum_{j\in\mathcal{S}} M_{ij}(t)\, p_j(t) \end{aligned}$$

where $M_{ij}(t) = \mathbb{P}\left(X(t+1)=i|X(t)=j\right)$ is a *transition matrix*. We can write this a bit more elegantly as a (linear) matrix equation

$$\boldsymbol{p}(t+1) = \mathbf{M}(t)\,\boldsymbol{p}(t). \tag{11.1}$$

Here I have used the (traditional) convention of denoting column vector in bold (e.g. $\boldsymbol{v}$) and row vector as their transposes (e.g. $\boldsymbol{v}^\mathsf{T}$). In the literature on Markov chains the dynamics is sometimes written as $\boldsymbol{p}^\mathsf{T}(t+1) = \boldsymbol{p}^\mathsf{T}(t)\,\mathbf{T}(t)$ where $\mathbf{T}(t) = \mathbf{M}^\mathsf{T}(t)$. We can visualise a Markov chain in terms of a directed graph. An example of this is shown in Figure 11.1.

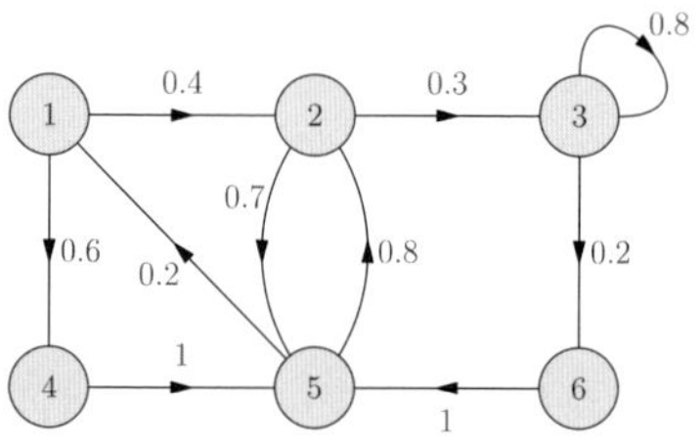

Figure 11.1 Example of a Markov chain depicted as a directed graph. The vertices correspond to states while the edges represent transition probabilities.

Since we are guaranteed that starting from some state j we will end up in some state then, for all j and all t,

$$\sum_{i\in\mathcal{S}} M_{ij}(t) = \sum_{i\in\mathcal{S}} \mathbb{P}\left(X(t+1)=i|X(t)=j\right) = 1 \qquad \text{or} \qquad \mathbf{1}^\mathsf{T}\mathbf{M}(t) = \mathbf{1}^\mathsf{T}$$

where $\mathbf{1}$ is the vector whose elements are all equal to 1. We also note that since the elements $M_{ij}(t)$ are transition probabilities they are non-negative (i.e. positive or zero). Matrices with non-negative elements whose columns (or rows) sum to 1 are known as *stochastic matrices*. Markov chains for discrete state spaces are therefore characterised by stochastic matrices. If the transition probabilities between states don't change over time (i.e. $\mathsf{M}(t) = \mathsf{M}$ for all t) then the system is said to be a time-homogeneous or a *stationary* Markov chain. In this case,

$$\boldsymbol{p}(t+1) = \mathsf{M}\,\boldsymbol{p}(t) = \mathsf{M}^2\,\boldsymbol{p}(t-1) = \cdots = \mathsf{M}^t\,\boldsymbol{p}(1).$$

Markov chains are ubiquitous. We have seen in Chapter 10 that random walks and branching processes are Markov chains. We will see in the next section that we can impose a dynamics on the Ising model which allows us to view them as Markov chains. We can use Markov chains to model any game where the next move only depends on the current state and ignores how we got there. Of course, some games might depend on the past history. Thus, in some variants of poker the likelihood of an unrevealed card might just depend on the current state of the game, but whether you judge a player to be bluffing is likely to depend on your past memory of how the player acted in the past.

Example 11.2 Population Genetics Revisited
Consider a population of organisms undergoing selection and mutation. For simplicity we assume the organisms are haploids (i.e. they carry only one copy of genes – the generalisation to diploids is easy, but unnecessary for this example). We also assume that the organisms undergo crossover, keeping the genes randomly shuffled (what population biologists call *linkage equilibrium*). This allows us to consider each gene in isolation from all the others. Further, it is assumed that the gene takes two forms (alleles), A and B. The probability of a mutation from A to B is denoted by u while the probability of a reverse mutation from B to A we denote by v. Finally, we assume that individuals with allele B produce on average $1 + s$ offspring for every offspring produced by individuals with allele A. This advantage is called by population biologists the *fitness*.

Assuming that there is a fixed population size, P, then the dynamics can be modelled by a Markov chain. It is helpful to consider mutation and selection taking place to create a large (effectively infinite) population of seeds – we will eventually sample from this seed population to obtain a new generation of organisms. If n is the number of individuals with the fitter allele B, then the proportion of the seed population in state B is given by

$$p_s = \frac{(1+s)n}{(1+s)n + P - n} = \frac{(1+s)n}{P + s\,n}.$$

After mutation the proportion, p_{sm}, of the seed population in allele B is

$$p_{sm} = (1-v)p_s + u(1-p_s) = \frac{(1-v)(1+s)n + u(P-n)}{P + s\,n}.$$

If we assume an infinite population then this would be equal to the proportion of individuals at the next generation

$$p(t+1) = \frac{(1-v)(1+s)p(t) + u\big(1-p(t)\big)}{1+s\,p(t)}$$

where $p(t) = n/P$. If the effect of selection and mutation is small then we can approximate the dynamics by a differential equation

$$\frac{\mathrm{d}p}{\mathrm{d}t} \approx p(t+1) - p(t) = \frac{\big(u - (u+v-s+u\,s)p(t) + s\,p^2(t)\big)}{1+s\,p(t)}.$$

Figure 11.2(a) shows the solution to this differential equation together with the exact solution to the difference equations.

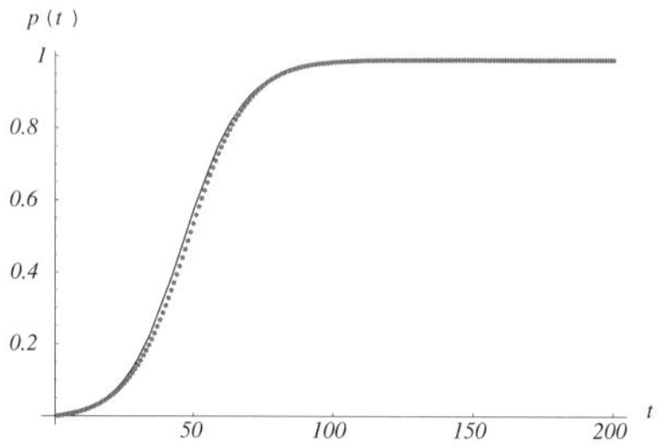

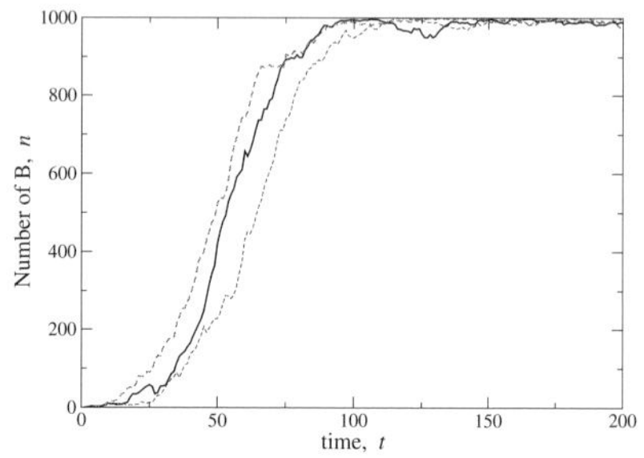

Figure 11.2 Illustration of the evolution of (a) an infinite population (the solid curve shows solution of approximate differential equation, while the dotted curve show exact solution of difference equation) and (b) three runs of a population of size 1000 under selection and mutation with $s = 0.1$, $u = v = 0.001$.

In finite populations (the situation we are trying to model) there will be random fluctuations resulting from sampling a new population of size P from the seed population. The probability of drawing n' mutants given a proportion p_{sm} of mutants in the seed population is given by a binomial distribution, $n' \sim \text{Bin}(P, p_{sm})$,

$$M_{n',n} = \mathbb{P}\left(n'|n\right) = \binom{P}{n'} p_{sm}^{n'} \left(1 - p_{sm}\right)^{P-n'}. \tag{11.2}$$

Using this probability distribution we can simulate the behaviour of a finite population. Figure 11.2(b) shows three realisations of the time evolution of the population using different streams of random numbers. The selection strength and mutation rate is the same as the infinite population solution shown in Figure 11.2(a). As discussed earlier, the dynamics of a finite population can be described by a Markov chain

$$\boldsymbol{p}(t+1) = \mathbf{M}\boldsymbol{p}(t) = \mathbf{M}^t \boldsymbol{p}(1)$$

where $\boldsymbol{p}(t)$ is the probability vector $(p_0(t), p_1(t), \ldots, p_P(t))$ and $p_n(t)$ is the probability of the population having n members with allele B. The matrix $\mathbf{M}$ is the transition matrix with probability $M_{n'n}$ given by Equation (11.2). For small populations we can use the Markov chain to solve the dynamics exactly. An example of this is shown in Figure 11.3 for a population of size 100.

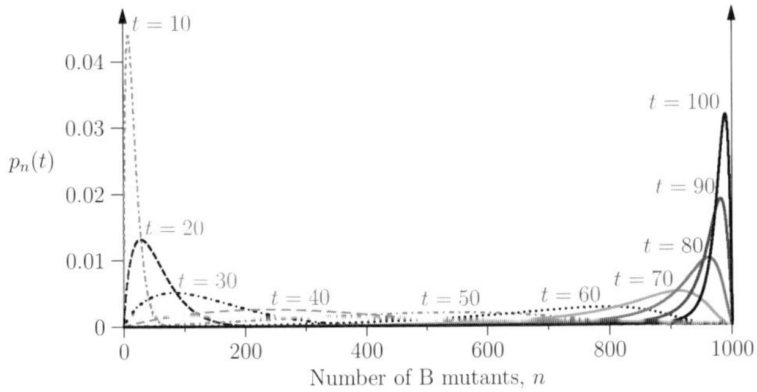

Figure 11.3 Markov chain simulation of the population showing the probability distribution at time steps $t = 10, 20, \cdots, 100$. This is for the same set of parameters as that used in Figure 11.2. Each curve represents a histogram showing the probability of n mutants being present at time t.

Unfortunately, the state space grows with the size of the population and we are often interested in very large populations; far larger than it is practical to compute a transition matrix. Fortunately, we will see in Chapter 12 that we can obtain an excellent approximation to this problem using the Fokker–Planck equation.

For systems with a small number of states, the dynamics of the system can be simulated numerically using the above matrix equation. This is numerically very stable and is often a very efficient means of obtaining a description of the dynamics. (A complete description of the dynamics would require assigning probabilities to all possible histories, $(X(t)|t = 1, \ldots, \infty)$. The Markov chain analysis gives the marginal distributions $\boldsymbol{p}(t)$ at each time step. Very often this is all we want to know.) Unfortunately, in many cases the state space is so large that it is not possible to write down the transition matrix, nevertheless, we can often deduce many properties about the dynamics of a system from a knowledge of the structure of its transition matrix.

11.1.2 Properties of Stochastic Matrices

Stochastic matrices have some interesting properties which are important for understanding the dynamics of Markov chains. These are valuable even when it is impractical to write down the transition matrix explicitly. We restrict our attention to the case when the search space is finite. Recall that elements of a stochastic matrix are non-negative and every column sums to one. (These are called column stochastic matrices. They need to be distinguished from the even more select set of doubly stochastic matrices where both the rows and columns sum to one.)

The important properties of stochastic matrices relate to their eigenvalues. A short description of the properties of eigenvectors for a general matrix can be found in Appendix 11.A. The first property of importance is that the modulus of any eigenvalue is less than or equal to 1. To show this we consider the eigenvalue equation of the $n \times n$ matrix, M,

$$\lambda\, v_i = \sum_{j=1}^{n} M_{ij} v_j .$$

Using this and the fact that M is a column stochastic matrix

$$|\lambda| \sum_{i=1}^{n} |v_i| \overset{(1)}{=} \sum_{i=1}^{n} \left| \sum_{j=1}^{n} M_{ij}\, v_j \right| \overset{(2)}{\leq} \sum_{i=1}^{n} \sum_{j=1}^{n} |M_{ij}|\, |v_j| \overset{(3)}{=} \sum_{j=1}^{n} |v_j| \sum_{i=1}^{n} |M_{ij}| \overset{(4)}{=} \sum_{j=1}^{n} |v_j|$$

(1) Taking the modulus of the equation above, using $|\lambda\, v_i| = |\lambda|\, |v_i|$, and summing over i.
(2) Using $|\sum_i a_i| \leq \sum_i |a_i|$ and $|M_{ij}\, v_j| = |M_{ij}|\, |v_j|$.
(3) Changing the order of summation and pulling out terms independent of i.
(4) Using $|M_{ij}| = M_{ij}$ since the transition probabilities are non-negative so that $\sum_{i=1}^{n} |M_{ij}| = \sum_{i=1}^{n} M_{ij} = 1$ which follows because M is a stochastic matrix.

Since $|\lambda| \sum_{i=1}^{n} |v_i| \leq \sum_j |v_j|$, it follows that $|\lambda| \leq 1$.[1]

The second property of a stochastic matrix is that it has at least one eigenvector with eigenvalue 1. It is easy to see that, because each column sums to 1, the vector $\boldsymbol{u}^{\mathsf{T}} = \mathbf{1}^{\mathsf{T}}$ (i.e. the row vector of all 1s) is a left eigenvector of any stochastic matrix, M, with eigenvalue 1 since

$$\sum_{i=1}^{n} u_i\, M_{ij} = \sum_{i=1}^{n} 1\, M_{ij} = 1 = u_j.$$

Any left-hand eigenvector has a corresponding right-hand eigenvector with the same eigenvalue. A right-hand eigenvector with eigenvalue 1 will correspond to a *stationary state* of a (stationary) Markov chain since it will be left unchanged from one time step to the next.[2] A stochastic matrix will have at least one eigenvector with eigenvalue 1, but can have more, corresponding to different stationary states.

Assuming an $n \times n$ stochastic matrix is not defective (see Appendix 11.A), we can expand the matrix as

$$\mathsf{M} = \sum_{i=1}^{n} \lambda^{(i)}\, \boldsymbol{v}^{(i)} \left(\boldsymbol{u}^{(i)}\right)^{\mathsf{T}}$$

where $\lambda^{(i)}$ is the ith eigenvector of M and $\boldsymbol{v}^{(i)}$ and $\boldsymbol{u}^{(i)}$ are the corresponding left- and right eigenvectors. Furthermore, due to the biorthogonality of the left- and right-hand eigenvectors (see Appendix 11.A) we have

$$\mathsf{M}^2 \overset{(1)}{=} \sum_{i=1}^{n} \lambda^{(i)}\, \boldsymbol{v}^{(i)}\, (\boldsymbol{u}^{(i)})^{\mathsf{T}} \sum_{j=1}^{n} \lambda^{(j)}\, \boldsymbol{v}^{(j)}\, (\boldsymbol{u}^{(j)})^{\mathsf{T}} \overset{(2)}{=} \sum_{i,j=1}^{n} \lambda^{(i)} \lambda^{(j)} \boldsymbol{v}^{(i)} \left((\boldsymbol{u}^{(i)})^{\mathsf{T}}\, \boldsymbol{v}^{(j)} \right) (\boldsymbol{u}^{(j)})^{\mathsf{T}}$$

$$\overset{(3)}{=} \sum_{i,j=1}^{n} \lambda^{(i)} \lambda^{(j)} \boldsymbol{v}^{(i)}\, [\![i = j]\!]\, (\boldsymbol{u}^{(j)})^{\mathsf{T}} \overset{(4)}{=} \sum_{i=1}^{n} (\lambda^{(i)})^2\, \boldsymbol{v}^{(i)}\, (\boldsymbol{u}^{(i)})^{\mathsf{T}}$$

(1) Using the eigenvector expansion of M given above.

[1] In fact, what we have just done is taken the 1-norm of $\lambda \boldsymbol{v} = \mathsf{M}\, \boldsymbol{v}$ giving $|\lambda|\, \|\boldsymbol{v}\|_1 = \|\mathsf{M}\, \boldsymbol{v}\|_1 \leq \|\mathsf{M}\|_1\, \|\boldsymbol{v}\|_1$ (which is true for any consistent norm). Now $\|\mathsf{M}\|_1 = 1$, implying that $|\lambda| \leq 1$.

[2] This is confusing because the word 'state' is being used in two distinct ways. The system takes a particular state, for example, in a random walk this would be the position of the walker. However, the 'stationary state' corresponds to a probability over states given by the eigenvector with eigenvalue equal to 1. In some cases the stationary state may be a single state, but this is not guaranteed.

(2) Taking the sums outside and rearranging the orders of terms.
(3) Using the biorthogonality of left and right eigenvectors $(\boldsymbol{u}^{(i)})^{\mathsf{T}} \boldsymbol{v}^{(j)} = [\![i = j]\!]$.
(4) Summing over j (the indicate function $[\![i = j]\!]$ means that we only keep the terms where $j = i$).

Similarly, taking any power of $\mathbf{M}$

$$\mathbf{M}^t = \sum_{i=1}^{n} (\lambda^{(i)})^t \, \boldsymbol{v}^{(i)} \, (\boldsymbol{u}^{(i)})^{\mathsf{T}}.$$

The notation here is rather ugly, with ^{T} *denoting taking the transpose, while* $(\lambda^{(i)})^t$ *denotes taking the eigenvalue,* $\lambda^{(i)}$*, to the power* t.

Thus, for a stationary Markov chain, starting from any initial distribution of states $\boldsymbol{p}(1)$, the distribution at time $t + 1$ is given by

$$\boldsymbol{p}(t+1) = \mathbf{M}^t \, \boldsymbol{p}(1) = \sum_{i=1}^{n} c_i \, (\lambda^{(i)})^t \, \boldsymbol{v}^{(i)} \tag{11.3}$$

where

$$c_i = (\boldsymbol{u}^{(i)})^{\mathsf{T}} \, \boldsymbol{p}(1).$$

Since $|\lambda^{(i)}| \leq 1$, the contributions of those terms in the above sum involving eigenvalues whose absolute value is less than 1 will diminish exponentially (i.e. as $\lambda^t = \mathrm{e}^{t \log(\lambda)}$). The Markov chain will reach a stationary distribution involving only those eigenvectors with eigenvalue $\lambda = 1$. If there is more than one such eigenvector then the final state will depend on the initial distribution, $\boldsymbol{p}(1)$.

The good news from Equation (11.3) is that most of the eigenvectors have only a transient effect and vanish exponentially quickly. The bad news is that there is nothing stopping an eigenvalue being very close to 1. If $|\lambda| = 1 - \epsilon$ for small ϵ then $|\lambda^t| \approx \mathrm{e}^{-\epsilon t}$ and the characteristic time for the contribution to decay by a factor $\mathrm{e} = \exp(1)$ is $1/\epsilon$. In general the convergence of a Markov chain depends on the gap between the eigenvalues equal to 1 and the next highest eigenvalue. Alas, in systems with a large number of states it is not uncommon for this gap to be very small.

In cases where it is feasible to write down the transition matrix and compute its eigenvectors then we can use Equation (11.3) to compute the full dynamics. However, some caution is required because the eigenvectors are not orthogonal, therefore the matrices of eigenvectors can be poorly conditioned so that this method of computing the evolution can be numerically unstable (in contrast to just applying the transition matrix which is numerically stable, but slow). For possible ways around this see the paper by Moler and Loan (2003).

Any state of a Markov chain which you cannot escape from is known as an *absorbing* state. In many problems there is one or a finite number of absorbing states. We are often interested in the expected time to reach the absorbing state starting from some known initial condition. This time is known as the *first-passage time* or *first-hitting time*. In Exercise 11.2 we ask you to calculate a closed formula for computing the first-passage times from the transition matrix, although this may be numerically difficult to compute (since it involves inverting matrices that can be very poorly conditioned).

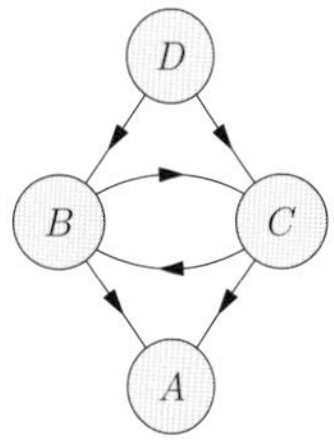

State A is an absorbing state.

11.1.3 Ergodic Stochastic Matrices

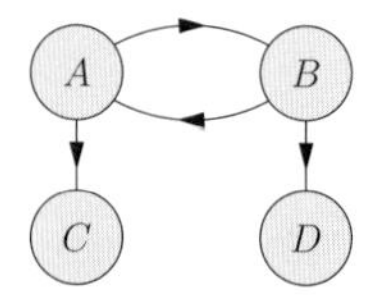

Markov chain with multiple stationary states.

In general, stochastic matrices can have multiple eigenvectors with eigenvalue 1, which means that the final state will depend on the initial conditions. However, there are special stochastic matrices which have a unique eigenvector with eigenvalue whose absolute value is 1. This guarantees that we will reach the same stationary distribution irrespective of the initial conditions. (Be aware, however, that this stationary distribution is, in general, a probability distribution over different possible states.) Fortunately, there are easy conditions for determining whether or not a matrix will have a unique stationary state. In fact, there are two properties that need to be checked. Firstly, the system must not get stuck in a different stationary state. This requires that for any state there is a finite probability of reaching the states in the stationary distribution in a finite number of steps. The second important condition is that the system does not go through a periodic cycle. For example, if we consider a random walk around a closed circuit where we take one step to the left or to the right with equal probability and there were an even number of sites, then if we started at site 1 at time $t = 1$, then on even time steps we would be at an even numbered site and at odd time steps we would be at an odd numbered site. This periodicity would be removed if we had a small probability of staying put, or if we had an odd number of sites around the circuit.

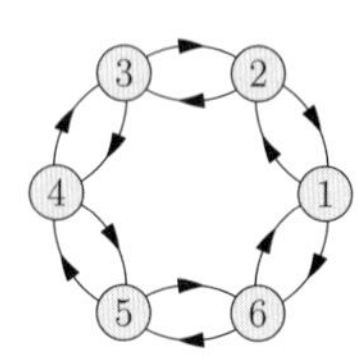

Periodic Markov chain.

A sufficient condition for a unique stationary distribution is that there is a finite probability of moving from one state to any other state in a finite number of steps and that the system is not periodic (that is, the return times to any particular state does not have any fixed period). In this case, the Markov chain is said to be *ergodic*. This result is formalised in the *Perron–Frobenius theorem*, which states for any irreducible stochastic matrix with period d the eigenvectors will have eigenvalues with values

$$\lambda_1 = 1, \quad \lambda_2 = e^{2\pi i/d}, \quad \ldots, \quad \lambda_d = e^{2\pi(d-1)i/d}$$

and all other eigenvalues satisfy $|\lambda_j| < 1$. For an aperiodic stochastic matrix there is therefore one eigenvalue equal to 1 and all other eigenvalues have magnitude strictly less than 1.

The condition that a matrix, $\mathbf{M}$, is irreducible is that there does *not* exist a permutation matrix $\mathbf{P}$ which will transform $\mathbf{M}$ so that

$$\mathbf{P}^{\mathsf{T}}\mathbf{M}\mathbf{P} = \begin{pmatrix} \mathbf{M}_1 & \mathbf{M}_2 \\ 0 & \mathbf{M}_3 \end{pmatrix}.$$

If such a matrix exist then

$$\mathbf{M}\left(\mathbf{P}\begin{pmatrix} \boldsymbol{v} \\ \mathbf{0} \end{pmatrix}\right) = \mathbf{P}\begin{pmatrix} \mathbf{M}_1\boldsymbol{v} \\ \mathbf{0} \end{pmatrix}$$

which implies there is a set of states which can never access a second set of states – thus the matrix is non-ergodic. Unfortunately, this definition is not very useful in practice. If all the elements of the transition matrix are strictly positive (i.e. all are greater than zero), then the matrix will be irreducible as there is a non-zero probability to move from any state to any other state of the system. This is a sufficient, but by no means necessary, condition for irreducibility.

Beware, though, that even for systems described by ergodic stochastic matrices, there can be transient states that take a time that is exponential in the size of the system to decay. Thus, even though there is a unique steady state in practice we may never reach it. Formally, in the limit of infinite size systems the time to visit configurations that make up the steady state may be infinite. This is a situation which physicists sometimes refer to as *ergodicity breaking*.

The importance of ergodic Markov chains is that they are relatively easy to construct and often provide an excellent model of real systems. It is not too difficult to construct a dynamic system with a unique stationary distribution – something that is useful for formally proving the convergence of many algorithms and that is also used in MCMC.

11.2 Markov Chain Monte Carlo

One of the cornerstones of modern probabilistic modelling is Markov Chain Monte Carlo, more usually referred to as MCMC. We recall from Chapter 3 that Monte Carlo procedures are ways of sampling a random variable X from some probability mass $P_X(x)$ or probability density $f_X(x)$. This allows us to obtain numerical approximations of expectations, which can be used to answer many of the questions we are interested in. When X is simple (e.g. a number) and the cumulative distribution function is easily invertible, then we can use the transformation method to generate random deviates. Otherwise we could use the rejection method. However, when X or its probability distribution becomes more complex (e.g. high dimensional) then the transformation method is usually impossible and the rejection method impractical because we cannot find a good approximating distribution. In this case, we end up nearly always rejecting a sample. MCMC overcomes this by visiting high probability states.

The idea is to build a Markov chain (i.e. a probabilistic dynamical system) that has a unique steady-state distribution, which is the distribution that we are trying to sample over. We use the heuristic that highly probable states are close together. Thus we explore the probability distribution over states by making small steps in the random-variable space. Although the sample points are correlated over a short time we are guaranteed that over a long enough time period we visit each state according to their probability (provided we have set up the dynamics correctly). However, whenever you hear claims such as this you need to be on your guard. The mathematical guarantee comes with the caveat that you have to wait 'long enough' and sometimes long enough is much longer than anyone is prepared to wait (it is easy to find systems where that wait could be much longer than the age of the universe). This being said, MCMC often gives you excellent results. We discuss the practicalities of MCMC before returning to explore its limitations and discussing improvements.

MCMC originated during the Second World War at Los Alamos where the first atom bomb was designed and built. The method was devised by Stanisław Ulam and taken up by von Neumann amongst others. The work was made public in a seminal paper by N. Metropolis, A. W. Rosenbluth, M. N. Rosenbluth, A. H. Teller, and E. Teller in 1953. It was initially regarded as a tool for computing physical quantities. In 1970 W. K. Hastings generalised MCMC to allow a non-symmetric function for generating candidate solutions – this is compensated for by modifying the transition probabilities. MCMC has now become a standard tool for Bayesian inference.

11.2.1 Detailed Balance

One apparent difficulty in using MCMC is to ensure that its stationary distribution is reached, however, this is surprisingly simple to achieve. Supposing we want a certain probability distribution, $\boldsymbol{\pi} = (\pi_i | i \in \mathcal{S})$ – where $\mathcal{S}$ is the set of states – to be the stationary distribution of a Markov chain. Then a sufficient (though not necessary) condition for this to happen is to choose the transition probability, M_{ij}, from state j to state i, to satisfy

$$M_{ij}\pi_j = M_{ji}\pi_i. \tag{11.4}$$

If we sum over j we find

$$\sum_{j \in \mathcal{S}} M_{ij}\pi_j = \sum_{j \in \mathcal{S}} M_{ji}\pi_i = \pi_i$$

since $\sum_{j \in \mathcal{S}} M_{ji} = 1$ (i.e. $\mathbf{M}$ is a stochastic matrix). Thus,

$$\mathbf{M}\boldsymbol{\pi} = \boldsymbol{\pi}$$

which is the condition we require for $\boldsymbol{\pi}$ to be a stationary state of our Markov chain. If $\mathbf{M}$ is ergodic this stationary distribution will be unique. Equation (11.4) is known as *detailed balance*. Any Markov chain satisfying detailed balance is said to be a *reversible Markov chain*.

The condition of detailed balance together with a stationary distribution $\boldsymbol{\pi}$ is not sufficient to uniquely determine the transition probability $\mathbb{P}\left(X(t+1) = i | X(t) = j\right) = M_{ij}$. Usually, starting from a state i we choose a neighbouring state j uniformly at random from a set of neighbours (note if i is a neighbour of j then j must be a neighbour of i). We then decide whether to accept the neighbour. There are different ways to achieve this. Two commonly used strategies are

1. Accept a move with probability $\min(1, \pi_i/\pi_j)$. (In practice you just draw a uniform deviate $U \sim \mathrm{U}(0,1)$ and accept the move if $U \leq \pi_i/\pi_j$.) To show that this satisfies the detailed balance condition we first consider the case then $\pi_i \geq \pi_j$ then $M_{ij} = 1$ and $M_{ji} = \pi_j/\pi_i$. In this case $M_{ij}\,\pi_j = \pi_j$ while $M_{ji}\,\pi_i = (\pi_j/\pi_i)\,\pi_i = \pi_j$, thus satisfying detailed balance. Similarly, if $\pi_i < \pi_j$ then

$M_{ji} = 1$ and $M_{ij} = \pi_i/\pi_j$. This also satisfies Equation (11.4). This procedure is often referred to as the *Metropolis* algorithm.

2. Accept a move from state j to state i with a probability

$$M_{ij} = \frac{\pi_i}{\pi_i + \pi_j}.$$

Direct substitution into Equation (11.4) shows that this satisfies detailed balance. This is called the *heat-bath* method, at least by physicists.

An important property of these algorithms is that they just depend on the ratio of the probabilities π_i. As a consequence we only need to know this quantity up to a constant. In practice, one frequently does not know the normalising factor. For example, in simulating physical systems we often know the energy function, $E(\boldsymbol{x})$, and we assume the system is in thermodynamic equilibrium with a large heat bath so that the probability of a configuration, $\boldsymbol{x}$, is given by the Boltzmann distribution

$$\mathbb{P}\left(\boldsymbol{x}\right) = \frac{e^{-\beta E(\boldsymbol{x})}}{Z}, \qquad Z = \sum_{\boldsymbol{x}} e^{-\beta E(\boldsymbol{x})},$$

but we cannot compute the normalisation factor (known as the partition function), Z, as it is computationally infeasible to sum over all configurations of the system. Similarly, in Bayesian inference we want to compute the posterior, $\mathbb{P}\left(\boldsymbol{x}|\mathcal{D}\right)$, where we know the joint distribution $\mathbb{P}\left(\boldsymbol{x}, \mathcal{D}\right) = \mathbb{P}\left(\mathcal{D}|\boldsymbol{x}\right) \mathbb{P}\left(\boldsymbol{x}\right)$, but to compute the evidence $\mathbb{P}\left(\mathcal{D}\right)$ requires marginalising out $\boldsymbol{x}$ which is again often intractable.

There remains a choice in what neighbourhood you choose. The smaller your neighbourhood the higher the probability of the move being accepted, but the longer it takes to obtain an independent sample. The neighbourhood has to be sufficiently large to ensure that there is a finite probability of visiting every part of the search space (i.e. the Markov chain is ergodic).

One choice of neighbourhoods for sampling from multivariate distributions,

$$\boldsymbol{X} = (X_1, X_2, \ldots, X_n) \sim f_{\boldsymbol{X}}$$

is to randomly choose a variable index $i \in \{1, 2, \ldots, n\}$ and then sample X_i from the marginal distribution, keeping all other components fixed. Stuart and Donald Geman in their seminal paper (Geman and Geman, 1984) called this *Gibbs' sampling* (in honour of the physicist Willard Gibbs). This is more limited than traditional MCMC sampling, but can be much quicker when the distribution $f_{\boldsymbol{X}}(\boldsymbol{x})$ is such that computing the change in probabilities due to changing a single variable is efficient.

MCMC provides a sequence of random samples (each of which may provide a description of the state of a complex multidimensional system). The first elements in the sequence are typically not correctly distributed and need to be thrown away. The notion of a random variable not having the correct distribution is rather meaningless. The correct way to think about this is to imagine an ensemble of MCMC simulations running the same algorithms but with different random numbers. At some time the probability distribution of the ensemble will converge to the probability $\boldsymbol{\pi}$. The time to reach this stationary distribution is sometimes called the burn-in time. We need to throw away the samples during this burn-in period. Of course, in practice we don't run an ensemble of MCMC simulations and it is difficult to know how long the burn-in takes (there are some heuristics for estimating this, but they are too detailed to discuss here). In practice, people often run an MCMC for as long as is practical and throw away some proportion of their data (e.g. the first half). Starting from a high probability state will often reduce the burn-in period.

Having thrown away the burn-in period we are left with a set of random deviates, but they are not independent. If we want independent deviates then we have to select samples separated by sufficient time so that the samples are no longer dependent of each other. Unfortunately, just as with burn-in there is no fixed rule for determining this. Fortunately, for computing expectations we don't care if nearby samples are independent provided we have run long enough that we have visited each part of space sufficiently often. One subtlety is that for the sample to be unbiased we have to consider an iteration as the current solution whether or not the candidate solution is accepted. That is, even when we reject a move so that $X(t+1) = X(t)$ we should use both $X(t)$ and $X(t+1)$ in computing the expectations empirically. If we don't we are being prejudiced against values of random variables that occur with high probability.

Example 11.3 One-Dimensional MCMC

We give a very simple one-dimensional example. Here we consider a continuous space. Suppose we want to sample from

This is a mixture of Gaussians (or normals). It is very different from the distribution of a sum of two normally distributed random variables $X = \frac{2}{3}Y + \frac{1}{3}Z$ *where* $Y \sim \mathcal{N}(0,1)$ *and* $Z \sim \mathcal{N}(4,9)$, *which is actually normally distributed* $X \sim \mathcal{N}(4/3, 37/81)$.

$$f_X(x) = \frac{2}{3}\mathcal{N}(x|0,1) + \frac{1}{3}\mathcal{N}(x|4,9)$$

using MCMC. To do this we start from some initial value $X(1)$ and then for $t = 1$ up to T we generate a candidate point $Y = X(t) + \mathcal{N}(0, \sigma^2)$ with step size $\sigma = 0.2$. We accept the candidate point (set $X(t+1)$ to Y) with a probability

$$\min\left(1, \frac{f_X(Y)}{f_X(X(t))}\right)$$

otherwise we set $X(t+1)$ to $X(t)$.

Although we have not chosen the neighbours of a point uniformly, nevertheless our transition probability satisfies detailed balance because the probability of choosing a point Y as a candidate point starting from X is the same as choosing X as a candidate point starting from Y. Also our process is clearly ergodic as there is a non-

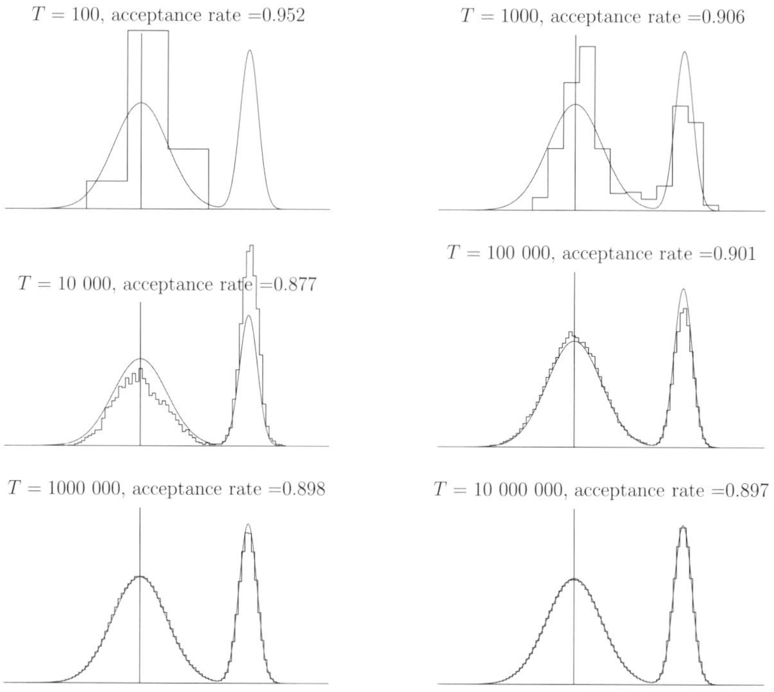

Figure 11.4 Example of MCMC where we draw samples from the distribution $f_X(x) = \frac{2}{3}\mathcal{N}(x|0,1) + \frac{1}{3}\mathcal{N}(x|4,9)$ (shown as a continuous curve). The curve consisting of a number of steps shows a histogram(with different bin sizes) of our samples from MCMC as we increase the number of samples, T.

zero probability of making a transition from any point to any other point.

Figure 11.4 shows a set of histograms from independent runs with a different number of sample points. In this problem the quality of the results depends on the step size. If the step size is too small then we have a high acceptance rate, but a low decorrelation rate (or high exploration time). Contrariwise if the step size is too large then we get a high decorrelation rate, but a low acceptance rate. The step size we have chosen ($\sigma = 0.2$) provides a good compromise.

The example above is nice to visualise, but is an unrealistic application of MCMC. As we pointed out in Chapter 3 the point of Monte Carlo is to compute expectations. In low dimensions we can compute the expectations by numerical integration with the distribution function as an added weight. We can usually obtain very high accuracy by sampling the domain of $f_X(x)$ at relatively few points. The difficulty comes when we have to integrate over many dimensions. The number of points we need to obtain a fixed accuracy grows exponentially with the number of dimensions. Thus, numerical integration simply becomes infeasible in high dimensions. However, Monte Carlo gives errors that fall off as $1/\sqrt{T}$ where T is the number of (independent) sample points. Thus, MCMC comes into its own in high dimensions.

Example 11.4 Ising Model
A classic high-dimensional problem in statistical physics is the Ising model described in Section 10.4. This is a discrete problem involving variables $\boldsymbol{S} = (S_i | i \in \mathcal{V})$ where $\mathcal{V}$ is a set of vertices and $S_i \in \{-1, 1\}$. The 'energy' of the Ising model is given by

$$E(\boldsymbol{S}) = -J \sum_{(i,j)\in\mathcal{E}} S_i\, S_j$$

where $\mathcal{E}$ is a set of edges linking the variables and J is a coupling constant. It is usual to consider the case where the variables live on a regular lattice. The probability of a configuration is given by the Boltzmann probability

$$\mathbb{P}\left(\boldsymbol{S}\right) \propto \mathrm{e}^{-\frac{E(\boldsymbol{S})}{kT}} = \exp\left(\frac{J}{k\,T} \sum_{(i,j)\in\mathcal{E}} S_i\, S_j\right)$$

where k and T are the Boltzmann constant and temperature respectively.

The difficulty in computing expectations is that the number of possible values the random variables can take is $2^{|\mathcal{V}|}$, which rapidly becomes intractable as the number of variables increase. MCMC is an extremely simple method for computing expectations. Starting from some initial configuration $\boldsymbol{S}$ we obtain a candidate set of variables, $\boldsymbol{S}'$, by choosing a variable S_i at random and 'flip' it (i.e. $S_i \leftarrow -S_i$) keeping all other variables constant (this is an example of Gibbs' sampling). We accept the update if $E(\boldsymbol{S}') < E(\boldsymbol{S})$, or else we accept the update with a probability

$$p = \mathrm{e}^{-(E(\boldsymbol{S}')-E(\boldsymbol{S}))/kT}.$$

This just depends on the difference in energy between variables $\boldsymbol{S}'$ and $\boldsymbol{S}$. However, to compute this difference we don't need to compute the complete energy function, but only need to consider the spin we have changed and its neighbours

$$E(\boldsymbol{S}') - E(\boldsymbol{S}) = 2\,J\,S_i \sum_{j\in\mathcal{N}_i} S_j$$

where $\mathcal{N}_i = \{j \in \mathcal{V} |\, (i,j) \in \mathcal{E}\}$ is the set of neighbours of site i. It is commonly the case in problems with a local energy function that the energy difference from changing a single variable is much faster to compute than the full energy. This makes using MCMC much faster than it would otherwise be. We will see, however, that we can simulate the Ising model even more effectively using a cluster MCMC algorithm. Incidentally, the physics community tends to call MCMC for these types of applications Monte Carlo.

Simulated Annealing

In passing it is perhaps worth mentioning simulated annealing because of its close connection with MCMC, even though it is not directly to do with probabilities. Many optimisation problems can be represented as a cost (or energy) function $E(\boldsymbol{X})$ which we want to minimise. In difficult problems this function varies quite rapidly as we change $\boldsymbol{X}$. One common technique for solving these problems is known as *simulated annealing* where we treat $E(\boldsymbol{X})$ as an energy function and perform MCMC while slowly lowering the temperature. The idea is that at higher temperatures the acceptance probability of making a change to $\boldsymbol{X}$ is relatively high, allowing the algorithms to find a good part of the solution space. As we lower the temperature, $\boldsymbol{X}$ spends more of its time at low-cost solutions. This is analogous to the physical process of annealing metals, where by slowly reducing the temperature a metal solidifies into its low-energy crystalline phase rather than a higher energy (suboptimal) glassy phase. There is even a guarantee that if we reduce the temperature slowly enough simulated annealing will find the global optimal solution, unfortunately, in some cases slowly enough would be much slower than performing a random search. In practice, one reduces the temperature much faster than needed to guarantee an optimal solution, and then hopes for the best.

11.2.2 Bayes and MCMC

MCMC is one of the most useful tools for carrying out Bayesian calculations when the models become too complex to be solved analytically. As we have mentioned already, one convenient feature of MCMC is that it does not require the probabilities to be normalised. Therefore to compute a posterior $f(\boldsymbol{\theta}|\mathcal{D})$ we need only to compute the joint probability $f(\boldsymbol{\theta}, \mathcal{D}) = f(\mathcal{D}|\boldsymbol{\theta})\, f(\boldsymbol{\theta})$ (that is, we can ignore the normalisation $f(\mathcal{D})$). To use the Metropolis algorithm we start with some guess for a probable $\boldsymbol{\theta}$ and then select a nearby value, $\boldsymbol{\theta}'$, according to some local probability distribution. We compute

$$r = \min\left(1, \frac{f(\mathcal{D}|\boldsymbol{\theta}')\, f(\boldsymbol{\theta}')}{f(\mathcal{D}|\boldsymbol{\theta})\, f(\boldsymbol{\theta})}\right)$$

and move to $\boldsymbol{\theta}'$ with probability r, otherwise we remain at $\boldsymbol{\theta}$. The initial values of $\boldsymbol{\theta}$ are discarded, the rest are taken to be samples from the posterior distribution which we can use, for example, to compute the posterior mean.

There is still some discretion in choosing the candidate or proposal parameters $\boldsymbol{\theta}'$. We want them to be sufficiently close to the current parameters $\boldsymbol{\theta}$ so that there is a good probability of the proposal being accepted, however, this has to be balanced against our desire to explore the full parameter space. We discuss this issues in more detail in the next section. If we are to use the Metropolis algorithm given above, another requirement on selecting our proposal parameters is that the proposal distribution, $p(\boldsymbol{\theta}'|\boldsymbol{\theta})$, is symmetric. That is, we require $p(\boldsymbol{\theta}'|\boldsymbol{\theta}) = p(\boldsymbol{\theta}|\boldsymbol{\theta}')$ in order that detailed balance is satisfied. This can be rather too restrictive, for example if our parameters are non-negative then some care is required to ensure that we have a symmetric proposal distribution. Even when it is symmetric it might be inefficient, e.g. by making large steps unlikely when a non-negative parameter becomes small. We can adapt the standard Metropolis algorithm to cope with any proposal distribution. Our detailed balance equation becomes

$$W(\boldsymbol{\theta} \to \boldsymbol{\theta}')\,\pi(\boldsymbol{\theta}) = W(\boldsymbol{\theta}' \to \boldsymbol{\theta})\,\pi(\boldsymbol{\theta}') \tag{11.5}$$

where a transition probability $W(\boldsymbol{\theta} \to \boldsymbol{\theta}')$ now equals the proposal distribution, $p(\boldsymbol{\theta}|\boldsymbol{\theta}')$, times the acceptance probability, $a(\boldsymbol{\theta}|\boldsymbol{\theta}')$,

$$W(\boldsymbol{\theta} \to \boldsymbol{\theta}') = p(\boldsymbol{\theta}'|\boldsymbol{\theta})\,a(\boldsymbol{\theta}'|\boldsymbol{\theta}).$$

(Equation (11.5) is just Equation (11.4) written for continuous rather than discrete variables – note that the function $W(\boldsymbol{\theta} \to \boldsymbol{\theta}')$ plays the same role as the transition elements M_{ij}.) Our more sophisticated detailed balance requirement is that

$$\frac{a(\boldsymbol{\theta}'|\boldsymbol{\theta})}{a(\boldsymbol{\theta}|\boldsymbol{\theta}')} = \frac{p(\boldsymbol{\theta}|\boldsymbol{\theta}')\,\pi(\boldsymbol{\theta}')}{p(\boldsymbol{\theta}'|\boldsymbol{\theta})\,\pi(\boldsymbol{\theta})}.$$

Similar to the Metropolis algorithm we compute

$$r = \min\left(1, \frac{a(\boldsymbol{\theta}'|\boldsymbol{\theta})}{a(\boldsymbol{\theta}|\boldsymbol{\theta}')}\right) = \min\left(1, \frac{p(\boldsymbol{\theta}|\boldsymbol{\theta}')\,\pi(\boldsymbol{\theta}')}{p(\boldsymbol{\theta}'|\boldsymbol{\theta})\,\pi(\boldsymbol{\theta})}\right)$$

and we accept with a probability r. For computing the posterior we can use $\pi(\boldsymbol{\theta}) \propto f(\mathcal{D}|\boldsymbol{\theta})\,f(\boldsymbol{\theta})$ so that r becomes

$$r = \min\left(1, \frac{p(\boldsymbol{\theta}|\boldsymbol{\theta}')\,f(\mathcal{D}|\boldsymbol{\theta}')\,f(\boldsymbol{\theta}')}{p(\boldsymbol{\theta}'|\boldsymbol{\theta})\,f(\mathcal{D}|\boldsymbol{\theta})\,f(\boldsymbol{\theta})}\right).$$

This modified Metropolis algorithm is known as the Metropolis–Hastings algorithm. It can be quite important in problems where the variables being estimated, θ_i, are constrained.

Example 11.5 Traffic Rate (Example 8.4 Revisited)

In Example 8.4 on page 200 we considered estimating the average traffic rate along a road from a series of observations, $\mathcal{D} = (N_1, N_2, \ldots, N_n)$. We assumed that the likelihood of the data is given by a Poisson distribution, $N_i \sim \text{Poi}(\mu)$. Our task is to estimate μ. We assume an uninformative prior $f_0(\mu) = \text{Gam}(\mu|0,0) \propto 1/\mu$. In Example 8.4 we showed that the posterior in this case is given by the gamma distribution $\text{Gam}(\mu|a_n, b_n)$ where $a_n = \sum_{i=1}^{n} N_i$ and $b_n = n$. However, let us compute this using MCMC instead.

Our variable μ has to be positive, which makes our proposal distribution more complicated. We want it to be quite close to our previous value. Thus we take our candidate to be $\mu' \sim \text{Gam}(a, b)$ where we choose a and b such that $\mathbb{E}\left[\mu'\right] = \frac{a}{b} = \mu$. We also want the variance to be close to the variance of our posterior (which unfortunately we don't know). This may take some parameter tuning, but in this example we take $\mathbb{V}\text{ar}\left[\mu'\right] = \frac{a}{b^2} = 1$. This implies that

$a = \mu^2$ and $b = \mu$ so that $\mu' \sim \text{Gam}(\mu^2, \mu)$. To perform an update we compute

$$r = \min\left(1, \frac{\text{Gam}(\mu|\mu'^2, \mu')\frac{1}{\mu'}\prod_{i=1}^n \text{Poi}(N_i|\mu')}{\text{Gam}(\mu'|\mu^2, \mu)\frac{1}{\mu}\prod_{i=1}^n \text{Poi}(N_i|\mu)}\right)$$

$$= \min\left(1, \frac{\mu\,\text{Gam}(\mu|\mu'^2, \mu')}{\mu'\,\text{Gam}(\mu'|\mu^2, \mu)} e^{-n(\mu'-\mu)+\sum_{i=1}^n N_i \log\left(\frac{\mu'}{\mu}\right)}\right).$$

We accept the update with a probability r.

In Figure 11.5 we show a histogram of the samples of the posterior distribution for the traffic rate μ produced by MCMC. Note that to compute the posterior histogram we used the current μ value at each step whether μ had changed or not.

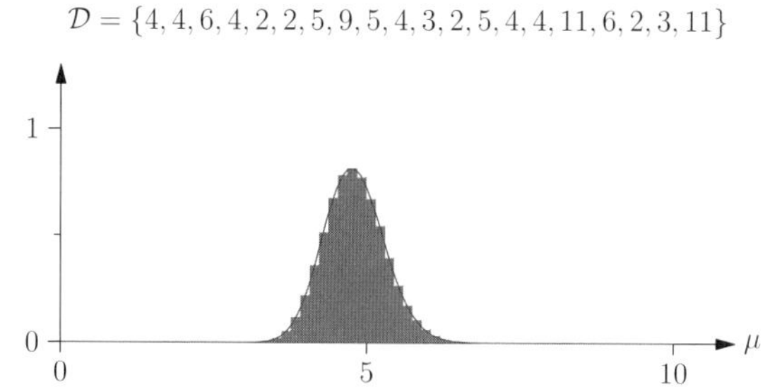

Figure 11.5 The histogram for the posterior distribution for the traffic rate computed using MCMC. The solid line shows the exact posterior (a gamma) distribution for this problem. The data is drawn from $\text{Poi}(N_i|5)$.

The efficiency of the MCMC depends on the proposal distribution. If we choose a proposal distribution which is identical to the posterior then $r = 1$ and we would always accept the update. Of course, as the whole point of performing MCMC is because we don't know the posterior we are unlikely to be able to make our proposal distribution identical to the posterior, so the practical goal is to try to design a proposal distribution that is not too different from the posterior and so improve the efficiency of the MCMC sampler.

In our example MCMC was unnecessary as for this case there is a nice conjugate prior. Also since we were estimating only a single parameter we could have computed the distribution numerically. Where MCMC comes into its own is for more complicated problems, particularly when we are trying to infer multivariate parameters (such as in hierarchical models with latent variables or even graphical models as described throughout Chapter 8). More realistic problems are described in many books specialising on Bayesian inference, e.g. Lee (2003); Gelman et al. (2003). In principle, using MCMC for Bayesian calculations is extremely straightforward, but it is such an important task that it has been very heavily developed. There are a host of books describing MCMC for Bayesian methods. Don't, however, become too intimidated by the sophistication of many MCMC methods. Computers are sufficiently fast that for many problems a naive implementation is usually good enough to get good estimates. By far the most important part of Bayesian inference is to build a good model. This means understanding the problem and being very clear about what you are doing. The actual computation is of secondary importance and efficiency only matters when you are pushing the envelope.

11.2.3 Convergence of MCMC

The time for MCMC to converge to the stationary distribution is generally very difficult even to estimate. Technically, the convergence speed depends on the size of the second-highest eigenvalue of the transition matrix – see Equation (11.3) –, but in practice if you could compute this you don't need to use MCMC at all. A number of different issues can affect the convergence time. In Example 11.3 on page 320, we had a multi-peaked distribution with a low probability region in between. This clearly slows down convergence as it is difficult to transition between the high probability regions.

This also occurs in the Ising model below its critical temperature when the probability of making a transition between having most variables equal to $+1$, to a situation where most variables equal -1, becomes exponentially small in the number of variables. If we interpret the Ising model as a simplistic model of a magnet where the variables represent magnetic spins which are either pointing up or down, then below the critical temperature we have a spontaneous magnetisation. The system has broken the symmetry of the energy function. This is an example of ergodicity breaking where the time average of a sample does not give the same result as averaging over all the states weighted by a Boltzmann factor (technically, this ergodicity breaking only occurs in the limit of infinite number of variables, but in practice the waiting time for a magnetic system to change its magnetisation is much larger than anyone cares to wait). MCMC provides a truer description of the physical system than performing the expectation – ergodicity breaking is observed in real magnets where the number of spins is around Avogadro's number (i.e. of order 10^{23}). The Ising model is fairly benign in that it only has two broken symmetry states (that is, the energy function has two minima). In many disordered systems (including the Ising model with random couplings) the energy function can have a very large number of minima, many of which have energies far from the global optimum energy. If MCMC is run at a low temperature it will, with high probability, gets 'stuck' in a local minima with an exponentially (in the number of variables) small probability of escaping. These systems are extremely hard to simulate using MCMC although there are many variants which have been developed to overcome these problems – there exists a large literature on performing Monte Carlo on disordered systems in the physics literature.

The Ising model also causes problems around its critical point (just where the symmetry breaking occurs). This is because at this point the system is so finely balanced that flipping a single variable has an effect which can ripple through the whole system. More formally, there are long-range correlations between variables. This again slows down convergences considerably. This is an entirely emergent property of the model that is not at all apparent from looking at the energy function. The lesson is that anticipating convergence times is tricky. There has been work to try to design MCMC with fast convergence times (this is often referred to as *rapid mixing Markov chains*). However, the problem is difficult.

Cluster Algorithms

For some classes of problem such as the Ising model we can achieve rapid mixing by carefully choosing a cluster of variables and then flipping the whole cluster. This has to be done so as to maintain detailed balance. If the cluster is badly chosen the probability of accepting the change (i.e. flipping all the spins) would be extremely low. However, for some problems we can choose the cluster so the probability of flipping it is 1! That is, we always flip the cluster. These algorithms have become state of the art for simulating many spin systems as they rapidly decorrelate the samples.

The clustering algorithm discussed in this section is completely different to the clustering discussed in Example 9.7.

Example 11.6 Cluster Algorithm for the Ising Model
For the Ising model there is a particularly simple clustering MCMC algorithm due to Wolff (1989) building on the ideas of Swendsen and Wang (1987). The idea is to construct a cluster of connected spins with the same orientation. This is done using Algorithm 11.1. The list can be a stack – a simple data structure on which we just add (push) sites and remove (pop) sites.

Algorithm 11.1 Cluster algorithm for the Ising model.

```
choose a site , i, at random
set X := S_i
set S_i := −S_i
add i to a list of sites
do until list of sites is empty
  remove a site , j, from list of sites
  for all neighbours, k, of the site j
    if S_k = X and U < p = 1 − e^{−2β} where U ~ U(0,1)
      set S_k := −S_k
      add k to list of sites
```

We add neighbouring sites with a probability $p = 1 - e^{-2\beta}$ if the spins at the site are the same orientation as the rest of the cluster. Suppose we have built a cluster of spins with n_s edges to neighbours outside the cluster with the same spin and n_o edges to neighbours outside the cluster with the opposite spin configuration. Let $\boldsymbol{S}$ be the original spin configuration and $\boldsymbol{S}'$ be the spin configuration after flipping all the spins in the cluster. Now the probability, $p(\boldsymbol{S}'|\boldsymbol{S})$ of choosing the candidate configuration $\boldsymbol{S}'$ is proportional to $(1-p)^{n_s}$ (the number of edges the algorithm decided not to include in the

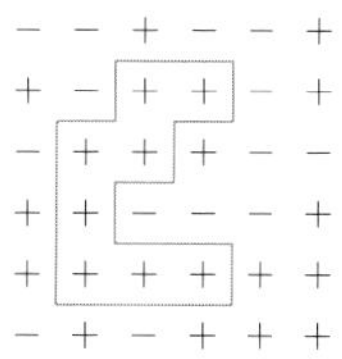

Cluster with $n_s = 8,\ n_o = 10$

cluster). If we built the cluster starting from $\boldsymbol{S}'$ then the probability of choosing a candidate configuration $\boldsymbol{S}$ would be proportional to $(1-p)^{n_o}$. Since $1-p = \mathrm{e}^{-2\beta}$,

$$\frac{p(\boldsymbol{S}'|\boldsymbol{S})}{p(\boldsymbol{S}|\boldsymbol{S}')} = \frac{(1-p)^{n_s}}{(1-p)^{n_o}} = \mathrm{e}^{-2\beta\,(n_s-n_o)}$$

(this, at least, looks plausible; you have to convince yourself that all the other terms that occur in the candidate probability cancel). The change in the Boltzmann probability when flipping the spins in the cluster is

$$\frac{\pi(\boldsymbol{S}')}{\pi(\boldsymbol{S})} = \mathrm{e}^{-\beta(E(\boldsymbol{S}')-E(\boldsymbol{S}))} = \mathrm{e}^{2\beta(n_o-n_s)}$$

since n_o edges are satisfied while n_s edges become unsatisfied. From the equation for detailed balance we require the acceptance ratio to be

$$r = \frac{a(\boldsymbol{S}'|\boldsymbol{S})}{a(\boldsymbol{S}|\boldsymbol{S}')} = \frac{p(\boldsymbol{S}|\boldsymbol{S}')\,\pi(\boldsymbol{S}')}{p(\boldsymbol{S}'|\boldsymbol{S})\,\pi(\boldsymbol{S})} = 1.$$

Thus by carefully choosing the candidate solution we end up always accepting the move. In Question 11.6 you are asked to simulate the Ising model using the clustering algorithm.

The great advantage of the clustering algorithm is that it changes many variables at each step so that the samples are much less correlated. This is particularly important close to criticality where traditional sampling methods are very slow. Interestingly, one feature of the clustering algorithm is that at low temperature it is quite capable of flipping most of the spins, thus transitioning between the states with positive and negative magnetisation. That is, it restores ergodicity. However, as we explained earlier, real systems suffer ergodicity breaking. Thus, our clustering algorithm does not model the dynamics observed in real systems. Fortunately, for the Ising model we can easily reinstate ergodicity breaking at low temperature by considering the absolute magnetisation.

Exact Sampling

Having stated how difficult it is to know whether a Markov chain has reached its stationary distribution, I now confess that it is possible to generate exact (perfect) random samples using a modified MCMC procedure. To understand the idea we start from an impractical algorithm and then work towards a practical algorithm. We consider running multiple MCMC simulations starting from all possible initial conditions, but using the same random number generator. If two runs ever reach exactly the same state they will remain the same from then on. This is known as coalescence. If all initial conditions coalesced then the resulting runs would not depend on the initial states. As a result the state would be an exact sample of the stationary distribution. Thus a simple algorithm would be to start the system off in every possible initial state and wait for all the states to coalesce. After this the system has reached its stationary state and we can obtain a perfectly random sample.

This algorithm is impractical since the number of initial conditions is huge. We can avoid the need to simulate every single initial condition by setting up the MCMC carefully so that it respects some (partial) ordering of the states. For example, the number of spins in state +1 in the Ising model provides a partial ordering of the states. If we use an update rule so that the partial ordering is maintained then we can be sure that all the initial states will coalesce provided that the maximum and minimum states coalesce. This provides a practical algorithm for drawing exact samples for a number of distributions including the Ising model. A more detailed review of exact sampling is given in MacKay (2003) or in the original paper by Propp and Wilson (1996).

Exact sampling is very elegant, but it remains a nice idea rather than a standard tool. It is too slow to be used regularly and is dominated by clustering algorithms. Both clustering and exact sampling are limited to particular types of problems – generally spin types. A much more general technique for speeding up MCMC is hybrid Monte Carlo

11.2.4 Hybrid Monte Carlo

One of the problems with MCMC is that it takes a considerable amount of time to obtain uncorrelated samples. If we increase the step size then neighbouring samples are less correlated, but this happens at the expense of having a higher rejection rate. Hybrid Monte Carlo is a technique developed in the late 1980s to overcome this. It is the method of choice in many applications involving continuous random variables. The idea grew out of simulations of molecular systems although the technique can be generalised to other systems. The dynamics of systems of molecules can be simulated using Newton's law of motion. These dynamics will move the system to an equally probable state quite different from where we started. (Newton's law ensures conservation of energy, which for a Boltzmann distribution means that the probability of the states before and after applying Newton's law is unchanged.) The idea behind hybrid Monte Carlo is to perform (inaccurate) molecular dynamics simulation for some time and then to use the resulting configuration of the molecules as a candidate sample in Monte Carlo. Using inaccurate dynamics insures that we visit states of different energy.

The idea can be generalised to problems involving continuous variables where there are no underlying dynamics. This is done by treating the random variables as if they were positions of particles. The log-probability distribution is then treated as an energy potential. These particles are given a mass and a momentum and then we simulate their 'dynamics' for some period of time. The position of the particles is used as a candidate which we accept or reject using the usual MCMC procedure. This is repeated with a new, randomly chosen, momentum to obtain the next candidate values for the random variable. The advantage of this over traditional MCMC is that the samples are much less correlated. The reversibility of classical mechanics ensures that detailed balance is satisfied.

To be more specific, supposing we wanted to draw a sample from the distribution $\pi(\boldsymbol{x})$, where $\boldsymbol{x} = (x_1, x_2, \ldots, x_n)$ is a high-dimensional random variable. We first define the potential $U(\boldsymbol{x})$ such that $\exp(-U(\boldsymbol{x})) \propto \pi(\boldsymbol{x})$. Next we define a Hamiltonian (potential plus kinetic energy)

$$H(\boldsymbol{x}, \boldsymbol{p}) = U(\boldsymbol{x}) + \sum_{i=1}^{n} \frac{p_i^2}{2\, m_i}$$

where $\boldsymbol{p} = (p_1, p_2, \ldots, p_n)$ is the fictitious momentum, with p_i being the momentum of variable x_i. Note that the distribution we are interested in sampling from is a marginal of $\exp(-H(\boldsymbol{x}, \boldsymbol{p}))$

$$\pi(\boldsymbol{x}) = \int \mathrm{e}^{-H(\boldsymbol{x},\boldsymbol{p})} \prod_{i=1}^{n} \frac{\mathrm{d}p_i}{\sqrt{2\pi\, m_i}}.$$

Rather than compute this integral we sample from it by choosing momenta $p_i \sim \mathcal{N}(0, m_i)$. We then simulate Newtonian dynamics. This is equivalent to updating the positions and momenta according to Hamilton's equations

$$\frac{\partial \boldsymbol{x}(t)}{\partial t} = \frac{\partial H(\boldsymbol{x}, \boldsymbol{p})}{\partial \boldsymbol{p}}, \qquad \frac{\partial \boldsymbol{p}(t)}{\partial t} = -\frac{\partial H(\boldsymbol{x}, \boldsymbol{p})}{\partial \boldsymbol{x}}.$$

If this was performed exactly it would preserve $H(\boldsymbol{x}(t), \boldsymbol{p}(t))$. However, the updates are performed numerically so the Hamiltonian is not exactly preserved. This, however, is not critical since the only purpose of following the dynamics is to obtain a new candidate point $\boldsymbol{x}$ which has a relatively high probability, $\pi(\boldsymbol{x})$.

Hybrid Monte Carlo (so-called because it is a hybrid between molecular dynamics and Monte Carlo) is sometimes referred to as Hamiltonian Monte Carlo because it is based on Hamiltonian dynamics. The technique was first proposed by Duane et al. (1987). Because of its huge importance hybrid Monte Carlo has seen a lot of recent advances. For example, rather than choosing the momenta independently they can be chosen from a multivariate normal distribution tailored to the problem so as to maximise the exploration. More details, including detailed numerical methods for simulating the dynamics, can be found in specialised books on MCMC.

11.3 Kalman and Particle Filtering

In Chapter 6 we showed that given a series of unbiased noisy measurements, X_i, we could obtain an estimate for the quantity of interest by taking the mean of the noisy measurements. Provided the measurements had a well-defined variance, the error in the estimate would reduce as $1/\sqrt{n}$, where n is the number of measurements. However, what happens when the quantity we are measuring is changing over time? Can we combine measurements taken at different times to get a superior estimate of the quantity of interest compared with the raw measurements? This is the aim of *filtering*. We are only going to get an improved estimate if we can model the dynamics of the quantity we are interested in reasonably accurately. This makes filtering challenging and often infeasible.

However, when it is feasible we can get improvements in our estimate that have very important practical applications. Thus the field of filtering is very well studied.

Filtering has many and very varied applications. As a result, its formulation is often rather abstract. We start by giving a concrete example. The details of the filter we use will be given later on.

Example 11.7 Tracking
Let us consider tracking an object through space. This might be a comet moving through the night sky or a tornado over the earth. We assume the trajectory is governed by Newton's laws and that the object experiences some weak unknown forces which slightly perturbs its motion from a dead straight line. We consider the position of the object at different times $t = 0, \Delta t, 2\Delta t, \ldots$ In Figure 11.6 we show the position of the object at these time intervals ($\times$) – the object is moving from left to right. We also show the noisy observations of its position (+). To estimate its position we use a Kalman filter (described below), which uses the noisy observations plus a model of the dynamics to obtain a more reliable prediction of the position of the object. We show the Kalman filter estimate by a dot and the error in the estimate by a circle. As we collect more data our estimate in the position improves (our estimator is usually more accurate than the noisy observation) and our error estimate shrinks.

Figure 11.6 An object ($\times$) moving through space. At each time interval we make a noisy observation of its position (+). We use a Kalman filter to estimate its position ($\odot$) where the size of the circle shows our estimated error in the position.

The example given shows filtering applied to tracking an object. This is a classic application. Kalman filters were used in the Apollo missions to the moon and are still used for tracking the international space station. However, Kalman filters are also used in many other applications. For example, it is the control mechanism used in phase-locked loops which is ubiquitous in FM radio. The purpose is to track the phase of a radio signal as it changes over time. The Kalman filter and its many extensions are widely used in many controllers. The somewhat confusing term *filtering* comes from applications in signal processing where the estimation procedure can be viewed as a way of reducing (filtering out) noise in the observations. Filtering is the term used when we try to predict the very next point. Sometimes we want to improve our estimation of the parameters over the whole signal. This is called *smoothing*. If we want to make predictions of the future the task is known as *prediction* (filtering is prediction immediately after making an observation).

One of the complications of filtering is that the underlying dynamical system governing the quantity we are interested in tracking is not generally fully known. The unknown part is modelled as independent noise. If this is a good model of the uncertainty and the modelling noise is small compared to the measurement noise, then filtering can be beneficial. In general we consider a system which evolves according to

$$\boldsymbol{X}(t+\Delta t) = \boldsymbol{a}(\boldsymbol{X}(t)) + \boldsymbol{\eta}(\boldsymbol{X}(t)).$$

where $\boldsymbol{\eta}(\boldsymbol{X}(t))$ models the noise. In addition to the dynamic equation for the variable we have a series of noisy observations that depend on $\boldsymbol{X}(t)$

$$\boldsymbol{Y}(t) = \boldsymbol{c}(\boldsymbol{X}(t)) + \boldsymbol{\epsilon}(\boldsymbol{X}(t)),$$

Our task is to make the best predictions we can for $\boldsymbol{X}(t)$. Finding the best prediction for $\boldsymbol{X}(t)$ can be extremely complicated, however, there is a case where things simplify substantially. That is, where the dynamics is linear so that the observations are a linear function of the quantities to be estimated, and the noise is normally distributed. In this case, the uncertainties can be modelled by a normal distribution. This only involves a relatively small number of variables (essentially the means and covariances). This problem was studied by Rudolf Kálmán in 1960.

11.3.1 The Kalman Filter

Although Kálmán received the glory, the development of the Kalman filter also involved Thorvald Thiele, Peter Swerling, Richard Bucy, Stanley Schmidt, and Ruslan Stratonovich. It was very much an idea whose time had come.

In the Kalman filter we assume that we have some set of states $\boldsymbol{X}(t)$ that update according to

$$\boldsymbol{X}(t+\Delta t) = \mathbf{A}(t)\,\boldsymbol{X}(t) + \boldsymbol{\eta}(t)$$

where $\mathbf{A}(t)$ is some matrix and $\boldsymbol{\eta}(t)$ is a normally distributed noise. In the two-dimensional tracking problem in Example 11.7, the random variable $\boldsymbol{X}(t)$ describes the position and velocity of the object being tracked and $\boldsymbol{\eta}(t)$ describes some uncertainties that enter due to unseen forces acting on the object. These uncertainties are assumed to be normally distributed with mean zero and covariance $\mathbf{Q}$, i.e.

$$\boldsymbol{\eta}(t) \sim \mathcal{N}(0, \mathbf{Q}(t)).$$

The second part of the Kalman filter is the set of noisy observations $\boldsymbol{Y}(t)$ where

$$\boldsymbol{Y}(t) = \mathbf{C}(t)\boldsymbol{X}(t) + \boldsymbol{\epsilon}(t)$$

$\mathbf{C}(t)$ is again some matrix and $\boldsymbol{\epsilon}(t)$ a normally distributed uncertainty

$$\boldsymbol{\epsilon}(t) \sim \mathcal{N}(0, \mathbf{R}(t)).$$

The noise $\boldsymbol{\eta}(t)$ and $\boldsymbol{\epsilon}(t)$ are assumed to be independent of each other (a fairly innocuous assumption), but also independent at different times steps (i.e. $\mathbb{E}\left[\eta_i(t)\,\eta_j(t')\right] = \mathbb{E}\left[\epsilon_i(t)\epsilon_j(t')\right] = 0$ when $t \neq t'$). However, this would be unlikely to be true if $\boldsymbol{\eta}(t)$ modelled some persistent unknown systematic error. If this is the case, then filtering is likely to give poor results. Note that we only require that our observations, $\boldsymbol{Y}(t)$, are some function of our states, $\boldsymbol{X}(t)$ – we don't need to observe all the state information. In the two-dimensional tracking example we might observe the position, but not the velocity. Nevertheless, we can use our series of observations to estimate both the position and velocity of the object. Note that filtering is only going to provide a significant pay-off when the typical errors in the model, $\mathbb{E}\left[|\boldsymbol{\eta}(t)|\right]$, are small compared to our observational errors, $\mathbb{E}\left[|\boldsymbol{\epsilon}(T)|\right]$.

Kalman filters are heavily used in control systems. Many control researchers seem unduly nervous of using probabilities, so often derive the Kalman filter without explicitly using probabilities.

There are a number of ways of deriving the Kalman filter. We give a probabilistic derivation. We assume that the initial uncertainty in the parameters, $\boldsymbol{X}(0)$, are normally distributed. Then, in the special case of a linear model with normal noise, the unknown state variable at all subsequent times, $\boldsymbol{X}(t)$, will be normally distributed, i.e. $\boldsymbol{X}(t) \sim \mathcal{N}(\boldsymbol{\mu}(t), \boldsymbol{\Sigma}(t))$. Using our updated equation we find that $\boldsymbol{X}(t+\Delta t) = \mathbf{A}(t)\,\boldsymbol{X}(t) + \boldsymbol{\eta}(t)$. Since $\boldsymbol{X}(t)$ is normally distributed so will the linear transform $\mathbf{A}(t)\,\boldsymbol{X}(t)$. Furthermore, as our uncertainty in $\boldsymbol{X}(t+\Delta t)$ is just the sum of two normally distributed terms, it will itself be normally distributed. The mean of the distribution is given by

$$\tilde{\boldsymbol{\mu}}(t+\Delta t) = \mathbb{E}\left[\boldsymbol{X}(t+\Delta t)\right] = \mathbf{A}(t)\,\mathbb{E}\left[\boldsymbol{X}(t)\right] = \mathbf{A}(t)\,\boldsymbol{\mu}(t)$$

where the expectation is on the probabilities of X given earlier observations. The covariance is

Recall that we use the notation $\mathbb{C}\text{ov}\left[\boldsymbol{X}\right]$ to denote the covariance matrix of a vector $\boldsymbol{X}$ with components $\mathbb{C}\text{ov}\left[X_i, X_j\right]$.

$$\begin{aligned}\tilde{\boldsymbol{\Sigma}}(t+\Delta t) &\overset{(1)}{=} \mathbb{C}\text{ov}\left[\mathbf{A}(t)\,\boldsymbol{X}(t) + \boldsymbol{\eta}(t)\right] \overset{(2)}{=} \mathbb{C}\text{ov}\left[\mathbf{A}(t)\,\boldsymbol{X}(t)\right] + \mathbb{C}\text{ov}\left[\boldsymbol{\eta}(t)\right] \\ &\overset{(3)}{=} \mathbf{A}(t)\,\boldsymbol{\Sigma}(t)\,\mathbf{A}^{\mathsf{T}}(t) + \mathbf{Q}(t)\end{aligned}$$

(1) Where we are measuring the covariance matrix for $\boldsymbol{X}(t+\Delta t) = \mathbf{A}(t)\,\boldsymbol{X}(t) + \boldsymbol{\eta}(t)$.
(2) Since $\mathbf{A}(t)\,\boldsymbol{X}(t)$ is independent of $\boldsymbol{\eta}(t)$.
(3) Since $\mathbf{A}(t)$ is constant

$$\begin{aligned}\mathbb{C}\text{ov}\left[\mathbf{A}\boldsymbol{X}\right] &= \mathbb{E}\left[(\mathbf{A}\boldsymbol{X})(\mathbf{A}\boldsymbol{X})^{\mathsf{T}}\right] - \mathbb{E}\left[\mathbf{A}\boldsymbol{X}\right]\,\mathbb{E}\left[\mathbf{A}\boldsymbol{X}\right]^{\mathsf{T}} \\ &= \mathbf{A}\left(\mathbb{E}\left[\boldsymbol{X}\boldsymbol{X}^{\mathsf{T}}\right] - \mathbb{E}\left[\boldsymbol{X}\right]\mathbb{E}\left[\boldsymbol{X}^{\mathsf{T}}\right]\right)\mathbf{A}^{\mathsf{T}} \\ &= \mathbf{A}\,\mathbb{C}\text{ov}\left[\boldsymbol{X}\right]\,\mathbf{A}^{\mathsf{T}} = \mathbf{A}\,\boldsymbol{\Sigma}\,\mathbf{A}^{\mathsf{T}}\end{aligned}$$

and $\mathbb{C}\text{ov}\left[\boldsymbol{\eta}(t)\right] = \mathbf{Q}(t)$ by definition of the model.

The probability density for $\boldsymbol{X}(t+\Delta t)$ given observation $\boldsymbol{Y}(t+\Delta t)$ is given by Bayes' rule as

$$\begin{aligned}f(\boldsymbol{X}(t+\Delta t) = \boldsymbol{x}|\boldsymbol{Y}(t+\Delta t) = \boldsymbol{y}) \\ = \frac{f(\boldsymbol{Y}(t+\Delta t) = \boldsymbol{y}|\boldsymbol{X}(t+\Delta t) = \boldsymbol{x})\,f(\boldsymbol{X}(t+\Delta t) = \boldsymbol{x})}{f(\boldsymbol{Y}(t+\Delta t) = \boldsymbol{y})}.\end{aligned}$$

Since $\boldsymbol{Y}(t) = \mathbf{C}(t)\boldsymbol{X}(t) + \boldsymbol{\epsilon}(t)$ with $\boldsymbol{\epsilon}(t) \sim \mathcal{N}(0, \mathbf{R}(t))$, our likelihood is given by

$$f(\boldsymbol{Y}(t+\Delta t) = \boldsymbol{y}|\boldsymbol{X}(t+\Delta t) = \boldsymbol{x}) = \mathcal{N}(\boldsymbol{y}|\mathbf{C}(t+\Delta t)\boldsymbol{x}, \mathbf{R}(t)).$$

We have already found our prior $f(\boldsymbol{X}(t+\Delta t) = \boldsymbol{x}) = \mathcal{N}(\boldsymbol{x}|\tilde{\boldsymbol{\mu}}(t+\Delta t), \tilde{\boldsymbol{\Sigma}}(t+\Delta t))$. The denominator $f(\boldsymbol{Y}(t+\Delta t) = \boldsymbol{y})$ is a normalisation factor which can be computed at the end. Combining the likelihood and prior we obtain a posterior

$$
\begin{aligned}
f(\boldsymbol{x}, t+\Delta t) &= f(\boldsymbol{X}(t+\Delta t) = \boldsymbol{x} | \boldsymbol{Y}(t+\Delta t) = \boldsymbol{y}) \\
&\propto \mathrm{e}^{-\frac{1}{2}(\boldsymbol{y}-\mathbf{C}\boldsymbol{x})^{\mathsf{T}}\mathbf{R}^{-1}(\boldsymbol{y}-\mathbf{C}\boldsymbol{x})-\frac{1}{2}(\boldsymbol{x}-\tilde{\boldsymbol{\mu}})^{\mathsf{T}}\tilde{\boldsymbol{\Sigma}}^{-1}(\boldsymbol{x}-\tilde{\boldsymbol{\mu}})}
\end{aligned}
$$

where we have dropped the index $t+\Delta t$ to make the equation readable. Collecting terms proportional to x_i^2 and x_i we get

$$
f(\boldsymbol{x}, t+\Delta t) \propto \mathrm{e}^{-\frac{1}{2}\boldsymbol{x}^{\mathsf{T}}(\mathbf{C}^{\mathsf{T}}\mathbf{R}^{-1}\mathbf{C}+\tilde{\boldsymbol{\Sigma}}^{-1})\boldsymbol{x}+\boldsymbol{x}^{\mathsf{T}}(\mathbf{C}^{\mathsf{T}}\mathbf{R}^{-1}\boldsymbol{y}+\tilde{\boldsymbol{\Sigma}}^{-1}\tilde{\boldsymbol{\mu}})}.
$$

We we have used $\boldsymbol{a}^{\mathsf{T}}\boldsymbol{x} = \boldsymbol{x}^{\mathsf{T}}\boldsymbol{a}$.

Finally completing the square we find

$$
f(\boldsymbol{x}, t+\Delta t) \propto \mathrm{e}^{-\frac{1}{2}(\boldsymbol{x}-\boldsymbol{\mu}(t+\Delta t))^{\mathsf{T}}\boldsymbol{\Sigma}^{-1}(t+\Delta t)(\boldsymbol{x}-\boldsymbol{\mu}(t+\Delta t))}
$$

with

$$
\begin{aligned}
\boldsymbol{\mu}(t+\Delta t) &= (\mathbf{C}^{\mathsf{T}}\mathbf{R}^{-1}\mathbf{C}+\tilde{\boldsymbol{\Sigma}}^{-1})^{-1}(\mathbf{C}^{\mathsf{T}}\mathbf{R}^{-1}\boldsymbol{y}+\tilde{\boldsymbol{\Sigma}}^{-1}\tilde{\boldsymbol{\mu}}) \\
\boldsymbol{\Sigma}(t+\Delta t) &= (\mathbf{C}^{\mathsf{T}}\mathbf{R}^{-1}\mathbf{C}+\tilde{\boldsymbol{\Sigma}}^{-1})^{-1}.
\end{aligned}
$$

Clearly, $f(\boldsymbol{x}, t+\Delta t)$ has the form of a normal distribution $\mathcal{N}(\boldsymbol{x}|\boldsymbol{\mu}(t+\Delta t), \boldsymbol{\Sigma}(t+\Delta t))$, thus the normalisation terms are those of a normal distribution.

These updated equations are not in their most convenient form. To obtain the standard form we use the Woodbury matrix identity (note that there are many different ways to express this identity – see Equation 5.5 on page 101). We use

$$
\left(\mathbf{C}^{\mathsf{T}}\mathbf{R}^{-1}\mathbf{C}+\tilde{\boldsymbol{\Sigma}}^{-1}\right)^{-1} = \tilde{\boldsymbol{\Sigma}} - \tilde{\boldsymbol{\Sigma}}\mathbf{C}^{\mathsf{T}}\left(\mathbf{R}+\mathbf{C}\tilde{\boldsymbol{\Sigma}}\mathbf{C}^{\mathsf{T}}\right)^{-1}\mathbf{C}\tilde{\boldsymbol{\Sigma}}. \tag{11.6}
$$

Using this identity we can write the updated equations (after some algebra) as

$$
\boldsymbol{\mu}(t+\Delta t) = \tilde{\boldsymbol{\mu}} + \mathbf{K}(\boldsymbol{y} - \mathbf{C}\tilde{\boldsymbol{\mu}}) \tag{11.7}
$$

$$
\boldsymbol{\Sigma}(t+\Delta t) = (\mathbf{I} - \mathbf{K}\mathbf{C})\tilde{\boldsymbol{\Sigma}} \tag{11.8}
$$

where $\mathbf{K}$ is known as the *optimal Kalman gain matrix* defined as

$$
\mathbf{K} = \tilde{\boldsymbol{\Sigma}}\mathbf{C}^{\mathsf{T}}\mathbf{S}^{-1}, \qquad \mathbf{S} = \mathbf{R} + \mathbf{C}\tilde{\boldsymbol{\Sigma}}\mathbf{C}^{\mathsf{T}}.
$$

$\mathbf{S}$ is known as the *innovation* or *residual* matrix. The advantage of this standard form is that it only requires taking the inverse of the *innovation* matrix $\mathbf{S}$. This leads to a computationally cheaper algorithm than if we had used the updated equations we initially obtained. The equivalence between the standard form and the result we derived is not obvious. For the sceptical I show this equivalence in the following page, although you will need pen, paper, and patience to verify all the steps. To show that the covariances are equivalent we substitute for $\mathbf{K}$ and $\mathbf{S}$ to obtain

$$
\begin{aligned}
\boldsymbol{\Sigma}(t+\Delta t) &\overset{(1)}{=} \tilde{\boldsymbol{\Sigma}} - \mathbf{K}\mathbf{C}\tilde{\boldsymbol{\Sigma}} \\
&\overset{(2)}{=} \tilde{\boldsymbol{\Sigma}} - \tilde{\boldsymbol{\Sigma}}\mathbf{C}^{\mathsf{T}}\left(\mathbf{C}\tilde{\boldsymbol{\Sigma}}\mathbf{C}^{\mathsf{T}}+\mathbf{R}\right)^{-1}\mathbf{C}\tilde{\boldsymbol{\Sigma}}
\end{aligned}
$$

(1) From Equation (11.8).
(2) Using the definition of the Kalman gain matrix, $\mathbf{K}$, and the innovation matrix, $\mathbf{S}$.

This then follows from the Woodbury formula, Equation (11.6). To prove the Woodbury formula we show that if we multiply the right-hand side, $\mathbf{R}$, of Equation (11.6) by the inverse of the left-hand side, $\mathbf{L}$, we obtain the identity

$$\begin{aligned}
\mathbf{L}^{-1}\mathbf{R} &\overset{(1)}{=} \left(\mathbf{C}^{\mathsf{T}}\mathbf{R}^{-1}\mathbf{C} + \tilde{\boldsymbol{\Sigma}}^{-1}\right)\left(\tilde{\boldsymbol{\Sigma}} - \tilde{\boldsymbol{\Sigma}}\mathbf{C}^{\mathsf{T}}\left(\mathbf{C}\tilde{\boldsymbol{\Sigma}}\mathbf{C}^{\mathsf{T}} + \mathbf{R}\right)^{-1}\mathbf{C}\tilde{\boldsymbol{\Sigma}}\right) \\
&\overset{(2)}{=} \mathbf{C}^{\mathsf{T}}\mathbf{R}^{-1}\mathbf{C}\tilde{\boldsymbol{\Sigma}} - \mathbf{C}^{\mathsf{T}}\mathbf{R}^{-1}\mathbf{C}\tilde{\boldsymbol{\Sigma}}\mathbf{C}^{\mathsf{T}}\left(\mathbf{C}\tilde{\boldsymbol{\Sigma}}\mathbf{C}^{\mathsf{T}} + \mathbf{R}\right)^{-1} \\
&\quad \mathbf{C}\tilde{\boldsymbol{\Sigma}} + \mathbf{I} - \mathbf{C}^{\mathsf{T}}\left(\mathbf{C}\tilde{\boldsymbol{\Sigma}}\mathbf{C}^{\mathsf{T}} + \mathbf{R}\right)^{-1}\mathbf{C}\tilde{\boldsymbol{\Sigma}} \\
&\overset{(3)}{=} \mathbf{I} + \mathbf{C}^{\mathsf{T}}\mathbf{R}^{-1}\mathbf{C}\tilde{\boldsymbol{\Sigma}} - \mathbf{C}^{\mathsf{T}}\mathbf{R}^{-1}\left(\mathbf{C}\tilde{\boldsymbol{\Sigma}}\mathbf{C}^{\mathsf{T}} + \mathbf{R}\right)\left(\mathbf{C}\tilde{\boldsymbol{\Sigma}}\mathbf{C}^{\mathsf{T}} + \mathbf{R}\right)^{-1}\mathbf{C}\tilde{\boldsymbol{\Sigma}} = \mathbf{I}
\end{aligned}$$

(1) From Equation (11.6) where the left-hand side is $\mathbf{L} = \left(\mathbf{C}^{\mathsf{T}}\mathbf{R}^{-1}\mathbf{C} + \tilde{\boldsymbol{\Sigma}}^{-1}\right)^{-1}$ and the right-hand side is $\mathbf{R} = \left(\tilde{\boldsymbol{\Sigma}} - \tilde{\boldsymbol{\Sigma}}\mathbf{C}^{\mathsf{T}}\left(\mathbf{C}\tilde{\boldsymbol{\Sigma}}\mathbf{C}^{\mathsf{T}} + \mathbf{R}\right)^{-1}\mathbf{C}\tilde{\boldsymbol{\Sigma}}\right)$.
(2) Multiplying out the terms.
(3) Rearranging and factoring terms involving $\left(\mathbf{C}\tilde{\boldsymbol{\Sigma}}\mathbf{C}^{\mathsf{T}} + \mathbf{R}\right)^{-1}$.
(4) The factored terms simplify, leading to a cancellation.

To show that the updated equations for $\boldsymbol{\mu}(t + \Delta t)$ are equivalent we note that we have just shown $\boldsymbol{\Sigma}(t + \Delta t) = (\mathbf{C}^{\mathsf{T}}\mathbf{R}^{-1}\mathbf{C} + \tilde{\boldsymbol{\Sigma}}^{-1})^{-1} = (\mathbf{I} - \mathbf{K}\mathbf{C})\tilde{\boldsymbol{\Sigma}}$, thus our updated equation is

$$\begin{aligned}
\boldsymbol{\mu}(t + \Delta t) &= \left(\mathbf{C}^{\mathsf{T}}\mathbf{R}^{-1}\mathbf{C} + \tilde{\boldsymbol{\Sigma}}^{-1}\right)^{-1}\left(\mathbf{C}^{\mathsf{T}}\mathbf{R}^{-1}\boldsymbol{y} + \tilde{\boldsymbol{\Sigma}}^{-1}\tilde{\boldsymbol{\mu}}\right) \\
&= (\mathbf{I} - \mathbf{K}\mathbf{C})\,\tilde{\boldsymbol{\Sigma}}\left(\mathbf{C}^{\mathsf{T}}\mathbf{R}^{-1}\boldsymbol{y} + \tilde{\boldsymbol{\Sigma}}^{-1}\tilde{\boldsymbol{\mu}}\right) \\
&= \tilde{\boldsymbol{\Sigma}}\mathbf{C}^{\mathsf{T}}\mathbf{R}^{-1}\boldsymbol{y} - \mathbf{K}\mathbf{C}\tilde{\boldsymbol{\Sigma}}\mathbf{C}^{\mathsf{T}}\mathbf{R}^{-1}\boldsymbol{y} + \tilde{\boldsymbol{\mu}} - \mathbf{K}\mathbf{C}\tilde{\boldsymbol{\mu}}.
\end{aligned}$$

But by definition $\mathbf{K} = \tilde{\boldsymbol{\Sigma}}\mathbf{C}^{\mathsf{T}}\mathbf{S}^{-1}$ or $\tilde{\boldsymbol{\Sigma}}\mathbf{C}^{\mathsf{T}} = \mathbf{K}\mathbf{S} = \mathbf{K}(\mathbf{C}\tilde{\boldsymbol{\Sigma}}\mathbf{C}^{\mathsf{T}} + \mathbf{R})$. Thus, the first two terms above are equal to

$$\tilde{\boldsymbol{\Sigma}}\mathbf{C}^{\mathsf{T}}\mathbf{R}^{-1}\boldsymbol{y} - \mathbf{K}\mathbf{C}\tilde{\boldsymbol{\Sigma}}\mathbf{C}^{\mathsf{T}}\mathbf{R}^{-1}\boldsymbol{y} = \mathbf{K}(\mathbf{C}\tilde{\boldsymbol{\Sigma}}\mathbf{C}^{\mathsf{T}} + \mathbf{R})\mathbf{R}^{-1}\boldsymbol{y} - \mathbf{K}\mathbf{C}\tilde{\boldsymbol{\Sigma}}\mathbf{C}^{\mathsf{T}}\mathbf{R}^{-1}\boldsymbol{y} = \mathbf{K}\boldsymbol{y}$$

so that

$$\boldsymbol{\mu}(t + \Delta t) = \tilde{\boldsymbol{\mu}} + \mathbf{K}(\boldsymbol{y} - \mathbf{C}\tilde{\boldsymbol{\mu}})$$

as advertised above.

Provided our initial estimate was normally distributed (i.e. $\boldsymbol{X}(0) \sim \mathcal{N}(\boldsymbol{\mu}(0), \boldsymbol{\Sigma}(0))$) then all remaining distributions are exact. Thus, the Kalman filter provides the true posterior probabilistic estimate for $\boldsymbol{X}(t)$. The underlying process is Markovian because the mean and covariance at each time step fully describe the uncertainty of the system. Thus the entire knowledge required to estimate the prior probability $f(\boldsymbol{x}, t + \Delta t)$ is contained in our posterior distribution at time t.

Knowledge of the observations made at previous times are irrelevant. Thus, the Kalman filter provides the best possible estimate of the state assuming that the underlying model is correct.

Our derivation of the Kalman filter looked intimidating, but all we did was take our current distribution $f(\boldsymbol{X}(t) = \boldsymbol{x}) = \mathcal{N}(\boldsymbol{x}|\boldsymbol{\mu}(t), \boldsymbol{\Sigma}(t))$ describing our uncertainty in the variables of interest, $\boldsymbol{X}(t)$, and update it according to our linear updated equation to obtain a prior for $\boldsymbol{X}(t+1)$. We then used the observation $\boldsymbol{Y}(t+1)$ to obtain a posterior distribution for $\boldsymbol{X}(t+1)$. The complications arose, as they often do when working with multivariate normal distribution, because the updated equation involves the inverse of the covariance matrix. Don't be put off using Kalman filters, they are extremely easy to implement in any computer language that allows you to invert matrices easily. Let us now return to our tracking problem.

Example 11.8 Tracking (Continuation of Example 11.7)
We consider tracking a two-dimensional object. We assume that the state of the system is given by its position and velocity. The velocity undergoes small perturbations due to hidden forces. The state updated equation is thus

$$\boldsymbol{x}(t+\Delta t) = \mathbf{A}\boldsymbol{x}(t) + \boldsymbol{\eta}$$

$$\begin{pmatrix} x_1(t+\Delta t) \\ x_2(t+\Delta t) \\ v_1(t+\Delta t) \\ v_2(t+\Delta t) \end{pmatrix} = \begin{pmatrix} 1 & 0 & \Delta t & 0 \\ 0 & 1 & 0 & \Delta t \\ 0 & 0 & 1 & 0 \\ 0 & 0 & 0 & 1 \end{pmatrix} \begin{pmatrix} x_1(t) \\ x_2(t) \\ v_1(t) \\ v_2(t) \end{pmatrix} + \begin{pmatrix} 0 \\ 0 \\ \eta_3(t) \\ \eta_4(t) \end{pmatrix}.$$

That is,

$$x_i(t+\Delta t) = x_i(t) + v_i(t)\,\Delta t, \qquad v_i(t+\Delta t) = v_i(t) + \eta_{i+1}(t).$$

Here we assume that the noise only affects the velocity. Clearly a perturbation in the velocity will also produce a small change in the position. This could be modelled and needs to be modelled if Δt is large. However, to keep the model simple I'm ignoring this. We assume our noise is uncorrelated

$$\mathbb{C}\text{ov}\left[\eta_3(t), \eta_4(t)\right] = \begin{pmatrix} 0.0001 & 0 \\ 0 & 0.0001 \end{pmatrix}.$$

We also assume that we making noisy observations of the position of our object. For simplicity we assume that these observations are the positions plus some noise

$$\boldsymbol{y}(t) = \mathbf{C}\boldsymbol{x}(t) + \boldsymbol{\epsilon}(t)$$

$$\begin{pmatrix} y_1(t) \\ y_2(t) \end{pmatrix} = \begin{pmatrix} 1 & 0 & 0 & 0 \\ 0 & 1 & 0 & 0 \end{pmatrix} \begin{pmatrix} x_1(t) \\ x_2(t) \\ v_1(t) \\ v_2(t) \end{pmatrix} + \begin{pmatrix} \epsilon_1(t) \\ \epsilon_2(t) \end{pmatrix}$$

where the noise is given by

$$\mathbb{C}\text{ov}\left[\epsilon_1(t), \epsilon_2(t)\right] = \begin{pmatrix} 0.16 & 0 \\ 0 & 0.16 \end{pmatrix}.$$

We have simulated the model and used a Kalman filter to predict its position and velocity. Figure 11.6 on page 331 shows the first few observations and the estimate in the position of the object produced by the Kalman filter. Figure 11.7 shows the root mean squared (RMS) error in the estimated position of the object as it evolves over time.

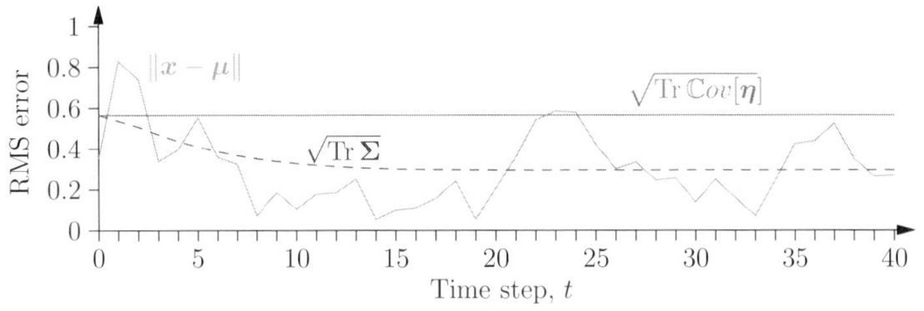

Figure 11.7 The RMS error in the estimated position of an object is plotted against time. The jagged curve shows the actual error, the dashed curve shows the Kalman filter's estimation of its error, while the horizontal line shows the typical errors in the observations.

We see for the example above that we rapidly achieve an error of around half the observed error. The size of the final error depends on the modelling error. If the modelling error $\mathbb{C}\text{ov}\left[\boldsymbol{\eta}\right] = \mathbf{Q}$ is small then the Kalman filter can provide a substantial improvement in the predicted position compared to the observations. If $\mathbf{Q}$ is around the same size as the observation error $\mathbb{C}\text{ov}\left[\boldsymbol{\epsilon}\right] = \mathbf{R}$, then the Kalman filter, or, in fact, any filter is unlikely to be of much use. Filtering relies on having a model giving accurate predictions of the next state – usually the error in the prediction will be proportional to the time since the last measurement, Δt, thus Kalman filters are very useful in instances where there are frequent, but noisy, measurements.

To be able to use the Kalman filter we require that our errors are normally distributed and that the update is linear. In this way, we can always treat our uncertainty as normally distributed. If our errors are non-normal or our model non-linear then we can no longer model our uncertainty as a normal distribution. If our updated equations are not too non-linear, then we can linearise our equations

$$\begin{aligned} \boldsymbol{X}(t+\Delta t) &= \boldsymbol{a}(\boldsymbol{X}(t)) + \boldsymbol{\eta}(\boldsymbol{X}(t)) \approx \boldsymbol{a}(\boldsymbol{\mu}(t)) + \left(\boldsymbol{X}(t) - \boldsymbol{\mu}(t)\right)^{\mathsf{T}} \boldsymbol{\nabla} \boldsymbol{a}(\boldsymbol{X}(t)) + \boldsymbol{\eta}(\boldsymbol{X}(t)) \\ \boldsymbol{Y}(t) &= \boldsymbol{c}(\boldsymbol{X}(t)) + \boldsymbol{\epsilon}(\boldsymbol{X}(t)) \approx \boldsymbol{c}(\boldsymbol{\mu}(t)) + \left(\boldsymbol{X}(t) - \boldsymbol{\mu}(t)\right)^{\mathsf{T}} \boldsymbol{\nabla} \boldsymbol{c}(\boldsymbol{X}(t)) + \boldsymbol{\epsilon}(\boldsymbol{X}(t)) \end{aligned}$$

this leads to the so-called *extended Kalman filter*. Extended Kalman filters can be quite successful, but if the non-linearities are too strong, or the model itself is inaccurate, then the prediction of the extended Kalman filter can be worse than the noisy observation. Probability theory doesn't provide us with any richer sets of distributions that remain closed under non-linear operations, thus if the Kalman filter fails we have to give up on trying to find a closed form solution to filtering. However, we still have Monte Carlo simulation to fall back on.

11.3.2 Particle Filtering

Sometimes we are given a noisy time series where the underlying dynamic model is Markovian but non-linear, or the noise is far from being normal. Our task might be to make a prediction at the next time step (discrete filtering) or we might want to remove noise from the signal (discrete smoothing). In this case we can perform *sequential Monte Carlo*, also known as *particle filtering* (sometimes known as the condensation algorithm). That is, we try to estimate the parameters of our model using the noisy observations, updating our distribution by characterising the uncertainty at each step.

Example 11.9 Repairing Clipped Signal

Suppose we have an old recording where we know that in addition to random noise in the signal, the signal also suffers by being clipped when the amplitude becomes too large. We illustrate this in Figure 11.8. A simple model of the true signal, $x(1), x(2), \ldots$ would be

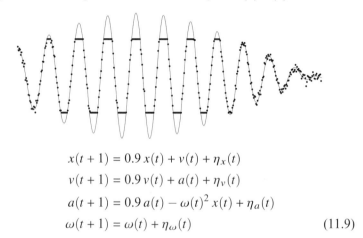

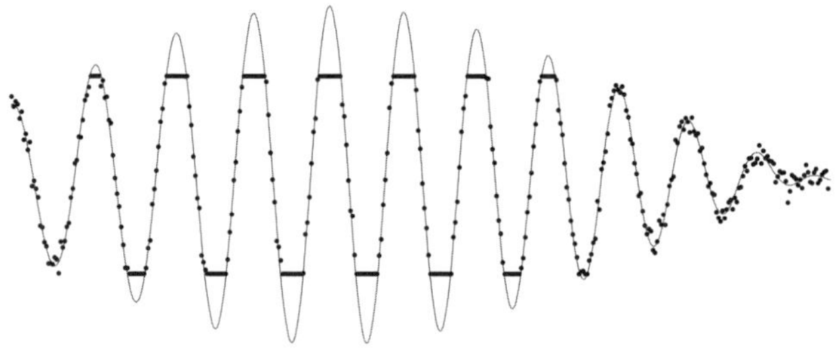

Figure 11.8 Example of a signal suffering non-linear distortions. The black dots show the distorted signal while the line shows the original signal.

This is a discretisation of the equation for simple harmonic motion

$$\frac{d^2 x}{dt^2} = -\omega^2 x$$

with added noise.

$$\begin{aligned} x(t+1) &= 0.9\,x(t) + v(t) + \eta_x(t) \\ v(t+1) &= 0.9\,v(t) + a(t) + \eta_v(t) \\ a(t+1) &= 0.9\,a(t) - \omega(t)^2\,x(t) + \eta_a(t) \\ \omega(t+1) &= \omega(t) + \eta_\omega(t) \end{aligned} \tag{11.9}$$

where the 'velocity', $v(t)$, 'acceleration', $a(t)$, and 'angular frequency', $\omega(t)$ are introduced to model the dynamics. The 'fudge'-factors 0.9 are introduced so that we get approximate simple harmonic motion – this isn't obvious but is a consequence of discretising a continuous time model. The errors, $\eta_\star(t)$, are included to model the relatively slow change in frequency and amplitude of the signal. The observed (noisy) signal, $\mathcal{D} = (y(1), y(2), \ldots, y(T))$, is related to the true signal by

$$y(t) = \text{clip}(x(t) + \epsilon(t)) \tag{11.10}$$

where $\epsilon(t) \sim \mathcal{N}(0, \mu)$ and

$$\text{clip}(z) = \begin{cases} \theta & \text{if } z > \theta \\ z & \text{if } -\theta \leq z \leq \theta \\ -\theta & \text{if } z < -\theta. \end{cases}$$

The signal restoration problem is a discrete smoothing problem, rather than a pure filtering problem. One solution to the smoothing problem is just to do filtering starting from the beginning of the signal and continue until you reach the end. This, however, only uses the Markov property moving forward in time. We can do better by running filtering backwards as well as forwards. We return to smoothing after discussing filtering.

The idea of particle filtering is to use Monte Carlo methods to evaluate all expectations we might be interested in (for example, the posterior mean of the signal and its variance). The idea is very simple. We use a population of particles which we evolve in time according to the stochastic dynamical equations. We weight the particles according to the likelihood that the observed data is generated by that particle. The expectations are then just weighted averages over the population of particles. Unfortunately as time increases many of the particles are likely to diverge from the actual process we are trying to model. We see this in the fact the likelihood for many particles will become very small. To overcome this, every so often we resample our population from the existing population according to their weights. This ensures that the population is concentrated around the true trajectory. Although it reduces the diversity of the population since the most likely particles in the old population will be selected many times in the new population, the population will quickly diverge again because of the stochastic nature of the dynamics. We discuss the details of this in the rest of this section.

Importance sampling. Particle filtering uses a type of Monte Carlo called *importance sampling*. This is rather like rejection sampling (see Section 3.3.2) where we generate samples, X_i, from a proposal distribution $q(x)$. The proposal distribution should have a finite probability of drawing a sample from any interval where there is a non-zero probability of drawing a sample from the distribution of interest $f(x)$. In rejection sampling we would obtain a deviate from f by accepting the sample $X_i = x$ with a probability $c\, f(x)/q(x)$ and otherwise rejecting the sample. The constant c has to be chosen so that $c\, f(x) \geq q(x)$ for all x. One drawback of this method is that it requires us to know $f(x)$ and $q(x)$ and in particular the normalisation constant (as we have already seen we often don't know this normalisation). In importance sampling we keep all the samples drawn from $q(x)$, but now we weight them by

$$w_i = \frac{f(X_i)}{q(X_i)}.$$

To compute an expectation of some function, $A(X)$, of our random variable, X, we use

$$\mathbb{E}\left[A(X)\right] \approx \sum_{i=1}^{n} \frac{w_i}{Z}\, A(X_i), \qquad Z = \sum_{j=1}^{n} w_j$$

where n is the number of samples. We see that the expectation is estimated by taking the ratio of two quantities that depend on the weights. Thus, we only need to know the weights up to a multiplicative factor. In particular we don't need to know the normalisation of $f(X)$.

Example 11.10 Importance Sampling of a Beta Distribution
Suppose we wanted to compute expectations over a beta distribution

$$\mathrm{Bet}(x|a,b) \propto x^{a-1}\,(1-x)^{b-1}$$

but we don't know the normalisation factor. One way to approximate such an expectation is to draw a number of samples uniformly at random $(U_1, U_2, \ldots, U_n)$ and then assign a weight

$$w_i = U_i^{a-1}\,(1-U_i)^{b-1}$$

to each deviate (note in this example the proposal distribution is $q(U) = \mathcal{U}(U|0,1)$ so that $q(U_i) = 1$). We show 20 random deviates and their weights together with the functions $\mathrm{Bet}(x|a,b)$ in Figure 11.9 for $(a,b) = (3,4)$ and $(a,b) = (30,40)$.

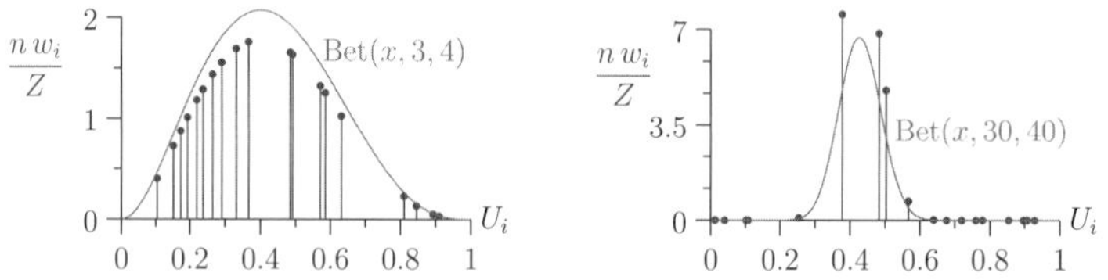

Figure 11.9 Examples of 20 random deviates and their weights for the case where $(a,b) = (3,4)$ and where $(a,b) = (30,40)$. Also shown are the beta distribution $\mathrm{Bet}(x|a,b)$. Note that because our samples don't perfectly capture the statistics of the distribution, the weights are not properly normalised.

To illustrate the accuracy of importance sampling we use it to estimate the mean, μ, and variance, σ^2, of the beta distribution. We show errors between these estimates and the true value for n equal 1 to 100 samples in Figure 11.10.

We note that the accuracy of estimate depends on how closely the proposal distribution (in our case a uniform distribution) is to the true distribution. Using a million deviates (which took around a second to generate) we computed the mean and variance correct to five decimal places for $\mathrm{Bet}(x|3,4)$ and four decimal places for $\mathrm{Bet}(x|30,40)$.

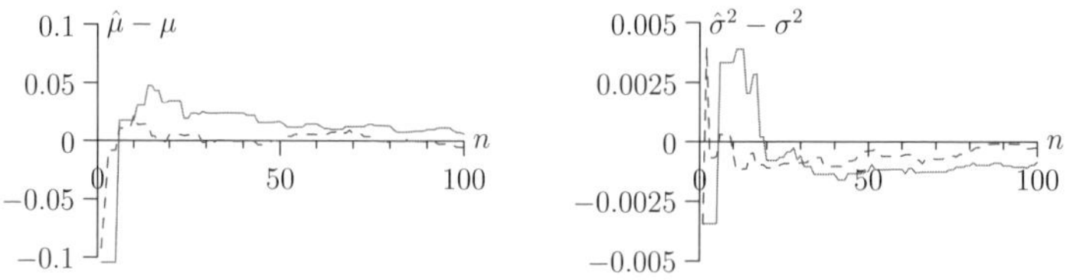

Figure 11.10 Errors in estimating the mean (left graph) and variance (right graph) of $\mathrm{Bet}(x|3,4)$ (dashed curve) and $\mathrm{Bet}(x|30,40)$ (continuous curve) using importance sampling plotted against sample size, n.

The accuracy of the estimate depends on how typical the samples are. That is, we want the weights $w_i = f(X_i)/q(X_i) \approx 1$. If $f(X_i)$ is large where $q(X_i)$ is small then the estimate is likely to be inaccurate, unless we have a very large sample.

Let us now return to our particle filtering problem where we want to predict the next value of $\boldsymbol{x}(t)$ given our observations $\boldsymbol{y}(1), \boldsymbol{y}(2), \ldots, \boldsymbol{y}(t)$. We cannot use a Kalman filter as our noise is highly non-Gaussian. Instead we use importance sampling. Let $\boldsymbol{x}(t) = (x(t), v(t), a(t), \omega(t))$ represent the state of the system (see Example 11.9 on page 338). We generate n samples (particles) from an initial distribution $f(\boldsymbol{X}_i(1) = \boldsymbol{x}) = f_1(\boldsymbol{x})$, encoding our initial uncertainty in the state. We can propagate these samples forward in time using the updated Equation (11.9), where we generate the noise $\eta_i(t)$ from a distribution chosen to capture the uncertainties in the model (e.g. $\mathcal{N}(0, \sigma_i^2)$ for appropriately chosen σ_i^2). We are left with n sample trajectories $\mathcal{T}_i = (\boldsymbol{X}_i(1), \boldsymbol{X}_i(2), \ldots, \boldsymbol{X}_i(T))$, drawn from the proposal distribution

$$q(\mathcal{T}_i) = f_1\left(\boldsymbol{X}_i(1)\right) \prod_{i=2}^{T} f\left(\boldsymbol{X}_i(t) | \boldsymbol{X}_i(t-1)\right)$$

where $f\left(\boldsymbol{X}_i(t) | \boldsymbol{X}_i(t-1)\right)$ is the probability of reaching $\boldsymbol{X}_i(t)$ from $\boldsymbol{X}_i(t-1)$ using the updated Equation (11.9). As with Kalman filter we are interested in computing the posterior probability of our sample given the observations $\mathcal{D}$,

$$f(\mathcal{T}_i | \mathcal{D}) = \frac{f(\mathcal{D} | \mathcal{T}_i)\, f(\mathcal{T}_i)}{f(\mathcal{D})}$$

where the prior $f(\mathcal{T}_i)$ is by construction equal to the proposal distribution $q(\mathcal{T}_i)$, and the likelihood of our data (in the example above the original noisy signal) is

$$f(\mathcal{D} | \mathcal{T}_i) = \prod_{i=1}^{T} f(\boldsymbol{y}(t) | \boldsymbol{X}_i(t)).$$

The probability of the data $f(\mathcal{D})$ (the evidence term) is a normalisation term that we don't know, but as it is the same for all trajectories $\mathcal{T}_i$ we don't need it to perform importance sampling (it would just cancel out when we normalised the weights). To compute the importance weights we use

$$w_i \propto \frac{f(\mathcal{D}, \mathcal{T}_i)}{q(\mathcal{T}_i)} \propto \frac{f(\mathcal{D} | \mathcal{T}_i)\, f(\mathcal{T}_i)}{q(\mathcal{T}_i)} = f(\mathcal{D} | \mathcal{T}_i)$$

since the probability of generating our sample trajectories, $\mathcal{T}_i$, is equal to $q(\mathcal{T}_i) = f(\mathcal{T}_i)$. Here we have been clever. Our prior (before seeing the data) is the probability of a trajectory. Because we are generating trajectories by iterating the stochastic dynamical Equation (11.9) our trajectories are just being drawn from the prior distribution. Thus, the weights only depend on the likelihood of the trajectories.

Figure 11.11 shows a schematic of the sequential importance sampling described above for a 10-particle system. The particles are generated by a stochastic updated equation. As time increases the variance of the particle position increases, leading to a lower accuracy solution. The importance weights of the particles depend on the whole trajectory.

Unfortunately, a naive sequential importance sampling is doomed, as the samples will rapidly loose track of the true state. The accuracy of the sampling

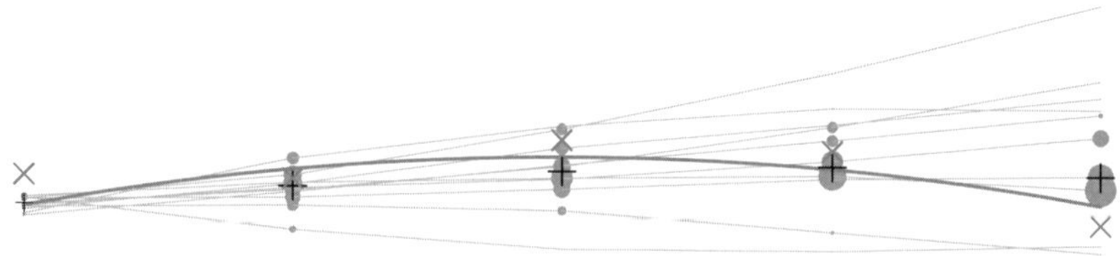

Figure 11.11 Illustration of sequential importance sampling. The blue curve shows the true time evolution, the crosses (×) the noisy observations, and the red curves the trajectories of the particles. The size of the particles represent their importance. The '+' symbol shows the weighted average of the particles at each time step.

method depends on samples being concentrated around the posterior. This concentration typically dissipates as the number of steps made by the algorithm increase, since in generating the trajectories we only use our model and are unguided by the data, $\mathcal{D}$. Furthermore, the likelihood of the samples is a product of the likelihoods $f(\boldsymbol{y}(t)|\boldsymbol{X}_i(t))$. But the variance of a product of numbers is likely to be huge (see the discussion of the log-normal distribution in Example 5.3) so that the vast majority of samples will have a negligible likelihood compared with the most probable. To overcome this, we can resample the set of particle points after a certain number of time steps. That is, we compute the posterior probability, $f(\mathcal{T}_i|\mathcal{D})$, associated with a partial trajectory $\mathcal{T}_i = (\boldsymbol{X}_i(1), \boldsymbol{X}_i(2), \ldots, \boldsymbol{X}_i(t))$. We then choose a new set of samples from the current set of samples $\{\boldsymbol{X}_i(t)\}$ with probabilities proportional to the posterior probabilities. Thus, we are likely to choose $\boldsymbol{X}_i(t)$ several times for those sequences $\mathcal{T}_i$ where $f(\mathcal{T}_i|\mathcal{D})$ is relatively large, while the less likely sequences may not be sampled at all. Because the noise associated with the updates are independent for each sample, the sample sequences will quickly diverge again after the resampling step. Thus, we have to resample again after a number of steps. This resampling is characteristic of the particle filter strategy.

Resampling produces a loss in the diversity of the samples (many points are no longer independent) so that the accuracy in the estimate from importance sampling is reduced. On the other hand, after resampling when we carry on propagating the particles the samples start to diverge and so become less correlated, but they are now much higher probability samples so we gain accuracy. There remains some debate about whether resampling adds bias in the estimates, but in most instances we have no choice; it is the only way to obtain high probability sample throughout the trajectory. In practice, one usually does not resample at each time step as we lose diversity and thus reduce the quality of the fit (it could also lead to over-fitting the current data – although true Bayesian methods should never over-fit, in this case we tune the model which breaks the part of the contract which guarantees that we cannot over-fit the data). It is common to resample only when some significant proportion of the particles have much smaller weights than the others (when this happens will depend on the problem and the data). In practice, provided we use a sufficiently large population and resample sufficiently often the estimates we obtain are usually good enough.

Example 11.11 Repairing Clipped Signal Continued
We continue Example 11.9 on page 338 where we considered repairing a clipped and noisy waveform shown in Figure 11.8.

A particle filter is used to predict the true signal. Some tuning of the model noise was required (in the end I set that standard deviation of $\eta_x = \eta_v = 0$ and $\eta_a = \eta_\omega > 0$ – its value depends on the characteristic of the signal). The likelihood of a sample at a particular time step when the signal is not being clipped is taken to be to $\mathcal{N}(x(t) - y(t), \sigma^2)$, where $x(t)$ is the position of the particle and σ^2 is the typical variance in the signal. For the clipped signal I added a weak prior, making large deviations from the mean less likely than smaller deviations. I resampled the population when the signal crossed zero. In Figure 11.12 we show the mean of the posterior as predicted by the particle filter (shown by $\times$). The results are not a perfect restoration – the model is rather crude and the particle filter used very primitive. Nevertheless, the resulting waveform is considerably closer to the true signal than the distorted signal.

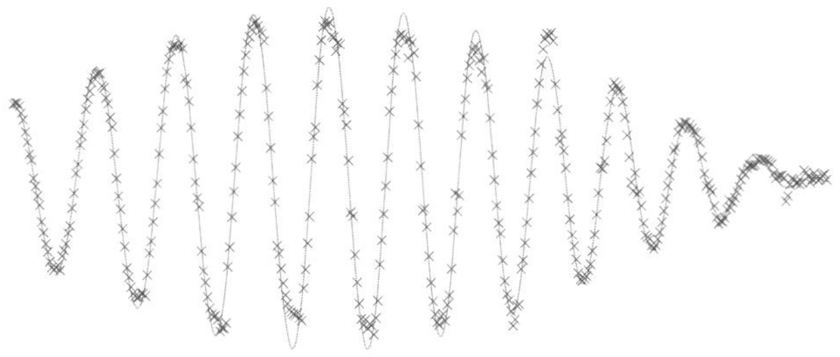

Figure 11.12 Particle filter prediction of the signal shown in Figure 11.8 on page 338. The reference signal is shown by the solid line.

Just as with MCMC, particle filtering has been heavily developed and can seem rather intimidating. However, the basic idea is simple, as is a basic implementation. More sophisticated approaches are used to improve the proposal distribution. Auxiliary particle filters use information from the data to improve the sampling. The resample-move algorithm and block sampling are other methods to improve the proposal distribution, while Rao–Blackwellised particle filters combine particle filters with analytically tractable methods, such as Kalman filters and their variants, to remove the need to sample so densely. The clipped-signal problem was not a filtering problem (except at its end point), but a smoothing problem. Smoothing is a more complicated problem than filtering as we have future as well as past data. Various forward–backward algorithms have been proposed as well as the use of a two-filter approach. There is a very extensive literature on particle filtering and smoothing that the interested reader can follow; see, for example, Crisan and Rozovskii (2011).

11.3.3 Approximate Bayesian Computation

A popular method for inferring unknown parameters of a dynamic system that has some of the flavour of particle filtering goes by the name of approximate Bayesian computation or ABC. Here a population of 'particles' is evolved forward in time using different parameters of the model. Instead of using the likelihood, trajectories that diverge too far from the data are simply rejected. After one iteration, a new population is constructed with slightly modified parameters to those that passed the first round. This is then repeated. Usually, the threshold for rejecting trajectories is reduced at each iteration. This has become a very popular for finding unknown parameters in dynamical systems. It has been argued that the thresholding acts as a crude approximation to a likelihood function. To me, it appears like a sensible and effective method for estimating parameters, however, its probabilistic justification is rather lost on me.

Markov chains describe the evolution of a random variable where its probability at the next time step depends only on the value of the random variable at the current time step. This situation occurs extremely often in all sorts of different fields. As a result, understanding Markov chains is often a good investment of time.

Markov chains have a second claim to importance. They can be used to generate random deviates from a distribution – so-called MCMC. This is particularly valuable in the case of high dimensions, for example, when we are interested in a set of coupled variables. This has found a huge number of applications in many different fields.

Many dynamical systems are well described using Markov chains and the problem of predicting the state of a moving object from a set of noisy observation, has many important applications. This problem is called filtering and has a classic closed form solution, the Kalman filter, applicable for linear systems with normal noise. When the dynamics cannot be described by linear equations or the noise is non-Gaussian the Kalman filter is no longer applicable. In these cases we can use particle filters, a sequential Monte Carlo technique for estimating the true state of a dynamical system.

Additional Reading

Markov chains are covered in most texts on probability. MacKay (2003) provides considerable insight into MCMC as well as much else. Radford Neal, one of the best exponents of MCMC, has produced a technical report which provides a classic and comprehensive (although possibly dated) reviews of MCMC which is available online (Neal, 1993). Geman and Geman (1984) is a classic paper introducing many of the ideas of how to apply MCMC in unusual situations. There are now a huge number of specialist books on MCMC. Many techniques for modelling more complex data including Kalman and particle filters are described in Cressie and Wikle (2011).

Exercise for Chapter 11

Exercise 11.1 (answer on page 432)

The figure below shows a diagrammatic representation of a Markov model consisting of six states. Write down the transition matrix and compute its eigenvalues and the steady state. What is the characteristic relaxation time to reach the steady state? Perform a Monte Carlo simulation to find the steady-state probabilities empirically and compare this with those obtained from considering the eigenvectors of the transition matrix.

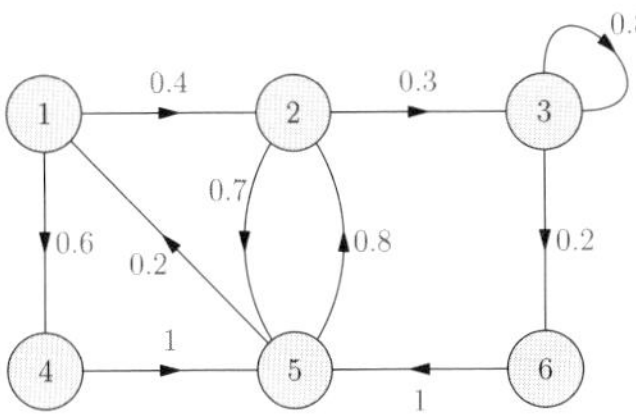

Exercise 11.2 (answer on page 433)

Consider an ergodic Markov chain described by the transition matrix $\mathbf{M}$ starting from an initial state $\boldsymbol{p}(1)$. To compute the expected time to reach a state i we can consider a modified transition matrix $\hat{\mathbf{M}}$ where $\hat{M}_{jk} = M_{jk}$ if $k \neq i$ and $\hat{M}_{ji} = 0$ otherwise. That is, in this modified system the dynamics are identical to the original until we reach state i, whereupon the walker disappears. The probability of being in state i at time t is equal to $\boldsymbol{\delta}_i \hat{\mathbf{M}}^{t-1} \boldsymbol{p}(1)$, where $\boldsymbol{\delta}_i$ is the vector with zero elements everywhere except in the ith position where the element is equal to 1. The expected first-passage time is thus given by

$$T_{\text{fpt}} = \sum_{t=1}^{\infty} (t-1)\, \boldsymbol{\delta}_i^{\mathsf{T}} \hat{\mathbf{M}}^{t-1} \boldsymbol{p}(1).$$

This just sums the number of steps taken times the probability of reaching state i in those steps. Show that we can write this as

$$T_{\text{fpt}} = \boldsymbol{\delta}_0^{\mathsf{T}} \left(\mathbf{I} - \hat{\mathbf{M}}\right)^{-1} \left(\mathbf{I} - \hat{\mathbf{M}}\right)^{-1} \boldsymbol{p}(1) - 1$$

where $\mathbf{I}$ is the identity matrix. Further show that this can be simplified to

$$T_{\text{fpt}} = \mathbf{1}^{\mathsf{T}} \left(\mathbf{I} - \hat{\mathbf{M}}\right)^{-1} \boldsymbol{p}(1) - 1$$

where $\mathbf{1}$ is the vector of all 1s. Note that you have to sum a geometric series of matrices and differentiate matrices so this exercise isn't for the faint-hearted.

Exercise 11.3 (answer on page 435)

Compute the expected first-passage time to reach state 6 starting from state 1 for the Markov chain described in Exercise 11.1 using the formula from Exercise 11.2, and compare this with empirical measurements.

Exercise 11.4 (answer on page 435)

Consider a random variable that can take values in the set $\mathbb{Z}_n = \{0, 1, 2, \ldots, n-1\}$. At each time step the value increases or decreases by 1 modulo n. That is, the system is performing a random walk around the ring. Construct the transition matrix and compute the eigenvalues for $n = 499$. Show that if n is even there are two eigenvalues

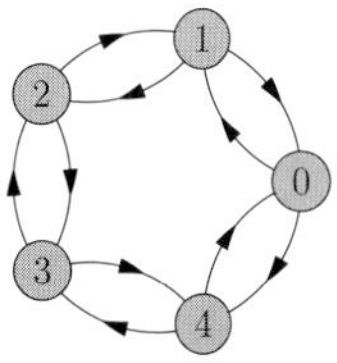

with modulus 1. Plot a histogram of the eigenvalues and show how the second-largest eigenvalue grows with n.

Exercise 11.5 (answer on page 436)

Use MCMC to generate normal deviates where the candidate solution is taken by adding $U \sim \mathrm{U}(-1/2, 1/2)$. Compute the time it takes to generate 10^6 deviates. Compare this to other methods for generating normal deviates (see Exercise 3.2).

Exercise 11.6 (answer on page 436)

Use the clustering algorithm (see Example 11.6) or, if you prefer, the Metropolis algorithm (see Example 11.4), to simulate the two-dimensional Ising model as described in Section 10.4. Compute the mean and variance of the energy and magnetisation per spin. (When using the clustering algorithm you should use the absolute magnetisation below the critical point because the clustering algorithm restores ergodicity which is broken in real magnets and by Metropolis.)

Exercise 11.7 (answer on page 437)

Use MCMC to obtain an estimate for the posterior given a normal likelihood, $X_i \sim \mathcal{N}(\mu, \sigma)$ using an uninformative prior. (Generate $n = 20$ data points from a normal distribution with $\mu = 1/2$, $\sigma = 2$.) Plot a histogram of the posterior for the mean and show that it is distributed according to a t-distribution (see Exercise 8.4 on page 255).

Exercise 11.8 (answer on page 438)

In Example 8.11 on page 228 and Example 8.12 on page 232 we considered the problem of estimating the parameters for a mixture of two Gaussians. Generate data consisting of 500 deviates from $\mathcal{N}(0, 1)$ and 200 deviates from $\mathcal{N}(3, 1)$. The task is to infer the means and variances of the two distributions assuming these distributions are unknown and we don't know which data point was generated by which distribution. We assume, however, that we know the likelihood of a data point is given by a mixture of Gaussians

$$f(X_i|\mu_1, \sigma_1, \mu_2, \sigma_2, p) = p\,\mathcal{N}(X_i|\mu_1, \sigma_1^2) + (1-p)\,\mathcal{N}(X_i|\mu_2, \sigma_2^2).$$

Use MCMC to estimate the unknown parameters μ_1, σ_1, μ_2, σ_2, and p, as well as their uncertainty.

Exercise 11.9 (answer on page 439)

Consider the problem of tracking a financial time series. Assume that a stock has a true value, $X(t)$. Due to market volatility, the price of the stock, $Y(t)$, is taken to be equal to the 'true value' plus some random noise

$$Y(t) = X(t) + \epsilon(t)$$

where $\epsilon(t) \sim \mathcal{N}(0, 1)$. The true value $X(t)$ varies much more slowly and we assume

$$X(t+1) = X(t) + V(t) \qquad V(t+1) = V(t) + \eta(t)$$

where $\eta(t) \sim \mathcal{N}(0, 10^{-4})$. Generate some time series data from this model. The figure below illustrates some time series data for 100 data points.

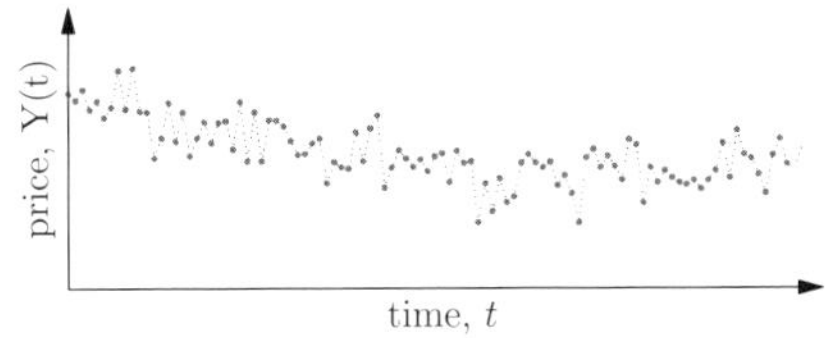

Use a Kalman filter to estimate the true price and its uncertainty. Also use a particle filter to perform the same calculation and compare the two estimates.

Appendix 11.A Eigenvalues and Eigenvectors of General Square Matrices

Recall that an eigenvector of a (square) matrix M is a vector, $\boldsymbol{v}$, with the property

$$\mathsf{M}\,\boldsymbol{v} = \lambda\,\boldsymbol{v}$$

where λ is the corresponding eigenvalue. We note that if $\boldsymbol{v}$ is an eigenvector so is $\boldsymbol{w} = c\,\boldsymbol{v}$ and $\boldsymbol{w}$ will have the same eigenvalue as $\boldsymbol{v}$.

For the special class of symmetric $n \times n$ matrices (i.e. matrices M where $\mathsf{M} = \mathsf{M}^{\mathsf{T}}$) there are n real eigenvalues, $\lambda^{(i)}$, and n independent real eigenvectors $\boldsymbol{v}^{(i)}$. If $\lambda^{(i)} \neq \lambda^{(j)}$ then the eigenvectors are orthogonal so that $\boldsymbol{v}^{(i)\mathsf{T}}\boldsymbol{v}^{(j)} = 0$. Furthermore, if we have m eigenvectors with the same eigenvalue then any linear combination of the eigenvectors are also eigenvectors and in consequence we can chose a set that are orthogonal to each other. We can also normalise the eigenvectors so that we can choose a set that form an orthonormal basis. That is, $\boldsymbol{v}^{(i)\mathsf{T}}\boldsymbol{v}^{(j)} = [\![i = j]\!]$. This is true for symmetric matrices with real components. The generalisation of symmetric matrices to matrices with complex components is the class of *normal matrices* that satisfy $\mathsf{M}^*\mathsf{M} = \mathsf{M}\mathsf{M}^*$ (where M^* denotes taking the conjugate transpose of the matrix). These all have an orthogonal set of eigenvectors with real eigenvalues. The most prominent type of complex normal matrices are the set of *Hermitian matrices* where the matrix equals the complex conjugate of the transpose. These matrices are ubiquitous in quantum mechanics, although, at least in my experience, don't turn up commonly elsewhere.

The situation is a bit more complicated when we deal with non-normal (real) matrices (i.e. asymmetric square matrices). In this case, the eigenvalues will either be real or exist in complex conjugate pairs. The eigenvectors are real if the eigenvalues are real and exist in complex conjugate pairs if the eigenvalues are complex. Unlike the case for symmetric matrices, for general matrices the eigenvectors are not generally orthogonal to each other, so that, commonly, $\boldsymbol{v}^{(i)\mathsf{T}}\boldsymbol{v}^{(j)} \neq 0$. Furthermore, the left eigenvectors

$$\boldsymbol{u}^{\mathsf{T}}\mathsf{M} = \lambda\,\boldsymbol{u}^{\mathsf{T}}$$

are generally different to the right eigenvectors, $\boldsymbol{v}$. However, for any right eigenvector, $\boldsymbol{v}$, with eigenvalue λ, there will be a corresponding left eigenvector, $\boldsymbol{u}$, with the same eigenvalue. Furthermore, the left eigenvectors will be orthogonal to all right eigenvectors with different eigenvalues. This is known as *biorthogonality*.

Note that if we wish we can choose a set of eigenvectors such that $\boldsymbol{u}^{(i)\mathsf{T}}\boldsymbol{v}^{(j)} = [\![i = j]\!]$, although when we do this, we cannot generally normalise both the left and right eigenvectors, $\boldsymbol{u}^{(i)}$ and $\boldsymbol{v}^{(i)}$.

A non-normal $n \times n$ matrix may not even have n independent eigenvectors, in which case it is said to be *defective*. An example of a defective matrix is

$$\begin{pmatrix} 1 & -1 \\ 0 & 1 \end{pmatrix}$$

which has an eigenvector $\begin{pmatrix} 1 \\ 0 \end{pmatrix}$ with an eigenvalue 1, but there are no other independent eigenvectors. Defective matrices are fortunately rare, but they do exist (and they include stochastic matrices). For non-defective matrices we can decompose the matrix as a product of three matrices

$$\mathbf{M} = \mathbf{V}\boldsymbol{\Lambda}\mathbf{V}^{-1}$$

Note that we use the convention that $\boldsymbol{a}$ represents a column vector while $\boldsymbol{b}^{\mathsf{T}}$ would represent a row vector. Thus, following the usual rules of matrix multiplication, $\boldsymbol{C} = \boldsymbol{a}\boldsymbol{b}^{\mathsf{T}}$ is a matrix with components $C_{ij} = a_i b_j$.

where $\mathbf{V}$ is a matrix whose columns are the right eigenvectors of $\mathbf{M}$ and $\boldsymbol{\Lambda}$ is a diagonal matrix whose diagonal elements are the eigenvalues of $\mathbf{M}$. The matrix $\mathbf{U} = \mathbf{V}^{-1}$ is a matrix whose rows are the left eigenvectors of the matrix $\mathbf{M}$. We can also write this decomposition as

$$\mathbf{M} = \sum_i \lambda^{(i)} \boldsymbol{v}^{(i)} \left(\boldsymbol{u}^{(i)}\right)^{\mathsf{T}}$$

where $\lambda^{(i)}$ is the ith eigenvector of $\mathbf{M}$, and $\boldsymbol{u}^{(i)}$ and $\boldsymbol{v}^{(i)}$ are the corresponding left and right eigenvectors. A consequence of this decomposition is that $\mathbf{M}^t = \mathbf{V}\boldsymbol{\Lambda}^t\mathbf{V}^{-1}$, where $\boldsymbol{\Lambda}^t$ will be a diagonal matrix with diagonal elements $(\lambda^{(i)})^t$.

12

Stochastic Processes

Contents

We often face situations where we have random variables extended over a continuous domain such as space or time, e.g. $\{X(t)|t \in \mathbb{R}\}$. These are modelled by stochastic processes. We describe properties shared by the most commonly used stochastic processes and discuss a particularly useful stochastic process, namely Gaussian processes. We then examine one of the most important processes, the diffusion process, also known as Wiener or Brownian process. This process comes with a powerful set of tools which allow it to be analysed relatively easily. Finally we consider point processes and in particular Poisson processes and their various applications.

Stochastic processes take a significantly longer time to master than the techniques we have covered in previous chapters. On the other hand, they can often provide a remarkably quick answer to otherwise hard problems (sometimes without the need to understand all the technical details). Being aware of these tools can often pay off.

12.1 Stochastic Processes

Stochastic processes generalise the idea of random variables to random functions. They provide an important tool set for modelling a large number of phenomena.

12.1.1 What Are Stochastic Processes?

In Chapter 11 we considered Markov chains which describe a stochastic system evolving over discrete time intervals. We could describe the system by a sequence or vector of random variables. By taking the limit where time becomes continuous we end up with a stochastic process, $X(t)$, which is now a function. *Process* in this context has a specific technical meaning: it is to a continuous function what a random variable is to an everyday variable. If the process has the Markov property (i.e. the value of $X(t)$ is conditionally independent of $X(t')$ given $X(t-\delta t)$ for all $t' < t-\delta t$ and any δt) then it is said to be a Markov process. (Although we will often consider the independent variable t to be time, we can, of course, have stochastic processes defined over any set of independent variables.)

Taking the continuum time limit opens up tools from calculus which can be used to quickly find properties of the system. However, considerable care is needed in taking the limit. Proving rigorous results in full generality can be challenging – although you can get a long way just considering convergence properties of the mean squared. For engineers, scientists, and even economists, stochastic calculus provides a fast route to getting useful results quickly.

Our approach to dealing with random variables was to consider their probability distribution. For stochastic processes the probability distribution is over an ensemble of possible functions. These are much less convenient objects to work with than random variables. Supposing we have a random process, $X(t)$, modelling a random function of time, then we can consider the probability density for stochastic process at a particular time, t,

$$f_{X(t)}(x,t) = \lim_{\delta x \to 0} \frac{1}{\delta x} \mathbb{P}\left(x \le X(t) \le x + \delta x\right).$$

This is a regular probability density function which we are very familiar with. Similarly, we could consider the joint probability density at two times

$$f_{X(t)}(x_1,t_1;x_2,t_2) = \lim_{\delta x \to 0} \frac{1}{\delta x^2} \mathbb{P}\left(x_1 \le X(t_1) \le x_1 + \delta x \wedge x_2 \le X(t_2) \le x_2 + \delta x\right).$$

This generalises to any number of points. The conditional probability density is the probability density at time t_2 given that $X(t_1) = x_1$,

$$f_{X(t)}(x_2,t_2|x_1,t_1) = \frac{f_{X(t)}(x_1,t_1;x_2,t_2)}{f_{X(t)}(x_1,t_1)}.$$

Again this can be generalised to any number of points.

12.1.2 Gaussian Processes

To get used to the idea of stochastic processes we consider a particularly simple class of stochastic process known as a *Gaussian process*. Such processes are characterised by the property that all their joint probabilities $f_{X(t)}(x_1, t_1)$, $f_{X(t)}(x_1, t_1; x_2, t_2)$, $f_{X(t)}(x_1, t_1; x_2, t_2; x_3, t_3)$, etc. are normal distributions of the variables x_1, x_2, etc. Vector-valued stochastic processes, $\boldsymbol{X}(t)$, are also Gaussian processes if all their joint probabilities are normally distributed. The distribution for a Gaussian process, $X(t)$, is fully specified by its mean and covariance

$$\mu(t) = \mathbb{E}\left[X(t)\right], \qquad K(t_1, t_2) = \mathbb{C}\mathrm{ov}\left[X(t_1), X(t_2)\right].$$

The covariance, $K(t_1, t_2)$, is sometimes known as a kernel function – it can often be seen as playing a similar role to kernel functions used in machine learning techniques such as support vector machines. We can think of the functions $X(t)$ having a probability density

$$f(X(t)) = \mathcal{GP}(X(t)|\mu(t), K(t_1, t_2)) \propto \mathrm{e}^{-\frac{1}{2}\int_{-\infty}^{\infty} (X(t_1)-\mu(t_1))\, K^{-1}(t_1,t_2)\, (X(t_2)-\mu(t_2))\, \mathrm{d}t_1\, \mathrm{d}t_2}$$

although some care is required in interpreting this 'density' as it is a density over random functions. One way to make sense of this is to discretise time so that the double integral in the exponent becomes a double sum

$$\frac{1}{2}\int_{-\infty}^{\infty} (X(t_1) - \mu(t_1))\, K^{-1}(t_1, t_2)\, (X(t_2) - \mu(t_2))\, \mathrm{d}t_1\, \mathrm{d}t_2$$
$$\approx \frac{1}{2}\sum_{i,j=1}^{n} (X(t_i) - \mu(t_i))\, K^{-1}(t_i, t_j)\, (X(t_i) - \mu(t_j)).$$

In this discretise case we are left with a well-defined multivariate normal distribution. A Gaussian process can be seen as the limit as $t_i - t_{i-1} \to 0$.

A random sample drawn from a Gaussian process would be some continuous function. In practice we cannot draw a random function as we would have to specify it at an uncountable number of points. However, we can generate points on the function. That is, we can discretise our space (or time) and generate values at the point which satisfy the Gaussian process properties. To do this we consider the matrix $\mathbf{K}$ with elements $K_{ij} = K(t_i, t_j)$ (our covariance or kernel function for times t_i and t_j). The kernel matrix must be positive definite (otherwise the Gaussian process's distribution would not be normalisable). A consequence of this is that we can compute a Cholesky decomposition, $\mathbf{K} = \mathbf{L}\mathbf{L}^{\mathsf{T}}$ where $\mathbf{L}$ is a lower diagonal matrix. Once we have done this we can generate a random deviate $\boldsymbol{X} = (X(t_1), X(t_2), \ldots, X(t_n)) = \boldsymbol{\mu} + \mathbf{L}\boldsymbol{Y}$ where $\boldsymbol{\mu} = (\mu(t_1), \mu(t_2), \ldots, \mu(t_n))$ and

$Y \sim \mathcal{N}(\mathbf{0}, \mathbf{I})$ is an n-dimensional normal deviate with zero mean and covariance $\mathbf{I}$ (so that, $Y_i \sim \mathcal{N}(0,1)$). Note that

Recall that we use the notation $\mathbb{C}\mathrm{ov}\left[\boldsymbol{X}\right]$ to denote the covariance matrix of a vector $\boldsymbol{X}$ with components $\mathbb{C}\mathrm{ov}\left[X_i, X_j\right]$.

$$\mathbb{E}\left[\boldsymbol{X}\right] = \boldsymbol{\mu}, \qquad \mathbb{C}\mathrm{ov}\left[\boldsymbol{X}\right] = \mathbf{L}\,\mathbb{E}\left[\boldsymbol{Y}\boldsymbol{Y}^{\mathsf{T}}\right]\mathbf{L}^{\mathsf{T}} = \mathbf{L}\,\mathbf{I}\,\mathbf{L}^{\mathsf{T}} = \mathbf{K},$$

or

$$\mathbb{E}\left[X(t_i)\right] = \mu(t_i), \qquad \mathbb{C}\mathrm{ov}\left[X(t_i), X(t_j)\right] = K(t_i, t_j).$$

Since $\boldsymbol{X} = \boldsymbol{\mu} + \mathbf{L}\boldsymbol{y}$ is a sum of normal deviates, $\boldsymbol{y}$, (and constants, $\boldsymbol{\mu}$) it will itself be normally distributed. Thus, the vector $\boldsymbol{X}(t)$ will correspond to points from a random deviate drawn from $\mathcal{GP}(X(t)|\mu(t), K(t_1,t_2))$. Figure 12.1 shows some samples of a Gaussian process for two different kernel functions, $K(t_1,t_2)$. The kernel function defines the auto-correlation of the function being fitted. By tuning the kernel function we can change the probability of different functions. Note that for the kernel $\exp(-b|t_2 - t_1|^a)$, the Gaussian processes will (with overwhelming probability) be smooth if $a = 2$, otherwise if $1 < a < 2$ it will be smooth, but non-analytic, or if $a \le 1$ it will be continuous, but no longer smooth.

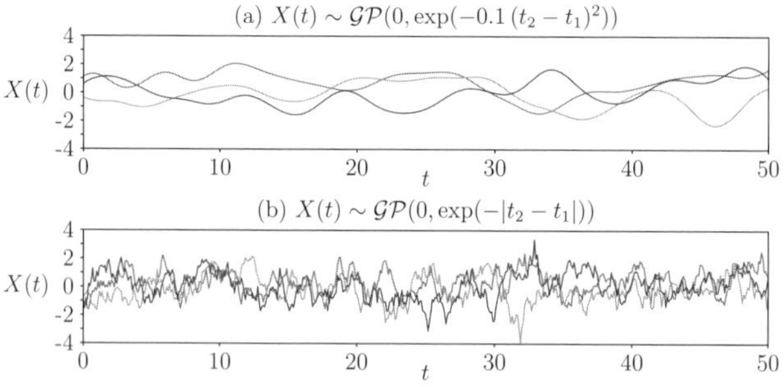

Figure 12.1 Three examples of samples $X(t) \sim \mathcal{GP}(0, K(t_1,t_2))$, for (a) $K(t_1,t_2) = \exp(-0.1\,(t_2 - t_1)^2)$ and (b) $K(t_1,t_2) = \exp(-0.1\,|t_2 - t_1|)$.

To generate deviates in higher dimensions we just have to discretise space on a high-dimensional lattice and use a high-dimensional kernel function. Figure 12.2 shows one particular example of a two-dimensional Gaussian process. Gaussian processes are often used to model very high-dimensional functions.

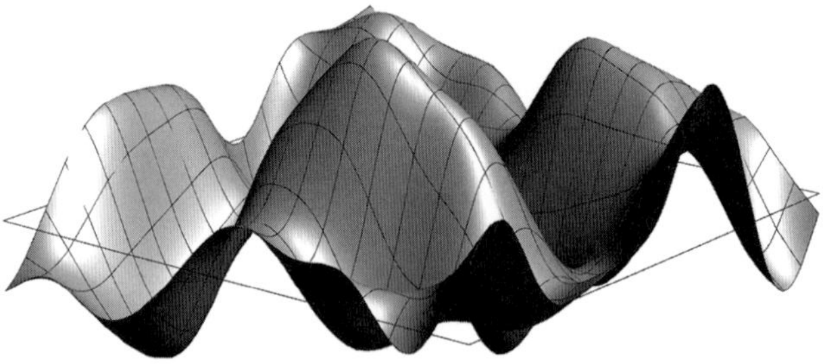

Figure 12.2 Example of an instance of a two-dimensional Gaussian process $X(\boldsymbol{x}) \sim \mathcal{GP}(0, K(\boldsymbol{x},\boldsymbol{y}))$ where $K(\boldsymbol{x},\boldsymbol{y}) = \exp(-\|\boldsymbol{x} - \boldsymbol{y}\|^2)$.

Gaussian Processes for Inference

Gaussian processes arise naturally in some of the diffusion processes we consider below. But they are also useful tools for modelling in their own right. For example, we can use them to provide a prior on smooth (or not so smooth) functions.

If we have some data points we wish to fit with a suitably smooth function then Gaussian processes allow us one route to do this in a reasonably principled fashion. This can be done for data with any number of independent variables. We assume we have n independent data points $\mathcal{D} = \big((\boldsymbol{x}_i, y_i)\big| i = 1, 2, \ldots, n\big)$ and we want to fit a function $a(\boldsymbol{x})$, where we take as a likelihood function

$$f(\mathcal{D}|a) = \prod_{i=1}^{n} \mathcal{N}(y_i|a(\boldsymbol{x}_i), \sigma^2) = \prod_{i=1}^{n} \frac{1}{\sqrt{2\pi}\,\sigma} \mathrm{e}^{-\frac{1}{2\sigma^2}(y_i - a(\boldsymbol{x}_i))^2}$$

with a prior on the functions $a(\boldsymbol{x})$ of $\mathcal{GP}\big(a(\boldsymbol{x})|0, K(\boldsymbol{x}_1, \boldsymbol{x}_2)\big)$. The posterior is proportional to the likelihood times the prior. We are not usually interested in generating deviates from this prior. The type of information we might be interested in is the probability distribution of $a(\boldsymbol{x}^*)$ at a particular point $\boldsymbol{x}^*$. To compute this we marginalise over all other points. Where we have no data points the marginalisation is trivial. After marginalising out the points where we have data we are left with a univariate distribution for $a(\boldsymbol{x}^*)$,

$$f\left(a(\boldsymbol{x}^*)|\mathcal{D}\right) = \int_{-\infty}^{\infty} \frac{f(\mathcal{D}|a(\boldsymbol{x}))\, f(a(\boldsymbol{x}))}{f(\mathcal{D})} \prod_{\boldsymbol{x} \neq \boldsymbol{x}^*} \mathrm{d}a(\boldsymbol{x}) \propto \int_{-\infty}^{\infty} \mathrm{e}^{G} \prod_{i=1}^{n} \mathrm{d}a(\boldsymbol{x}_i)$$

where

$$\begin{aligned} G = &-\frac{1}{2\sigma^2}\sum_{i=1}^{n}\big(a(\boldsymbol{x}_i) - y_i\big)^2 - \frac{1}{2}\sum_{i,j=1}^{n} a(\boldsymbol{x}_i)\, K^{-1}(\boldsymbol{x}_i, \boldsymbol{x}_j)\, a(\boldsymbol{x}_j) \\ &-\sum_{i=1}^{n} a(\boldsymbol{x}_i)\, K^{-1}(\boldsymbol{x}_i, \boldsymbol{x}^*)\, a(\boldsymbol{x}^*) - \frac{1}{2} a(\boldsymbol{x}^*)\, K^{-1}(\boldsymbol{x}_i, \boldsymbol{x}^*)\, a(\boldsymbol{x}^*). \end{aligned}$$

It is now a purely algebraic exercise to compute the posterior density for $a(\boldsymbol{x}^*)$. This direct method leads to an answer in a non-standard form; later we will rederive the result using a less direct method, but one which gives the answer in a more convenient form. We proceed with the direct derivation as it is a more straightforward application of Bayesian inference. We follow the standard technique when faced with a Gaussian integral of completing the square. To help to do this we introduce: the matrix $\mathbf{L}$ with components $L_{ij} = K^{-1}(\boldsymbol{x}_i, \boldsymbol{x}_j)$, the vectors $\boldsymbol{l}$ and $\boldsymbol{a}$ with elements $l_i = K^{-1}(\boldsymbol{x}_i, \boldsymbol{x}^*)$ and $a_i = a(\boldsymbol{x}_i)$, and finally the scalars $a = a(\boldsymbol{x}^*)$ and $l = K^{-1}(\boldsymbol{x}^*, \boldsymbol{x}^*)$ (here $\mathbf{L}$ is a sub-matrix of $\mathbf{K}^{-1}$ *not* the Cholesky decomposition of a matrix). Thus,

$$\mathbf{K}^{-1} = \begin{pmatrix} \mathbf{L} & \boldsymbol{l} \\ \boldsymbol{l}^\mathsf{T} & l \end{pmatrix}$$

$$\begin{aligned} G &\overset{(1)}{=} -\frac{1}{2\sigma^2}\|\boldsymbol{y} - \boldsymbol{a}\|^2 - \frac{1}{2}\boldsymbol{a}^\mathsf{T}\mathbf{L}\boldsymbol{a} - a\,\boldsymbol{a}^\mathsf{T}\boldsymbol{l} - \frac{1}{2}a^2 l \\ &\overset{(2)}{=} -\frac{1}{2}\boldsymbol{a}^\mathsf{T}\left(\mathbf{L} + \sigma^{-2}\mathbf{I}\right)\boldsymbol{a} + \boldsymbol{a}^\mathsf{T}\left(\sigma^{-2}\boldsymbol{y} - a\,\boldsymbol{l}\right) - \frac{1}{2\sigma^2}\|\boldsymbol{y}\|^2 - \frac{1}{2}a^2 l \\ &\overset{(3)}{=} -\frac{1}{2}\left(\boldsymbol{a} + (\mathbf{L} + \sigma^{-2}\mathbf{I})^{-1}\left(\sigma^{-2}\boldsymbol{y} - a\,\boldsymbol{l}\right)\right)^\mathsf{T} \\ &\quad \left(\mathbf{L} + \sigma^{-2}\mathbf{I}\right)\left(\boldsymbol{a} + (\mathbf{L} + \sigma^{-2}\mathbf{I})^{-1}\left(\sigma^{-2}\boldsymbol{y} - a\,\boldsymbol{l}\right)\right) \\ &\quad + \frac{1}{2}\left(\sigma^{-2}\boldsymbol{y} - a\,\boldsymbol{l}\right)^\mathsf{T}\left(\mathbf{L} + \sigma^{-2}\mathbf{I}\right)^{-1}\left(\sigma^{-2}\boldsymbol{y} - a\,\boldsymbol{l}\right) - \frac{1}{2}a^2 l - \frac{1}{2\sigma^2}\|\boldsymbol{y}\|^2 \end{aligned}$$

(1) Rewriting the equation above in matrix form.
(2) Expanding $\|\boldsymbol{y} - \boldsymbol{a}\|^2$ and collecting together terms.
(3) Completing the square in terms of $\boldsymbol{a}$.

The first term is now in quadratic form and by making a change of variables $\boldsymbol{a} \to \boldsymbol{a}' = \boldsymbol{a} + (\mathsf{L} + \sigma^{-2}\mathsf{I})^{-1}\left(\sigma^{-2}\boldsymbol{y} - a\,\boldsymbol{l}\right)$ we can integrate it out. The terms left in the exponent that involve a are

$$
\begin{aligned}
G' &\overset{(1)}{=} -\frac{1}{2}a^2 l + \frac{1}{2}\left(\sigma^{-2}\boldsymbol{y} - a\,\boldsymbol{l}\right)^{\mathsf{T}}\left(\mathsf{L} + \sigma^{-2}\mathsf{I}\right)^{-1}\left(\sigma^{-2}\boldsymbol{y} - a\,\boldsymbol{l}\right) + \text{const} \\
&\overset{(2)}{=} -\frac{1}{2}a^2\left(l + \boldsymbol{l}^{\mathsf{T}}\left(\mathsf{L} - \sigma^{-2}\mathsf{I}\right)^{-1}\boldsymbol{l}\right) - a\,\sigma^{-2}\boldsymbol{l}^{\mathsf{T}}\left(\mathsf{L} + \sigma^{-2}\mathsf{I}\right)^{-1}\boldsymbol{y} + \text{const} \\
&\overset{(3)}{=} -\frac{1}{2}\left(l - \boldsymbol{l}^{\mathsf{T}}\left(\mathsf{L} - \sigma^{-2}\mathsf{I}\right)^{-1}\boldsymbol{l}\right)\left(a + \frac{\boldsymbol{l}^{\mathsf{T}}\left(\mathsf{L} + \sigma^{-2}\mathsf{I}\right)^{-1}\boldsymbol{y}}{\left(l - \boldsymbol{l}^{\mathsf{T}}\left(\mathsf{L} - \sigma^{-2}\mathsf{I}\right)^{-1}\boldsymbol{l}\right)}\right)^2 + \text{const}
\end{aligned}
$$

(1) Showing the terms involving a after integrated out $\boldsymbol{a}'$.
(2) Multiplying out the last term and keeping terms proportional to a^2 and a.
(3) Completing the square in terms of a.

As the remaining exponent is quadratic in a we can read off the mean and variance, giving

$$
f\left(a(\boldsymbol{x}^*)\middle|\mathcal{D}\right) = \mathcal{N}\left(a(\boldsymbol{x}^*)\middle|\frac{-\boldsymbol{l}^{\mathsf{T}}\left(\mathsf{L} + \sigma^{-2}\mathsf{I}\right)^{-1}\boldsymbol{y}}{l - \boldsymbol{l}^{\mathsf{T}}\left(\mathsf{L} - \sigma^{-2}\mathsf{I}\right)^{-1}\boldsymbol{l}}, \frac{1}{l - \boldsymbol{l}^{\mathsf{T}}\left(\mathsf{L} - \sigma^{-2}\mathsf{I}\right)^{-1}\boldsymbol{l}}\right). \tag{12.1}
$$

This result is in terms of the inverse of the correlation matrix, $\mathsf{L} = \mathsf{K}^{-1}$, which is inefficient to work with. We now give the more traditional derivation. In our model

$$
y_i = a(\boldsymbol{x}_i) + \epsilon_i
$$

where $\epsilon_i \sim \mathcal{N}(0, \sigma^2)$. Since ϵ_i is independent of $a(\boldsymbol{x}_i)$ we have $\mathbb{E}\left[y_i\right] = a(\boldsymbol{x}_i)$ and

$$
\mathbb{C}\text{ov}\left[y_i, y_j\right] = \mathbb{C}\text{ov}\left[a(\boldsymbol{x}_i), a(\boldsymbol{x}_j)\right] + \mathbb{C}\text{ov}\left[\epsilon_i, \epsilon_j\right] = K(\boldsymbol{x}_i, \boldsymbol{x}_j) + \sigma^2 \left[\!\left[i = j \right]\!\right].
$$

The joint distribution of the vector $\binom{\boldsymbol{y}}{a(\boldsymbol{x}^*)}$ will therefore be distributed according to

$$
\begin{pmatrix} \boldsymbol{y} \\ a(\boldsymbol{x}^*) \end{pmatrix} \sim \mathcal{N}\left(\boldsymbol{0}, \begin{pmatrix} \mathsf{K} + \sigma^2\,\mathsf{I} & \boldsymbol{k} \\ \boldsymbol{k}^{\mathsf{T}} & k \end{pmatrix}\right)
$$

where K is a matrix with elements $K_{ij} = K(\boldsymbol{x}_i, \boldsymbol{x}_j)$, I is the identity matrix, $\boldsymbol{k}$ is a vector with elements $k_i = K(\boldsymbol{x}_i, \boldsymbol{x}^*)$ and $k = K(\boldsymbol{x}^*, \boldsymbol{x}^*)$. Using one of those mysterious matrix identities involving the inverse of partitioned matrices (see Equations (5.3) and (5.4) on page 101),

$$
\begin{pmatrix} \mathsf{K} + \sigma^2\,\mathsf{I} & \boldsymbol{k} \\ \boldsymbol{k}^{\mathsf{T}} & k \end{pmatrix}^{-1} = \begin{pmatrix} \left(\mathsf{K} + \sigma^2\,\mathsf{I} - \frac{\boldsymbol{k}\boldsymbol{k}^{\mathsf{T}}}{k}\right)^{-1} & \frac{\left(\mathsf{K}+\sigma^2\,\mathsf{I}\right)^{-1}\boldsymbol{k}}{k+\boldsymbol{k}^{\mathsf{T}}(\mathsf{K}+\sigma^2\,\mathsf{I})^{-1}\boldsymbol{k}} \\ \frac{\boldsymbol{k}^{\mathsf{T}}\left(\mathsf{K}+\sigma^2\,\mathsf{I}\right)^{-1}}{k+\boldsymbol{k}^{\mathsf{T}}(\mathsf{K}+\sigma^2\,\mathsf{I})^{-1}\boldsymbol{k}} & \frac{1}{k+\boldsymbol{k}^{\mathsf{T}}(\mathsf{K}+\sigma^2\,\mathsf{I})^{-1}\boldsymbol{k}} \end{pmatrix}.
$$

Writing out the probability distribution for the vector $\binom{\boldsymbol{y}}{a(\boldsymbol{x}^*)}$ explicitly (ignoring terms not involving $a(\boldsymbol{x}^*)$), we find

$$f\big(a(\boldsymbol{x}^*)\big|\mathcal{D}\big) \propto \mathrm{e}^{-\frac{a(\boldsymbol{x}^*)^2}{2\left(k+\boldsymbol{k}^\mathsf{T}(\mathsf{K}+\sigma^2\mathsf{I})^{-1}\boldsymbol{k}\right)}+\frac{a(\boldsymbol{x}^*)\boldsymbol{k}^\mathsf{T}(\mathsf{K}+\sigma^2\mathsf{I})^{-1}\boldsymbol{y}}{k+\boldsymbol{k}^\mathsf{T}(\mathsf{K}+\sigma^2\mathsf{I})^{-1}\boldsymbol{k}}}.$$

Completing the square we find

$$f\big(a(\boldsymbol{x}^*)\big|\mathcal{D}\big) \propto \mathrm{e}^{\frac{-1}{2\left(k+\boldsymbol{k}^\mathsf{T}(\mathsf{K}+\sigma^2\mathsf{I})^{-1}\boldsymbol{k}\right)}\left(a(\boldsymbol{x}^*)-\boldsymbol{k}^\mathsf{T}(\mathsf{K}+\sigma^2\mathsf{I})^{-1}\boldsymbol{y}\right)^2}$$

or

$$f\big(a(\boldsymbol{x}^*)\big|\mathcal{D}\big) = \mathcal{N}\Big(a(\boldsymbol{x}^*)\Big|\boldsymbol{k}^\mathsf{T}(\mathsf{K}+\sigma^2\mathsf{I})^{-1}\boldsymbol{y}, k-\boldsymbol{k}^\mathsf{T}(\mathsf{K}+\sigma^2\mathsf{I})^{-1}\boldsymbol{k}\Big). \tag{12.2}$$

Rather remarkably, this is equivalent to Equation (12.1) – see Exercise 12.1. It is usually more convenient to work with Equation (12.2) as it involves the kernel function $K(\boldsymbol{x},\boldsymbol{x}')$ rather than its inverse – recall that the kernel function is equal to the correlation function of the prior.

As all quantities are normally distributed we can compute the evidence $f(\mathcal{D})$ in closed form

$$f(\mathcal{D}) = -\frac{1}{2}\boldsymbol{y}^\mathsf{T}\left(\mathsf{K}+\sigma^2\mathsf{I}\right)^{-1}\boldsymbol{y} - \frac{1}{2}\log\left(\left|\mathsf{K}+\sigma^2\mathsf{I}\right|\right) - \frac{n}{2}\log(2\pi).$$

(This can also be seen as the probability of the vector or targets $\boldsymbol{y}$ given the features $\mathbf{X}$ where the columns of $\mathbf{X}$ correspond to the feature vector $\boldsymbol{x}_i$.) The evidence allows us to select between different models. That is, between different mean functions, $\mu(t)$, and in different covariance functions $K(t,t')$. This is particularly useful, as there are often parameters in the covariance function that are difficult to choose a priori.

Example 12.1 Gaussian Process Curve Fitting

To illustrate using a Gaussian process prior in regression we consider a one-dimensional example since this is the easiest to plot – we considered curve fitting with polynomials previously in Example 9.5 on page 278. We have generated some data (in this case by choosing deviates $y_i \sim \cos(x_i) + \mathcal{N}(0, 0.05^2)$ for $x_i = 0, 1, 2, \ldots$). As our prior we used the Gaussian process $\mathcal{GP}(0, K(x,x'))$ with the kernel function $K(x,x') = \exp(-(x-x')^2/(2\ell^2))$. Figure 12.3 shows the mean of the posterior for the fitting curve for (a) $\ell = 1$ and (b) $\ell = \frac{1}{2}$. The grey area shows deviation up to one standard deviation from the mean.

The outcome depends critically on the kernel function. From the Bayesian point of view, the Gaussian process we choose as our prior should encode our prior beliefs. Although we may have some vague belief that the function we are fitting should be relatively smooth, encoding this belief as a kernel function is non-trivial. Of course, we could choose some of the parameters in our kernel function

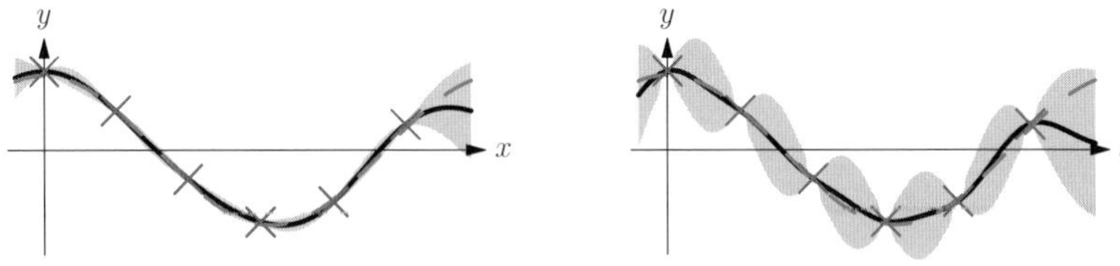

(a) $\ell = 1$ (b) $\ell = \frac{1}{2}$

Figure 12.3 Fitting data using a Gaussian process prior with kernel $K(x, x') = \exp(-(x - x')^2/(2\,\ell^2))$ with (a) $\ell = 1$, (b) $\ell = \frac{1}{2}$. The data points are shown as red crosses. The data points were taken from the cosine function (shown by the blue dashed line) plus normally distributed noise. The mean of the posterior is shown by a thick black line and predicted errors of ± 1 standard deviation are shown by the grey area.

(for example, the scale parameter ℓ) as a hyperparameter which we learn in a hierarchical model. We could just take the more pragmatic approach of testing a few kernels on some unseen data and choosing the best – for example, as measured by the evidence. We lose the moral high ground of following a truly Bayesian philosophy by doing this, but Gaussian processes can give very good results all the same.

The use of Gaussian processes for machine learning has a long history. It was used to model spatial distributions where it is known as kriging (Cressie and Wikle, 2011). It was independently reinvented in the machine learning community where it has been highly developed. The interested reader should consult the specialist literature, e.g. Rasmussen and Williams (2006).

In recent years Gaussian processes have been heavily used for optimising continuous parameters of algorithms. A Gaussian process is fitted through a number of points where the performance of the algorithm has been evaluated. For stochastic algorithms there is usually some uncertainty in the expected performance of the algorithm that can be captured well by Gaussian processes. The fit is then used to choose a point with a strong likelihood of being better than any other point, or of providing more information about the landscape. In very high dimensions finding optimal solutions for the Gaussian process model can be computationally demanding in itself. However, in cases where evaluating the model is very costly, this approach, known as *Bayesian optimisation*, is very attractive.

12.1.3 Markov Processes

By far the most commonly used stochastic processes are Markov processes, where the conditional probabilities satisfy the usual Markov property

$$f_{X(t)}(x_n, t_n | x_{n-1}, t_{n-1};\ x_{n-2}, t_{n-2};\ \ldots;\ x_1, t_1) = f_{X(t)}(x_n, t_n | x_{n-1}, t_{n-1})$$

assuming $t_1 < t_2 < \cdots < t_{n-1} < t_n$. A consequence of the Markov property is that the conditional probabilities satisfy the *Chapman–Kolmogorov equation*

$$f_{X(t)}(x_3,t_3|x_1,t_1) = \int f_{X(t)}(x_3,t_3|x_2,t_2)\, f_{X(t)}(x_2,t_2|x_1,t_1)\,\mathrm{d}x_2 \tag{12.3}$$

where $t_1 < t_2 < t_3$ and the integral is over the whole domain of $X(t_2)$.

For a sufficiently smooth Markov process we can derive an equation describing the evolution of the conditional probability density. To compute this we Taylor expand the conditional density

$$f_{X(t)}(x_2,t+\Delta t|x_1,t) = f_{X(t)}(x_2,t|x_1,t) + \Delta t\,\frac{\partial f_{X(t)}(x_2,t|x_1,t)}{\partial t} + O\left((\Delta t)^2\right).$$

The first term is equal to the Dirac delta function, $f_{X(t)}(x_2,t|x_1,t) = \delta(x_2-x_1)$, since $X(t)$ is a single valued function. The derivative can be written as

$$\frac{\partial f_{X(t)}(x_2,t|x_1,t)}{\partial t} = W_t(x_2|x_1) - \delta(x_2-x_1)\int W_t(x_2|x_1)\,\mathrm{d}x_2.$$

The term $W_t(x_2|x_1)$ gives the probability of moving to x_2 from x_1 while the second term gives the probability of leaving x_1. Note that the second term is necessary to ensure that (at order Δt)

$$\int f_{X(t)}(x_2,t+\Delta t|x_1,t)\,\mathrm{d}x_2 = 1.$$

Using the Chapman–Kolmogorov Equation (12.3) together with the Taylor expansion

$$\begin{aligned}
&f_{X(t)}(x_2,t+\delta t)\\
&\quad= \int f_{X(t)}(x_2,t+\Delta t|x_1,t)\, f_{X(t)}(x_1,t)\,\mathrm{d}x_1\\
&\quad= \int \left(\delta(x_2-x_1) + \Delta t\,\frac{\partial f_{X(t)}(x_2,t|x_1,t)}{\partial t} + O\left((\Delta t)^2\right)\right) f_{X(t)}(x_1,t)\,\mathrm{d}x_1\\
&\quad= f_{X(t)}(x_2,t) + \Delta t \int \big(W_t(x_2|x_1) f_{X(t)}(x_1,t)\\
&\qquad - W_t(x_1|X_2)\, f_{X(t)}(x_2,t)\big)\,\mathrm{d}x_1 + O\left((\Delta t)^2\right).
\end{aligned}$$

Rearranging and taking the limit $\Delta t \to 0$

$$\frac{\partial f_{X(t)}(x_2,t)}{\partial t} = \int \big(W_t(x_2|x_1)\, f_{X(t)}(x_1,t) - W_t(x_1|x_2)\, f_{X(t)}(x_2,t)\big)\,\mathrm{d}x_1. \tag{12.4}$$

Equation (12.4) is known as the *master equation*. Just as for a Markov chain if the transition probabilities don't depend on time (i.e. $W_t(x_1|x_2) = W(x_1|x_2)$) then we say the process is *stationary*. For a system where $X(t)$ can only take discrete values the master equation is

$$\frac{\mathrm{d}p_i(t)}{\mathrm{d}t} = \sum_j \big(W(i|j)\,p_j(t) - W(j|i)\,p_i(t)\big).$$

Note that the term $W(i|i)$ does not contribute to the master equation.

We can sometimes obtain the master equation directly. We illustrate this in the next example.

Example 12.2 Decay Process
Consider some number of radioactive particles where the probability of any particle decaying in time δt is equal to $\gamma\,\delta t$. Assuming that there are j particles then the probability, in an interval δt, of there being i particles is

$$W(i|j)\,\delta t = \begin{cases} 0 & \text{if } i > j \\ \gamma\, j\, \delta t & \text{if } i = j-1 \\ O\left((\delta t)^2\right) & \text{if } i < j-1. \end{cases}$$

We don't need to specify $W(i|i)$. The discrete master equation becomes

$$\begin{aligned}\frac{\mathrm{d}\,p_i(t)}{\mathrm{d}t} &= W(i|i+1)\,p_{i+1}(t) - W(i-1|i)\,p_i(t)\\ &= (i+1)\,\gamma\,p_{i+1}(t) - \gamma\, i\, p_i(t).\end{aligned}$$

All other terms vanish. This is still an awkward equation to work with; however, we can quickly obtain a differential equation for the expected number of remaining radioactive particles

$$\begin{aligned}\sum_{i=0}^{\infty} i\,\frac{\mathrm{d}\,p_i(t)}{\mathrm{d}t} &= \gamma \sum_{i=0}^{\infty} i\,(i+1)\,p_{i+1}(t) - \gamma \sum_{i=0}^{\infty} i^2\, p_i(t)\\ &= \gamma \sum_{i=0}^{\infty} \left((i-1)\,i - i^2\right) p_i(t) = \gamma \sum_{i=0}^{\infty} i\, p_i(t)\end{aligned}$$

where we changed the dummy index in the first term (in doing this we added the term $0 \times 1 \times p_0(t)$ into the sum – fortunately this equals 0). Now as the expected number of particles $\mathbb{E}\left[N(t)\right]$ is given by $\sum_{i=0}^{\infty} i\, p_i(t)$ we find

$$\frac{\mathrm{d}\,\mathbb{E}\left[N(t)\right]}{\mathrm{d}t} = -\gamma \mathbb{E}\left[N(t)\right]$$

or $\mathbb{E}\left[N(t)\right] = N_0 \mathrm{e}^{-\gamma t}$.

Sometimes we have a Markov chain (e.g. a random walk) where the time is discretised

$$\boldsymbol{p}(t+1) = \mathbf{M}\,\boldsymbol{p}(t)$$

but we know that the change in the probabilities is rather slow. We then consider the continuous time Markov process as an approximation to the discrete system

$$\frac{\mathrm{d}\,\boldsymbol{p}(t)}{\mathrm{d}t} \approx \boldsymbol{p}(t+1) - \boldsymbol{p}(t) = (\mathbf{M} - \mathbf{I})\,\boldsymbol{p}.$$

This is equivalent to

$$\frac{\mathrm{d}\, p_i(t)}{\mathrm{d}t} = \sum_j \big(M_{ij} p_j(t) - M_{ji} p_i(t)\big)$$

since, as probability is conserved, $\sum_j M_{ji} = 1$ (i.e. M is a stochastic matrix). This is identical to the master Equation (12.4) with $W(i|j) = M_{ij}$.

Deriving stochastic processes through the master equation is not always the most natural route. In many situations a more natural way of deriving stochastic processes is through a modified differential equation known as Itô's equation. This is particularly useful for diffusion-type processes which we discuss below. Diffusion processes tend to overshadow other stochastic processes, but there are many stochastic processes which are not diffusion processes (e.g. in Section 12.3 we consider point processes).

12.2 Diffusion Processes

Diffusion processes are useful for describing a huge group of phenomena that experiences small random perturbations on a short time scale. The prototypical example is a particle diffusing through a material, where its motion is caused by many small (molecular) collisions. The applications of diffusion processes are vast, ranging from physical models of heat diffusion to genetics and finance.

12.2.1 Brownian Motion

In Section 10.1 we introduced random walks. Consider a one-dimensional walker stating from the origin, $X(0) = 0$, that at each time interval he/she takes a step $\Delta X \sim \mathcal{N}(0, 1)$. After t times the position of the walker is

$$X(t) \sim \sum_{t'=1}^{t} \Delta X(t').$$

The sum of t independent normally distributed random variables is itself normally distributed (we showed this when discussing the central limit theorem in Section 5.3). Thus $X(t) \sim \mathcal{N}(0, t)$. Note that the root mean squared (RMS) distance from the origin grows as $\sqrt{t}$. Now, consider defining a Markov process that has the same large t behaviour as the discrete random walk described here. We can do this by introducing time intervals Δt such that a random walker makes a step

$$X(t + \Delta t) = X(t) + \Delta W(t)$$

where $\Delta W(t)$ are independent at each time interval and

$$\Delta W(t) \sim \mathcal{N}(0, \Delta t).$$

That is,

$$\mathbb{E}\left[\Delta W(t)\right] = 0, \qquad \mathbb{E}\left[\Delta W(t) \Delta W(t')\right] = \Delta t \left[\!\left[t = t' \right]\!\right].$$

In the limit $\Delta t \to 0$, we can write

$$\mathrm{d}X(t) = X(t + \mathrm{d}t) - X(t) = \mathrm{d}W(t)$$

where $\mathrm{d}W(t)$ is known as the *Brownian* or *Wiener measure* with the properties

$$\mathbb{E}\left[\mathrm{d}W(t)\right] = 0, \qquad \mathbb{E}\left[\mathrm{d}W(t)\;\mathrm{d}W(t')\right] = \mathrm{d}t\,\delta(t - t')$$

where $\delta(t - t')$ is the Dirac delta function. The solution of this random walk, $X(t) = W(t)$ (assuming $X(0) = 0$), has the property that

$$X(t') - X(t) = W(t') - W(t) \sim \mathcal{N}(0, |t' - t|).$$

Furthermore, $W(t') - W(t)$ is independent of $W(t)$ since $W(t') - W(t)$ describes future fluctuations. This model is known by mathematicians as 'Brownian motion' and by physicists as a Wiener process. Physicists use the term Brownian motion to refer to a different process which mathematicians call an Ornstein–Uhlenbeck process (which can be viewed as a noisy particle in a quadratic potential).

If the walker had a bias so that in expectation $X(t)$ increases by a after 1 second then the walker's dynamics can be described by

$$\mathrm{d}X(t) = a\,\mathrm{d}t + \mathrm{d}W(t). \tag{12.5}$$

This is an example of a *stochastic differential equation*. The first term is known by physicists as the drift term as it describes the general drift of a particle. The second term is the fluctuation term – it is equal to zero on average. (Unfortunately, stochastic differential equations are also used in population dynamics where the term genetic drift is used to describe the fluctuation term.)

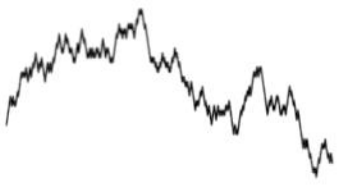

The stochastic differential equation, $\mathrm{d}X(t) = \mathrm{d}W(t)$, defines a continuous random walk, $X(t)$. The walk depends on the Wiener measure $\mathrm{d}W(t)$, which is a random variable. Consequently, the stochastic differential equation describes a family of possible walks. As the magnitude of $\Delta W(t)$ goes to zero as $\Delta t \to 0$, the walk is a continuous function. This is not true of the derivative. To see this, consider the case of zero drift for the discrete random walk. The mean squared of the discrete derivative is

$$\mathbb{E}\left[\left(\frac{X(t+\Delta t) - X(t)}{\Delta t}\right)^2\right] = \mathbb{E}\left[\left(\frac{\Delta W(t)}{\Delta t}\right)^2\right] = \frac{1}{(\Delta t)^2}\mathbb{E}\left[(\Delta W(t))^2\right] = \frac{1}{\Delta t}.$$

Thus in the limit $\Delta t \to 0$, the magnitude of the derivative diverges (at least, in expectation). To understand this recall that the typical size of $\Delta W(t)$ is $\sqrt{\Delta t}$. In the limit $\Delta t \to \infty$ the fluctuations diverge, but much slower than any drift term. Conversely, when $\Delta t \to 0$, the fluctuation goes to zero, but also much slower than any drift term. In consequence our particle seems to have infinite velocity on very short time scales. Physically we don't have to alarmed about this because we are only using the limiting process as a trick. In real systems the fluctuations are usually caused by a set of small knocks to the system at finite intervals.

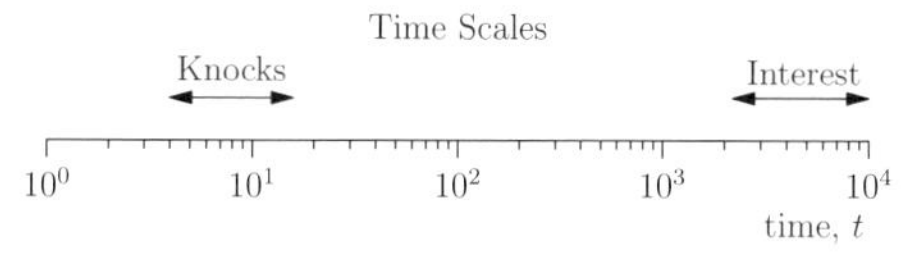

We are interested in the behaviour of the system over long time periods. For any interval at the time scale of interest there will have been many knocks. Assuming the knocks are independent (which is the key assumption we make about the noise) then by the central limit theorem the sum of all the knocks will lead to a noise which is normally distributed. This is true irrespective of the distribution of knocks provided they satisfy the conditions of the central limit theorem (i.e. their distribution is not so long tailed that it has infinite variance). It is this fact which makes diffusion models so ubiquitous. As we don't care about the distribution of the knocks (other than its mean and variance) we can pretend that the origin of the noise is a set of infinitely small, normally distributed knocks occurring infinitely often – this allows us to use tricks from differential calculus.

Mathematically though, the fact that our stochastic process is non-differentiable is more troubling, since it is unclear what we mean by Equation (12.5) as the derivative doesn't exist. This equation can be viewed as a shorthand for the limiting process defined above. We discuss this in more detail in the next section where we consider general stochastic differential equations.

12.2.2 Stochastic Differential Equations

We can generalise the stochastic differential equation for a random walk with drift to the more complex situation

$$\mathrm{d}X(t) = a(t, X(t))\,\mathrm{d}t + b(t, X(t))\,\mathrm{d}W(t) \tag{12.6}$$

where the drift term, $a(t, X(t))$, and the fluctuation term, $b(t, X(t))$, may depend on both the current value of $X(t)$ and on the time, t. Physicists frequently write this as

$$\frac{\mathrm{d}X(t)}{\mathrm{d}t} = a(t, X(t)) + b(t, X(t))\,\eta(t)$$

where $\eta(t) = \mathrm{d}W(t)/\mathrm{d}t$. This is often called a *Langevin equation*. We might like to think of this equation as a shorthand for some limit, but be aware that that the derivative diverges so we shouldn't think of it as particularly meaningful. An alternative view is to think of the stochastic differential equation as a shorthand for the integral equation

$$X(t) - X(0) = \int_0^t a(t', X(t'))\,\mathrm{d}t' + \int_0^t b(t', X(t'))\,\mathrm{d}W(t')$$

where the first integral is a normal well-defined integral, but the second integral is a special stochastic integral. The stochastic integral can be interpreted in terms of a limiting processes. However, in general the result of the expectation of this

The Riemann Integral

$$\int_a^b f(t)\,\mathrm{d}t = \lim_{\Delta_i \to 0} \sum_{i=0}^{n-1} f_i\,\Delta_i$$

$\Delta_i = t_{i+1} - t_i$

$f_i = f(t)$ for $t \in [t_i, t_{i+1}]$

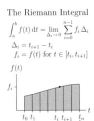

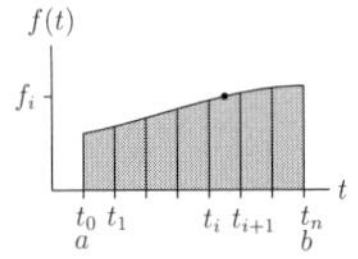

integral depends on how we take the limit! That is, we can approximate the stochastic integral as a sum

$$\int_0^t b(t', X(t'))\,\mathrm{d}W(t') \approx \sum_{i=0}^{n-1} b_i \left(W(t_{i+1}) - W(t_i)\right)$$

where $(t_i | i = 0, 1, \ldots, n)$ is an ordered set of times from 0 to t and b_i depends on $b(t', X(t'))$ in the interval $[t_i, t_{i+1}]$. For normal integrals the limit (usually) does not depend on how we choose b_i (e.g. it might be the initial point, the mid-point, or the end point). However, in stochastic integrals the result we get does. To see this we consider the case when $b(t, X(t)) = W(t)$. We consider the two common choices

Itô's stochastic integral where $b_i = b(t_i, X(t_i))$, i.e. it is the first point in the time interval. Then when $b(t, X(t)) = W(t)$

$$I_1 = \int_0^t W(t)\,\mathrm{d}W(t) \approx \sum_{i=0}^{n-1} W(t_i) \left(W(t_{i+1}) - W(t_i)\right).$$

In expectation $\mathbb{E}\left[I_1\right] = 0$ since $\left(W(t_{i+1}) - W(t_i)\right)$ is a normally distributed random variable, $\mathcal{N}(0, t_{i+1} - t_i)$ that is uncorrelated with $W(t_i)$.

Stratonovich stochastic integral where b_i is taken as the average of the two end-point values

$$b_i = \frac{b(t_i, X(t_i)) + b(t_{i+1}, X(t_{i+1}))}{2}.$$

If we take $b(t, X(t)) = W(t)$ then

$$\begin{aligned} I_2 = \int_0^t W(t)\,\mathrm{d}W(t) &\overset{(1)}{=} \approx \sum_{i=0}^{n-1} \frac{1}{2} \left(W(t_{i+1}) + W(t_i)\right) \left(W(t_{i+1}) - W(t_i)\right) \\ &\overset{(2)}{=} \frac{1}{2} \sum_{i=0}^{n-1} \left(W^2(t_{i+1}) - W^2(t_i)\right) \overset{(3)}{=} \frac{1}{2} \left(W^2(t_n) - W^2(t_0)\right) \overset{(4)}{=} \frac{1}{2} W^2(t) \end{aligned}$$

(1) From the definition of the Stratonovich stochastic integral we choose $W(t) = \frac{1}{2}\left(W(t_{i+1}) + W(t_i)\right)$ and $\mathrm{d}W(t) = W(t_{i+1}) - W(t_i)$.
(2) From the difference of two squares.
(3) Follows from the cancellation of all the intermediate terms in the sum.
(4) Using $t_n = t$, $t_0 = 0$ and $W^2(0) = 0$.

In expectation $\mathbb{E}\left[I_2\right] = \mathbb{E}\left[W^2(t)/2\right] = t/2$.

This seems rather unsatisfactory, as the result depends on how we interpret the constant $b(t, X(t))$! This is a consequence of the fact that the infinitesimal noise is large ($\sqrt{\mathrm{d}t}$) compared to the interval itself ($\mathrm{d}t$). Which interpretation should we use? This depends on the actual process we are trying to model. Remember that we tend to use stochastic processes to model processes where randomness usual occurs at discrete time points. The way that this randomness influences the random variables in the system determines which interpretation we should choose. It is nearly always Itô's interpretation which correctly describes the process people try to model, and this interpretation is by far the most commonly used.

Unfortunately, Itô's interpretation leads to a slightly strange calculus. Note that the Stratonovich interpretation of I_2 gave an expectation that you would expect from applying the normal rules of calculus. This is generally the case: you can treat Stratonovich's stochastic integrals as if they are normal integrals. However, Itô's interpretation nearly always gives the answer you are after. It is possible to map from Itô's interpretation to Stratonovich, but in practice most people just use Itô's calculus despite its strange rules.

Itô's calculus appeared so weird that it was initially treated with some degree of scepticism.

Itô's calculus lets you obtain results for simple stochastic differential equations in terms of functions of the Wiener measure. A useful result known as *Itô's isometry* tells you about the variance of an Itô integral. We note that in expectation

$$\begin{aligned}\mathbb{E}\left[\int_{t_0}^{t_1} X(t)\mathrm{d}W(t)\right] &= \lim_{(t_{i+1}-t_i)\to 0} \mathbb{E}\left[\sum_i X(t_i)\left(W(t_{i+1}) - W(t_i)\right)\right] \\ &= \sum_i \mathbb{E}\left[X(t_i)\left(W(t_{i+1}) - W(t_i)\right)\right]\end{aligned}$$

but $W(t_{i+1}) - W(t_i)$ is by definition independent of $X(t_i)$ and has zero mean so the expectation of this integral is zero. The variance is

$$\begin{aligned}&\mathbb{V}\mathrm{ar}\left[\int_{t_0}^{t_1} X(t)\,\mathrm{d}W(t)\right] \\ &\stackrel{(1)}{=} \mathbb{E}\left[\left(\int_{t_0}^{t_1} X(t)\,\mathrm{d}W(t)\right)^2\right] \\ &\stackrel{(2)}{=} \mathbb{E}\left[\int_{t_0}^{t_1} X(t)\,\mathrm{d}W(t)\int_{t_0}^{t_1} X(t')\,\mathrm{d}W(t')\right] \\ &\stackrel{(3)}{=} \lim_{(t_{i+1}-t_i)\to 0} \mathbb{E}\left[\sum_i X(t_i)\left(W(t_{i+1}) - W(t_i)\right)\sum_j X(t_j)\left(W(t_{j+1}) - W(t_j)\right)\right] \\ &\stackrel{(4)}{=} \lim_{(t_{i+1}-t_i)\to 0} \sum_{i,j} \mathbb{E}\left[X(t_i)\,X(t_j)\left(W(t_{i+1}) - W(t_i)\right)\left(W(t_{j+1}) - W(t_j)\right)\right] \\ &\stackrel{(5)}{=} \lim_{(t_{i+1}-t_i)\to 0} \sum_i \mathbb{E}\left[X^2(t_i)\right]\mathbb{E}\left[\left(W(t_{i+1}) - W(t_i)\right)^2\right] \\ &\stackrel{(6)}{=} \lim_{(t_{i+1}-t_i)\to 0} \sum_i \mathbb{E}\left[X^2(t_i)\right](t_{i+1} - t_i) \\ &\stackrel{(7)}{=} \int_{t_0}^{t_1} \mathbb{E}\left[X^2(t)\right]\,\mathrm{d}t \stackrel{(8)}{=} \mathbb{E}\left[\int_{t_0}^{t_1} X^2(t)\,\mathrm{d}t\right]\end{aligned}$$

(1) Since the mean is zero, the variance is the expectation of the square.
(2) Rewriting the squared integral as a double integral.
(3) Replacing the integral by the limit of the Itô summation.
(4) Simple rearranging (pulling the sums out of the expectation).
(5) Using the fact that the Wiener measures are independent on all time intervals so that $\mathbb{E}\left[X(t_i)\,X(t_j)\left(W(t_{i+1})-W(t_i)\right)\left(W(t_{j+1})-W(t_j)\right)\right]=0$ if $i\neq j$. We are also using that $\left(W(t_{i+1})-W(t_i)\right)$ is independent of $X(t_i)$.
(6) Using $\mathbb{E}\left[\left(W(t_{i+1})-W(t_i)\right)^2\right]=t_{i+1}-t_i$.
(7) Identifying the sum as a normal Riemann integral (there are no longer any fluctuating quantities).
(8) Reversing the order of the integral and expectation.

A second basic building block of Itô's calculus is *Itô's lemma*, which describes how Itô processes transform under a change of variable. This is different to the normal change of variable for differentials (i.e. the chain rule) because we cannot neglect terms of order $(\mathrm{d}X(t))^2$. In general, consider a stochastic process, $X(t)$, defined by the stochastic differential equation

$$\mathrm{d}X(t)=a(t)\,\mathrm{d}t+b(t)\,\mathrm{d}W(t)$$

interpreted in the sense of Itô ($a(t)$ and $b(t)$ may be functions of $X(t)$ we just haven't bothered to write this explicitly). Suppose we wish to make a change of variables $X(t)\to Y(t)=g(X(t))$ where we assume g is a twice differentiable function. Taylor expanding $g(X(t)+\mathrm{d}X(t))$ around $X(t)$ we find

$$\begin{aligned}g(X(t)+\mathrm{d}X(t))&=g(X(t))+g'(X(t))\,\mathrm{d}X(t)\\&\quad+\frac{1}{2}g''(X(t))\,(\mathrm{d}X(t))^2+O\left((\mathrm{d}X(t))^3\right).\end{aligned}$$

But

$$\begin{aligned}(\mathrm{d}X(t))^2&=\left(a(t)\,\mathrm{d}t+b(t)\,\mathrm{d}W(t)\right)^2\\&=a^2(t)\,(\mathrm{d}t)^2+2\,a(t)\,b(t)\,\mathrm{d}W(t)\,\mathrm{d}t+b^2(t)\,(\mathrm{d}W(t))^2\end{aligned}$$

involves the term $b^2(t)\,(\mathrm{d}W(t))^2$ which in expectation is equal to $b^2(t)\,\mathrm{d}t$ (the other two terms are all of order $(\mathrm{d}t)^{3/2}$ and $(\mathrm{d}t)^2$ so are safe to ignore). Itô showed that the correct behaviour is obtained by replacing $(\mathrm{d}W(t))^2$ by its expected value $\mathrm{d}t$. This leads to Itô's lemma

$$\begin{aligned}\mathrm{d}g(X(t))&=g(X(t)+\mathrm{d}X(t))-g(X(t))\\&=g'(X(t))\,\mathrm{d}X(t)+\frac{1}{2}g''(X(t))\,b^2(t)\,\mathrm{d}t.\end{aligned}$$

The first term is the usual change of variable term, but now we have a new correction to the drift coming from the fluctuations. If $g(X(t))$ is linear the second term vanishes and the transformation of variables follows the normal rules, but for more complex transformations we need to include the additional term $\frac{1}{2}g''(X(t))\,b^2(t)\,\mathrm{d}t$.

Example 12.3 Stock Price
A very simple model for stock prices is to assume that the return is equal to a long-term growth term plus a fluctuation term. The return is the growth per stock, so that

$$\frac{\mathrm{d}S(t)}{S(t)} = \mu \,\mathrm{d}t + \sigma \,\mathrm{d}W(t)$$

or

$$\mathrm{d}S(t) = \mu \, S(t) \,\mathrm{d}t + \sigma \, S(t) \,\mathrm{d}W(t)$$

where μ describes the long-term rise (or fall) in the values of stocks while σ is a measure of the volatility of the price. This is a somewhat unexpected model. In a world where fluctuations were caused by many small random events we might expect the variance to grow in proportion to $S(t)$ (so that the typical fluctuations grow as $\sqrt{S(t)}$). For example, if the value of the company depends on the performance of the individual employees we might expect the variance to grow with the size of the company. However, empirically the variance in the stock price seems to grow as $S(t)^2$, suggesting that the fluctuations are *not* caused by many independent random components, but are just a fixed proportion of the company size. This might be the case if the stock price reflected the effectiveness of the spending decisions of the CEO. Of course, many of the fluctuations in stock price might be caused by the stochastic behaviour of investors and have little to do with the true value of the company. However, one hopes that at some level the stock price provides some approximation to the true value of the company. Let us see the consequence of our model.

We can obtain a simpler equation to analyse by making the change of variables $X(t) = g(S(t)) = \log(S(t))$. Applying Itô's lemma we find

$$\begin{aligned}
\mathrm{d}X(t) &\overset{(1)}{=} g'(S(t))\mathrm{d}S(t) + g''(S(t))(\sigma \, S(t))^2 \,\mathrm{d}t \\
&\overset{(2)}{=} \frac{\mathrm{d}S(t)}{S(t)} - \frac{1}{S^2(t)}(\sigma \, S(t))^2 \,\mathrm{d}t \\
&\overset{(3)}{=} \left(\mu - \frac{\sigma^2}{2}\right) \mathrm{d}t + \sigma \,\mathrm{d}W(t)
\end{aligned}$$

(1) From Itô's lemma.
(2) Using $g'(S(t)) = 1/S(t)$ and $g''(S(t)) = -1/S(t)^2$.
(3) From the stochastic differential equation $\mathrm{d}S(t)/S(t) = \mu \,\mathrm{d}t + \sigma \,\mathrm{d}W(t)$, cancelling $S(t)^2$ in the second term and rearranging.

We can make a further change of variables that $Y(t) = X(t) - (\mu - \sigma^2/2)t$ in this case

$$\mathrm{d}Y(t) = \sigma \,\mathrm{d}W(t).$$

As this was a linear change of variables (in $X(t)$), we get no correction term from Itô's lemma. We observe in passing that $Y(t)$ is a martingale (since in expectation $\mathbb{E}\left[\mathrm{d}Y(t)\right] = 0$). It is trivial to solve this equation giving $Y(t) = \sigma W(t) + c$. Thus, $X(t) = c + (\mu - \sigma^2/2)t + \sigma W(t)$ and finally

$$S(t) = S(0)\,\mathrm{e}^{(\mu-\sigma^2/2)t+\sigma W(t)}$$

where $S(0) = \mathrm{e}^c$. Note that our solution, $S(t)$, depends on the normally distributed stochastic variable $W(t)$. Since $\log(S(t))$ is normally distributed with mean $\log(S(0)) + (\mu - \sigma^2/2)t$ and variance $\sigma^2 t$, the stock price, $S(t)$, is log-normally distributed

$$f_{S(t)}(S(t) = s) = \mathrm{LogNorm}\left(s \middle| \log(S(0)) + (\mu - \sigma^2/2)t, \sigma^2 t\right).$$

The mean and variance of a log-normal distribution LogNorm $(x|\mu, \sigma^2)$ are $\mathrm{e}^{\mu+\sigma^2/2}$ and $(\mathrm{e}^{\sigma^2} - 1)\mathrm{e}^{2\mu+\sigma}$, respectively. Thus $\mathbb{E}\left[S(t)\right] = \mathrm{e}^{\mu t} S(0)$ as we might expect, and $\mathbb{V}\mathrm{ar}\left[S(t)\right] = (\mathrm{e}^{\sigma^2 t} - 1)\,\mathbb{E}\left[S(t)\right]^2$. For large t, the standard deviation grows approximately as $\mathrm{e}^{\sigma^2 t/2}\,\mathbb{E}\left[S(t)\right]$. That is, for stocks with identical expected growth rates and volatility (variance) the standard deviation grows exponentially (see Exercise 12.2). That is, some stocks do spectacularly well, while others fail. Yet within this model this is entirely down to luck!

Because the gains and losses on investments are compounded, the returns are typically (at least, approximately) log-normally distributed and the fluctuations can be very large. This tends to make markets inherently unstable, with bubbles, crashes, grossly inflated bonuses, and massive inequalities. We can deplore this unfairness at the same time as asking whether there is a way we can make a little extra money.

Example 12.4 Investing to Win (Example 5.4 Revisited)
In Example 5.4 on page 87 we discussed how, when betting on multiple outcomes it might be more rational to try to maximise your expected log-winnings rather than your expected winnings. The argument being that your expected winnings are dominated by the rare event of winning every time, whereas by maximising your log-winnings you tend to maximise what you will typically win.

For investing on a stock whose value changes according to

$$S(t+1) = S(t) + \mu\, S(t)\,\Delta t + \sigma\, S(t)\,\Delta W(t) \tag{12.7}$$

we can follow a strategy similar to that of Kelly as proposed by Edward Thorp for continuous investments. Suppose we have an initial capital of 1 pound which we can either invest in a bank with fixed interest rate r or in a stock. We consider a scenario where at each time interval Δt we rebalance our portfolio of stock and cash in the bank so that we have a fraction f of our capital in stocks. The fractional gain in our capital at round t is

$$\Delta X_t = (1 - f)\, r\, \Delta t + f\, \big(\mu\, \Delta t + \sigma\, \Delta W(t)\big).$$

After n rounds our capital will be worth

$$G_n(f) = \prod_{t=1}^{n} (1 + \Delta X_t).$$

The expected log-gain is

$$\begin{aligned}\log\big(G_n(f)\big) &= \sum_{t=1}^{n} \mathbb{E}\left[\log\big(1 + \Delta X_t\big)\right] = n\, \mathbb{E}\left[\log\big(1 + \Delta X_t\big)\right] \\ &\approx n\, \mathbb{E}\left[\Delta X_t - \frac{(\Delta X_t)^2}{2} + \ldots\right]\end{aligned}$$

$\log(1 + \epsilon) = \epsilon - \frac{\epsilon}{2} + \cdots$

where we have expanded the logarithm in terms of the small fractional gains. Now

$$\begin{aligned}\mathbb{E}\left[\Delta X_t\right] &= \mathbb{E}\left[(1 - f)\, r\, \Delta t + f\, \big(\mu\, \Delta t + \sigma\, \Delta W(t)\big)\right] \\ &= \big((1 - f)\, r + f\, \mu\big)\, \Delta t \\ \mathbb{E}\left[(\Delta X_t)^2\right] &= f^2\, \sigma^2 \mathbb{E}\left[(\Delta W(t))^2\right] + O((\Delta t))^2 = f^2\, \sigma^2\, \Delta t + O((\Delta t)^2)\end{aligned}$$

all higher-order terms are at least of order $O((\Delta t)^2)$. Thus

$$\log\big(G_n(f)\big) \approx n\, \Delta t\, \left((1 - f)\, r + f\, \mu - \frac{f^2\, \sigma^2}{2}\right) + O((\Delta t)^2).$$

Maximising the first-order term we find that the optimal strategy is to choose f, such that

$$f = \frac{\mu - r}{\sigma^2}.$$

Let us see what this means. Suppose we played this strategy daily for a year. We take $\Delta t = 1/365$. We assume an expected growth rate of $\mu = 365(1.05^{1/365} - 1) \approx 4.88\%$ corresponding to an expected yearly growth rate of 5% (the small difference is because the growth is compounded daily). Let us assume that $\sigma^2 = 0.06$. In Figure 12.4 we show two time series evolving according to Equation (12.7).

Now suppose we can invest our capital in a risk-free investment with interest rate $r = 365(1.02^{1/365} - 1) \approx 1.98\%$, which equates to a yearly interest rate of 2%. Plugging this into our continuous Kelly formula we find $f \approx 0.48$. In other words, each day we ensure that just under half our capital is invested in stocks. Consider using three strategies, we can choose the risk-free investment and increase our capital by a 21.9% after 10 years. We can put all our money into the

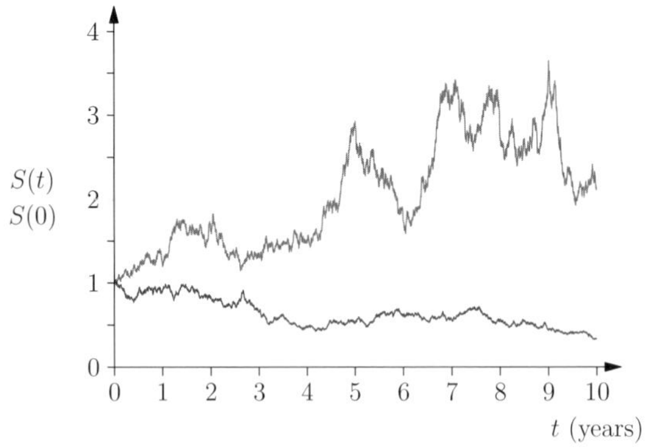

Figure 12.4 Two 10-year times series showing simulated stock prices evolving according to Equation (12.7).

stock. This gives an expected return on our capital of 62.9%, but with very high fluctuations. Finally we can follow the Kelly criterion and rebalance our capital daily, giving us an expected return of 40.2%, but with less fluctuations. Figure 12.5 shows the capital after 10 years for 1000 investors, each betting on a different stock with an expected yearly growth rate of 5%. Using Kelly's strategy reduces the chances of a small return, although it also reduces the chances of making a very large profit if you are lucky enough to invest in the right stock.

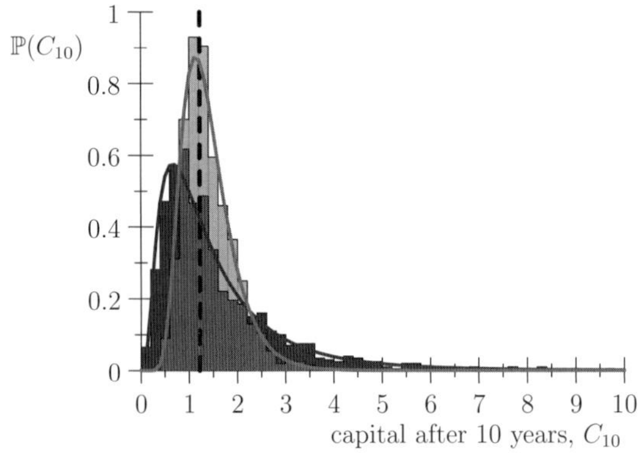

Figure 12.5 Capital after 10 years' investment for 1000 different stocks. The dark histogram is when you invest completely in stocks, the light histogram follows Kelly strategy, while the vertical dashed line shows the return on the risk-free investment. We have cut off the axis; the best return on capital in this simulation was 16.3. The solid line shows the continuous time solution.

We can get an analytic approximation of our distribution of returns by assume that we rebalance our portfolio continuously. In this case our combined holdings of stock and bank evolves according to

$$\mathrm{d}S(t) = S(t)\,\big((1-f)\,r + f\,\mu + f\,\sigma\,\mathrm{d}W(t)\big).$$

Here we use $\mu = \log(1.05)$ and $r = \log(1.02)$ corresponding to expected yearly gains of 5% and 2% per year. Following the same change of variables that we used in Example 12.3, the distribution of stock prices is

$$f_{S(t)}(S(t)=s)$$
$$= \mathrm{LogNorm}\left(s\,\middle|\,\log(S(0)) + ((1-f)\,r + f\mu - f^2\sigma^2/2)t,\ f^2\sigma^2 t\right).$$

We also show the continuous time solution for Kelly's solution ($f = (\mu - r)/\sigma^2$) and for investing entirely in the stock (i.e. $f = 1$) in Figure 12.5.

Of course, to mitigate against bad luck, one would typically invest in many different stocks. Kelly's proposal to maximise the log-gain can again help to reduce risk in this situation. However, before using Kelly's proposal to invest your life savings be aware of the limitation of the model. In our simplified world we assumed that the expected growth and volatility (variance) are fixed (stationary). In real markets, the growth rates and volatility typically change over time. If you could estimate these quantities accurately you could still follow a Kelly-type strategy, but the estimation process is non-trivial. It is also somewhat counter-intuitive. We have seen that the price of stocks with identical growth rates and volatility diverge exponentially, thus using the change in price over a long period is a hopelessly poor measure of the expected growth of a stock on a short time window. Furthermore, stock prices are not independent of each other, thus hedging your bets by investing in many different stocks is not guaranteed to prevent you from losing money. Investment banks tend to use a more reliable strategy to prevent large losses. They invest other people's money and charge fees whether or not the investment make money.

If you are a gambler then you might prefer to take a risk on getting a high return and put all your money in stocks; however, if you are investing your pension following Kelly would seem to have some virtue. Rather surprisingly Kelly is not much appreciated in financial economics. One cause of this is a vitriolic paper by two Nobel laureates in economics, Robert Merton and Paul Samuelson, who ripped to pieces Henry Latané, who independently proposed a similar strategy for portfolio selection – Kelly was not even mentioned, presumably because he discusses betting, which is completely different from asset management. Merton and Samuelson's argument is entirely theoretical and very difficult to follow, but seems to rest on the fact that the central limit theorem is only valid in the large n limit so Latané's and Kelly's strategy is provably suboptimal in some sense. However, in practice the central limit theorem is usually a good approximation for even a small number of random variables (provided the random variables being summed don't have very long tails), so Kelly's criterion may not be optimal, but it will be a pretty good approximation to the optimal strategy. Nevertheless, the fear of being ridiculed by two Nobel laureates has kept Kelly firmly outside of main stream economics (see Ziemba, MacLean, and Thorp (2012) for a comprehensive set of papers covering the use of the Kelly criterion).

We don't pursue Itô's calculus any further. Instead we study the distribution of a stochastic process at a particular time. This gives rise to the Fokker–Planck equation.

12.2.3 Fokker–Planck Equation

Stochastic processes are unfamiliar mathematical objects. To return to more familiar territory we can ask about the distribution of a stochastic process. We start from the identity

$$f_X(x,t) = \mathbb{E}\left[\delta(X(t) - x)\right] = \int f_X(y,t)\,\delta(y-x)\mathrm{d}y$$

where $\delta(y-x)$ is the Dirac delta function defined in Appendix 5.A on page 103. Now consider the distribution at a slightly later time, $t + \mathrm{d}t$ (we assume for the moment $\mathrm{d}t$ is finite and take the limit $\mathrm{d}t \to 0$ later on)

$$\begin{aligned} f_X(x,t+\mathrm{d}t) &\overset{(1)}{=} \mathbb{E}\left[\delta(X(t+\mathrm{d}t) - x)\right] \overset{(2)}{=} \mathbb{E}\left[\delta(X(t) + \mathrm{d}X(t) - x)\right] \\ &\overset{(3)}{=} \left[\delta(X(t) - x) + \delta'(X(t) - x)\,\mathrm{d}X(t)\right. \\ &\quad \left. + \frac{1}{2}\delta''(X(t) - x)\,(\mathrm{d}X(t))^2 + O((\mathrm{d}X(t))^3)\right] \\ &\overset{(4)}{=} f_X(x,t) + \mathbb{E}\left[\delta'(X(t) - x)\,\mathrm{d}X(t)\right] \\ &\quad + \frac{1}{2}\mathbb{E}\left[\delta''(X(t) - x)\,(\mathrm{d}X(t))^2\,\mathrm{d}t\right] + O((\mathrm{d}t)^2). \end{aligned}$$

(1) Using the identity above (except at time $t + \mathrm{d}t$).
(2) Writing $X(t+\mathrm{d}t) = X(t) + \mathrm{d}X(t)$.
(3) Taylor expanding the delta function to second order.
(4) Using the linearity of the expectation operator.

Where we assume our process satisfies an Itô equation

$$\mathrm{d}X(t) = a(t, X(t))\,\mathrm{d}t + b(t, X(t))\,\mathrm{d}W(t).$$

Averaging over the noise in the next time step, $\mathrm{d}W(t)$, so that

$$\begin{aligned} \mathbb{E}_{\mathrm{d}W(t)}\left[\mathrm{d}X(t)\right] &= a(t, X(t))\,\mathrm{d}t \\ \mathbb{E}_{\mathrm{d}W(t)}\left[(\mathrm{d}X(t))^2\right] &= b^2(t, X(t))\,\mathrm{d}t + O((\mathrm{d}t)^2) \end{aligned}$$

then up to corrections of order $(\mathrm{d}t)^2$

$$\begin{aligned} f_X(x,t+\mathrm{d}t) = f_X(x,t) &+ \mathbb{E}\left[\delta'(X(t) - x)\,a(t, X(t))\mathrm{d}t\right] \\ &+ \frac{1}{2}\mathbb{E}\left[\delta''(X(t) - x)\,b^2(t, X(t))\,\mathrm{d}t\right] \end{aligned}$$

(we are allowed to average $\mathrm{d}W(t)$ independently of $\delta'(X(t) - x)$ since it corresponds to fluctuations from time t to $t + \delta t$). Writing the expectation in integral form we find

$$\frac{f_X(x,t+\mathrm{d}t)-f_X(x,t)}{\mathrm{d}t} = \int f(y,t)\,\delta'(y-x)\,a(t,y)\mathrm{d}y + \frac{1}{2}\int f(y,t)\,\delta''(y-x)\,b^2(t,y)\mathrm{d}y \qquad (12.8)$$

where we have ignored terms of order $\mathrm{d}t$ which vanish in the limit $\mathrm{d}t \to 0$. Now we use integration by parts

$$\int_{-\infty}^{\infty} f(y,t)\,\delta'(y-x)\,a(t,y)\mathrm{d}y = \left[f(y,t)\,\delta(y-x)\,a(t,y)\right]_{-\infty}^{\infty} - \int_{-\infty}^{\infty}\delta(y-x)\frac{\partial}{\partial y}\left(f(y,t)\,a(t,y)\right)\mathrm{d}y = -\frac{\partial f(x,t)\,a(t,x)}{\partial x}.$$

Similarly, applying integration by parts twice to the second term in Equation (12.8) we obtain

$$\int f(y,t)\,\delta''(y-x)\,b^2(t,y)\mathrm{d}y = \frac{\partial^2 f(x,t)\,b^2(t,x)}{\partial x^2}.$$

Substituting these back and taking the limit $\mathrm{d}t \to 0$ we obtain

$$\frac{\partial f(x,t)}{\partial t} = -\frac{\partial a(t,x)\,f(x,t)}{\partial x} + \frac{1}{2}\frac{\partial^2 b^2(t,x)\,f(x,t)}{\partial x^2}. \qquad (12.9)$$

This is the *Fokker–Planck equation*. It is a generalised form of the diffusion equation. Diffusion is an example of a stochastic process where $a = 0$ and b is a constant (b^2 is sometimes known as the diffusion coefficient).

The Fokker–Planck equation is also known as Kolmogorov's forward equation. Kolmogorov provided a mathematical foundation for stochastic processes in terms of measure theory and an independent derivation of the Fokker–Planck equation. He also derived a backward equation which describes the evolution of a stochastic process going backwards in time. The backward equation is given by

$$\frac{\partial f(x,t)}{\partial t} = -a(t,x)\,\frac{\partial f(x,t)}{\partial x} - \frac{1}{2}b^2(t,x)\,\frac{\partial^2 f(x,t)}{\partial x^2}.$$

This is the adjoint equation of the forward equation. The backward equation is useful for calculating properties such as the expected probability of reaching an absorbing state or the mean time to reach such a state.

The Fokker–Planck (Kolmogorov) equations can be generalised to multi-component vectors. Consider the case where $\boldsymbol{X}(t)$ is an n-component random vector satisfying the Itô equation

$$\mathrm{d}\boldsymbol{X}(t) = \boldsymbol{a}(t,\boldsymbol{X}(t))\,\mathrm{d}t + \mathbf{B}(t,\boldsymbol{X}(t))\,\mathrm{d}\boldsymbol{W}(t)$$

with $\boldsymbol{a}(t,\boldsymbol{X}(t))$ as an n-component drift vector, $\mathrm{d}\boldsymbol{W}(t)$ as an m component Wiener measure, and $\mathbf{B}(t,\boldsymbol{X}(t))$ as a matrix measuring the size of the fluctuations. The corresponding Fokker–Planck equation is

$$\frac{\partial f(\boldsymbol{x},t)}{\partial t} = -\sum_{i=1}^{n} \frac{\partial}{\partial x_i}\left(a_i(t,\boldsymbol{x}(t)) f(\boldsymbol{x},t)\right) + \sum_{i=1}^{n}\sum_{j=1}^{n} \frac{\partial^2}{\partial x_i x_j{}^2}\left(\sum_{k=1}^{n} B_{ik}(t,\boldsymbol{X}(t))\, B_{ik}(t,\boldsymbol{x}(t)) f(\boldsymbol{x},t)\right).$$

Fokker–Planck equations are familiar objects (i.e. they are partial differential equations), however, partial differential equations are often difficult to work with. If you are lucky, you can use the method of separation of variables to obtain a solution (often in terms of a sum of polynomials).

Example 12.5 Noisy Particle in a Quadratic Potential
Consider a particle experiencing a restoring force towards the origin and a stochastic force in a random direction. In one dimension the equation of motion is given by the Itô equation

$$\mathrm{d}X(t) = -a\, X(t)\, \mathrm{d}t + \sigma\, \mathrm{d}W(t).$$

This is known as an Ornstein–Uhlenbeck process. The corresponding Fokker–Planck equation is

$$\frac{\partial f_X(x,t)}{\partial t} = a\, \frac{\partial \left(x\, f_X(x,t)\right)}{\partial x} + \frac{\sigma^2}{2} \frac{\partial^2 f_X(x,t)}{\partial x^2}.$$

Given an initial condition $f_X(x,t) = \delta(x - x_0)$, the full solution is given by

$$f_X(x,t) = \sqrt{\frac{a}{\pi\, \sigma^2 \left(1 - \mathrm{e}^{-2at}\right)}}\; \mathrm{e}^{-\frac{a(x - x_0\, \mathrm{e}^{-at})^2}{\sigma^2 (1 - \mathrm{e}^{-2at})}}.$$

The easiest way to see this is a solution to the Fokker–Planck equation is to substitute it into the differential equation (although this is a slow, thankless exercise). We observe that the Ornstein–Uhlenheck process is a Gaussian process. In the limit $t \to \infty$ this becomes

$$f_X(x,\infty) = \sqrt{\frac{1}{\pi\, \sigma^2}}\; \mathrm{e}^{-a\, x^2/\sigma^2}.$$

Ornstein–Uhlenbeck processes are important models in many realms, ranging from physical systems to fluctuations in the price of stocks.

Kramer–Moyal's Expansion

Rather than deriving the Fokker–Planck equation through a stochastic differential equation we can also obtain it through an expansion of the master equation.

Recall that the master equation is a first-order equation describing the evolution of the distribution of a random process at time t

$$\frac{\partial f_{X(t)}(x,t)}{\partial t} = \int \left(W_t(x|x')\, f_{X(t)}(x',t) - W_t(x'|x)\, f_{X(t)}(x,t) \right) \mathrm{d}x'$$

where $W_t(x|x')\,\mathrm{d}t$ is the probability of moving from x' to x at time $\mathrm{d}t$. Defining $W_t(x|x') = W_t(x';\Delta) = W_t(x-\Delta;\Delta)$ where $\Delta = x - x'$, then we can write the master equation as

$$\frac{\partial f_{X(t)}(x,t)}{\partial t} = \int W_t(x-\Delta;\Delta)\, f_{X(t)}(x-\Delta,t)\,\mathrm{d}\Delta - f_{X(t)}(x,t) \int W_t(x;-\Delta)\,\mathrm{d}\Delta.$$

Although this is just a change of variables from x to Δ and a redefinition of the transition probability, it is nevertheless important to get all the signs correct. We assume that the density $f_{X(t)}(x-\Delta,t)$ is heavily concentrated around x (that is, large jumps are unlikely) and that $f_{X(t)}(x,t)$ is a relatively smooth function of x. These are natural conditions in many situations. Under these assumptions we can Taylor expand the product $W_t(x-\Delta;\Delta)\, f_{X(t)}(x-\Delta,t)$ around x

$$W_t(x-\Delta;\Delta)\, f_{X(t)}(x-\Delta,t) = \sum_{i=0} \frac{1}{i!}(-\Delta)^i \frac{\partial^m}{\partial x^i} W_t(x;\Delta)\, f_{X(t)}(x,t).$$

Substituting this back into the master equation (and separating off the $i = 0$ term) we get

$$\begin{aligned}\frac{\partial f_{X(t)}(x,t)}{\partial t} &= \int W_t(x;\Delta)\, f_{X(t)}(x,t)\,\mathrm{d}\Delta \\ &\quad + \int \sum_{i=1} \frac{1}{i!}(-\Delta)^i \frac{\partial^i}{\partial x^i} W_t(x;\Delta)\, f_{X(t)}(x,t)\,\mathrm{d}\Delta \\ &\quad - f_{X(t)}(x,t) \int W_t(x;-\Delta)\,\mathrm{d}\Delta.\end{aligned}$$

The first and third term cancel. Taking the integral inside the derivative we get

$$\frac{\partial f_{X(t)}(x,t)}{\partial t} = \sum_{i=1} \frac{(-1)^i}{i!} \frac{\partial^i}{\partial x^i} \left(f_{X(t)}(x,t) \int \Delta^i\, W_t(x;\Delta)\mathrm{d}\Delta \right) \tag{12.10}$$

$$= \sum_{i=1} \frac{(-1)^i}{i!} \frac{\partial^i}{\partial x^i} \left(f_{X(t)}(x,t)\, M^{(i)}(x,t) \right) \tag{12.11}$$

where $M^{(i)}(x,t)$ is the ith *jump moment* of the transition probability

$$M^{(i)}(x,t) = \int \Delta^i\, W_t(x;\Delta)\,\mathrm{d}\Delta = \int \Delta^i\, W_t(x+\Delta|x)\,\mathrm{d}\Delta.$$

Equation (12.11) is known as the Kramer–Moyal's expansion of the master equation. If the transition probability is suitable local the higher-order jump moments will be negligible and we obtain the second order Kramer–Moyal's expansion

$$\frac{\partial f_{X(t)}(x,t)}{\partial t} = -\frac{\partial M^{(1)}(x,t)\, f_{X(t)}(x,t)}{\partial x} + \frac{1}{2}\frac{\partial^2 M^{(2)}(x,t)\, f_{X(t)}(x,t)}{\partial x^2}$$

but this is just the familiar Fokker–Planck equation—we identify $M^{(1)}(x,t) = a(t, x(t))$ and $M^{(2)}(x,t) = b(t,x)$. Occasionally it is more convenient to obtain the Fokker–Planck equation through the master equation rather than a stochastic differential equation and very occasionally (when larger jumps occur) the higher-order terms in the Kramer–Moyal's expansion may provide a more accurate approximation for the dynamics of the density $f_{X(t)}(x,t)$.

When analytical solutions to the Fokker–Planck equation are difficult to obtain one can always seek numerical solutions. This loses some of the insight that comes from having an analytical solution, nevertheless it can provide useful results quite easily. Numerical solutions are reasonably doable in low (one or two) dimensions, but they become increasingly time-consuming in higher dimensions where Monte Carlo simulation often gives results more accurately and a lot more easily. The problem simplifies considerably when seeking the stationary state distribution.

12.2.4 Stationary Distribution of Stochastic Processes

Sometimes stochastic processes have a well-defined stationary distribution. This is often a quantity of considerable interest as it describes the long-term behaviour of a process. However, not all stochastic processes will have a stationary distribution. This can be due to the fact that $a(t,x)$ and $b(t,x)$ change continuously with t. However, even when there is no explicit t dependence in the stochastic differential equation there may still be no stationary distribution. For example, the simple random walk

$$\mathrm{d}X(t) = \mathrm{d}W(t)$$

has a solution $X(t) = W(t) \sim \mathcal{N}(0,t)$, which is a normal deviate with an ever-increasing variance. If the process is non-ergodic the stationary distribution may depend on the initial conditions.

Nevertheless, in many cases there is a well-defined and unique stationary distribution, $f(x) = \lim_{t\to\infty} f(x,t)$, which is the distribution of interest. To obtain this distribution set

$$\frac{\partial f(x,t)}{\partial t} = 0$$

in the Fokker–Planck Equation (12.9). For a one-dimensional stochastic process we are left with an ordinary differential equation

$$-\frac{\mathrm{d}\,a(x)\,f(x)}{\mathrm{d}x} + \frac{1}{2}\frac{\mathrm{d}^2 b^2(x)\,f(x)}{\mathrm{d}x^2} = 0.$$

Integrating, we find

$$-a(t,x)\,f(x) + \frac{1}{2}\frac{\mathrm{d}\,b^2(t,x)\,f(x)}{\mathrm{d}x} = c_1,$$

however, $c_1 = 0$ (otherwise $f(x)$ would be unnormalisable). Dividing through by $b^2(x)\, f(x)/2$ (and rearranging)

$$\frac{2\,a(x)}{b^2(x)} = \frac{1}{b^2(x)\, f(x)} \frac{\mathrm{d}\, b^2(x)\, f(x)}{\mathrm{d}x} = \frac{\mathrm{d} \log(b^2(x)\, f(x))}{\mathrm{d}x}.$$

Integrating, exponentiating, and dividing through by $b^2(x)$ we obtain

$$f(x) \propto \frac{1}{b^2(x)} \exp\left(2 \int_{-\infty}^{x} \frac{a(y)}{b^2(y)} \mathrm{d}y\right). \tag{12.12}$$

This is a simple equation, and can provide a quick solution to problems that look difficult to solve.

Example 12.6 Population Genetics Rides Again

In Example 11.2 (on page 311) we considered the evolution of two possible alleles, A and a fitter mutant B, at a single site. The site undergoes mutation and selection, such that at the next generation the expected proportion of individuals with the mutant gene is

$$p_{sm} = \frac{(1-v)\,(1+s)\,X + u\,(P-X)}{P + s\,X}$$

where P is the population size, X the number of mutants at the previous generation, s is the fitness (selective advantage) of the mutant, u the probability of undergoing a mutation from A to B, and v the probability of a mutation from B to A.

For a large population, using the Markov chain description to compute the dynamics becomes intractable (as the state space and transition matrix becomes too large to work with). However, we can approximate the system by a stochastic differential equation

$$X(t + \Delta t) = X(t) + a(X(t))\Delta t + b(X(t))\,\Delta W(t)$$

where on average

$$\mathbb{E}\left[X(t+1)\right] = X(t) + a(X(t))\,\Delta t$$

while

$$\mathbb{E}\left[X^2(t+1)\right] - \mathbb{E}\left[X(t+1)\right]^2 = b^2(X(t))\,\Delta t.$$

For the model we are considering

$$\mathbb{P}\left(X(t+1) = x\right) = \mathbb{P}\left(n' = Px\right) = \mathrm{Bin}(P\,x | P, p_{sm})$$

where we have taken $\Delta t = 1$ (generation). Taking averages

$$\begin{aligned} a(X(t)) &= \mathbb{E}\left[X(t+1)\right] - X(t) = p_{sm} - X(t) \\ &= \frac{(u - (u + v - s + u\,s)X(t) + sX^2(t))}{1 + s\,X(t)}. \end{aligned}$$

The variance of a binomial is given by

$$b^2(X(t)) = \mathbb{E}\left[X^2(t+1)\right] - \mathbb{E}\left[X(t+1)\right]^2 = \frac{p_{sm}(1-p_{sm})}{P}.$$

Assuming s, u, and v are small, then expanding to first order

$$a(X) = \mathbb{E}\left[\Delta X\right] \approx -v\,X + u(1-X) + sX(1-X)$$

$$b(X) = \mathbb{E}\left[X^2\right] - \mathbb{E}\left[X\right]^2 \approx \frac{X(1-X)}{P}.$$

This model is known as the Fisher–Wright model (predominantly in the UK) or the Wright–Fisher model (predominantly in the USA). It was first proposed by Ronald Fisher and Sewell Wright around 1930. The dynamics are given by the Fokker–Planck equation

It was studied independently by Fisher and Wright. Each made a mistake in their derivation which was corrected by the other author.

$$\frac{\partial f(x,t)}{\partial t} = -\frac{\partial\left(\left(-v\,x + u(1-x) + sx(1-x)\right)\, f(x,t)\right)}{\partial x} + \frac{1}{2P}\frac{\partial^2\left(x\,(1-x)\,f(x,t)\right)}{\partial x^2}.$$

This is a bit of a pain to solve numerically, but the stationary state has a relatively simple form

$$f_{eq}(x) \propto \frac{1}{x\,(1-x)} \mathrm{e}^{2P\int_0^x \frac{-v\,y+u\,(1-y)+s\,y\,(1-y)}{y\,(1-y)}\mathrm{d}y} = \frac{x^{2Pu-1}\,(1-x)^{2Pv-1}\,\mathrm{e}^{2Psx}}{B(2Pu, 2Pv)F(2Pu; 2P(u+v); 2Ps)}$$

where $B(2Pu, 2Pv)\,F(2Pu; 2P(u+v); 2Ps)$ are normalisation constants (which happen to be expressible in terms of standard special functions – the beta function and the hypergeometric function). In 1930 it would have been impossible to solve the Markov chain even for a small system numerically. The stochastic differential equation provides a quick answer to a difficult problem.

In higher dimensions the stationary distribution is given by a partial differential equation. For all but the simplest stochastic processes it is not possible to solve these equations analytically.

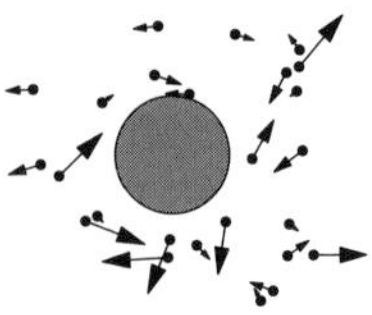

Example 12.7 Physical Brownian Motion

Real Brownian motion as observed in nature concerns a small particle in a viscous liquid experiencing fluctuations in its velocity caused by random knocks. Physicists describe this using the Langevin equation

$$m\frac{\mathrm{d}\boldsymbol{V}}{\mathrm{d}t} = -\beta\,\boldsymbol{V} + \sigma\boldsymbol{\eta}$$

where $\boldsymbol{V}$ is a three-dimensional velocity vector, m is a mass, β is a resistive force, and $\boldsymbol{\eta}$ is a three-dimensional Wiener measure describing random knocks of the particle. In the Itô form

$$\mathrm{d}\boldsymbol{V} = -\frac{\beta}{m}\boldsymbol{V}\,\mathrm{d}t + \frac{\sigma}{m}\mathrm{d}\boldsymbol{W}(t).$$

The drift coefficient is β/m, and the size of the fluctuations is σ/m. The probability of the velocity being close to $\boldsymbol{v}$ is $\mathbb{P}\left(\boldsymbol{V}(t) - \boldsymbol{v} \in B(\epsilon)\right) \approx f(\boldsymbol{v},t)\,|B(\epsilon)|$, where $B(\epsilon)$ is a small (Borel) ball of radius ϵ and volume $|B(\epsilon)|$. The density $f(\boldsymbol{v},t)$ is governed by the Fokker–Planck equation

$$\frac{\partial f(\boldsymbol{v},t)}{\partial t} = \frac{\beta}{m}\sum_{i=1}^{n}\frac{\partial v_i f(\boldsymbol{v},t)}{\partial v_i} + \frac{\sigma^2}{2\,m^2}\frac{\partial^2 f(\boldsymbol{v},t)}{\partial {v_i}^2}.$$

This is hard to solve in complete generality but it has a stationary distribution

$$f(\boldsymbol{v}) = \left(\frac{2\beta\, m}{\pi\,\sigma^2}\right)^{3/2} \mathrm{e}^{-\frac{\beta\, m}{\sigma^2}\sum_{i=1}^{3} v_i^2}$$

(which is easily checked by substitution into the Fokker–Planck equation). Thus the velocities of a particle in a viscous fluid undergoing random knocks are normally distributed. Note, that the form of the equation for Brownian motion in one dimension is identical to that of an Ornstein–Uhlenbeck process.

In general we can solve the equation for the stationary distribution numerically by discretising the variable and replacing the partial derivatives by difference equations. The boundary conditions must be defined in such a way that no probability is lost. This is achieved by so-called *reflecting boundary conditions*, which can be visualised as having an identical but reflected system living at the boundary.

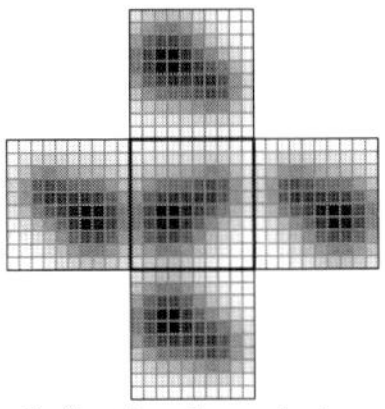

Reflecting boundaries

12.2.5 Pricing Options

Stochastic processes have taken on significant interest due to applications in financial markets. For engineers and scientists with higher aspirations, such grubby monetary applications are, no doubt, of little interest. Nevertheless, it provides a different use of stochastic methods so may be of academic interest.

The classic application of stochastic processes is to obtain a fair price for options. This was proposed by Fischer Black and Myron Scholes in 1973. Robert C. Merton extended the analysis to option pricing and won the 1997 Nobel Prize for Economics with Scholes (Black missed out as he died in 1995). We consider here a European option, which is an agreement to have the right to buy or sell an amount of a commodity or shares at some future date at some agreed price (the strike price). This right does not need to be taken up. We imagine some commodity with price $S(t)$ at time t. We assume that the value of this commodity is governed by the stochastic differential equation

$$\mathrm{d}S(t) = \mu\, S(t)\,\mathrm{d}t + \sigma\, S(t)\,\mathrm{d}\,W(t) \tag{12.13}$$

where μ is the long-term drift in the price and σ measures the volatility in the market (which is determined empirically). $S(t)$ is the spot price (the market price at the current time t). It is an assumption of the model that the spot price has some trend (the drift) and some fluctuations proportional to the spot price – this implies that the dynamics of the price is independent of the units it is measured in – we discussed this model earlier in Example 12.3 on page 365.

Now let $V(S,t)$ be the value of an option (derivative) at time t. At the time when the option matures, T, the value $V(S,T)$ would be the difference between the spot price and the strike price if this was positive and zero otherwise (you would not take up the option if you lose money on it – i.e. the spot price is less than the agreed or strike price). Assuming $V(S,t)$ is doubly differentiable in S and singularly differentiable in t, then by Itô's lemma

$$\begin{aligned} \mathrm{d}V(S,t) &= \frac{\partial V(S,t)}{\partial S}\mathrm{d}S(t) + \frac{\partial V(S,t)}{\partial t}\mathrm{d}t + \frac{1}{2}\frac{\partial^2 V(S,t)}{\partial S^2}\mathbb{E}\left[(\mathrm{d}S)^2\right] \\ &= \frac{\partial V(S,t)}{\partial S}\mathrm{d}S(t) + \frac{\partial V(S,t)}{\partial t}\,\mathrm{d}t + \frac{1}{2}\sigma^2\, S^2(t)\frac{\partial^2 V(S,t)}{\partial S^2}\,\mathrm{d}t. \end{aligned} \tag{12.14}$$

Black and Scholes imagined owning one option $V(S,t)$ and then trading in the commodity in such a way as to remove the fluctuations in the value of $V(S,t)$. In particular, they imagined buying and selling the commodity so that at time t you had $-\partial V(S,t)/\partial S$ shares of the commodity. In that way the asset of the option plus shares would be equal to

$$\Pi(S,t) = V(S,t) - S(t)\,\frac{\partial V(S,t)}{\partial S}$$

where the last term is equal to the price of the commodity times the amount of the commodity. The profit from time t to $t + \delta t$ is

$$\begin{aligned} \mathrm{d}\Pi(S,t) &= \mathrm{d}V(S,t) - \frac{\partial V(S,t)}{\partial S}\,\mathrm{d}S(t) \\ &= \left(\frac{\partial V(S,t)}{\partial t} + \frac{1}{2}\sigma^2\, S^2(t)\frac{\partial^2 V(S,t)}{\partial S^2}\right)\,\mathrm{d}t \end{aligned}$$

which is obtained by substituting Equation (12.14) for $\mathrm{d}V(S,t)$ in the equation above. We note that this has no fluctuations by construction, i.e. $\mathrm{d}\Pi(S,t)$ does not depend on $\mathrm{d}S(t)$ (which in turn depends on $\mathrm{d}W(t)$) but only on non-fluctuating quantities. The combination of option plus stock would be risk free. Black and Scholes argued that in an efficient market any risk-free gain should be equal to the mean interest rate in the market r. If this were not so it would be possible to make money in the market without taking any risk (this is known as arbitrage).

According to classical economic theory, in an efficient market there would be no arbitrage. (There are financial companies that make their livelihoods through identifying and exploiting arbitrage, often across different markets, but they correct market prices, reducing the remaining arbitrage.) If r is the mean interest rate in the market then the profit from a risk-free asset $\Pi(S,t)$ in time $\mathrm{d}t$ should be $r\,\Pi(S,t)\,\mathrm{d}t$. Putting this equal to $\mathrm{d}\Pi(S,t)$ and substituting in the value of $\Pi(S,t)$ we arrive at the *Black–Scholes partial differential equation*

$$\frac{\partial V(S,t)}{\partial t} + \frac{1}{2}\sigma^2\,S^2(t)\frac{\partial^2 V(S,t)}{\partial S^2} + r\,S(t)\,\frac{\partial V(S,t)}{\partial S} - r\,V(S,t) = 0. \tag{12.15}$$

We can use this partial differential equation to obtain a fair price for the option. There are options to be allowed to buy a commodity (a call option) and options for being allowed to sell (a call option). We consider here only the call option. We want to determine the price, $V(S,t)$ of this option at time t. To determine this from Equation (12.15) we need to determine the boundary conditions. These are

$$V(0,t) = 0, \qquad V(S,T) = \max(S(T) - K, 0), \qquad \lim_{S\to\infty} V(S,t) \to S,$$

where K is the spot price. The first of these equations says that the option is valueless if the price of the commodity is zero. The second equation says that at the time of maturity the value of the option is equal to the difference between the spot price at that time and the agreed (strike) price of the commodity K (if the spot price is lower than strike price then we don't buy the commodity and the option is worthless). The last equation says that the value of the option tends to the value of commodity when the price goes through the roof.

Finding solutions to partial differential equations is a black art. Fortunately Equation (12.15) is a linear equation which is closely related to the diffusion equation. For this problem (the call option) the value of the asset (and hence a fair price for the option) is given by

$$\begin{aligned} V(S,t) = {} & S\,\Phi\!\left(\frac{1}{\sigma\sqrt{T-t}}\left(\log\!\left(\frac{S}{K}\right) + \left(r + \frac{\sigma^2}{2}\right)(T-t)\right)\right) \\ & + K\mathrm{e}^{-r(T-t)}\,\Phi\!\left(\frac{1}{\sigma\sqrt{T-t}}\left(\log\!\left(\frac{S}{K}\right) + \left(r - \frac{\sigma^2}{2}\right)(T-t)\right)\right) \end{aligned}$$

where

$$\Phi(x) = \int_{-\infty}^{x} \mathcal{N}(z|0,1)\,\mathrm{d}z$$

is the familiar cumulative probability density for the zero mean, unit variance normal distribution (the easiest way to verify that this is a solution of Equation (12.15) is through direct substitution – a rather tedious exercise, but greatly helped using a symbolic manipulation language such as Mathematica). A similar formula can be obtained for a put option.

Options are used by companies to try to hedge against the vagaries of future prices. As such, they can be useful instruments. The Black–Scholes equations mark the limits of what can be solved in closed forms. They are extremely useful for providing a quick pricing formula for a rather complex investment. They make a number of simplifying assumptions, for example that the price of a stock is governed by Equation (12.13). To go beyond such assumptions usually involves using Monte Carlo techniques. Many otherwise respectable scientists have been siphoned off into finance to perform such calculations. Whether this is to the greater benefit of humankind is open to some debate.

12.3 Point Processes

Although diffusion processes are very important there are many other stochastic processes of importance. Figure 12.6 shows some points in a two-dimensional plane. They may be users in a mobile telephone network or stars in the sky. We assume that the points are in some way random. Sets of points drawn from some random distribution are known as point processes. We can define point processes in any number of dimensions. For example, in one dimension they may be events happening at different times. These processes are important in a large number of applications.

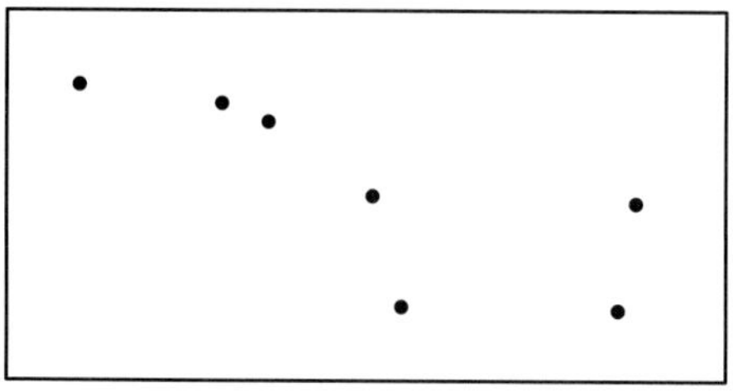

Figure 12.6 Example of a point process in a plane.

12.3.1 Poisson Processes

The simplest point process is the Poisson process, where we assume that the number of points in any two non-overlapping regions are independent of each other. In this case, the probability of k points occurring in any region will be distributed according to a Poisson distribution. Poisson processes may not be a perfect model. For example, it may be impossible to have two events occurring very close to each other. If we were modelling the distribution of stars in three-dimensional space then, due to their gravitational influence, it would be incorrect to assume that their positions are independent of each other when they are very close. However, in many cases the inaccuracy that occurs in assuming a point process is a Poisson process will often not be worth worrying about.

Poisson processes have some very nice properties, making them easy to handle. These follow from the properties of the Poisson distribution. The first important property is that the sum of Poisson distributed variables is also Poisson distributed. For example, let $X \sim \text{Poi}(\lambda)$ and $Y \sim \text{Poi}(\mu)$ then

$$P(X+Y=k) = \sum_{i=0}^{k} \text{Poi}(i|\lambda)\,\text{Poi}(k-i|\mu) = \sum_{i=0}^{k} \frac{\lambda^i\,\mu^{k-i}}{i!\,(k-i)!}\text{e}^{-\lambda-\mu}$$

$$= \frac{\text{e}^{-\lambda-\mu}}{k!}\sum_{i=0}^{k}\binom{k}{i}\lambda^i\,\mu^{k-i} = \frac{\text{e}^{-\lambda-\mu}}{k!}(\lambda+\mu)^k = \text{Poi}(k|\lambda+\mu).$$

To describe a Poisson process, Π_λ, we define the intensity function $\lambda(\boldsymbol{x})$ such that in any region of space A, if $N(A)$ denotes the number of points in that region A, then

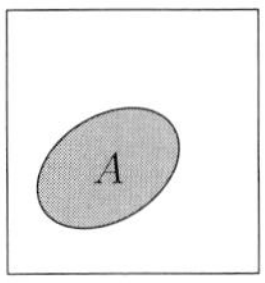

$$\mathbb{P}\left(N(A)=k\right) = \text{Poi}\left(k \,\middle|\, \int_A \lambda(\boldsymbol{x})\,\text{d}\boldsymbol{x}\right)$$

where the integral is over the region A. Note that because of the additive property of the Poisson distribution, Poisson processes inherit a superposition property. That is, if we have two Poisson processes with intensity functions $\lambda(\boldsymbol{x})$ and $\mu(\boldsymbol{x})$ then the sum of the Poisson processes is itself a Poisson process with intensity function $\lambda(\boldsymbol{x})+\mu(\boldsymbol{x})$. This property extends to any number of Poisson processes (which is easy to see from an induction argument).

A second nice feature of Poisson processes is that if we make a change of variables to a Poisson process the new processes is again a Poisson process (provided we don't map some region to a single point, which would result in there being a finite probability of two events occurring at the same point). This mapping property follows since any region in the new coordinate system can be mapped to a region in the old coordinate system, and this region will satisfy the Poisson property.

To compute quantities of interest we consider the moment or cumulant generating function (CGF). Suppose we are interested in

$$Q = \sum_{\boldsymbol{X}\in\Pi} g(\boldsymbol{X})$$

where the sum is over points in the Poisson process. The function $g(\boldsymbol{X})$ is some function of interest that depend on the point process. We assume that $g(\boldsymbol{x})$ is sufficiently smooth that if we partition the region of interest into many small subregion, A_i, then we can approximate Q by

g_0 N_0	g_4 N_4	g_8 N_8	g_{12} N_{12}
g_1 N_1	g_5 N_5	g_9 N_9	g_{13} N_{13}
g_2 N_2	g_6 N_6	g_{10} N_{10}	g_{14} N_{14}
g_3 N_3	g_7 N_7	g_{11} N_{11}	g_{15} N_{15}

$$Q \approx \sum_i g_i\,N_i$$

where the sum is over the subregions, g_i is a representative value of $g(\boldsymbol{X})$ in subregion A_i, and $N_i = N(A_i)$ is the number of points that fall in region A_i. Provided $g(\boldsymbol{x})$ is sufficiently smooth we would expect this approximation to become more accurate as the subregions shrink. We now consider the moment generation function

$$\mathbb{E}\left[\mathrm{e}^{\beta Q}\right] \approx \prod_i \mathbb{E}\left[\mathrm{e}^{\beta g_i N_i}\right]$$

where the product is over the subregions. Since we are considering a Poisson process, $N_i = N(A_i)$ will be Poisson distributed with mean $\lambda_i = \int_{A_i} \lambda(\boldsymbol{x})\,\mathrm{d}\boldsymbol{x}$ so that

$$\mathbb{E}_{N_i}\left[\mathrm{e}^{\beta g_i N_i}\right] = \sum_{N_i=0}^{\infty} \mathrm{e}^{\beta g_i N_i} \frac{\lambda_i^{N_i} \mathrm{e}^{-\lambda_i}}{N_i!} = \mathrm{e}^{\lambda_i (\mathrm{e}^{\beta g_i} - 1)}$$

and thus

$$\mathbb{E}\left[\mathrm{e}^{\beta Q}\right] \approx \prod_i \mathrm{e}^{\lambda_i (\mathrm{e}^{\beta g_i} - 1)} = \mathrm{e}^{\sum_i \lambda_i (\mathrm{e}^{\beta g_i} - 1)}.$$

Taking the limit where the size of the subregions go to zero then the sum in the exponent can be replaced by an integral and we have

$$\log\left(\mathbb{E}\left[\mathrm{e}^{\beta Q}\right]\right) = \int \lambda(\boldsymbol{x}) \left(\mathrm{e}^{\beta g(\boldsymbol{x})} - 1\right) \mathrm{d}\boldsymbol{x}.$$

This is the CGF for the quantity Q. The integral may not always be convergent. This can often be remedied, for example, by making β imaginary (i.e. using the logarithm of the characteristic function). The mean and variance of Q is found simply by taking the derivatives of the CGF

$$\mathbb{E}\left[Q\right] = \int \lambda(\boldsymbol{x})\, g(\boldsymbol{x})\, \mathrm{d}\boldsymbol{x} \qquad \mathbb{V}\mathrm{ar}\left[Q\right] = \int \lambda(\boldsymbol{x})\, g^2(\boldsymbol{x})\, \mathrm{d}\boldsymbol{x}.$$

Example 12.8 Gravitational Force

A typical example of using a Poisson process is to compute the gravitational force due to a cluster of stars. The force $\boldsymbol{F}$ at the origin produced by a mass m at position $\boldsymbol{X}$ is

$$\boldsymbol{F} = \frac{G\, m\, \boldsymbol{X}}{\|\boldsymbol{X}\|^3}.$$

To compute the typical gravitational force for a cluster of stars we assume that the stars are distributed in space according to a Poisson process, Π_λ, with intensity $\lambda(\boldsymbol{x})$ describing the cluster. However, we would expect the masses also to follow some distribution. Here the superposition of Poisson processes comes in. We can think of our stars as drawn from distinct Poisson processes at each mass level. That is, we consider an intensity function $\lambda(\boldsymbol{x}, m)$ as defining the intensity of stars of mass m, such that the expected number of stars of mass m in region A is $\int_A \lambda(\boldsymbol{x}, m)\, \mathrm{d}\boldsymbol{x}$. We can then add the Poisson processes at different masses. The CGF for this process is

$$\log\left(\mathbb{E}\left[\mathrm{e}^{\boldsymbol{\beta}^{\mathsf{T}} \boldsymbol{F}}\right]\right) = \int \mathrm{d}\boldsymbol{x} \int \mathrm{d}m\, \lambda(\boldsymbol{x}, m) \left(\mathrm{e}^{\frac{G\, m\, \boldsymbol{x}}{\|\boldsymbol{x}\|^3}} - 1\right)$$

with

$$\mathbb{E}\left[\boldsymbol{F}\right] = G \int \mathrm{d}\boldsymbol{x} \int \mathrm{d}m\, \lambda(\boldsymbol{x}, m) \frac{m\, \boldsymbol{x}}{\|\boldsymbol{x}\|^3}.$$

For a spherically symmetric intensity function the integral is zero by symmetry. More interesting would be the covariance

$$\mathbb{C}\mathrm{ov}\left[\boldsymbol{F}\right] = G^2 \int \mathrm{d}\boldsymbol{x} \int \mathrm{d}m\, \lambda(\boldsymbol{x}, m) \frac{m^2\, \boldsymbol{x}\, \boldsymbol{x}^{\mathsf{T}}}{\|\boldsymbol{x}\|^6}.$$

For an infinite uniform universe the integrals will diverge. This would be a somewhat surprising and unpleasant universe to live in (the gravitational force would vary everywhere).

12.3.2 Poisson Processes in One Dimension

One of the most important types of point processes are one-dimensional processes. These might, for example, be used to model the position of cars along a road. Frequently we use one-dimensional point processes to model systems evolving in time. That is, we consider the set of times when events occur. The one-dimensional Poisson process inherits all the properties of general Poisson processes, but also has some distinct properties. In particular, due to the fact that any change of variables to a Poisson process gives a Poisson process, we can map any one-dimensional Poisson process to a uniform Poisson process, i.e. a Poisson process with a constant intensity function. In this case, the expected number of events in any interval is just proportional to the length of the interval.

One-dimensional Poisson processes also allow us to order the events. If we have a set of events happening at times $T_1 < T_2 < T_3 < \cdots < T_n$, then for a Poisson process with a constant intensity function, λ, the interval between events is exponentially distributed, $T_i - T_{i-1} \sim \mathrm{Exp}(\lambda)$. To show this, we note that the probability of no event occurring in a time interval t is given by $\mathrm{Poi}(0|\lambda\, t) = \exp(-\lambda\, t)$. The probability of at least one event occurring in time, t, is $1 - \mathrm{e}^{-\lambda t}$. Thus, the probability of the time between two events, $T_i - T_{i-1}$ being less than t is just the probability that one or more event has happened in this time, so that

$$\mathbb{P}\left(T_i - T_{i-1} < t\right) = 1 - \mathrm{e}^{-\lambda\, t}.$$

Thus the probability density of the interval is equal to

$$f_{T_i - T_{i-1}}(y) = \frac{\mathrm{d}}{\mathrm{d}t} \mathbb{P}\left(T_i - T_{i-1} < t\right) = \lambda\, \mathrm{e}^{-\lambda\, t} = \mathrm{Exp}(t|\lambda).$$

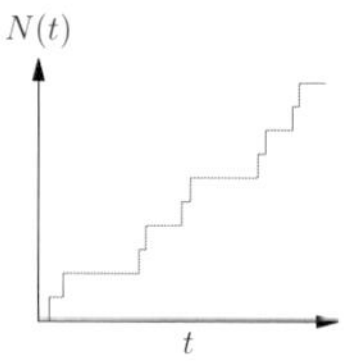

The staircase function, defined by $N(t) = \max\{n|T_n \leq t\}$ (i.e. the number of events that has occurred up to time t), is known as a *renewal processes* (the length between events being random variables). It can be used to model the replacement of light bulbs or the decay of a particle. The simplest renewal processes are Poisson processes where the time period between events is exponentially distributed. A slightly more elaborate problem is that of queues. Here the arrival time of a customer at a queue will be described by a renewal process, but now there is also a service provider clearing the queue. We consider a very simple queuing problem in the following example.

Example 12.9 Queues

For a call centre, a post office, or a hospital, a reasonable (zeroth order) approximation is to assume that the arrivals of clients are independent of each other so that they can be modelled using a Poisson process. The intensity may well change depending, for example, on the time of day. We consider here a very simple queue where the customers arrive according to Poisson process Π_λ with constant intensity λ and they are dealt with in time $T \sim \text{Exp}(\mu)$. Provided there is a customer in the queue, the time it takes for the customer to be served can also be modelled as a Poisson process. We can model queues using a master equation. Let $p_n(t)$ be the probability of there being n customers in the queue at time t. Then if $n \geq 1$

$$\begin{aligned}p_n(t+\delta t) = p_n(t)(1 - \lambda\,\delta t - \mu\,\delta t) + \lambda\,p_{n-1}(t)\,\delta t \\ + \mu\,p_{n+1}(t)\,\delta t + O\left((\delta t)^2\right).\end{aligned}$$

Subtracting $p_n(t)$ from both sides, dividing by δt, and taking the limit $\delta t \to 0$ we obtain the master equation

$$\frac{\mathrm{d}\,p_n(t)}{\mathrm{d}t} = -(\lambda+\mu)\,p_n(t) + \lambda\,p_{n-1}(t) + \mu\,p_{n+1}.$$

If $n = 0$ then

$$\frac{\mathrm{d}\,p_0(t)}{\mathrm{d}t} = -\lambda p_0(t) + \mu\,p_1(t).$$

To obtain a full solution we can use generating function techniques. A much simpler problem is to consider the steady state where $\mathrm{d}p_n(t)/\mathrm{d}t = 0$ – note that we will only reach a steady state if $\mu > \lambda$, i.e. the queue-clearing rate is shorter than the customer arrival rate. From the equation $\mathrm{d}p_0(t)/\mathrm{d}t = 0$ we find that $p_1 = (\lambda/\mu)\,p_0$ and from $\mathrm{d}p_1(t)/\mathrm{d}t = 0$ we find $p_2 = (\lambda/\mu)^2 p_0$; repeating this we find $p_n = (\lambda/\mu)^n p_0$ (which we can also prove by induction). To find p_0 we note that

$$1 = \sum_{n=1}^{\infty} p_n = p_0 \sum_{n=1}^{\infty} \left(\frac{\lambda}{\mu}\right)^n = \frac{p_0}{1-\lambda/\mu}$$

so that $p_0 = 1 - \lambda/\mu$ and

$$p_n = \left(1 - \frac{\lambda}{\mu}\right)\left(\frac{\lambda}{\mu}\right)^n = (1-\rho)\,\rho^n = \text{Geo}(n|\rho)$$

where $\rho = \lambda/\mu$ (recall that for there to be a steady-state solution $\lambda < \mu$) and $\text{Geo}(n|\rho)$ is a geometric distribution. The expected queue length is $1/(1-\rho)$. The probability of the queue being empty, i.e. the cost of the queue to the service provider is $(1-\rho)$. Figure 12.7 show the equilibrium distribution of queue lengths for a relatively benign queue with $\rho = \lambda/\mu = 0.5$ so that the expected queue length is 2, and a more economical queue with $\rho = 0.9$ with an expected length of 10.

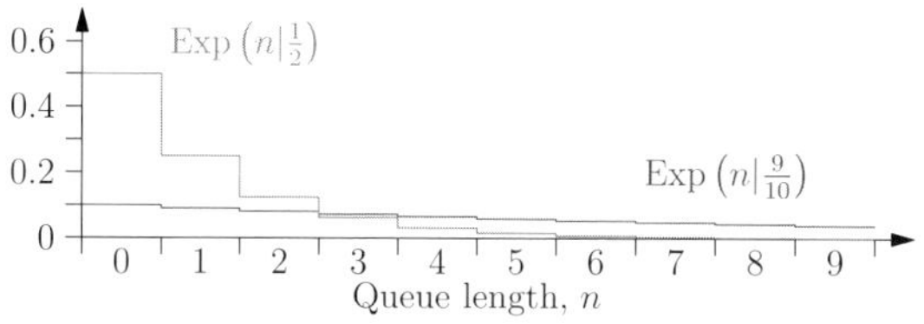

Figure 12.7 Equilibrium distribution of queue lengths for $\rho = \lambda/\mu = 0.5$ and 0.9.

Queue theory has very important applications and has been much studied. The queue which we looked at goes by the name of $M/M/1$ where M (Markovian) indicates exponential waiting times and the 1 denotes that there is one service provider. Queues are covered in many advanced texts on probability, e.g. Grimmett and Stirzaker (2001a), as well as many specialised books just on queues.

12.3.3 Chemical Reactions

An important area of mathematical modelling is in studying chemical reactions. The starting point for modelling chemical reactions is a set of reaction equations together with the reaction rates, e.g.

$$\emptyset \overset{k_1}{\rightarrow} X \qquad X \overset{k_2}{\rightarrow} Y \qquad 2X + Y \overset{k_3}{\rightarrow} 3X \qquad X \overset{k_4}{\rightarrow} \emptyset \tag{12.16}$$

where in the first reaction $\emptyset \overset{k_1}{\rightarrow} X$ implies that X is spontaneously produced and in the last equation that X spontaneously decays (or, at least, we are not interested in what X turns into). Chemically such equations look very strange, but in modelling the biology of the cell we often use such equations as a shorthand for a more complex process. When there are a large number of molecules we can model the dynamics of the concentrations of the reactants as a set of couple differential equations

$$\frac{\mathrm{d}[X]}{\mathrm{d}t} = k_1[X] - k_2[X] + k_3[X]^2[Y] - k_4[X], \qquad \frac{\mathrm{d}[Y]}{\mathrm{d}t} = k_2[X] - k_3[X]^2[Y],$$

where $[X]$ and $[Y]$ denote the concentration of reactants X and Y, respectively. As reactions are usually stochastic events there will be fluctuations, but in many situations the number of atoms are so large that these fluctuations are negligible. Note that the speed of a reaction depends on its reaction rate, k_i, and the product of the concentrations of the reactants on the left-hand side of the reaction equations. This is because these reactants all have to collide at a particular location. Thus, for example, the rate of the reactions $2X + Y \rightarrow 3X$ will be proportional to the concentration of X squared times the concentration of Y (two X molecules and a Y molecule must collide – we assume that the position of reactants are independent of each other). In this reaction an X molecule gets transformed to a Y molecule increasing the number of X molecules by 1 and decreasing the number of Y molecules by 1 (hence the constant factors in the terms of the differential equations – in this case ± 1 as each reaction either increases or decreases the number of molecules by one). The model is often extremely accurate in the case when we have a large number of reactants. In some situations however (often when modelling biological systems) the number of reactants can be quite small and differential equations provide a poor description of the observed dynamics. In this case, we can build a more accurate model by considering individual reactions between molecules.

The probability per unit time of a reaction occurring is known as the *propensity*. For the system described in Equation (12.16) the propensities are given by

$$\alpha_1 = k_1 \qquad \alpha_2 = k_2\, n_X \qquad \alpha_3 = \frac{k_3}{\Omega^2}\, n_X^2\, n_Y \qquad \alpha_4 = k_4\, n_X$$

where n_X and n_Y are the number of reactants, X and Y respectively, and where $\Omega = V\, N_{\mathrm{A}}$, V is the volume of the reaction vessel and N_{A} is Avogadro's number. For a reactant Z, the concentration (in moles per unit volume) is given by $[Z] = n_Z/\Omega$. If the change of concentration of a reactant Z (in the limit of large n_Z) is $\mathrm{d}[Z]/\mathrm{d}t$, then the expected change in the number of molecules of reactant Z in a time interval $\mathrm{d}t$ is given by

$$\mathrm{d}\mathbb{E}\left[n_Z\right] = \Omega\, \frac{\mathrm{d}\,[Z]}{\mathrm{d}t}\, \mathrm{d}t.$$

Thus, to obtain the propensities we multiply the term in the differential equation corresponding to a particular reaction by Ω and divide by Ω for each concentration – hence the factor Ω^{-2} in the propensity α_3[1]. (When we write $\emptyset \rightarrow X$ we implicitly assume that the rate of producing X is proportional to the volume of the reaction vessel – this would be true if X was being created by some precursor P whose concentration remains constant over time.) In what follows I am going to assume that concentrations are measured in nano-moles per litre. In this case $\Omega = 10^{-9}\, V\, N_{\mathrm{A}}$. I will also assume the reaction vessels has size $V = 10^9\, N_{\mathrm{A}}^{-1}$ litres so that $\Omega = 1$ (for some cells this is a reasonable assumption).

[1] If we are being even more careful then we would write $\alpha_3 = k_2\, n_X\, (n_X - 1)\, n_Y$ since the probability of two molecules of species X being in the same location is proportional to $n_X\, (n_X - 1)$. You might think there should be a combinatorial factor of a half, but by convention all such factors are absorbed into the definition of reaction rate – k_2 in this case.

We assume the reactions occur independently with probabilities given by the propensities, α_i. We can write down a master equation for the probabilities $p_{n,m}(t) = \mathbb{P}\left(n_X(t) = n,\ n_Y(t) = m\right)$. We consider a small time interval δt, so that the probability of two reactions occurring in the time period δt is $(\delta t)^2$, which can be ignored in the limit $\delta t \to 0$. For the above example,

$$\begin{aligned} p_{n,m}(t+\delta t) = (1 - (k_1 + k_2\, n + k_3\, n^2\, m + k_4\, n)\, \delta t) p_{n,m}(t) + k_1\, p_{n-1,m}(t)\, \delta t \\ + k_2\, (n+1) p_{n+1,m-1}(t)\, \delta t + k_3\, (n-1)^2 (m+1) p_{n-1,m+1}(t)\, \delta t \\ + k_4\, (n+1)\, p_{n+1,m}(t)\, \delta t \end{aligned}$$

subtracting $p_{n,m}(t)$ from both sides of the equation, dividing by δt, and taking the limit $\delta t \to 0$ we obtain the chemical master equation

$$\begin{aligned} \frac{\mathrm{d}\, p_{n,m}(t)}{\mathrm{d}t} &= -\,(k_1 + k_2\, n + k_3\, n^2\, m + k_4\, n)\, p_{n,m}(t) \\ &\quad + k_1\, p_{n-1,m}(t) + k_2\, (n+1) p_{n+1,m-1}(t) \\ &\quad + k_3\, (n-1)^2 (m+1) p_{n-1,m+1}(t) + k_4\, (n+1)\, p_{n+1,m}(t). \end{aligned}$$

Although the chemical master equation provides a full description of the system it is often impractical to solve. The state space is often far too large to enumerate (it grows exponentially in the number of species involved in the reactions). Instead it is often more convenient to perform a Monte Carlo simulation.

An efficient method for performing a Monte Carlo simulation of chemical reactions was proposed by Dan Gillespie in 1977 and is known as the *Gillespie algorithm*. Under the assumption that the chemical reactions are independent of each other then the reaction events can be treated as a Poisson process (although the propensities change after each reaction). The probability of the ith reaction occurring in time t is just given by the exponential distribution $\mathrm{Exp}(t|\alpha_i)$ where α_i is the propensity (probability per unit time) of reaction i occurring. If we have r reactions then

$$\alpha_0 = \sum_{i=1}^{r} \alpha_i$$

is the propensity of any of the reactions happening. The Gillespie algorithm is a real-time Monte Carlo simulation where at each step we compute the propensities (which depends on the number of molecules for each species and the reaction rates). We then generate a random deviate $T \sim \mathrm{Exp}(\alpha_0)$ (recall from Example 3.2 on page 52 that to we can generate an exponential deviate from a uniform deviate $U \sim \mathrm{U}(0,1)$ using the transformation $T = -\log(1-U)/\alpha_0$). Next we select a reaction with probability α_i/α_0. Finally we update the species number according to the reaction that is selected.

Example 12.10 Brusselator

As an example of a Gillespie simulation we take the system given by

$$\emptyset \xrightarrow{k_1} X \qquad X \xrightarrow{k_2} Y \qquad 2X + Y \xrightarrow{k_3} 3X \qquad X \xrightarrow{k_4} \emptyset$$

where $k_1 = 1$, $k_2 = 2$, $k_3 = 0.02$, and $k_4 = 0.04$ (with $\Omega = 1$). As initial conditions we take $n_X = n_Y = 0$. In Figure 12.8 we show a Gillespie simulation for this system. The system is highly non-linear and has a interesting flip-flop behaviour. It was first proposed by Ilya Prigogine and colleagues at the Free University of Brussels and is sometimes known as a Brusselator. The flip-flop behaviour is a consequence of the fluctuations in a finite system – they are not observed in the solution to the differential equations.

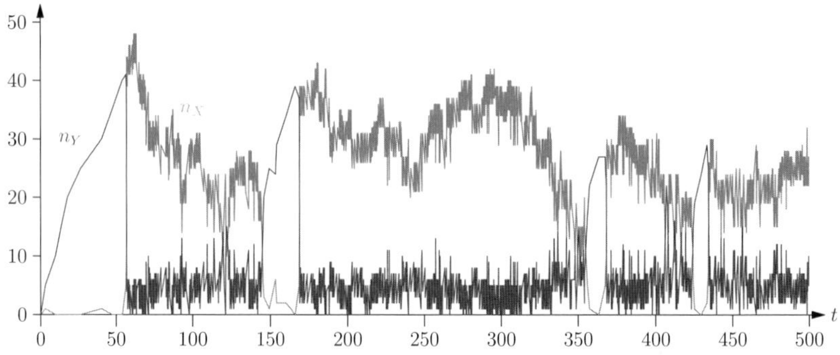

Figure 12.8 Gillespie simulation of reactions described above, run for 500 time units. We have slightly smoothed the curves by only plotting the curve after the total change in the two species is at least five – otherwise the printer goes on strike.

Stochastic processes describe a sequence of random variables that evolve in some random manner. These might be continuous functions such as Gaussian processes or diffusion processes or they may involve points or events such as in Poisson processes. They have a huge number of applications. We saw the use of Gaussian processes as providing useful priors on functions used for regression. Diffusion processes are useful for modelling functions that experience many small kicks. They are particularly helpful as they allow us to use many techniques from calculus and thus often supply a short cut to obtaining useful solutions. Provided we are interested in scales much larger than the small kicks they give excellent approximations. Thus, they are also used to model discrete Markov chains (which we saw in Example 12.6) as well as many continuous time problems such as those found in finance or real diffusion. In addition to diffusion processes there are a lot of other important stochastic processes. One that occurs in many problem are point processes, of which the simplest type are Poisson processes. These are particularly important in one dimension where they can be used to model the occurrence of independent events. These have important applications in modelling queues and simulating chemical reactions.

Additional Reading

There are so many books on stochastic processes catering for different tastes. The book by Grimmett and Stirzaker (2001a) covers probability and random processes and is an established text. Gardiner (2009) is one of the classic texts on stochastic processes with a very comprehensive and up-to-date coverage of the field. Another classic much loved by physical scientists is by van Kampen (2001). This is a nice read although missing some of the more recent developments.

Exercise for Chapter 12

Exercise 12.1 (answer on page 441)

We want to show that for a positive-definite matrix K' and its inverse $\mathsf{L}' = \mathsf{K}'^{-1}$ given by

$$\mathsf{K}' = \begin{pmatrix} \mathsf{K} & \boldsymbol{k} \\ \boldsymbol{k}^\mathsf{T} & k \end{pmatrix} \qquad \mathsf{L}' = \mathsf{K}'^{-1} = \begin{pmatrix} \boldsymbol{L} & \boldsymbol{l} \\ \boldsymbol{l}^\mathsf{T} & l \end{pmatrix}$$

the Equations (12.1) on page 354, and (12.2) on page 354 are identical. That is,

$$\boldsymbol{k}^\mathsf{T}(\mathsf{K} + \sigma^2 \mathsf{I})^{-1}\boldsymbol{y} = \frac{-\boldsymbol{l}^\mathsf{T}\left(\mathsf{L} + \sigma^{-2}\mathsf{I}\right)^{-1}\boldsymbol{y}}{l - \boldsymbol{l}^\mathsf{T}\left(\mathsf{L} - \sigma^{-2}\mathsf{I}\right)^{-1}\boldsymbol{l}}$$

$$k - \boldsymbol{k}^\mathsf{T}(\mathsf{K} + \sigma^2 \mathsf{I})^{-1}\boldsymbol{k} = \frac{1}{l - \boldsymbol{l}^\mathsf{T}\left(\mathsf{L} - \sigma^{-2}\mathsf{I}\right)^{-1}\boldsymbol{l}}.$$

This is difficult to show using matrix identities. Instead, for a randomly generated positive-definite matrix K' and for a particular value of σ and $\boldsymbol{y}$ show, using your favourite matrix manipulation language (e.g. MATLAB, Octave, python), that these two expressions give the same answer (up to machine accuracy).

Exercise 12.2 (answer on page 442)

Simulate a discrete time model for the evolution of stock prices

$$\Delta S(t) = \mu\, S(t)\, \Delta t + \sigma\, S(t)\, \sqrt{\Delta t}\, \eta$$

where $\eta \sim \mathcal{N}(0, 1)$ with $\mu = \sigma = \Delta t = 0.1$, starting from $S(0) = 1$ up to $t = 10$ and $t = 20$. Repeat this multiple times and plot histograms of $S(10)$, $S(20)$. Compare this with the theoretical result for the system

$$\mathrm{d}S(t) = \mu\, S(t)\, \mathrm{d}t + \sigma\, S(t)\, \mathrm{d}W(t)$$

given by

$$\begin{aligned} f_{S(t)}(S(10) = t) &= \mathrm{LogNorm}\left(s \middle| \left(\mu - \tfrac{\sigma^2}{2}\right) t, \sigma^2 t\right) \\ &= \frac{1}{s\, \sigma\, \sqrt{2\pi t}} \mathrm{e}^{-\left(\log(s) - (\mu - \sigma^2/2)t\right)/(2\sigma^2 t)}. \end{aligned}$$

Exercise 12.3 (answer on page 442)

Consider a particle experiencing normally distributed perturbations in a quartic potential

$$\mathrm{d}X(t) = -X^3(t)\, \mathrm{d}t + \mathrm{d}W(t).$$

Compute the stationary distribution $f(x) = \lim_{t\to\infty} f_{X_t}(x, t)$ using Equation (12.12) on page 375, and compare it with a simulation of a discrete time system.

Exercise 12.4 (answer on page 443)

Gillespie simulations are extremely easy to code. Simulate the set of reactions

$$\emptyset \overset{k_1=1}{\longrightarrow} X \qquad \emptyset \overset{k_2=0.5}{\longrightarrow} Y \qquad X+Y \overset{k_3=0.0001}{\longrightarrow} X \qquad X \overset{k_4=0.01}{\longrightarrow} \emptyset$$

starting from thc initial condition $n_X = n_Y = 0$ (assuming $\Omega = 1$). Write down approximate stochastic differential equations for $\mathrm{d}n_X(t)$ and $\mathrm{d}n_Y(t)$ assuming that the number of reactions that occur in a short time interval are given by a Poisson distribution.

Appendix A

Answers to Exercises

Contents

We provide a full set of answers to the exercises given at the end of each chapter.

A.1 Answers to Chapter 1. Introduction

Exercise 1.1 (on page 22): *Consider an honest dice and the two events* $L = \{1, 2, 3\}$ *and* $M = \{3, 4\}$. *Compute the joint probabilities* $\mathbb{P}(L, M)$, $\mathbb{P}(\neg L, M)$, $\mathbb{P}(L, \neg M)$, *and the conditional probabilities* $\mathbb{P}(L|M)$, $\mathbb{P}(M|L)$, $\mathbb{P}(\neg M|L)$, *and* $\mathbb{P}(M|\neg L)$. *Compute (i)* $\mathbb{P}(L, M) + \mathbb{P}(L, \neg M)$, *(ii)* $\mathbb{P}(M|L) + \mathbb{P}(\neg M|L)$, *and (iii)* $\mathbb{P}(M|L) + \mathbb{P}(M|\neg L)$.

Because the dice is honest $\mathbb{P}(A) = |A|/6$, where $|A|$ is the number of elements (cardinality) of the set A. Thus $\mathbb{P}(L) = 1/2$ and $\mathbb{P}(M) = 1/3$. Then the joint probabilities are given by

$$\mathbb{P}(L, M) = \mathbb{P}(\{3\}) = \tfrac{1}{6}, \quad \mathbb{P}(\neg L, M) = \mathbb{P}(\{4\}) = \tfrac{1}{6}, \quad \mathbb{P}(L, \neg M) = \mathbb{P}(\{1, 2\}) = \tfrac{1}{3}.$$

The joint probabilities are symmetric (i.e. $\mathbb{P}(A, B) = \mathbb{P}(B, A)$), but, beware, the conditional probabilities are not necessarily symmetric. Now the conditional probabilities are given by

$$\mathbb{P}(L|M) = \frac{\mathbb{P}(L, M)}{\mathbb{P}(M)} = \frac{1/6}{1/3} = \tfrac{1}{2}, \qquad \mathbb{P}(M|L) = \frac{\mathbb{P}(L, M)}{\mathbb{P}(L)} = \frac{1/6}{1/2} = \tfrac{1}{3},$$

$$\mathbb{P}(\neg M|L) = \frac{\mathbb{P}(\neg M, L)}{\mathbb{P}(L)} = \frac{1/3}{1/2} = \tfrac{2}{3}, \quad \mathbb{P}(M|\neg L) = \frac{\mathbb{P}(\neg L, M)}{\mathbb{P}(\neg M)} = \frac{1/3}{2/3} = \tfrac{1}{2}.$$

Finally

i. $\mathbb{P}(L, M) + \mathbb{P}(L, \neg M) = \frac{1}{6} + \frac{1}{3} = \frac{1}{2} = \mathbb{P}(L)$
ii. $\mathbb{P}(M|L) + \mathbb{P}(\neg M|L) = \frac{1}{3} + \frac{2}{3} = 1$
iii. $\mathbb{P}(M|L) + \mathbb{P}(M|\neg L) = \frac{1}{3} + \frac{1}{2} = 5/6$.

While for any events A and B, $\mathbb{P}(A, B) + \mathbb{P}(A, \neg B) = \mathbb{P}(A)$ and $\mathbb{P}(A|B) + \mathbb{P}(\neg A|B) = 1$, the sum $\mathbb{P}(A|B) + \mathbb{P}(A|\neg B)$ may take different values depending on the events. Sometimes this might be 1 (in fact, $\mathbb{P}(L|M) + \mathbb{P}(L|\neg M) = 1$), but as we have seen in part (iii) this is just chance.

(There is a choice in the singular form for dice of either 'dice' or 'die'. Following my own personal usage, I've gone with the newfangled 'dice' rather than the more traditional 'die'.)

Exercise 1.2 (on page 22): *Let* D_1 *and* D_2 *denote the number rolled by two independent and honest dice. Enumerate all the values of* $S = D_1 + D_2$ *and compute their probabilities. Compute* $\mathbb{E}[S]$, $\mathbb{E}[S^2]$, *and* $\mathbb{V}\mathrm{ar}[S]$. *Also compute* $\mathbb{E}[D_1] = \mathbb{E}[D_2]$ *and* $\mathbb{V}\mathrm{ar}[D_1] = \mathbb{V}\mathrm{ar}[D_2]$ *and verify that* $\mathbb{E}[S] = \mathbb{E}[D_1] + \mathbb{E}[D_2]$ *and* $\mathbb{V}\mathrm{ar}[S] = \mathbb{V}\mathrm{ar}[D_1] + \mathbb{V}\mathrm{ar}[D_2]$.

This is a rather tedious but straightforward exercise. Its serves the purpose of illustrating that using properties of independence can significantly speed up calculations.

The only non-trivial part is evaluating the probability of S. We can compute this by listing all pairs of dice rolls that add up to S. Since each dice has 6 possible outcomes, the total number of distinct pairs is $6^2 = 36$. The pairs adding up to 2 are $\{(1,1)\}$, to 3 is $\{(1,2),(2,1)\}$, to 4 is $\{(1,3),(2,2),(3,1)\}$, etc.

s	2	3	4	5	6	7	8	9	10	11	12
s^2	4	9	16	25	36	49	64	81	100	121	144
$\mathbb{P}(S=s)$	$\frac{1}{36}$	$\frac{2}{36}$	$\frac{3}{36}$	$\frac{4}{36}$	$\frac{5}{36}$	$\frac{6}{36}$	$\frac{5}{36}$	$\frac{4}{36}$	$\frac{3}{36}$	$\frac{4}{36}$	$\frac{1}{36}$

A straightforward if boring calculation shows that $\mathbb{E}[S] = 7$, while $\mathbb{E}[S^2] = 1974/36$ so that $\mathbb{V}\mathrm{ar}[S] = 1974/36 - 7^2 = 210/36 = 35/6$. For an individual dice

$$\mathbb{E}[D_1] = \sum_{i=1}^{6} \frac{i}{6} = \frac{7}{2}, \qquad \mathbb{E}\left[D_1^2\right] = \sum_{i=1}^{6} \frac{i^2}{6} = \frac{91}{6}$$

so that $\mathbb{V}\mathrm{ar}[D_1] = \frac{91}{6} - \left(\frac{7}{2}\right)^2 = 35/12$. Thus $\mathbb{E}[S] = \mathbb{E}[D_1] + \mathbb{E}[D_2] = 2\,\mathbb{E}[D_1]$ while $\mathbb{V}\mathrm{ar}[S] = \mathbb{V}\mathrm{ar}[D_1] + \mathbb{V}\mathrm{ar}[D_2] = 2\,\mathbb{V}\mathrm{ar}[D_1]$.

Exercise 1.3 (on page 22): *Repeat the proof given on page 15 of the unconscious statistician theorem*

$$\mathbb{E}_X[g(X)] = \sum_{X\in\mathcal{X}} g(X)\, f_X(X)$$

starting from $\mathbb{E}_Y[Y]$ *using indicator functions.*
As before we start from

$$\mathbb{E}_X[g(X)] = \mathbb{E}_Y[Y] = \sum_{Y\in\mathcal{Y}} Y\, f_Y(Y)$$

where we have

$$f_Y(Y) = \sum_{X\in\mathcal{X}} [\![Y = g(X)]\!]\, f_X(X).$$

Substituting this in, we get

$$\begin{aligned}
\mathbb{E}_X[g(X)] = \mathbb{E}_Y[Y] &= \sum_{Y\in\mathcal{Y}} Y \sum_{X\in\mathcal{X}} [\![Y = g(X)]\!]\, f_X(X) \\
&= \sum_{X\in\mathcal{X}} \sum_{Y\in\mathcal{Y}} Y\, [\![Y = g(X)]\!]\, f_X(X) \\
&= \sum_{X\in\mathcal{X}} f_X(X) \sum_{Y\in\mathcal{Y}} Y\, [\![Y = g(X)]\!] = \sum_{X\in\mathcal{X}} f_X(X)\, g(X)
\end{aligned}$$

where we have exchanged the order of summation and used

$$\sum_{Y\in\mathcal{Y}} Y\, [\![Y = g(X)]\!] = g(X).$$

Although this follows the same steps as the proof given on page 22, it feels more algebriac and requires less thinking to convince yourself that each step is correct.

Exercise 1.4 (on page 22): *A classic result in machine learning is known as the* bias-variance dilemma, *where it is shown that the expected generalisation error can be viewed as the sum of a bias term and a variance term. The derivation is an interesting exercise in simplifying expectations. We consider a regression problem where we have a feature vector,* $\boldsymbol{X}$, *and we wish to predict some underlying function,* $g(\boldsymbol{X})$. *The function,* $g(\boldsymbol{X})$, *is unknown, although we are given a finite training set* $\mathcal{D} = \big((\boldsymbol{X}_i, Y_i)|i = 1, 2, \ldots, n\big)$, *where* $Y_i = g(\boldsymbol{X}_i)$. *We use* $\mathcal{D}$ *to train a learning machine* $m(\boldsymbol{X}, \mathcal{D})$ *(often the function* m *will depend on a set of weights that is chosen in order to minimise the prediction error for the training set). The (mean-squared) generalisation error is given by*

$$E(\mathcal{D}) = \mathbb{E}_{\boldsymbol{X}}\left[\left(g(\boldsymbol{X}) - m(\boldsymbol{X}, \mathcal{D})\right)^2\right].$$

That is, it is the mean squared difference between the true result, $g(\boldsymbol{X})$, *and the prediction of learning machine,* $m(\boldsymbol{X}, \mathcal{D})$. *The expectation is with respect to some (usually unknown) underlying probability density. The generalisation error depends on the training set,* $\mathcal{D}$. *We are interested in the expected error averaged over all data sets of a given size* n,

$$E = \mathbb{E}_{\mathcal{D}}\left[E(\mathcal{D})\right] = \mathbb{E}_{\mathcal{D}}\left[\mathbb{E}_{\boldsymbol{X}}\left[\left(g(\boldsymbol{X}) - m(\boldsymbol{X}, \mathcal{D})\right)^2\right]\right].$$

In the bias-variance dilemma, we consider the prediction of the mean machine, $\mathbb{E}_{\mathcal{D}}\left[m(\boldsymbol{X}, \mathcal{D})\right]$. *The task in this exercise is to show that the expected error,* E, *is the sum of two terms known as the* bias *and the* variance. *The bias is given by*

$$\mathbb{E}_{\boldsymbol{X}}\left[\left(g(\boldsymbol{X}) - \mathbb{E}_{\mathcal{D}}\left[m(\boldsymbol{X}, \mathcal{D})\right]\right)^2\right]$$

and measures the difference between the true function and the response of the mean machine (i.e. the response we get by averaging the responses of an ensemble of learning machines trained on every possible training set of size n*). The variance is equal to*

$$\mathbb{E}_{\boldsymbol{X}}\left[\mathbb{E}_{\mathcal{D}}\left[\left(m(\boldsymbol{X}, \mathcal{D}) - \mathbb{E}_{\mathcal{D}}\left[m(\boldsymbol{X}, \mathcal{D})\right]\right)^2\right]\right]$$

and measures the expected variance in the responses of the machines. Hint: add and subtract the response of the mean machine in the definition of the expected generalisation. You may assume that the expectations over $\boldsymbol{X}$ *and* $\mathcal{D}$ *commute (i.e.* $\mathbb{E}_{\mathcal{D}}\left[\mathbb{E}_{\boldsymbol{X}}\left[\cdots\right]\right] = \mathbb{E}_{\boldsymbol{X}}\left[\mathbb{E}_{\mathcal{D}}\left[\cdots\right]\right]$*).*

This follows from straightforward algebra and using the properties of expectations.

$$\begin{aligned}
E &\overset{(1)}{=} \mathbb{E}_{\mathcal{D}}\left[\mathbb{E}_{\boldsymbol{X}}\left[\left(g(\boldsymbol{X}) - m(\boldsymbol{X}, \mathcal{D})\right)^2\right]\right] \\
&\overset{(2)}{=} \mathbb{E}_{\mathcal{D}}\left[\mathbb{E}_{\boldsymbol{X}}\left[\left(g(\boldsymbol{X}) - \mathbb{E}_{\mathcal{D}}\left[m(\boldsymbol{X}, \mathcal{D})\right] + \mathbb{E}_{\mathcal{D}}\left[m(\boldsymbol{X}, \mathcal{D})\right] - m(\boldsymbol{X}, \mathcal{D})\right)^2\right]\right] \\
&\overset{(3)}{=} \mathbb{E}_{\mathcal{D}}\left[\mathbb{E}_{\boldsymbol{X}}\left[a^2 + 2\,a\,b + b^2\right]\right] = B + C + V
\end{aligned}$$

where $a = g(\boldsymbol{X}) - \mathbb{E}_{\mathcal{D}}\big[m(\boldsymbol{X}, \mathcal{D})\big]$, $b = \mathbb{E}_{\mathcal{D}}\big[m(\boldsymbol{X}, \mathcal{D})\big] - m(\boldsymbol{X}, \mathcal{D})$, $B = \mathbb{E}_{\mathcal{D}}\big[\mathbb{E}_{\boldsymbol{X}}\big[a^2\big]\big]$, $C = 2\,\mathbb{E}_{\mathcal{D}}\big[\mathbb{E}_{\boldsymbol{X}}\big[a\,b\big]\big]$, and $V = \mathbb{E}_{\mathcal{D}}\big[\mathbb{E}_{\boldsymbol{X}}\big[b^2\big]\big]$.

(1) By the definition of the expected generalisation given in the question.
(2) Adding and subtracting $\mathbb{E}_{\mathcal{D}}\big[m(\boldsymbol{X}, \mathcal{D})\big]$.
(3) Expanding out the square, $(a+b)^2 = a^2 + 2\,a\,b + b^2$.

Tackling each term

$$\begin{aligned} B &= \mathbb{E}_{\mathcal{D}}\Big[\mathbb{E}_{\boldsymbol{X}}\Big[\big(g(\boldsymbol{X}) - \mathbb{E}_{\mathcal{D}}\big[m(\boldsymbol{X}, \mathcal{D})\big]\big)^2\Big]\Big] \\ &= \mathbb{E}_{\boldsymbol{X}}\Big[\big(g(\boldsymbol{X}) - \mathbb{E}_{\mathcal{D}}\big[m(\boldsymbol{X}, \mathcal{D})\big]\big)^2\Big] \end{aligned}$$

because $\mathbb{E}_{\boldsymbol{X}}\Big[\big(g(\boldsymbol{X}) - \mathbb{E}_{\mathcal{D}}\big[m(\boldsymbol{X}, \mathcal{D})\big]\big)^2\Big]$ doesn't depend on $\mathcal{D}$ so we can use $\mathbb{E}_{\mathcal{D}}\big[c\big] = c$ (although $m(\boldsymbol{X}, \mathcal{D})$ does depend on $\mathcal{D}$ we have already taken the expectation so it no longer depends on the data set). We note that B is equal to the bias defined in the question. We consider next the term

$$\begin{aligned} V &= \mathbb{E}_{\mathcal{D}}\Big[\mathbb{E}_{\boldsymbol{X}}\Big[\big(\mathbb{E}_{\mathcal{D}}\big[m(\boldsymbol{X}, \mathcal{D})\big] - m(\boldsymbol{X}, \mathcal{D})\big)^2\Big]\Big] \\ &= \mathbb{E}_{\boldsymbol{X}}\Big[\mathbb{E}_{\mathcal{D}}\Big[\big(m(\boldsymbol{X}, \mathcal{D}) - \mathbb{E}_{\mathcal{D}}\big[m(\boldsymbol{X}, \mathcal{D})\big]\big)^2\Big]\Big] \end{aligned}$$

where we have used $(u-v)^2 = (v-u)^2$ with $u = \mathbb{E}_{\mathcal{D}}\big[m(\boldsymbol{X}, \mathcal{D})\big]$ and $v = m(\boldsymbol{X}, \mathcal{D})$, and we have exchanged the order of the expectation. Changing the order of expectations is nearly always innocuous, but there can be times when exchanging the order of expectations gives different answers – a careful mathematician would check whether this is allowed. We assume our expectations are suitably well behaved. We observe that V is equal to the variance term given in the question. Finally we consider

$$\begin{aligned} C &\overset{(1)}{=} 2\,\mathbb{E}_{\mathcal{D}}\Big[\mathbb{E}_{\boldsymbol{X}}\Big[\big(g(\boldsymbol{X}) - \mathbb{E}_{\mathcal{D}}\big[m(\boldsymbol{X}, \mathcal{D})\big]\big)\big(\mathbb{E}_{\mathcal{D}}\big[m(\boldsymbol{X}, \mathcal{D})\big] - m(\boldsymbol{X}, \mathcal{D})\big)\Big]\Big] \\ &\overset{(2)}{=} 2\,\mathbb{E}_{\boldsymbol{X}}\Big[\mathbb{E}_{\mathcal{D}}\Big[\big(g(\boldsymbol{X}) - \mathbb{E}_{\mathcal{D}}\big[m(\boldsymbol{X}, \mathcal{D})\big]\big)\big(\mathbb{E}_{\mathcal{D}}\big[m(\boldsymbol{X}, \mathcal{D})\big] - m(\boldsymbol{X}, \mathcal{D})\big)\Big]\Big] \\ &\overset{(3)}{=} 2\,\mathbb{E}_{\boldsymbol{X}}\Big[\big(g(\boldsymbol{X}) - \mathbb{E}_{\mathcal{D}}\big[m(\boldsymbol{X}, \mathcal{D})\big]\big)\,\mathbb{E}_{\mathcal{D}}\Big[\big(\mathbb{E}_{\mathcal{D}}\big[m(\boldsymbol{X}, \mathcal{D})\big] - m(\boldsymbol{X}, \mathcal{D})\big)\Big]\Big] \\ &\overset{(4)}{=} 2\,\mathbb{E}_{\boldsymbol{X}}\Big[\big(g(\boldsymbol{X}) - \mathbb{E}_{\mathcal{D}}\big[m(\boldsymbol{X}, \mathcal{D})\big]\big)\big(\mathbb{E}_{\mathcal{D}}\big[m(\boldsymbol{X}, \mathcal{D})\big] - \mathbb{E}_{\mathcal{D}}\big[m(\boldsymbol{X}, \mathcal{D})\big]\big)\Big] \overset{(5)}{=} 0 \end{aligned}$$

(1) By definition of C.
(2) Exchanging the order of taking expectations (assuming this is allowed).
(3) Noting that the term $g(\boldsymbol{X}) - \mathbb{E}_{\mathcal{D}}\big[m(\boldsymbol{X}, \mathcal{D})\big]$ is independent of $\mathcal{D}$ so can be taken out of the expectation.
(4) Using the linearity of expectations ($\mathbb{E}\big[f(X) + g(X)\big] = \mathbb{E}\big[f(X)\big] + \mathbb{E}\big[g(X)\big]$) and $\mathbb{E}_{\mathcal{D}}\big[\mathbb{E}_{\mathcal{D}}\big[m(\boldsymbol{X}, \mathcal{D})\big]\big] = \mathbb{E}_{\mathcal{D}}\big[m(\boldsymbol{X}, \mathcal{D})\big]$.
(5) Observing that the last terms cancel.

In consequence, $E = B + V$, which is what we are asked to show. The question illustrates that using the linearity of the expectation operator and exchanging the order of expectations can often be very powerful in simplifying complex expressions. We also used the fact that once we have taken an expectation of a random variable we are left with a quantity which is then independent of the variable and so can be treated as a constant if we take a second expectation with respect to that random variable.

Having taken the trouble to do this rather elaborate calculation, you deserve an explanation of why this is a 'classic result in machine learning'. Firstly, it has to be admitted it is of little practical use as both the bias and the variance are usually incomputable (if we knew $g(\boldsymbol{X})$ then there would be little point trying to estimate it using machine learning). However, the result provides a lot of intuition about the difficulty of machine learning. If we use a very simple learning machine then the bias would typically be very large while the variance would be small (we would learn a similar machine given a slightly different data set). On the other hand, if we use a complicated machine we might be able to approximate a complex function $g(\boldsymbol{X})$ well (giving a low bias), but we are likely to have a large variance (the function we learn, $m(\boldsymbol{X}, \mathcal{D})$, is likely to be very sensitive to the data set). This leads to the *bias-variance dilemma*. Most of the very successful learning machines find some way of finessing this dilemma.

A.2 Answers to Chapter 2. Survey of Distributions

Exercise 2.1 (on page 40): *What distribution might you use to model the following situations:*

i. *the proportion of the gross national product (GDP) from different sectors of the economy*
   This could be modelled with a Dirichlet distribution.
ii. *the probability of three buses arriving in the next five minutes*
   The answer to this is country specific. In the UK a reasonable assumption might be that there are a large number of buses each with a small random probability of arriving in the next five minutes. In this case, the Poisson distribution would be an appropriate model. In other countries where buses run to timetables this might not require a probabilistic answer.
iii. *the length of people's stride*
   These kinds of measurements are often well modelled by normal distributions.
iv. *the salary of people*
   Given these are positive with a long tail a reasonable distribution might be a Gamma or Weibull (or even a log-normal) distribution.
v. *the outcome of a roulette wheel spun many times*
   As there are a number of different outcomes a multinomial distribution seems reasonable.
vi. *the number of sixes rolled in a fixed number of trials*

As there are two possibilities (six or not six) a binomial distribution is reasonable. A Poisson distribution would also provide a crude approximation.

vii. *the odds of a particular horse winning the Grand National*
A beta distribution could be used.

Exercise 2.2 (on page 41): *Assume that one card is chosen from an ordinary pack of cards. The card is then replaced and the pack shuffled. This is repeated 10 times. What is the chance that an ace is drawn exactly three times?*
Here we have 10 independent trials with a success probability of $1/13$ so that the probability of three successes is given by $\text{Bin}(3|10, 1/13) = 0.0312$.

Exercise 2.3 (on page 41): *Assume that a pack of cards is shuffled and 10 cards are dealt. What is the probability that exactly three of the cards are aces?*
In this case the trials are no longer independent. The probability is given by the hypergeometric distribution $\text{Hyp}(3|52, 4, 10) = 0.0186$.

Exercise 2.4 (on page 41): *Show that for the hypergeometric distribution,* $\text{Hyp}(k|N, m, n)$*, converges to a binomial distribution,* $\text{Bin}(k|n, p)$*, in the limit where N and m go to infinity in such a way that $m/N = p$. Explain why this will happen in words.*
Writing out the hypergeometric distribution in full

$$\text{Hyp}(k|N, m, n) = \frac{\binom{m}{k}\binom{N-m}{n-k}}{\binom{N}{n}} = \frac{m!\,(N-m)!\,(N-n)!\,n!}{(m-k)!\,k!\,(n-k)!\,(N-m-n+k)!\,N!}.$$

Judiciously collecting together terms we can write this as

$$\text{Hyp}(k|N, m, n) = \frac{n!}{(n-k)!\,k!}\,\frac{m!}{(m-k)!}\,\frac{(N-m)!}{(N-m-n+k)!}\,\frac{(N-n)!}{N!}.$$

As N and m go to infinity we observe

$$\lim_{m\to\infty}\frac{m!}{(m-k)!} = m^k, \qquad \lim_{N,m\to\infty}\frac{(N-m)!}{(N-m-n+k)!} = (N-m)^{n-k},$$

$$\lim_{N\to\infty}\frac{(N-n)!}{N!} = \frac{1}{N^n},$$

thus

$$\begin{aligned}\lim_{\substack{N,m\to\infty\\ m/N=p}} \text{Hyp}(k|N, m, n) &= \lim_{\substack{N,m\to\infty\\ m/N=p}} \frac{n!}{(n-k)!\,k!}\,\frac{m^k\,(N-m)^{n-k}}{N^n} \\ &= \frac{n!}{(n-k)!\,k!}\,p^k\,(1-p)^{n-k}\end{aligned}$$

which is equal to $\text{Bin}(k|n, p)$.

This should be no surprise. If we consider the example of m red balls in a bag of N balls, then if we sample k we would expect in the limit as N and m become very large that, irrespective of which balls we have already sampled, we barely change the probability of choosing a red ball. Thus sampling without replacement (described by the hypergeometric distribution) becomes identical to sampling with replacement (described by the binomial distribution).

Exercise 2.5 (on page 41): *In the UK National Lottery players choose six numbers between 1 to 59. On draw day six numbers are chosen and the players who correctly guess two or more of the drawn numbers win a prize. The prizes increase substantially as you guess more of the chosen numbers. Write down the probability of guessing k balls correctly using the hypergeometric distribution and compute the probabilities for k equal 2 to 6.*

The probability of choosing k of the six drawn numbers in six attempts is $\text{Hyp}(k|59,6,6)$. The probability of guessing: two correct numbers is 1 454 125/15 019 158=0.0975, three correct numbers is 234 260/22 528 737=0.0104, four correct numbers is $3\,445/7\,509\,579 = 4.587 \times 10^{-4}$, five correct numbers is $53/7\,509\,579 = 7.058 \times 10^{-6}$, and six correct numbers is $1/45\,057\,474 = 2.219 \times 10^{-8}$.

Exercise 2.6 (on page 41): *Show that if* $Y \sim \text{Exp}(1)$ *then the random variable* $X = \lambda \sqrt[k]{Y}$ *(or* $Y = (X/\lambda)^k$*) is distributed according to the Weibull density*

$$\text{Wei}(x|\lambda,k) = \frac{k}{\lambda}\left(\frac{x}{\lambda}\right)^{k-1} \mathrm{e}^{-(x/\lambda)^k}.$$

Plot the Weibull densities $\text{Wei}(x|1,k)$ *and the gamma distribution with the same mean and variance*

$$\text{Gam}\left(\frac{\Gamma^2(1+1/k)}{\Gamma(1+2/k)-\Gamma^2(1+1/k)}, \frac{\Gamma(1+2/k)}{\Gamma(1+2/k)-\Gamma^2(1+1/k)}\right)$$

for $k = 1/2, 1, 2,$ *and 5.*

Using the formula for the change of variables (see page 13)

$$\text{Wei}(x|\lambda,k) = \frac{\partial\,(x/\lambda)^k}{\partial x}\text{Exp}\left(\left(\frac{x}{\lambda}\right)^k \middle| 1\right) = \frac{k}{\lambda}\left(\frac{x}{\lambda}\right)^{k-1} \mathrm{e}^{-(x/\lambda)^k},$$

as advertised.

The densities for the Weibull distribution with $\lambda = 1$ and $k = 1/2$, 1, 2, and 5 are shown in Figure A.1, together with the gamma densities with the same mean and variance. We observe that the curves are subtly different (except for $k = 1$ where both densities are equal to $\text{Exp}(x|1)$). For many empirical uses there is often insufficient data to choose which of these two distributions is a more accurate approximation. However, there are cases where there is sufficient data. For example, in modelling wind speeds the Weibull distribution is often preferred. Although the differences are small, they can make a considerable difference when trying to infer the probability of large rare events (i.e. events that occur in the tail of the distribution).

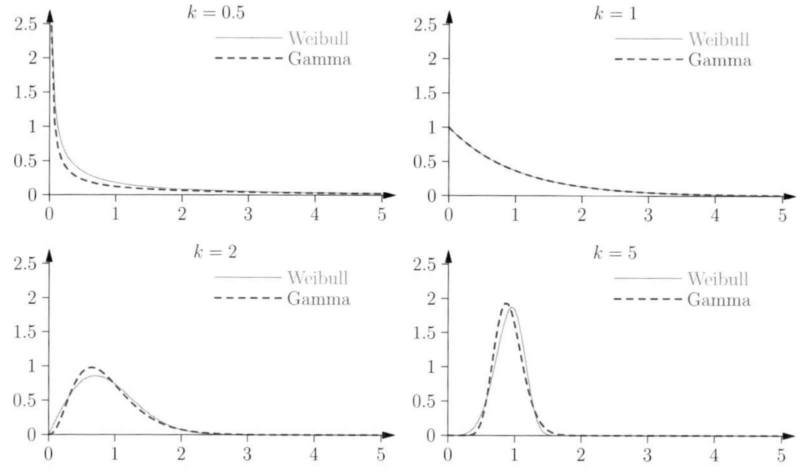

Figure A.1 Curves for the Weibull distribution.

A.3 Answers to Chapter 3. Monte Carlo

Exercise 3.1 (on page 58): *Using the transformation method show how you could generate a Cauchy deviate. The density of the Cauchy distribution is given by*

$$\mathrm{Cau}(x) = \frac{1}{\pi(1+x^2)}$$

and the cumulative distribution is

$$F(x) = \int_{-\infty}^{x} \mathrm{Cau}(y)\, dy = \frac{1}{2} + \frac{\arctan(x)}{\pi}.$$

Writing $y = F(x)$ and solving for x we find

$$F^{-1}(y) = \tan\left(\pi\left(y - \tfrac{1}{2}\right)\right).$$

To generate a Cauchy deviate we first generate a uniform deviate $U \sim \mathrm{U}(0,1)$ and then set $X = \tan(\pi(U - 1/2))$.

Exercise 3.2 (on page 58): *Implement a normal deviate generator:*

i. *using the transformation method by numerically inverting the erfc function (see Section 5.4 on page 90) using Newton–Raphson's method (alternatively the third edition of numerical recipes (Press et al., 2007) provides an inverse of the cumulative distribution function (CDF) for a normal distribution);*
ii. *using the standard Box–Muller method;*
iii. *using the rejection method from Cauchy deviates.*

Time how long it takes each implementation to generate 10^6 *random deviates.*

On my laptop I got timings of:

i. 365 ms (using the inverse-CDF function in numerical recipes);
ii. 51 ms;
iii. 115 ms.

I made no attempt to optimise the code.

Exercise 3.3 (on page 58): *In simulating a simple model of evolution, one is often faced with mutating each allele in a gene with a small probability* $p \ll 1$ *(this task also occurs in running a genetic algorithm). It is far quicker to generate a random deviate giving the distance between mutations than to decide for each allele whether to mutate it. The distance between mutations is given by the geometric distribution. Denoting by* $K > 0$ *the distance to the next allele that is mutated then*

$$\mathbb{P}\left(K = k\right) = \text{Geo}(k|p) = p\,(1-p)^{k-1}.$$

Using the transformation method show how to generate a deviate $K \sim \text{Geo}(p)$ *starting from a uniform deviate* $U \sim \text{U}(0,1)$.

The cumulative probability that $K \leq i$ is given by

$$\mathbb{P}\left(K \leq i\right) = \sum_{k=1}^{i} \text{Geo}(k|p) = p \sum_{k=1}^{i} (1-p)^{k-1} = 1 - (1-p)^i$$

where we have used that the sum of a geometric series is

$$\sum_{k=1}^{i} x^{k-1} = 1 + x + \ldots + x^{i-1} = \frac{1-x^i}{1-x}.$$

In the transformation method (see Figure A.2) we generate a uniform deviate $U \sim \text{U}(0,1)$ and then generate a geometric deviate $I \in \mathbb{N}$ if

$$\begin{aligned}
\mathbb{P}\left(K < I - 1\right) &\leq U < \mathbb{P}\left(K < I\right) \\
1 - (1-p)^{I-1} &\leq U < 1 - (1-p)^I \\
(1-p)^{I-1} &\geq 1 - U > (1-p)^I \\
(I-1)\log(1-p) &\geq \log(1-U) > I\log(1-p) \\
I - 1 &\leq \frac{\log(1-U)}{\log(1-p)} < I
\end{aligned}$$

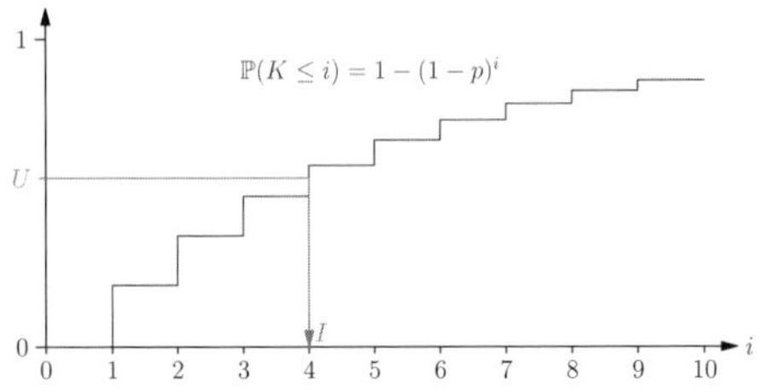

Figure A.2 Illustration of using the transformation method to obtain a geometric deviate. Note that i can be arbitrarily large.

or

$$I = \left\lfloor \frac{\log(1-U)}{\log(1-p)} \right\rfloor + 1$$

where $\lfloor x \rfloor$ denotes the floor function, i.e. the largest integer less than or equal to x.

A.4 Answers to Chapter 4. Discrete Random Variables

Exercise 4.1 (on page 72): *Consider a succession of Bernoulli trials with a success probability of p. The probability that the first success occurs after k trials is given by the* geometric distribution

$$\mathbb{P}\left(\textit{First success after k trial}\right) = \text{Geo}(k|p) = p\,(1-p)^{k-1}$$

(The geometric distribution is sometimes defined to be the probability of success after k failures so is defined as $p\,(1-p)^k$ *where* $k = 0, 1, 2, \ldots$*) Compute the cumulant generating function (CGF) for the geometric distribution and from this compute the mean and variance. If the probability of winning a lottery is* 10^{-6}*, how many attempts do you have to make on average before you win?*

The moment generating function is

$$\begin{aligned} M(k) &= \mathbb{E}\left[e^{l\,k}\right] = \sum_{k=1}^{n} e^{l\,k}\,p\,(1-p)^{k-1} = p\,e^{l} \sum_{k=1}^{n} \left((1-p)e^{l}\right)^{k-1} \\ &= \frac{p\,e^{l}}{1-(1-p)\,e^{l}} \end{aligned}$$

where we have used the sum of a geometric progression. Thus, the CGF is

$$G(k) = \log\bigl(M(k)\bigr) = \log(p) + l - \log\bigl(1-(1-p)\,e^{l}\bigr).$$

Taking derivatives

$$\begin{aligned} G'(k) &= 1 + \frac{(1-p)\,e^{l}}{1-(1-p)\,e^{l}} \\ G''(k) &= \frac{(1-p)\,e^{l}}{1-(1-p)\,e^{l}} + \frac{(1-p)^2\,e^{2l}}{\left(1-(1-p)\,e^{l}\right)^2} \end{aligned}$$

and recalling $\kappa_n = G^{(n)}(0)$ we find

$$\begin{aligned} \kappa_1 &= G'(0) = 1 + \frac{1-p}{p} = \frac{1}{p} \\ \kappa_2 &= G''(0) = \frac{1-p}{p} + \frac{(1-p)^2}{p^2} = \frac{1-p}{p^2}. \end{aligned}$$

To win a lottery with a probability of winning of one in a million, the expected number of trials before a success would be 1 million, which fits with many people's intuition. The standard deviation is fractionally less than 1 million.

Exercise 4.2 (on page 72): *Show that the probability of k successes until r failures have occurred in a sequence of Bernoulli trials with success probability p given by the* negative binomial distribution

$$\mathrm{NegBin}(k|r,p) = \binom{k+r-1}{k} p^k\,(1-p)^r$$

(note that the last trial will have been a failure, hence the combinatorial factor). Compute the CGF for the negative binomial and compute the mean and variance. If you are playing a game of pure chance with four friends how many times do you expect to win before you lose nine times. (Note that NegBin$(k|r,p)$ *is a properly normalised distribution so that*

$$\sum_{k=0}^{\infty} \binom{k+r-1}{k} p^k = \frac{1}{(1-p)^r}$$

which is true for any p. This is exactly the sum we need to compute the CGF.)

Consider a set of Bernoulli trials with k successes and r failures. A typical sequence of k successes before the rth failure would be

$$\underbrace{s\,f\,f\,s\,s\,f\,s\,s\,f\,f\,s\,f\,s\,s\,s\,s\,f\,s\,f\,s\,f\,f\,s\,s\,s\,s\,f\,f\,s\,s}_{k \text{ successes, } r-1 \text{ failures}}\ \underbrace{f}_{r^{th}\text{ failure}}$$

The probability of any particular sequence of k success and r failures is $p^k\,(1-p)^r$, but the number of such sequences is $\binom{k+r-1}{k}$. Multiplying these together we get the negative binomial distribution.

The moment generating function is

$$\begin{aligned} M(l) &= \sum_{k=0}^{\infty} \binom{k+r-1}{k} \mathrm{e}^{lk} p^k\,(1-p)^r \\ &= (1-p)^r \sum_{k=0}^{\infty} \binom{k+r-1}{k} (p\,\mathrm{e}^l)^k = \left(\frac{1-p}{1-p\,\mathrm{e}^l}\right)^r \end{aligned}$$

where we used the summation formula that was deduced from the normalisation condition. The CGF is

$$G(l) = \log\bigl(M(l)\bigr) = r\,\log(1-p) - r\,\log\bigl(1-p\,\mathrm{e}^l\bigr).$$

Thus

$$\begin{aligned} G'(l) &= \frac{r\,p\,\mathrm{e}^l}{1-p\,\mathrm{e}^l}, & \kappa_1 &= G'(0) = \frac{r\,p}{1-p}, \\ G''(l) &= \frac{r\,p\,\mathrm{e}^l}{1-p\,\mathrm{e}^l} + \frac{r\,p^2\,\mathrm{e}^{2l}}{(1-p\,\mathrm{e}^l)^2} = \frac{r\,p\,\mathrm{e}^l}{(1-p\,\mathrm{e}^l)^2}, & \kappa_2 &= G''(0) = \frac{r\,p}{(1-p)^2}. \end{aligned}$$

In a game of chance with a probability of winning of $p = 1/4$ then the expected number of wins before you lose $r = 9$ times is equal to $\kappa_1 = 3$. The standard deviation is $\sqrt{\kappa_2} = 2$.

Exercise 4.3 (on page 72): *Consider a series of Bernoulli trials occurring at a time interval δt with a success probability $\mu\,\delta t$ (so the expected number of successes in one second is μ). Show that, in the limit $\delta t \to 0$, the probability of the first successful event occurring at time t is distributed according to the* exponential distribution

$$f(t) = \mathrm{Exp}(t|\mu) = \mu\,\mathrm{e}^{-\mu\,t}.$$

(This is the waiting time between Poisson distributed events. It is used, for example, in simulating chemical reactions at a molecular level – see Section 12.3.2.)

Consider a set of Bernoulli trials with success probability $p = \mu\,\delta t$. A sequence where the first success occurs after k trial would be

$$\underbrace{f\,f\,f\,f\,f\,f\,f\,f\,f\,f\,f\,f\,f\,f\,f\,f\,f\,f\,f\,f\,f\,f\,f\,f\,f\,f\,f\,f\,f}_{k-1 \text{ failures}}\underbrace{s}_{\text{first success}}$$

The probability of a success on the kth trial is given by the geometric distribution

$$\mathbb{P}\left(\text{Success at time } t = k\,\delta t\right) = \mathrm{Geo}(k|\mu\,\delta t) = \mu\,\delta t\,(1-\mu\,\delta t)^{k-1}.$$

In the limit $\delta t \to 0$

$$\lim_{\delta t\to 0}(1-\mu\,\delta t)^{k-1} = \lim_{\delta t\to 0} \mathrm{e}^{(k-1)\,\log(1-\mu\,\delta t)} = \lim_{\delta t\to 0} \mathrm{e}^{-(k-1)\,\mu\,\delta t + O(\delta t^2)} = \mathrm{e}^{-\mu\,t}$$

where $t = k\,\delta t$. The probability of the event occurring in the kth time interval is thus

$$f(t)\,\delta t = \mathbb{P}\left(\text{Success at time } t = k\,\delta t\right) = \mu\,\mathrm{e}^{-\mu\,t}\,\delta t$$

where $f(t)$ is the probability density. Thus,

$$f(t) = \mu\,\mathrm{e}^{-\mu\,t} = \mathrm{Exp}(t|\mu).$$

Exercise 4.4 (on page 72): *The cumulative probability mass function for a Poisson distribution* $\mathrm{Poi}(\mu)$ *is given by*

$$\mathbb{P}\left(\textit{number of success} < k\right) = \sum_{i=0}^{k-1} \frac{\mu^i}{i!}\mathrm{e}^{-\mu}.$$

Starting from the normalised incomplete gamma function (see Appendix 2.A) defined by

$$Q(k,\mu) = \frac{1}{(k-1)!}\int_{\mu}^{\infty} z^{k-1}\,\mathrm{e}^{-z}\,dz$$

obtain a relationship between $Q(k,\mu)$ and $Q(k-1,\mu)$ using integration by part and hence prove that

$$\mathbb{P}\left(\textit{number of success} < k\right) = Q(k,\mu).$$

Using integration by part we find

$$
\begin{aligned}
Q(k,\mu) &\overset{(1)}{=} \frac{1}{(k-1)!}\int_\mu^\infty z^{k-1}\,\mathrm{e}^{-z}\,\mathrm{d}z \\
&\overset{(2)}{=} \frac{1}{(k-1)!}\left[-z^{k-1}\,\mathrm{e}^{-z}\right]_\mu^\infty - \frac{1}{(k-1)!}\int_\mu^\infty -(k-1)\,z^{k-2}\,\mathrm{e}^{-z}\,\mathrm{d}z \\
&\overset{(3)}{=} \frac{\mu^{k-1}\,\mathrm{e}^{-\mu}}{(k-1)!} + \frac{1}{(k-2)!}\int_\mu^\infty z^{k-2}\,\mathrm{e}^{-z}\,\mathrm{d}z \overset{(4)}{=} \frac{\mu^{k-1}\,\mathrm{e}^{-\mu}}{(k-1)!} + Q(k-1,\mu)
\end{aligned}
$$

(1) From the definition given in the question.
(2) Using integration by parts

$$
\int_a^b u\,\frac{\mathrm{d}v}{\mathrm{d}z}\,\mathrm{d}z = [u\,v]_a^b - \int_a^b v\,\frac{\mathrm{d}u}{\mathrm{d}z}\,\mathrm{d}z
$$

where $u = z^{k-1}$ and $\frac{\mathrm{d}v}{\mathrm{d}z} = \mathrm{e}^{-z}$.
(3) Substituting in the limits (the upper limit vanishes) and cancelling the factor $k-1$ in the second term.
(4) By definition of the normalised incomplete gamma function.

Repeating this recursion

$$
\begin{aligned}
Q(k,\mu) &= \frac{\mu^{k-1}\,\mathrm{e}^{-\mu}}{(k-1)!} + \frac{\mu^{k-2}\,\mathrm{e}^{-\mu}}{(k-2)!} + Q(k-2,\mu) \\
&= \frac{\mu^{k-1}\,\mathrm{e}^{-\mu}}{(k-1)!} + \cdots + \frac{\mu^{1}\,\mathrm{e}^{-\mu}}{1!} + Q(1,\mu)
\end{aligned}
$$

but

$$
Q(1,\mu) = \int_\mu^\infty \mathrm{e}^{-z}\,\mathrm{d}z = \mathrm{e}^{-\mu}.
$$

Thus

$$
Q(k,\mu) = \sum_{i=0}^{k-1} \frac{\mu^i\,\mathrm{e}^{-\mu}}{i!}
$$

which is what we set out to prove.

Exercise 4.5 (on page 73): *The probability of k or more success in n Bernoulli trials with a success probability of p is given by*

$$
\mathbb{P}\left(\textit{number of success} \geq k\right) = \sum_{i=k}^{n} \binom{n}{i} p^i\,(1-p)^{n-i} = I_p(k, n-k+1).
$$

Starting from the normalised incomplete beta function (see Appendix 2.B) defined by

$$
I_p(k, n-k+1) = \frac{n!}{(k-1)!\,(n-k)!}\int_0^p z^{k-1}\,(1-z)^{n-k}\,dz
$$

obtain a relationship between $I_p(k, n-k+1)$ and $I_p(k-1, n-k+2)$ using integration by part and hence prove

$$\mathbb{P}\left(\textit{number of success} \geq k\right) = I_p(k, n-k+1).$$

Using integration by parts

$$\begin{aligned}
&I_p(k, n-k+1)\\
&\stackrel{(1)}{=} \frac{n!}{(k-1)!\,(n-k)!}\left[-\frac{z^{k-1}\,(1-z)^{n-k+1}}{n-k+1}\right]_0^p\\
&\qquad - \frac{n!}{(k-1)!\,(n-k)!}\int_0^p -\frac{(k-1)}{n-k+1} z^{k-2}\,(1-z)^{n-k+1}\,\mathrm{d}z\\
&\stackrel{(2)}{=} -\frac{n!\,p^{k-1}\,(1-p)^{n-k+1}}{(k-1)!\,(n-k+1)!} + \frac{n!}{(k-2)!\,(n-k+1)!}\int_0^p z^{k-2}\,(1-z)^{n-k+1}\,\mathrm{d}z\\
&\stackrel{(3)}{=} -\binom{n}{k-1} p^{k-1}\,(1-p)^{n-k+1} + I_p(k-1, n-k-2)
\end{aligned}$$

(1) Using integration by parts

$$\int_a^b u\,\frac{\mathrm{d}v}{\mathrm{d}z}\,\mathrm{d}z = [u\,v]_a^b - \int_a^b v\,\frac{\mathrm{d}u}{\mathrm{d}z}\,\mathrm{d}z$$

with $u = z^{k-1}$ and $\frac{\mathrm{d}\,v}{\mathrm{d}\,z} = (1-z)^{n-k}$.

(2) Substituting in the limits (the lower limit vanishes – assuming $k > 1$) and simplify the constant in the second term.

(3) Using the definition of the binomial coefficient and the definition of the normalised incomplete beta function.

Using this recursion $k-1$ times we get

$$\begin{aligned}
I_p(k, n-k+1) = &-\binom{n}{k-1} p^{k-1}\,(1-p)^{n-k+1} - \cdots\\
&-\binom{n}{1} p\,(1-p)^{n-1} + I_p(1, n)
\end{aligned}$$

but

$$I_p(1, n) = n\int_0^p (1-z)^{n-1}\mathrm{d}z = 1 - (1-p)^n.$$

Thus

$$I_p(k, n-k+1) = 1 - \sum_{i=0}^{k-1}\binom{n}{i} p^i\,(1-p)^{n-i} = \sum_{i=k}^{n}\binom{n}{i} p^i\,(1-p)^{n-i}.$$

□

Exercise 4.6 (on page 73): *In Example 4.1 on page 70 we calculated the probability that there are no empty boxes when we randomly place n balls in k boxes. This probability was*

$$p_{neb} = \left(\prod_{j=1}^{k} \sum_{n_j>0}\right) \text{Mult}(\boldsymbol{n}|n, \boldsymbol{p}) \left[\!\left[\sum_{j=1}^{k} n_j = n \right]\!\right]$$

where $p_j = 1/k$ for, $j = 1, 2, \ldots, k$. Perform this sum using the integral representation of the indicator function

$$\left[\!\left[\sum_{j=1}^{k} n_j = n \right]\!\right] = \int_{-\infty}^{\infty} \mathrm{e}^{i\omega\left(\sum_j n_j - n\right)} \frac{d\omega}{2\pi}. \tag{A.1}$$

Note that the indicator function ensures that the total number of balls equals n, so we can treat the sums as unconstrained. Having performed the sums we can expand out the expression similar to what was done in Example 4.1 and then use Equation (A.1) to get a simplified answer. (Be warned the answer is not in an identical form to that given in Example 4.1, although it can be made so by a change of variables.)

The probability of no empty boxes is

$$p_{neb} \overset{(1)}{=} \sum_{\substack{\boldsymbol{n}\in\Lambda_n^k \\ \forall i,\, n_i>0}} \text{Mult}(\boldsymbol{n}|n, \boldsymbol{p}) \overset{(2)}{=} n! \prod_{j=1}^{k} \sum_{\{n_j>0\}} \frac{p_i^{n_j}}{n_j!} \left[\!\left[\sum_{j=1}^{k} n_j = n \right]\!\right]$$

$$\overset{(3)}{=} \frac{n!}{k^n} \prod_{j=1}^{k} \sum_{\{n_j>0\}} \frac{1}{n_j!} \left[\!\left[\sum_{j=1}^{k} n_j = n \right]\!\right]$$

(1) From the definition of the problem.
(2) Using the definition of the multinomial distribution and using an indicator function to allow us to replace the sum by an unbounded sum.
(3) Using $p_i = \frac{1}{k}$ and taking all constant terms outside the summation.

Using the integral representation of the indicator function, Equation (A.1), we can write this as

$$p_{neb} \overset{(1)}{=} \frac{n!}{k^n} \prod_{j=1}^{k} \sum_{\{n_j>0\}} \frac{1}{n_j!} \int_{-\infty}^{\infty} \mathrm{e}^{\mathrm{i}\omega\left(\sum_j n_j - n\right)} \frac{\mathrm{d}\omega}{2\pi}$$

$$\overset{(2)}{=} \frac{n!}{k^n} \int_{-\infty}^{\infty} \mathrm{e}^{\mathrm{i}\omega n} \prod_{j=1}^{k} \left(\sum_{n_j>0} \frac{\mathrm{e}^{\mathrm{i}\omega n_j}}{n_j!} \right) \frac{\mathrm{d}\omega}{2\pi} = \frac{n!}{k^n} \int_{-\infty}^{\infty} \mathrm{e}^{\mathrm{i}\omega n} \prod_{j=1}^{k} \left(\mathrm{e}^{\mathrm{e}^{\mathrm{i}\omega}} - 1 \right) \frac{\mathrm{d}\omega}{2\pi}$$

$$\overset{(3)}{=} \frac{n!}{k^n} \int_{-\infty}^{\infty} \mathrm{e}^{\mathrm{i}\omega n} \left(\mathrm{e}^{\mathrm{e}^{\mathrm{i}\omega}} - 1 \right)^k \frac{\mathrm{d}\omega}{2\pi}$$

(1) Replacing the indicator function by the integral representation.
(2) Using $\sum_{n=1} a^n/n! = \exp(a) - 1$ where $a = \mathrm{e}^{\mathrm{i}\omega}$.
(3) Using $\prod_{i=1}^{k} A = A^k$ where $A = \mathrm{e}^{\mathrm{e}^{\mathrm{i}\omega}} - 1$.

The integral representation has allowed us to treat the sums as decoupled from each other. This then allows us to sum the series for each n_i. We now expand the term we have just summed (this may look like going backwards, but we will have reduced k sums to two sums):

$$\begin{aligned}
p_{neb} &\overset{(1)}{=} \frac{n!}{k^n}\int_{-\infty}^{\infty} e^{-i\omega n}\left(\sum_{j=0}^{k}\binom{k}{j} e^{j\,e^{i\omega}}(-1)^{k-j}\right)\frac{d\omega}{2\pi}\\
&\overset{(2)}{=} \frac{n!}{k^n}\int_{-\infty}^{\infty} e^{-i\omega n}\left(\sum_{j=0}^{k}\binom{k}{j}(-1)^{k-j}\sum_{l=0}^{\infty}\frac{\left(j\,e^{i\omega}\right)^l}{l!}\right)\frac{d\omega}{2\pi}\\
&\overset{(3)}{=} \frac{n!}{k^n}\sum_{j=0}^{k}\binom{k}{j}(-1)^{k-j}\sum_{l=0}^{\infty}\frac{j^l}{l!}\int_{-\infty}^{\infty} e^{i\omega(l-n)}\frac{d\omega}{2\pi}
\end{aligned}$$

(1) Using the binomial expansion for $(a-b)^k$ with $a = e^{e^{i\omega}}$ and $b = 1$.
(2) Using the Taylor expansion of the exponential.
(3) Interchanging the order of integration of summation.

Using the integral

$$\int_{-\infty}^{\infty} e^{i\omega(l-n)}\frac{d\omega}{2\pi} = [\![l = n]\!]$$

we have

$$p_{neb} \overset{(1)}{=} \frac{n!}{k^n}\sum_{j=0}^{k}\binom{k}{j}(-1)^{k-j}\frac{j^n}{n!} \overset{(2)}{=} \sum_{j=0}^{k}\binom{k}{j}(-1)^{k-j}\left(\frac{j}{k}\right)^n$$

(1) Using $\sum_l f(l)\,[\![l = n]\!] = f(n)$.
(2) Cancelling factors of $n!$ and bringing k^{-n} inside the sum.

To show that this is identical to the answer we obtained in Example 4.1 we make a change of variables from j to $i = k - j$; using $\binom{k}{k-i} = \binom{k}{i}$ we find

$$p_{neb} = \sum_{i=0}^{k}\binom{k}{i}(-1)^i\left(1-\frac{i}{k}\right)^n.$$

To my mind this is a more natural calculation than Poissonisation, although I admittedly have a lot of familiarity with this approach. The approach also readily generalises and can easily provide good approximate answers through saddle-point evaluation of the integral (which is often exact in some asymptotic limit).

A.5 Answers to Chapter 5. The Normal Distribution

Exercise 5.1 (on page 102): *Let*

$$S_n = \frac{1}{\sqrt{n/12}}\sum_{i=1}^{n}(U_i - 1/2)$$

where $U_i \sim U(0,1)$. That is, S_n is the sum of n uniform deviates normalised so that it has zero mean and unit variance. Generate a histogram of deviates S_5 and

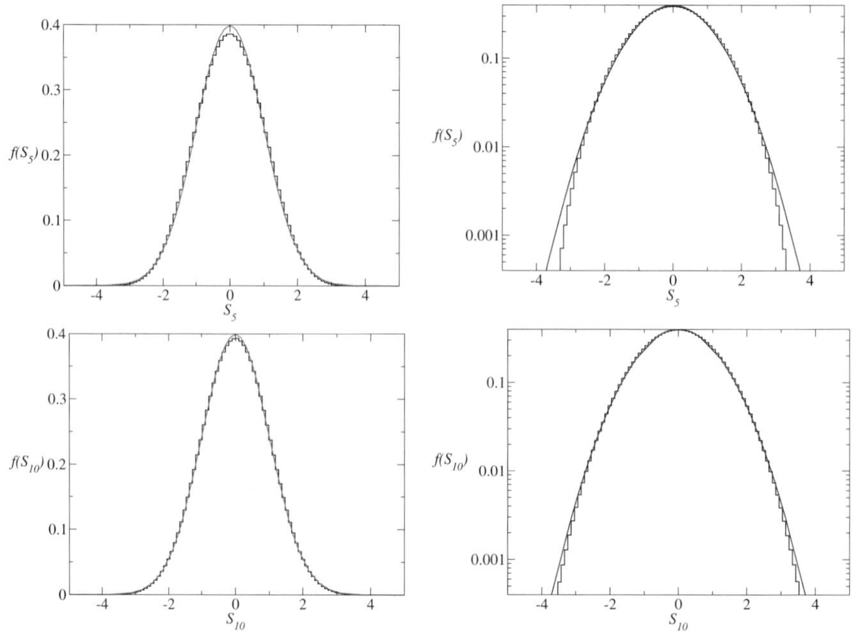

Figure A.3
Histogram of S_5 and S_{10} and logarithm of the histogram. The histogram is collected from 10^8 samples. Note that the error in the histogram is of the order of the width of a line. The smooth line shows the probability density function of a normal distribution.

compare it to the density $\mathcal{N}(x|0,1)$ (see Section 6.2 on page 115 for details on generating histograms). Repeat this for deviates S_{10}.

The histogram of S_5 and S_{10} is shown in Figure A.3. We also show the logarithm of the distribution to show the deviation in the tails of the distribution. As we increase n we see that S_n converges to the normal distribution although the tails clearly differ. We could use S_5 or S_{10} as an approximation for a normal deviate. The time to generate 10^6 deviates S_5 was around 55 ms while the time to generate 10^6 deviates S_{10} was around 105 ms (cf. Exercise 3.2 where we can see that S_5 is quite competitive).

Exercise 5.2 (on page 103): *Let*

$$T_n = \frac{1}{n}\sum_{i=1}^{n} C_i$$

where $C_i \sim$ Cau are Cauchy deviates. Generate a histogram T_{10} and compare this with the density $\text{Cau}(x)$.

We show a histogram of T_{10} in Figure A.4. We note that this is also Cauchy distributed.

Exercise 5.3 (on page 103): *Show that the CDF for normal random variables*

$$\Phi(x) = \int_{-\infty}^{x} e^{y^2/2} \frac{dy}{\sqrt{2\pi}}$$

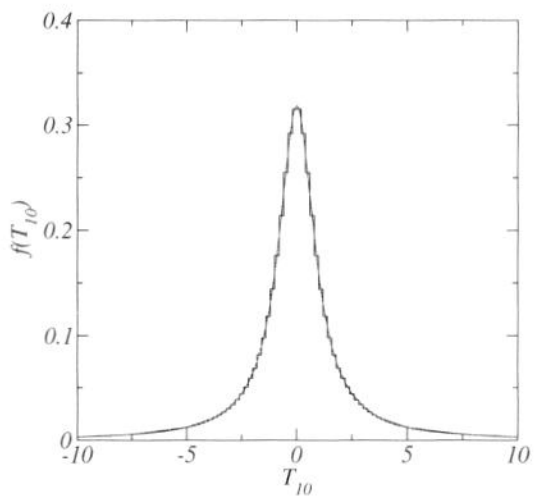

Figure A.4 Histogram of T_{10} collected on 10^8 samples. The smooth line shows the probability density for a Cauchy distribution.

can be written in terms of an incomplete gamma function by making the change of variables $y^2/2 = t$.

Under this change of variables $y = \sqrt{2t}$ and $\mathrm{d}y = \mathrm{d}t/\sqrt{2t}$, however, we have to treat the case $x \geq 0$ separately from the case $x < 0$. For the case $x \geq 0$

$$\begin{aligned}\Phi(x) &= \frac{1}{2} + \int_0^x \mathrm{e}^{-y^2/2} \frac{\mathrm{d}y}{\sqrt{2\pi}} \\ &= \frac{1}{2} + \int_0^{x^2/2} \mathrm{e}^{-t}\, t^{-1/2} \frac{\mathrm{d}t}{2\sqrt{\pi}} = \frac{1}{2} + \frac{1}{2\sqrt{\pi}}\gamma(1/2, x^2/2)\end{aligned}$$

where $\gamma(1/2, x^2/2)$ is the incomplete gamma function (see Section 2.A on page 41). But $\Gamma(1/2) = \sqrt{\pi}$ so

$$\Phi(x) = \frac{1}{2} + \frac{P(1/2, x^2/2)}{2}$$

where $P(a, x)$ is the normalised incomplete gamma function. The error function is equal to $\mathrm{erf}(x) = 2\Phi(\sqrt{2}\,x) - 1 = P(1/2, x^2)$. For the case $x < 0$ we use

$$\begin{aligned}\Phi(x) &= \int_{-\infty}^x \mathrm{e}^{-y^2/2} \frac{\mathrm{d}y}{\sqrt{2\pi}} = \int_{x^2/2}^{\infty} \mathrm{e}^{-t}\, t^{-1/2} \frac{\mathrm{d}t}{2\sqrt{\pi}} \\ &= \frac{1}{2\sqrt{\pi}}\Gamma(1/2, x^2/2) = \frac{Q(1/2, x^2/2)}{2}.\end{aligned}$$

The complementary error function for $(x > 0)$ is equal to $\mathrm{erfc}(x) = 2\Phi(-\sqrt{2}\,x) = Q(1/2, x^2)$.

Interestingly, the incomplete gamma function is the CDF for the Poisson (see Question 4.4), normal, and gamma distributions.

Exercise 5.4 (on page 103): *Marginalise out the variables* $\boldsymbol{y}$ *from the multivariate normal distribution*

$$\begin{aligned}\begin{pmatrix}\boldsymbol{x}\\ \boldsymbol{y}\end{pmatrix} &\sim \mathcal{N}\left(\begin{pmatrix}\boldsymbol{\mu}_x\\ \boldsymbol{\mu}_y\end{pmatrix}, \begin{pmatrix}\mathsf{A} & \mathsf{B}\\ \mathsf{B}^\mathsf{T} & \mathsf{C}\end{pmatrix}^{-1}\right) \\ &= \left|2\pi\begin{pmatrix}\mathsf{A} & \mathsf{B}\\ \mathsf{B}^\mathsf{T} & \mathsf{C}\end{pmatrix}\right|^{-1/2} \mathrm{e}^{-\frac{1}{2}(\boldsymbol{x}-\boldsymbol{\mu}_x)^\mathsf{T}\mathsf{A}(\boldsymbol{x}-\boldsymbol{\mu}_x) - (\boldsymbol{x}-\boldsymbol{\mu}_x)^\mathsf{T}\mathsf{B}(\boldsymbol{y}-\boldsymbol{\mu}_y) - \frac{1}{2}(\boldsymbol{y}-\boldsymbol{\mu}_y)^\mathsf{T}\mathsf{C}(\boldsymbol{y}-\boldsymbol{\mu}_y)}\end{aligned}$$

to obtain a density for the variables $\boldsymbol{x}$.

This is an exercise in completing squares

$$
\begin{aligned}
-\frac{1}{2}&(\boldsymbol{x}-\boldsymbol{\mu}_x)^{\mathsf{T}}\mathbf{A}(\boldsymbol{x}-\boldsymbol{\mu}_x)-(\boldsymbol{x}-\boldsymbol{\mu}_x)^{\mathsf{T}}\mathbf{B}\,(\boldsymbol{y}-\boldsymbol{\mu}_y)-\frac{1}{2}(\boldsymbol{y}-\boldsymbol{\mu}_y)^{\mathsf{T}}\mathbf{C}(\boldsymbol{y}-\boldsymbol{\mu}_y)\\
&=-\frac{1}{2}(\boldsymbol{x}-\boldsymbol{\mu}_x)^{\mathsf{T}}\mathbf{A}(\boldsymbol{x}-\boldsymbol{\mu}_x)\\
&\quad-\frac{1}{2}(\boldsymbol{y}-\boldsymbol{\mu}_y+\mathbf{C}^{-1}\mathbf{B}^{\mathsf{T}}(\boldsymbol{x}-\boldsymbol{\mu}_x))^{\mathsf{T}}\mathbf{C}(\boldsymbol{y}-\boldsymbol{\mu}_y+\mathbf{C}^{-1}\mathbf{B}^{\mathsf{T}}(\boldsymbol{x}-\boldsymbol{\mu}_x))\\
&\quad+\frac{1}{2}(\boldsymbol{x}-\boldsymbol{\mu}_x)^{\mathsf{T}}\mathbf{B}\,\mathbf{C}^{-1}\mathbf{B}^{\mathsf{T}}(\boldsymbol{x}-\boldsymbol{\mu}_x)\\
&=-\frac{1}{2}(\boldsymbol{x}-\boldsymbol{\mu}_x)^{\mathsf{T}}(\mathbf{A}-\mathbf{B}\,\mathbf{C}^{-1}\mathbf{B}^{\mathsf{T}})(\boldsymbol{x}-\boldsymbol{\mu}_x)\\
&\quad-\frac{1}{2}(\boldsymbol{y}-\boldsymbol{\mu}_y+\mathbf{C}^{-1}\mathbf{B}^{\mathsf{T}}(\boldsymbol{x}-\boldsymbol{\mu}_x))^{\mathsf{T}}\mathbf{C}(\boldsymbol{y}-\boldsymbol{\mu}_y+\mathbf{C}^{-1}\mathbf{B}^{\mathsf{T}}(\boldsymbol{x}-\boldsymbol{\mu}_x)).
\end{aligned}
$$

To integrate over $\boldsymbol{y}$, we make a change of variables $\boldsymbol{z}=\boldsymbol{y}-\boldsymbol{\mu}_y+\mathbf{C}^{-1}\mathbf{B}^{\mathsf{T}}(\boldsymbol{x}-\boldsymbol{\mu}_x)$ so we are left with an integral

$$\int_{-\infty}^{\infty}\mathrm{e}^{-\frac{1}{2}\boldsymbol{z}^{\mathsf{T}}\mathbf{C}\boldsymbol{z}}\,\mathrm{d}\boldsymbol{z}=|2\pi\,\mathbf{C}|^{1/2}$$

where we have used the result for the Schur complement

$$\left|2\pi\begin{pmatrix}\mathbf{A} & \mathbf{B}\\ \mathbf{B}^{\mathsf{T}} & \mathbf{C}\end{pmatrix}\right|=|2\pi\,\mathbf{C}|\times|2\pi\,(\mathbf{A}-\mathbf{B}\mathbf{C}^{-1}\mathbf{B}^{\mathsf{T}})|.$$

Then

$$
\begin{aligned}
f_{\boldsymbol{X}}(\boldsymbol{x})&=\int_{\infty}^{\infty}f_{\boldsymbol{X},\boldsymbol{Y}}(\boldsymbol{x},\boldsymbol{y})\,\mathrm{d}\boldsymbol{y}=\int_{-\infty}^{\infty}\mathcal{N}\left(\begin{pmatrix}\boldsymbol{x}\\ \boldsymbol{y}\end{pmatrix}\middle|\begin{pmatrix}\boldsymbol{\mu}_x\\ \boldsymbol{\mu}_y\end{pmatrix},\begin{pmatrix}\mathbf{A} & \mathbf{B}\\ \mathbf{B}^{\mathsf{T}} & \mathbf{C}\end{pmatrix}^{-1}\right)\mathrm{d}\boldsymbol{y}\\
&=|2\pi\,(\mathbf{A}-\mathbf{B}\mathbf{C}^{-1}\mathbf{B}^{\mathsf{T}})|^{-1/2}\mathrm{e}^{-\frac{1}{2}(\boldsymbol{x}-\boldsymbol{\mu}_x)^{\mathsf{T}}(\mathbf{A}-\mathbf{B}\,\mathbf{C}^{-1}\mathbf{B}^{\mathsf{T}})(\boldsymbol{x}-\boldsymbol{\mu}_x)}\\
&=\mathcal{N}\left(\boldsymbol{x}|\boldsymbol{\mu}_x,(\mathbf{A}-\mathbf{B}\,\mathbf{C}^{-1}\mathbf{B}^{\mathsf{T}})^{-1}\right).
\end{aligned}
$$

Using the formula for the inverse of a partitioned symmetric matrix, Equation (5.6),

$$\boldsymbol{\Sigma}^{-1}=\begin{pmatrix}\mathbf{A} & \mathbf{B}\\ \mathbf{B}^{\mathsf{T}} & \mathbf{C}\end{pmatrix}$$

then

$$\boldsymbol{\Sigma}=\begin{pmatrix}(\mathbf{A}-\mathbf{B}\,\mathbf{C}^{-1}\mathbf{B}^{\mathsf{T}})^{-1} & -(\mathbf{A}-\mathbf{B}\,\mathbf{C}^{-1}\mathbf{B}^{\mathsf{T}})^{-1}\mathbf{B}\mathbf{C}^{-1}\\ -\mathbf{C}^{-1}\mathbf{B}^{\mathsf{T}}(\mathbf{A}-\mathbf{B}\,\mathbf{C}^{-1}\mathbf{B}^{\mathsf{T}})^{-1} & (\mathbf{C}-\mathbf{B}\,\mathbf{A}^{-1}\mathbf{B}^{\mathsf{T}})^{-1}\end{pmatrix}.$$

Thus for a distribution

$$f_{\boldsymbol{X},\boldsymbol{Y}}(\boldsymbol{x},\boldsymbol{y})=\mathcal{N}\left(\begin{pmatrix}\boldsymbol{x}\\ \boldsymbol{y}\end{pmatrix}\middle|\begin{pmatrix}\boldsymbol{\mu}_x\\ \boldsymbol{\mu}_y\end{pmatrix},\begin{pmatrix}\boldsymbol{\Sigma}_x & \boldsymbol{\Sigma}_{xy}\\ \boldsymbol{\Sigma}_{xy}^{\mathsf{T}} & \boldsymbol{\Sigma}_y\end{pmatrix}\right)$$

the marginal distribution is

$$f_{\boldsymbol{X}}(\boldsymbol{x})=\int_{-\infty}^{\infty}f_{\boldsymbol{X},\boldsymbol{Y}}(\boldsymbol{x},\boldsymbol{y})\,\mathrm{d}\boldsymbol{y}=\mathcal{N}(\boldsymbol{x}|\boldsymbol{\mu}_x,\boldsymbol{\Sigma}_x).$$

Exercise 5.5 (on page 103): *Let $U = S^2 = \sum_{i=1}^{n} X_i^2$ where $X_i \sim \mathcal{N}(0,1)$, show that U is chi-squared distributed*

$$f_U(u) = \chi^2(n) = \text{Gam}\left(u \middle| \frac{n}{2}, \frac{1}{2}\right) = \frac{u^{n/2-1}\,\mathrm{e}^{-u/2}}{2^{n/2}\,\Gamma(n/2)}.$$

We consider the probability of S being between s and $s + \delta s$ (i.e. the probability of having a radius from the origin between s and $s + \delta s$)

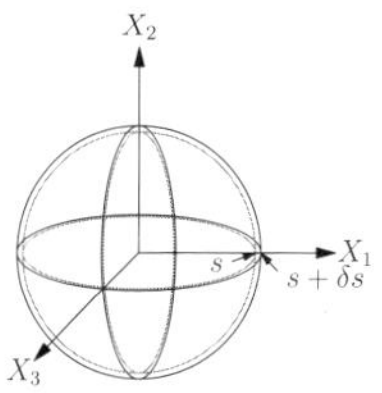

$$f_S(s)\delta s = \mathbb{P}\left(s \le S < s + \delta s\right) = \delta s \int \delta\left(s^2 - \sum_{i=1}^{n} X_i^2\right) \mathrm{e}^{-\frac{1}{2}\sum_{i=1}^{n} X_i^2} \prod_{i=1}^{n} \frac{\mathrm{d}X_i}{\sqrt{2\pi}}.$$

The difficulty of making the change of variables is to compute the Jacobian. However, this is just equal to the surface area of the n-dimensional hypersphere of radius s times δs. The surface area of the n-dimensional hypersphere of radius s is equal to $S_n s^{n-1}$ where S_n is the surface area of the n-dimensional hypersphere of radius 1. Thus,

$$f_S(s)\,\mathrm{d}s = \frac{S_n\, s^{n-1}\,\mathrm{e}^{-s^2/2}}{(2\pi)^{n/2}}\,\mathrm{d}s;$$

and making the change of variables $u = s^2$ so that $\mathrm{d}u = 2s\,\mathrm{d}s$ we find

$$f_u(u)\,\mathrm{d}u = f_S(s)\,\mathrm{d}s = f_S(\sqrt{u})\,\frac{\mathrm{d}u}{2\sqrt{u}} = \frac{S_n}{2\,(2\pi)^{n/2}}\,u^{n/2-1}\,\mathrm{e}^{-u/2}\,\mathrm{d}u.$$

We note that

$$\int_0^\infty u^{n/2-1}\,\mathrm{e}^{-u/2}\,\mathrm{d}u = 2^{n/2}\,\Gamma(n/2)$$

so that for $f_u(u)$ to be normalised requires $S_n = 2\pi^{n/2}/\Gamma(n/2)$. This is a relatively painless way of finding the surface area of a unit hypersphere in n dimensions.

A.6 Answers to Chapter 6. Handling Experimental Data

Exercise 6.1 (on page 125): *Generate 50 deviates $X_i \sim \mathcal{N}(0,1)$ and 50 deviates $Y_i \sim \mathcal{N}(0,4)$ and compute the empirical means and variances for the X and Y variables. From this compute the t-value and the corresponding p-value using Welch's t-test. Repeat this $n = 10^6$ times recording the p-values. Sort the p-values such that the pair (t_r, p_r) has*

$$t_1 \le t_2 \le t_3 \le \cdots \le t_r \le \cdots \le t_n.$$

Note that the p-value is a monotonic function of the t-value so it is sufficient to sort out the t-values. Recall that the p-value is defined such that, if the null hypothesis is true (i.e. the two sets of deviates have the same mean, which in our case is true), then $\mathbb{P}\left(|T| \ge t_r\right) = p_r$. Thus, we would expect that a proportion p_r of the sample have t-values greater than t_r. Empirically this proportion is $(n-r)/n$. Thus, if we plot p_r versus $1 - r/n$ where r is the rank of the t-value we should get, roughly, a straight line. Plot this (it helps to do this on a log-log plot).

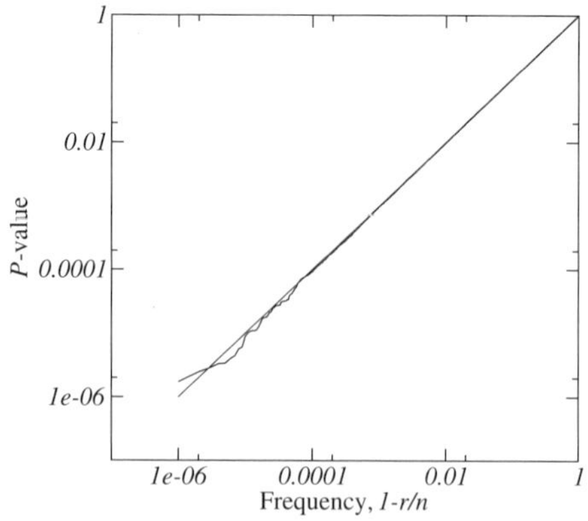

Figure A.5 The p-values are plotted against the proportion of t-values greater than the observed t-value for normally distributed data.

Figure A.5 shows the results of the experiment described. We can see that the data follows more or less a straight line. Note that on the right of the figure we have a lot more data and the line is straight. On the left we only have a small amount of data to determine the ranking. The wiggles are due to sampling errors (if we use 10 times as much data we get a straight line down to a p-value of 10^{-6}).

It is worth taking time to understand what this graph means. We are simulating the null hypothesis a million times and plotting the p-value versus its rank (i.e. the frequency with which it happens). This is, as it should be, a dead straight line (up to statistical fluctuations) – note that this is a log-log plot, but it has gradient 1 so this would also be a straight line in a linear plot. It clearly shows that low p-values will occur by chance even if the null hypothesis is true. One paper in 100 that claims a significant result because of a p-value of 0.01 or less is going to be wrong! This, however, is what the scientific community has agreed as acceptable. However, as soon as you do any kind of selection (e.g. collecting just enough data to get a required p-value) then your significance claim is invalid. It's a good job that never happens.

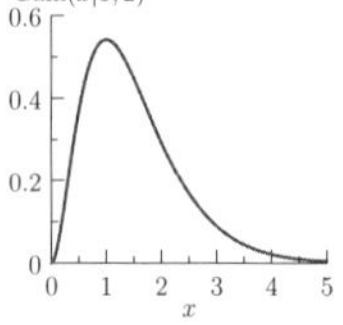

Exercise 6.2 (on page 126): *Repeat the experiment described in Exercise 6.1, but for* X_i, $Y_i \sim \text{Gam}(3, 2)$. *As our random variables are not normally distributed we have no right to believe in the p-value for the t-test. Interestingly, the p-value is not a bad estimate of the probability of the t-value occurring. This gives some confidence that even when the data is not exactly normal the p-values of the t-test are not always too misleading. Of course, it is always possible to cook up examples where the p-values are significantly underestimated, particularly for thick-tailed distributions.*

Using gamma deviates and repeating the t-test we get the results shown in Figure A.6. In this case we find the p-values are not too inaccurate at least when they are large. Note that in this case the p-values are conservative in that they predict a slightly larger p-value than the true probability $\mathbb{P}\left(|T| \geq t\right)$.

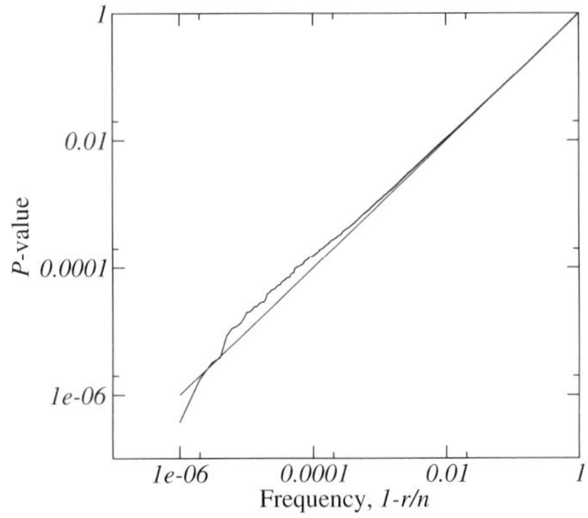

Figure A.6 The p-values are plotted against the proportion of t-values greater than the observed t-value for gamma distributed data.

Exercise 6.3 (on page 126): *Recall in Section 2.4 on page 38 we observed that many distributions, including the normal and gamma distributions, could be written in the form*

$$f_{\boldsymbol{X}}(\boldsymbol{x}|\boldsymbol{\eta}) = g(\boldsymbol{\eta})\, h(\boldsymbol{x})\, \mathrm{e}^{\boldsymbol{\eta}^{\mathsf{T}} \boldsymbol{u}(\boldsymbol{x})}$$

where $\boldsymbol{u}(\boldsymbol{x})$ *are some functions of our random variables and* $\boldsymbol{\eta}$ *are 'natural parameters'. We showed that the maximum likelihood estimators for the natural parameters satisfy*

$$-\boldsymbol{\nabla} \log\big(g(\boldsymbol{\eta})\big) = \frac{1}{n} \sum_{i=1}^{n} \boldsymbol{u}(\boldsymbol{x}_n). \tag{A.2}$$

Compute $\boldsymbol{\nabla} \log\big(g(\boldsymbol{\eta})\big)$ *and* $\boldsymbol{u}(\boldsymbol{x}_n)$ *for the normal and gamma distributions and show that they are consistent with the maximum likelihood estimators we obtained in Section 6.4.*

The normal distribution, $\mathcal{N}(x|\mu, \sigma^2)$, can be written in the form of the exponential family as

$$\begin{pmatrix} \eta_1 \\ \eta_2 \end{pmatrix} = \begin{pmatrix} \mu/\sigma^2 \\ -1/(2\sigma^2) \end{pmatrix}, \quad \begin{pmatrix} u_1(x) \\ u_2(x) \end{pmatrix} = \begin{pmatrix} x \\ x^2 \end{pmatrix},$$
$$g(\boldsymbol{\eta}) = \sqrt{-2\,\eta_2}\,\mathrm{e}^{\eta_1^2/(4\,\eta_2)}, \quad h(\boldsymbol{x}) = \frac{1}{\sqrt{2\,\pi}}.$$

Thus

$$\log\big(g(\boldsymbol{\eta})\big) = \frac{1}{2} \log(-2\,\eta_2) + \frac{\eta_1^2}{4\,\eta_2}$$
$$\boldsymbol{\nabla} \log\big(g(\boldsymbol{\eta})\big) = \begin{pmatrix} \dfrac{\eta_1}{2\,\eta_2} \\ \dfrac{1}{2\,\eta_2} - \dfrac{\eta_1^2}{4\,\eta_2^2} \end{pmatrix} = \begin{pmatrix} -\mu \\ -\sigma^2 - \mu^2 \end{pmatrix}.$$

Using Equation (A.2)

$$\hat{\mu} = \frac{1}{n}\sum_{i=1}^{n} x_i, \qquad \hat{\sigma}^2 + \hat{\mu}^2 = \frac{1}{n}\sum_{i=1}^{n} x_i^2$$

in agreement with the result we obtained in Section 2.4.

The gamma distribution, $\Gamma(x|a,b)$, can be written in the form of an exponential family with

$$\begin{pmatrix}\eta_1\\ \eta_2\end{pmatrix} = \begin{pmatrix}-b\\ a-1\end{pmatrix}, \quad \begin{pmatrix}u_1(x)\\ u_2(x)\end{pmatrix} = \begin{pmatrix}x\\ \log(x)\end{pmatrix}, \quad g(\boldsymbol{\eta}) = \frac{(-\eta_1)^{\eta_2+1}}{\Gamma(\eta_2+1)}, \quad h(\boldsymbol{x}) = 1.$$

Thus

$$\log\big(g(\boldsymbol{\eta})\big) = (\eta_2+1)\,\log(-\eta_1) - \log\big(\Gamma(\eta_2+1)\big)$$

$$\boldsymbol{\nabla}\log\big(g(\boldsymbol{\eta})\big) = \begin{pmatrix}\frac{\eta_2+1}{\eta_1}\\ \log(-\eta_1) - \psi(\eta_2+1)\end{pmatrix} = \begin{pmatrix}-\frac{a}{b}\\ \log(b) - \psi(a)\end{pmatrix}.$$

Using Equation (A.2) we have

$$\frac{\hat{a}}{\hat{b}} = \frac{1}{n}\sum_{i=1}^{n} x_i, \qquad \psi(a) - \log(b) = \frac{1}{n}\sum_{i=1}^{n}\log(x_i);$$

again in agreement with the result we obtained in Section 2.4.

A.7 Answers to Chapter 7. Mathematics of Random Variables

Exercise 7.1 (on page 176): *Use the Cauchy–Schwarz inequality to show that Pearson's correlation*

$$\rho = \frac{\mathbb{E}\left[(X-\mathbb{E}[X])\,(Y-\mathbb{E}[Y])\right]}{\sqrt{\mathbb{V}\mathrm{ar}[X]\ \mathbb{V}\mathrm{ar}[Y]}} = \frac{\mathbb{C}\mathrm{ov}[X,Y]}{\sqrt{\mathbb{V}\mathrm{ar}[X]\ \mathbb{V}\mathrm{ar}[Y]}}$$

satisfies $-1 \le \rho \le 1$.

Using the Cauchy–Schwarz inequality

$$\begin{aligned}\mathbb{C}\mathrm{ov}[X,Y]^2 &= \mathbb{E}\left[(X-\mathbb{E}[X])\,(Y-\mathbb{E}[Y])\right]^2\\ &\le \mathbb{E}\left[(X-\mathbb{E}[X])^2\right]\,\mathbb{E}\left[(Y-\mathbb{E}[Y])^2\right] = \mathbb{V}\mathrm{ar}[X]\ \mathbb{V}\mathrm{ar}[Y].\end{aligned}$$

Thus, $\rho^2 \le 1$ or $-1 \le \rho \le 1$. Equality holds when X is proportional to Y in which case $\rho = \pm 1$ (i.e. X is either completely correlated or completely anti-correlated with Y).

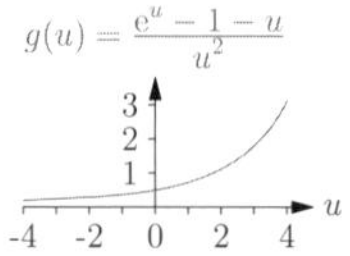

Exercise 7.2 (on page 176): *Using the fact that* $g(u) = (\mathrm{e}^u - 1 - u)/u^2$ *is monotonically increasing, show that if* $X_i - \mathbb{E}[X_i] \le b$ *then*

$$\mathbb{E}\left[\mathrm{e}^{\lambda(X_i - \mathbb{E}[X_i])}\right] \le 1 + \frac{\mathbb{V}\mathrm{ar}[X_i]}{b^2}(\mathrm{e}^{b\lambda} - 1 - b\lambda).$$

Assuming X_i are independent variables and defining $S_n = \sum_{i=1}^{n} X_i$ and $\mu = \mathbb{E}\left[S_n\right]$ and using the fact that

$$\mathbb{P}\left(S_n \leq \mu + t\right) \leq \mathrm{e}^{-\psi(t)}, \qquad \psi(t) = \max_{\lambda} \lambda\, t - \log\Big(\mathbb{E}\left[\mathrm{e}^{\lambda\,(S_n - \mu)}\right]\Big)$$

derive Bennett's inequality

$$\mathbb{P}\left(S_n \leq \mu + t\right) \leq \exp\left(-\frac{\sigma^2}{b^2}\, h\left(\frac{b\,t}{\sigma^2}\right)\right)$$

where $\sigma^2 = \mathbb{V}\mathrm{ar}\left[S_n\right]$ and $h(u) = (1+u)\log(1+u) - u$.

To prove Bennett's inequality we start (as usual) from the Markov inequality

$$\mathbb{P}\left(S_n - \mu \geq t\right) = \mathbb{P}\left(\mathrm{e}^{\lambda\,(S_n-\mu)} \geq \mathrm{e}^{\lambda\, t}\right) \leq \frac{\mathbb{E}\left[\mathrm{e}^{\lambda\,(S_n-\mu)}\right]}{\mathrm{e}^{\lambda\, t}} = \mathrm{e}^{-\tilde{\psi}(\lambda,t)}$$

where $\lambda > 0$ and

$$\tilde{\psi}(\lambda, t) = \lambda\, t - \bar{G}(\lambda), \qquad \bar{G}(\lambda) = \log\Big(\mathbb{E}\left[\mathrm{e}^{\lambda\,(S_n-\mu)}\right]\Big).$$

As this is true for any $\lambda \geq 0$ we have that $\mathbb{P}\left(S_n - \mu \geq t\right) \leq \exp(-\psi(t))$ where

$$\psi(t) = \max_{\lambda \geq 0} \tilde{\psi}(\lambda, t).$$

(The key to proving Bennett's inequality is the observation that $g(u) = (\mathrm{e}^u - 1 - u)/u^2$ is monotonically increasing. To show this we note that

$$g'(u) = \frac{\mathrm{e}^u - 1}{u^2} - 2\frac{\mathrm{e}^u - u - 1}{u^3} = \frac{\mathrm{e}^u + 1}{u^2} - 2\frac{\mathrm{e}^u - 1}{u^3}$$

but

$$\frac{\mathrm{e}^u + 1}{u^2} \geq 2\frac{\mathrm{e}^u - 1}{u^3}.$$

To see this we multiply both sides by $u^3/(2(\mathrm{e}^u + 1))$ giving

$$\frac{u}{2} \geq \frac{\mathrm{e}^u - 1}{\mathrm{e}^u + 1} \quad \text{if } u \geq 0 \qquad \frac{u}{2} \leq \frac{\mathrm{e}^u - 1}{\mathrm{e}^u + 1} \quad \text{if } u \leq 0.$$

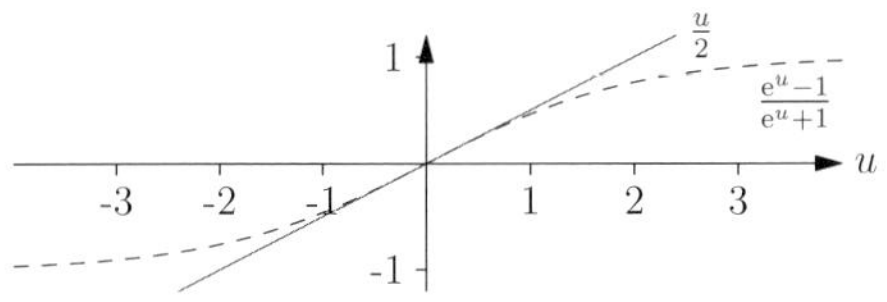

Thus $g'(u) \geq 0$, showing that $g(u)$ is monotonically increasing.)

Getting on with the proof we notice that as $X_i - \mathbb{E}\left[X_i\right] \leq b$ then by the monotonicity of $g(u)$

$$g(\lambda\,(X_i - \mathbb{E}\left[X_i\right])) \leq g(\lambda\, b)$$

or

$$\frac{e^{\lambda(X_i - \mathbb{E}[X_i])} - 1 - \lambda(X_i - \mathbb{E}[X_i])}{\lambda^2 (X_i - \mathbb{E}[X_i])^2} \le \frac{e^{\lambda b} - 1 - \lambda b}{\lambda^2 b^2}.$$

Multiplying both sides by $\lambda^2 (X_i - \mathbb{E}[X_i])^2$ and adding 1 we find

$$e^{\lambda(X_i - \mathbb{E}[X_i])} - \lambda(X_i - \mathbb{E}[X_i]) \le 1 + (X_i - \mathbb{E}[X_i])^2 \left(\frac{e^{\lambda b} - 1 - \lambda b}{b^2}\right).$$

Taking expectations with respect to X_i we find

$$\mathbb{E}\left[e^{\lambda(X_i - \mathbb{E}[X_i])}\right] \le 1 + \mathbb{V}\mathrm{ar}\left[X_i\right] \left(\frac{e^{\lambda b} - 1 - \lambda b}{b^2}\right)$$

or

$$\begin{aligned} \log\left(\mathbb{E}\left[e^{\lambda(X_i - \mathbb{E}[X_i])}\right]\right) &\le \log\left(1 + \mathbb{V}\mathrm{ar}\left[X_i\right] \left(\frac{e^{\lambda b} - 1 - \lambda b}{b^2}\right)\right) \\ &\le \mathbb{V}\mathrm{ar}\left[X_i\right] \left(\frac{e^{\lambda b} - 1 - \lambda b}{b^2}\right) \end{aligned}$$

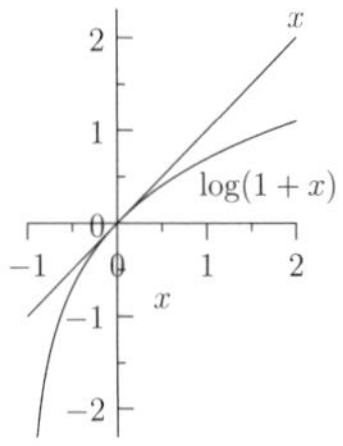

where we have used the fact that $\log(1 + x) \le x$. Summing both sides

$$\sum_{i=1}^{n} \log\left(\mathbb{E}\left[e^{\lambda(X_i - \mathbb{E}[X_i])}\right]\right) \le \sum_{i=1}^{n} \mathbb{V}\mathrm{ar}\left[X_i\right] \left(\frac{e^{\lambda b} - 1 - \lambda b}{b^2}\right)$$

and using the independence of X_i and the definition $S_n = \sum_{i=1}^{n} X_i$

$$\log\left(\mathbb{E}\left[e^{\lambda(S_n - \mu)}\right]\right) = \bar{G}(\lambda) \le \frac{\sigma^2}{b^2}\left(e^{\lambda b} - 1 - \lambda b\right)$$

where $\mu = \sum_{i=1}^{n} \mathbb{E}\left[X_i\right]$ and $\sigma^2 = \sum_{i=1}^{n} \mathbb{V}\mathrm{ar}\left[X_i\right]$ is the variance of S_n.

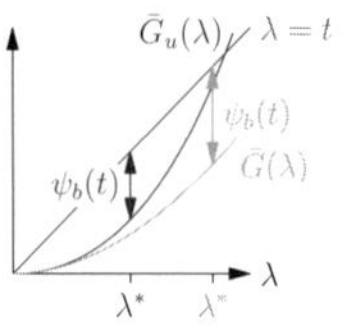

The rest is simple algebra. By replacing $\bar{G}(\lambda)$ by an upper bound, $\bar{G}_u(\lambda)$, and choosing λ to maximise

$$\tilde{\psi}_u(\lambda, t) = \lambda t - \bar{G}_u(\lambda) = \lambda t - \frac{\sigma^2}{b^2}\left(e^{\lambda b} - 1 - \lambda b\right)$$

we obtain a lower bound for $\psi(t)$. Thus we set the derivative of $\tilde{\psi}_u(\lambda, t)$ to zero

$$\tilde{\psi}'_u(\lambda, t) = t - \frac{\partial \bar{G}_u(\lambda)}{\partial \lambda} = t - \frac{\sigma^2}{b}\left(e^{\lambda b} - 1\right) = 0$$

giving

$$\lambda^* = \frac{1}{b}\log\left(1 + \frac{b\,t}{\sigma^2}\right).$$

Thus

$$\begin{aligned}\tilde{\psi}_u(\lambda^*, t) &= t\,\lambda^* - \frac{t}{b} + \frac{\sigma^2\,\lambda^*}{b} \\ &= \frac{\sigma^2}{b^2}\left(1 + \frac{t\,b}{\sigma^2}\right)\log\left(1 + \frac{t\,b}{\sigma^2}\right) - \frac{t}{b} = \frac{\sigma^2}{b^2}\,h\left(\frac{t\,b}{\sigma^2}\right)\end{aligned}$$

where $h(u) = (1 + u)\log(1 + u) - u$. So,

$$\mathbb{P}\left(S_n \geq \mu + t\right) \leq \exp\left(-\frac{\sigma^2}{b^2}\,h\left(\frac{t\,b}{\sigma^2}\right)\right)$$

which is what we set out to prove.

Exercise 7.3 (on page 176): *Use Taylor's expansion to second order to show that*

$$g(u) = (1 + u)\log(1 + u) - u - \frac{u^2}{2(1 + u/3)} \geq 0,$$

hence derive Bernstein's inequality from Bennett's inequality.

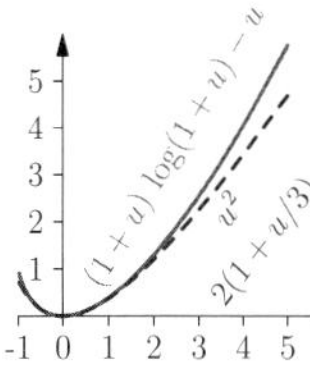

To understand where this inequality comes from we note that $\log(1 + u) = u - u^2/2 + u^3/3 - \ldots$ so that the Taylor expansion of $h(u) = (1 + u)\log(1 + u) - u$ equals

$$h(u) = \frac{u^2}{2} - \frac{u^2}{6} + \frac{u^4}{12} + O(u^5) = \frac{u^2}{2}\left(1 - \frac{u}{3} + \frac{u^2}{6} + O(u^3)\right).$$

Thus, we are asked to prove the term in the bracket can be lower bounded by $1/(1 + u/3)$. To show this we note that

$$g'(u) = \log(1 + u) - \frac{u^2/3 + 2u}{2\,(1 + u/3)^2} \qquad g''(u) = \frac{1}{1 + u} - \frac{1}{(1 + u/3)^3}.$$

Now $g(0) = g'(0) = 0$ and using the Taylor expansion to second order

$$g(u) = g(0) + u\,g'(0) + \frac{u^2}{2}g''(\zeta) = \frac{u^2}{2}g''(\zeta)$$

where ζ lies in the interval $[0, u]$. But again using Taylor's expansion to second order

$$(1 + u/3)^3 = 1 + u + u^2\frac{1}{3}\left(1 + \frac{\zeta}{3}\right)$$

where once again ζ lies in the interval $[0, u]$. As a consequence for $u > -3$ (note that $h(u)$ is only define for $u > -1$) we have that $(1 + u/3)^3 > 1 + u$, which implies that $g''(u) > 0$, thus $g(u) \geq 0$ or $h(u) \geq u^2/(2(1 + u/3))$. Substituting this into Bennett's inequality we obtain Bernstein's inequality.

This question illustrates the power of Taylor's expansion in proving bounds between functions. Admittedly, the algebra is tedious and error-prone. Using a package such as Mathematica can be a great aid if you struggle with accurately performing algebraic manipulations.

Exercise 7.4 (on page 176): *In Example 7.3 on page 143, we showed, using the Cauchy–Schwarz inequality, that variances have to be positive. Use Jensen's inequality to obtain the same result.*

As x^2 is a convex-up function, Jensen's inequality implies that

$$\mathbb{E}\left[X\right]^2 \leq \mathbb{E}\left[X^2\right].$$

Since the variance is given by

$$\mathbb{V}\text{ar}\left[X\right] = \mathbb{E}\left[X^2\right] - \mathbb{E}\left[X\right]^2$$

it is always positive (or zero). The condition for Jensen's inequality being an equality is that X is a constant, which is the condition for the variance to vanish.

Exercise 7.5 (on page 176): *One way of defining the median, m, of a distribution is as the value that minimises the mean absolute error*

$$m = \underset{c}{\operatorname{argmin}}\, \mathbb{E}\left[|X - c|\right].$$

Use Jensen's inequality for the absolute function and for the square to show that the median satisfies the inequality

$$|\mu - m| \leq \sigma$$

where $\mu = \mathbb{E}\left[X\right]$ is the mean and σ the standard deviation of the distribution.

From the definition of the mean

$$|\mu - m| = \left|\mathbb{E}\left[X - m\right]\right| \leq \mathbb{E}\left[|X - m|\right]$$

where we have used Jensen's inequality for the absolute function (which is a convex-up function) so $\left|\mathbb{E}\left[X\right]\right| \leq \mathbb{E}\left[|X|\right]$. But by the definition of the median, m, we have that for any c

$$\mathbb{E}\left[|X - m|\right] \leq \mathbb{E}\left[|X - c|\right]$$

so this inequality holds when $c = \mu$

$$\mathbb{E}\left[|X - m|\right] \leq \mathbb{E}\left[|X - \mu|\right].$$

Finally, by Jensen's inequality $\mathbb{E}\left[X\right]^2 \leq \mathbb{E}\left[X^2\right]$ or $\mathbb{E}\left[X\right] \leq \sqrt{\mathbb{E}\left[X^2\right]}$ so we have

$$|\mu - m| \leq \mathbb{E}\left[|X - \mu|\right] \leq \sqrt{\mathbb{E}\left[(X - \mu)^2\right]} = \sigma.$$

In passing, it is not at all obvious (at least to me) that the median should minimise the mean absolute error. To show that this is the case for a density $f_X(x)$, let

$$\begin{aligned}
\text{MAE}(m) = \mathbb{E}\left[|X - m|\right] &= \int_{-\infty}^{\infty} |x - m|\, f_X(x)\, \mathrm{d}x \\
&= \int_{m}^{\infty} (x - m)\, f_X(x)\, \mathrm{d}x - \int_{-\infty}^{m} (x - m)\, f_X(x)\, \mathrm{d}x.
\end{aligned}$$

To find the minimum of $\text{MAE}(m)$ we consider its derivative

$$\begin{aligned}\frac{\mathrm{d}\,\text{MAE}(m)}{\mathrm{d}m} &= -\left[(x-m)\,f_X(x)\right]_{x=m} - \int_m^\infty f_X(x)\,\mathrm{d}x - \left[(x-m)\,f_X(x)\right]_{x=m} \\ &\quad + \int_{-\infty}^m f_X(x)\,\mathrm{d}x \\ &= -\int_m^\infty f_X(x)\,\mathrm{d}x + \int_{-\infty}^m f_X(x)\,\mathrm{d}x.\end{aligned}$$

Setting $\mathrm{d}\text{MAE}(m)/\mathrm{d}m = 0$ implies

$$\int_m^\infty f_X(x)\,\mathrm{d}x = \int_{-\infty}^m f_X(x)\,\mathrm{d}x$$

which is the defining equation for the median. As we can represent the probability mass function for a discrete distribution as a density with a sum of Dirac deltas, this shows that the result holds also for discrete distributions.

Exercise 7.6 (on page 176): *By considering the second derivative show that the generating function $G(\lambda) = \log\left(\mathbb{E}\left[e^{\lambda X}\right]\right)$ is a convex-up function of λ.*

The first two derivative of $G(\lambda)$ are

$$G'(\lambda) = \frac{\mathbb{E}\left[X\,e^{\lambda X}\right]}{\mathbb{E}\left[e^{\lambda X}\right]}, \qquad G''(\lambda) = \frac{\mathbb{E}\left[X^2\,e^{\lambda X}\right]}{\mathbb{E}\left[e^{\lambda X}\right]} - \left(\frac{\mathbb{E}\left[X\,e^{\lambda X}\right]}{\mathbb{E}\left[e^{\lambda X}\right]}\right)^2.$$

Using the Cauchy–Schwarz inequality, $\mathbb{E}\left[A\,B\right]^2 \leq \mathbb{E}\left[A^2\right]\,\mathbb{E}\left[B^2\right]$, we have

$$\mathbb{E}\left[X\,e^{\lambda X}\right]^2 = \mathbb{E}\left[X\,e^{\lambda X/2} \times e^{\lambda X/2}\right]^2 \leq \mathbb{E}\left[X^2\,e^{\lambda X}\right]\,\mathbb{E}\left[e^{\lambda X}\right].$$

Thus

$$\left(\frac{\mathbb{E}\left[X\,e^{\lambda X}\right]}{\mathbb{E}\left[e^{\lambda X}\right]}\right)^2 \leq \frac{\mathbb{E}\left[X^2\,e^{\lambda X}\right]}{\mathbb{E}\left[e^{\lambda X}\right]};$$

from this we deduce $G''(\lambda) \geq 0$. This is the condition required for $G(\lambda)$ to be a convex-up function.

Exercise 7.7 (on page 177): *Show that the negative logarithm of the partition function*

$$G(\beta) = \log(Z), \qquad\qquad Z = \sum_{\boldsymbol{X}} e^{-\beta\,E(\boldsymbol{X})}$$

is a convex-up function of β, where the Boltzmann probability of the random variable, $\boldsymbol{X}$, is given by

$$p(\boldsymbol{X}) = \frac{e^{-\beta\,E(\boldsymbol{X})}}{Z}.$$

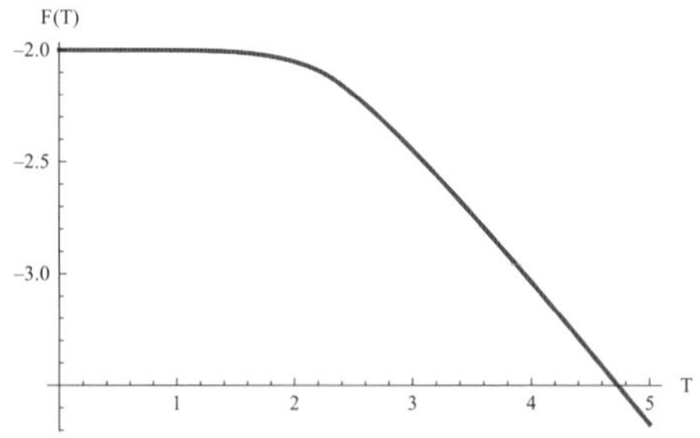

Figure A.7 Free energy per spin of the two-dimensional Ising model (with the Boltzmann constant k set to 1).

Show that for any function $G(\beta)$ (where $\beta = 1/(kT)$ – it is common to work in a systems of units where $k = 1$ so that $\beta = 1/T$, i.e. the inverse temperature)

$$\frac{\partial^2 T\, G(\frac{1}{T})}{\partial T^2} = \frac{1}{T^3} G''(\tfrac{1}{T})$$

and hence show that the free energy defined by

$$f = -k\,T \log\left(\sum_{X} e^{-E(X)/k\,T} \right)$$

is a concave (convex-down) function of T.

Define $G(\beta) = \log(Z)$ then

$$G'(\beta) = -\frac{\sum_X E(X)\, e^{-\beta E(X)}}{\sum_X e^{-\beta E(X)}} = -\sum_X E(X)\, p(X) = -\mathbb{E}\left[E(X)\right]$$

while

$$\begin{aligned} G''(\beta) &= \frac{\sum_X E(X)^2\, e^{-\beta E(X)}}{\sum_X e^{-\beta E(X)}} - \left(\frac{\sum_X E(X)\, e^{-\beta E(X)}}{\sum_X e^{-\beta E(X)}} \right)^2 \\ &= \sum_X E(X)^2\, p(X) - \left(\sum_X E(X)\, p(X) \right)^2 = \mathbb{V}\mathrm{ar}\left[E(X)\right] \geq 0. \end{aligned}$$

Thus, $G''(\beta)$ is convex up. Taking a derivative of $T\,G(1/T)$ we find

$$\begin{aligned} \frac{\partial T\, G\left(\frac{1}{T}\right)}{\partial T} &= G\left(\tfrac{1}{T}\right) + T\, G'\left(\tfrac{1}{T}\right) \frac{\partial \frac{1}{T}}{\partial T} = G\left(\tfrac{1}{T}\right) - \tfrac{1}{T} G'\left(\tfrac{1}{T}\right) \\ \frac{\partial^2 T\, G\left(\frac{1}{T}\right)}{\partial T^2} &= -\frac{1}{T^2} G'\left(\tfrac{1}{T}\right) + \frac{1}{T^2} G'\left(\tfrac{1}{T}\right) + \frac{1}{T^3} G''\left(\tfrac{1}{T}\right) = \frac{1}{T^3} G''\left(\tfrac{1}{T}\right). \end{aligned}$$

Therefore, taking $f(T) = -k\,T\,G(\frac{1}{kT})$ we have

$$\frac{\partial^2 f}{\partial T^2} = -\frac{1}{k\,T^3} \mathbb{V}\mathrm{ar}\left[E(X)\right] \leq 0,$$

implying that the free energy is a concave (convex-down) function of T. We show the free energy for the two-dimensional Ising model discussed in Chapter 10.

Exercise 7.8 (on page 177): *Use Jensen's inequality to show that the Kullback–Leibler (KL) divergence*

$$KL\left(f \,\|\, q\right) = -\sum_i f_i \log\left(\frac{q_i}{f_i}\right)$$

is greater or equal to zero (the variables f_i and q_i are probabilities).

Since the logarithm is a convex-down function, $\mathbb{E}\left[\log(x)\right] \leq \log(\mathbb{E}\left[x\right])$ or $\mathbb{E}\left[-\log(x)\right] \geq -\log(\mathbb{E}\left[x\right])$, thus

$$\begin{aligned}\mathrm{KL}\left(f \,\|\, q\right) &= \mathbb{E}_f\left[\log\left(\frac{q_i}{f_i}\right)\right] \geq \log\left(\mathbb{E}_f\left[\frac{q_i}{f_i}\right]\right) = -\log\left(\sum_i f_i \frac{q_i}{f_i}\right)\\ &= -\log\left(\sum_i g_i\right) = -\log(1) = 0.\end{aligned}$$

A.8 Answers to Chapter 8. Bayes

Exercise 8.1 (on page 255): *Harry the Axe is one of ten suspects found at the murder scene. On searching Harry a rare watch was found on him identical to that owned by the victim. The victim's watch is missing. Harry claimed it was his, but only 0.1% of the population has a similar watch. What is the probability that Harry is the murderer given the new evidence?*

Assuming the prior probability of Harry the Axe being guilty is $\mathbb{P}\left(\text{guilty}\right) = 0.1$, the probability of Harry having the watch given he is the murderer is $\mathbb{P}\left(\text{watch}|\text{guilty}\right) = 1$ (you can argue about this, but it does not make much difference what value you use here provided it's not too small), while $\mathbb{P}\left(\text{watch}|\neg\text{guilty}\right) = 0.001$, then by Bayes

$$\begin{aligned}\mathbb{P}\left(\text{guilty}|\text{watch}\right) &= \frac{\mathbb{P}\left(\text{watch}|\text{guilty}\right)\mathbb{P}\left(\text{guilty}\right)}{\mathbb{P}\left(\text{watch}\right)}\\ &= \frac{\mathbb{P}\left(\text{watch}|\text{guilty}\right)\mathbb{P}\left(\text{guilty}\right)}{\mathbb{P}\left(\text{watch}|\text{guilty}\right)\mathbb{P}\left(\text{guilty}\right) + \mathbb{P}\left(\text{watch}|\neg\text{guilty}\right)\mathbb{P}\left(\neg\text{guilty}\right)}\\ &= \frac{1 \times 0.1}{1 \times 0.1 + 0.001 \times 0.9} = 0.917.\end{aligned}$$

That is, the evidence has increased the chance of his guilt from 10% to 91.7%.

Exercise 8.2 (on page 255): *Show that the Dirichlet distribution is a conjugate prior for a multinomial likelihood and derive the update equations.*

We consider a likelihood $\text{Mult}(\boldsymbol{N}|n, \boldsymbol{p})$ where $\boldsymbol{N} = (N_1, N_2, \ldots, N_k)$ are our observations, n is the total number of observations, and $\boldsymbol{p} = (p_1, p_2, \ldots, p_k)$ is the probability of the observations happening in a particular class. Our task is to estimate $\boldsymbol{p}$ from our observations. We use the Dirichlet prior $\text{Dir}(\boldsymbol{p}|\boldsymbol{\alpha})$, where

$\boldsymbol{\alpha} = (\alpha_1, \alpha_2, \ldots, \alpha_k)$ is a set of parameters encoding our prior belief. The posterior will be

$$\begin{aligned} f(\boldsymbol{p}|\boldsymbol{N}, \boldsymbol{\alpha}) &\propto \text{Mult}(\boldsymbol{N}|n, \boldsymbol{p})\text{Dir}(\boldsymbol{p}|\boldsymbol{\alpha}) \\ &\propto \left(\prod_{i=1}^{k} p_i^{N_i}\right)\left(\prod_{i=1}^{k} p_i^{\alpha_i - 1}\right) = \prod_{i=1}^{k} p_i^{N_i+\alpha_i-1} \propto \text{Dir}(\boldsymbol{p}|\boldsymbol{N} + \boldsymbol{\alpha}). \end{aligned}$$

In the derivation we have ignored all terms *not* involving p_i. The update equations are thus $\boldsymbol{\alpha}' = \boldsymbol{N} + \boldsymbol{\alpha}$.

The reason for ignoring the terms not involving p_i is that the normalisation will sort that all out. However, it is sometimes worthwhile seeing this explicitly. So this is a repeat of the previous calculation where we keep all the terms. Bayes' rule tells us

$$f(\boldsymbol{p}|\boldsymbol{N}, \boldsymbol{\alpha}) = \frac{f(\boldsymbol{p}, \boldsymbol{N}|\boldsymbol{\alpha})}{\mathbb{P}\left(\boldsymbol{N}|\boldsymbol{\alpha}\right)}$$

where the joint probability is the product of the likelihood and prior

$$\begin{aligned} f(\boldsymbol{p}, \boldsymbol{N}|\boldsymbol{\alpha}) &= \text{Mult}(\boldsymbol{N}|n, \boldsymbol{p})\text{Dir}(\boldsymbol{p}|\boldsymbol{\alpha}) \\ &= \left(n! \prod_{i=1}^{k} \frac{p_i^{N_i}}{N_i!}\right)\left(\Gamma\left(\sum_{i=1}^{k} \alpha_i\right) \prod_{i=1}^{k} \frac{p_i^{\alpha_i-1}}{\Gamma(\alpha_i)}\right). \\ &= n!\, \Gamma\left(\sum_{i=1}^{k} \alpha_i\right) \prod_{i=1}^{k} \frac{p_i^{N_i+\alpha_i-1}}{N_i!\, \Gamma(\alpha_i)}. \end{aligned}$$

To compute the probability of the data we use the integral

$$\int_{\boldsymbol{p}\in\Lambda^k} \prod_{i=1}^{k} p_i^{x_i-1} \mathrm{d}\boldsymbol{p} = \frac{\prod_{i=1}^{k} \Gamma(x_i)}{\Gamma\left(\sum_{i=1}^{k} x_i\right)} \tag{A.3}$$

where Λ^k is the simplex representing all allowed values of $\boldsymbol{p}$. This integral has to be true for the Dirichlet distribution to be normalised. For readers who want to see all the tiny details we derive the integral below. Continuing for the moment with computing the probability of the data we observe that

$$\begin{aligned} \mathbb{P}\left(\boldsymbol{N}|\boldsymbol{\alpha}\right) &= \int_{\boldsymbol{p}\in\Lambda^k} f(\boldsymbol{p}, \boldsymbol{N}|\boldsymbol{\alpha})\, \mathrm{d}\boldsymbol{p} \\ &= n!\, \Gamma\left(\sum_{i=1}^{k} \alpha_i\right)\left(\prod_{i=1}^{k} \frac{1}{N_i!\, \Gamma(\alpha_i)}\right) \int_{\boldsymbol{p}\in\Lambda^k} \prod_{i=1}^{k} p_i^{N_i+\alpha_i-1} \mathrm{d}\boldsymbol{p} \\ &= n!\, \frac{\Gamma\left(\sum_{i=1}^{k} \alpha_i\right)}{\Gamma\left(\sum_{i=1}^{k} (N_i + \alpha_i)\right)} \prod_{i=1}^{k} \frac{\Gamma(N_i + \alpha_i)}{N_i!\, \Gamma(\alpha_i)}. \end{aligned}$$

Thus the posterior is

$$f(\boldsymbol{p}|\boldsymbol{N},\boldsymbol{\alpha}) = \frac{f(\boldsymbol{p},\boldsymbol{N}|\boldsymbol{\alpha})}{\mathbb{P}\left(\boldsymbol{N}|\boldsymbol{\alpha}\right)} = \Gamma\left(\sum_{i=1}^{k}(N_i+\alpha_i)\right)\prod_{i=1}^{k}\frac{p_i^{N_i+\alpha_i-1}}{\Gamma(N_i+\alpha_i)} = \mathrm{Dir}(\boldsymbol{p}|\boldsymbol{N}+\boldsymbol{\alpha})$$

where many of the normalisation terms cancel.

We return now to the integral in Equation (A.3). This can be proved directly through a rather fiddly set of multiple integrals (for the case $k = 2$ this is just the beta function integral which we solved in Appendix 2.B on page 43). We present a different derivation. We can rewrite the integral as

Here we use results from integration in the complex plane. If you are not familiar with this then you have to take the integral on faith.

$$\int_{\boldsymbol{p}\in\Lambda^k}\prod_{i=1}^{k}p_i^{x_i-1}\mathrm{d}\boldsymbol{p} = \int_0^\infty \mathrm{d}p_1\cdots\int_0^\infty \mathrm{d}p_k\prod_{i=1}^{k}p_i^{x_i-1}\delta\left(\sum_{i=1}^{k}p_i-1\right)$$

where $\delta(\cdot)$ represents the Dirac delta function. We using the integral representation of the Dirac delta function (see Appendix 5.A on page 103)

$$\delta(x) = \int_{-\mathrm{i}\infty}^{\mathrm{i}\infty}\mathrm{e}^{-\omega x}\frac{\mathrm{d}\omega}{2\pi\mathrm{i}}$$

(note that the integral is along the imaginary axis). Using this representation we can rewrite our integral as

$$\begin{aligned}\int_{\boldsymbol{p}\in\Lambda^k}\prod_{i=1}^{k}p_i^{x_i-1}\mathrm{d}\boldsymbol{p} &= \int_0^\infty \mathrm{d}p_1\cdots\int_0^\infty \mathrm{d}p_k\prod_{i=1}^{k}p_i^{x_i-1}\int_{-\mathrm{i}\infty}^{\mathrm{i}\infty}\mathrm{e}^{-\omega\left(\sum_{i=1}^{k}p_i-1\right)}\frac{\mathrm{d}\omega}{2\pi\mathrm{i}}\\ &= \int_{-\mathrm{i}\infty}^{\mathrm{i}\infty}\mathrm{e}^{\omega}\frac{\mathrm{d}\omega}{2\pi\mathrm{i}}\prod_{i=0}^{k}\left(\int_0^\infty p_i^{x_i-1}\mathrm{e}^{-\omega p_i}\mathrm{d}p_i\right).\end{aligned}$$

Use the standard integral representation for the gamma function

$$\int_0^\infty p_i^{x_i-1}\mathrm{e}^{-\omega p_i}\mathrm{d}p_i = \frac{\Gamma(x_i)}{\omega^{x_i}}$$

(see Appendix 2.A on page 41) so that

$$\int_{\boldsymbol{p}\in\Lambda^k}\prod_{i=1}^{k}p_i^{x_i-1}\mathrm{d}\boldsymbol{p} = \left(\prod_{i=1}^{k}\Gamma(x_i)\right)\int_{-\mathrm{i}\infty}^{\mathrm{i}\infty}\frac{\mathrm{e}^{\omega}}{\omega^{x_0}}\frac{\mathrm{d}\omega}{2\pi\mathrm{i}}$$

where $x_0 = \sum_{i=1}^{k}x_i$. Finally we use the integral representation of the reciprocal of the gamma function (see Appendix 2.A on page 41 once again)

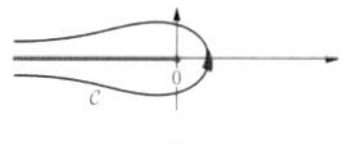

$$\frac{1}{\Gamma(z)} = \frac{1}{2\pi\mathrm{i}}\int_C x^{-z}\,\mathrm{e}^{x}\,\mathrm{d}x$$

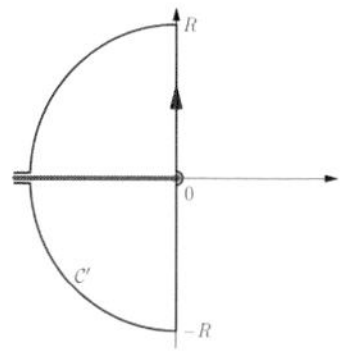

where C is the Hankel contour. Note that as there are no poles away from the branch cut so we can distort the Hankel contour to C' that runs along the imaginary axis. As the integrand vanishes in the limit $|x| \to \infty$ this integral is equal to

$$\frac{1}{\Gamma(z)} = \frac{1}{2\pi\,\mathrm{i}} \int_{-\mathrm{i}\infty}^{\mathrm{i}\infty} x^{-z}\,\mathrm{e}^{x}\,\mathrm{d}x.$$

Using this then leads to the result given in Equation (A.3).

Exercise 8.3 (on page 255): *Suppose we have an experiment with k possible outcomes. Consider performing n independent experiments. What is the likelihood of the n experiments have outcomes $\boldsymbol{N} = (N_1, N_2, \cdots, N_k)$ if the probability of a single outcome is given by $\boldsymbol{p} = (p_1, p_2, \cdots, p_k)$? Suppose we want to estimate $\boldsymbol{p}$ from our observations $\boldsymbol{N}$. What are the maximum likelihood probabilities? Show how using a conjugate prior to encode our prior beliefs about $\boldsymbol{p}$ will modify the expected value of $\boldsymbol{p}$ given our observation (the answer to Exercise 8.2 will be of use).*

We consider making n independent observations of an experiment with k possible outcomes. We denote the result of the experiments by $\boldsymbol{N} = (N_1, N_2, \cdots, N_k)$ where N_i is the number of times outcome i occurs. We assume that the probability of the outcomes are given by $\boldsymbol{p} = (p_1, p_2, \cdots, p_k)$ where p_i is the probability of the outcome i occurring. Then the likelihood of $\boldsymbol{N}$ given $\boldsymbol{p}$ is given by the multinomial distribution, $\mathrm{Mult}(\boldsymbol{N}|n, \boldsymbol{p})$ (see Section 2.3.1 on page 35).

To compute the maximum likelihood probabilities we have to maximise the multinomial distribution subject to the constraint $\sum_{i=1}^{k} p_i = 1$. To compute this we consider the Lagrangian

$$L = \mathrm{Mult}(\boldsymbol{N}|n, \boldsymbol{p}) + \lambda \left(\sum_{i=1}^{k} p_i - 1 \right)$$

where λ is a Lagrange multiplier which has to be chosen to ensure that the constraint is satisfied (see Appendix 7.C on page 182). The maximum likelihood is given by

$$\frac{\partial L}{\partial p_i} = \frac{\partial}{\partial p_i} \left(n! \prod_{j=1}^{k} \frac{p_j^{N_j}}{N_j!} + \lambda \left(\sum_{j=1}^{k} p_j - 1 \right) \right) = n! \frac{N_i}{p_i} \prod_{j=1}^{k} \frac{p_j^{N_j}}{N_j!} + \lambda = 0$$

or

$$p_i = -\frac{N_i}{\lambda} n! \prod_{j=1}^{k} \frac{p_j^{N_j}}{N_j!}.$$

Summing both sides from $i = 1$ to k and using the constraint $\sum_{i=1}^{k} p_i = 1$ and the fact that there are n observations so $\sum_{i=1}^{k} N_i = n$ we find

$$\lambda = -n\, n! \prod_{j=1}^{k} \frac{p_j^{N_j}}{N_j!}$$

so that

$$p_i = \frac{N_i}{n}.$$

In other words, the maximum likelihood probability is just equal to the factions of the samples with outcome i.

The maximum likelihood probability, although very natural, suffers from the fact that we would infer $p_i = 0$ if we had never seen the event occur. This is actually a sensible answer if we really have no prior information. The probability that if we toss a coin it turns into an elephant (or any other random outcome you think of) should probably be given zero probability. However, we usually consider outcomes that we have some suspicion might happen. In this case we would want to assign a prior probability to the outcomes. The most convenient prior to use is the Dirichlet prior $\text{Dir}(\boldsymbol{p}|\boldsymbol{\alpha})$. For example, we might choose $\alpha_i = 1$ for all i. This corresponds to the case where we consider all probabilities $\boldsymbol{p} \in \Lambda^k$ to be equally likely. (It is not the uninformative prior $\alpha_i = 0$, but then we believe it is worthwhile to consider these outcomes.) As we showed in the answer to Exercise 8.2, using a Dirichlet prior the posterior probability for the probabilities will be $\text{Dir}(\boldsymbol{p}|\boldsymbol{N}+\boldsymbol{\alpha})$. Thus, after making the observations, $\boldsymbol{N}$, the expected value of $\boldsymbol{p}$ is given by

$$\mathbb{E}\left[p_i\right] = \frac{N_i + \alpha_i}{\sum_i^k (N_i + \alpha_i)}.$$

If we compare this to the maximum likelihood values for p_i we see that the counts N_i have been increased by α_i, which are often referred to as *pseudo-counts*. In the case when we used $\alpha_i = 1$ for all i then the number of times we observe event i is increased by 1. In many algorithms using pseudo-counts this can make a very significant difference since if $n\,p_i$ is small it is quite likely that $N_i = 0$ and the maximum likelihood estimator would give $p_i = 0$ – this is often a much stronger statement than we want to make. (If we have never seen a spam email containing the word 'banana' we don't necessarily want to conclude that the probability of a new email containing the word 'banana' cannot be spam – something that many models may conclude if we don't use pseudo-counts.)

Exercise 8.4 (on page 255): *Equation (8.4) gives the posterior probability for the mean and variance given a set of data $\mathcal{D} = (X_1, X_2, \ldots, X_n)$ and assuming an uninformative prior. Show that the marginal created by averaging over the variance is distributed according to Student's t-distribution*

$$f(\mu|\mathcal{D}) = \int_0^\infty f(\mu, \tau|\mathcal{D})\, d\tau = \mathrm{T}\left(\frac{n\,(\mu - \hat{\mu})}{\sqrt{S}}\middle| n\right)$$

with

$$\hat{\mu} = \frac{1}{n}\sum_{i=1}^n X_i, \quad S = \sum_{i=1}^n (X_i - \hat{\mu})^2, \quad \mathrm{T}(t|\nu) = \frac{1}{\sqrt{\nu}\, B\left(\frac{1}{2}, \frac{\nu}{2}\right)} \left(1 + \frac{t^2}{\nu}\right)^{-\frac{\nu+1}{2}}.$$

From Equation (8.4):

$$\begin{aligned} f(\mu,\tau|\mathcal{D}) &= \mathcal{N}(\mu|\hat{\mu},(n\,\tau)^{-1})\,\mathrm{Gam}\big(\tau\big|\tfrac{n}{2},\tfrac{S}{2}\big) \\ &= \sqrt{\frac{n\,\tau}{2\,\pi}}\mathrm{e}^{-n\,\tau\,(\mu-\hat{\nu})^2/2}\times\left(\frac{S}{2}\right)^{\frac{n}{2}}\frac{\tau^{\frac{n}{2}-1}\,\mathrm{e}^{-\frac{S\,\tau}{2}}}{\Gamma(\frac{n}{2})} \\ &= \frac{\sqrt{n}}{\sqrt{2\,\pi}\,\Gamma\big(\frac{n}{2}\big)}\left(\frac{S}{2}\right)^{\frac{n}{2}}\tau^{\frac{n-1}{2}}\,\mathrm{e}^{-\frac{n\,\tau}{2}(\mu-\hat{\mu})^2-\frac{\tau}{2}S}. \end{aligned}$$

Marginalising out τ (using $\sqrt{\pi}=\Gamma(\frac{1}{2})$)

$$\begin{aligned} \int_0^\infty f(\mu,\tau|\mathcal{D})\,\mathrm{d}\tau &= \frac{\sqrt{n}}{\sqrt{2\,\pi}\,\Gamma\big(\frac{n}{2}\big)}\left(\frac{S}{2}\right)^{\frac{n}{2}}\int_0^\infty \tau^{\frac{n-1}{2}}\,\mathrm{e}^{-\tau\left(\frac{n}{2}(\mu-\hat{\mu})^2+\frac{S}{2}\right)}\,\mathrm{d}\tau \\ &= \frac{\sqrt{n}\,\Gamma\big(\frac{n+1}{2}\big)}{\Gamma\big(\frac{1}{2}\big)\,\Gamma\big(\frac{n}{2}\big)}\left(\frac{S}{2}\right)^{\frac{n}{2}}\left(\frac{S}{2}+\frac{n}{2}(\mu-\hat{\mu})^2\right)^{-\frac{n+1}{2}} \\ &= \sqrt{\frac{n}{S}}\frac{1}{B\big(\frac{1}{2},\frac{n}{2}\big)}\left(1+\frac{n\,(\mu-\hat{\mu})^2}{S}\right)^{-\frac{n+1}{2}} = \mathrm{T}\left(\frac{n\,(\mu-\hat{\mu})}{\sqrt{S}}\bigg|\,n\right) \end{aligned}$$

where $\mathrm{T}(X|\mu)$ is (Student's) t-distribution. Note that, for small n, the t-distribution has very large tails (reflecting the fact that we initially know nothing about the variance). For larger n the t-distribution quite rapidly converges towards a normal distribution.

Exercise 8.5 (on page 255): *Obtain some data, e.g. the area of different countries, their population, physical constants, etc., and plot the frequency of the most significant figure. Compare this with Benford's law.*

In Figure A.8 I have done this for the population of 238 countries. Note that it shouldn't really matter which set of data you have used.

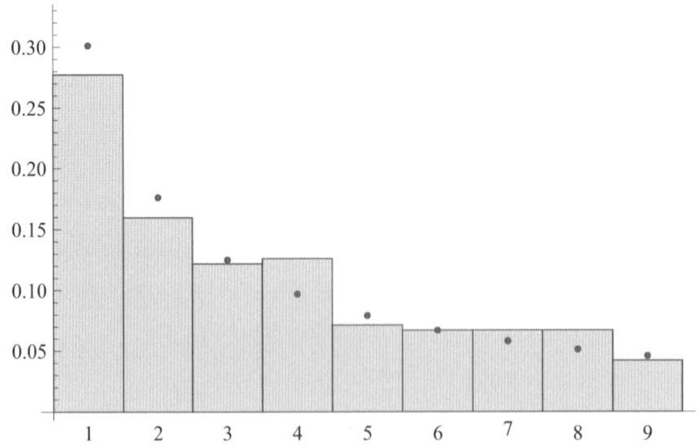

Figure A.8 Frequency of occurrence of the most significant figure for the population of 238 countries. The red dots show Benford's law.

Exercise 8.6 (on page 256): *Show that a scale parameter, σ, of the form $f_X(x|\sigma)=\frac{1}{\sigma}g_X(\frac{x}{\sigma})$ turns into a location parameter under the change of variables $X=\mathrm{e}^Y$.*

Use the usual change of variables

$$f_Y(y) = \frac{\mathrm{d}x}{\mathrm{d}y} f_X(x(y)|\sigma) = \frac{\mathrm{e}^y}{\sigma} g_X\left(\tfrac{\mathrm{e}^y}{\sigma}\right) = \mathrm{e}^{y-\log(\sigma)} g_X\left(\mathrm{e}^{y-\log(\sigma)}\right) = g_Y(y-\mu)$$

where $g_Y(y) = \mathrm{e}^y g_X(\mathrm{e}^y)$ and $\mu = \log(\sigma)$.

If we have a variable $X \in (0, \infty)$ it is sometimes more convenient to work with the unconstrained variable $Y = \log(X) \in (-\infty, \infty)$.

Exercise 8.7 (on page 256): *Show that the uninformative prior for a probability, P, transforms to a uniform prior for the log-odds (logit) parameter $Y = \log\left(\frac{P}{1-P}\right)$.*

The uninformative prior for a variable, P, in the range $(0, 1)$ is $f_P(p) \propto p^{-1}(1-p)^{-1}$ (recall this is an improper prior). Again use the usual change of variables

$$f_Y(y) = \frac{\mathrm{d}p}{\mathrm{d}y} f_P(p)$$

but invert the logit function

$$p = \mathrm{logit}^{-1}(y) = \frac{1}{1+\mathrm{e}^{-y}}, \qquad \frac{\mathrm{d}p}{\mathrm{d}y} = p(1-p)$$

so that

$$f_Y(y) \propto p(1-p) \times p^{-1}(1-p)^{-1} = 1.$$

If we have a variable $P \in (0, 1)$ it is sometimes more convenient to work with the unconstrained variable $Y = \mathrm{logit}(P) \in (-\infty, \infty)$. This possibly provides a more succinct argument for believing that $f_P(p) \propto p^{-1}(1-p)^{-1}$ is an uninformative prior for random variables confined to the interval $[0, 1]$.

A.9 Answers to Chapter 9. Entropy

Exercise 9.1 (on page 290): *Which has the most uncertainty: rolling an honest dice or tossing three fair coins assuming*

i. you care about the order of the results;
ii. you only care about the number of heads.

We use the entropy as our measure of uncertainty. For an honest dice

$$H_{dice} = -\sum_{i=0}^{6} p_i \log_2(p_i) = -\sum_{i=0}^{6} \frac{1}{6} \log_2\left(\frac{1}{6}\right) = \log_2(6) = 2.585 \text{ bits}.$$

For two coins we have

i. The outcome of each coin is 1 bit. As there are three independent coins the total uncertainty is 3 bits.
ii. If we only care about the number of heads then

$$\begin{aligned}
H_{heads} &= -p_{3h}\log(p_{3h}) - p_{2h}\log(p_{2h}) - p_{1h}\log(p_{1h}) - p_{0h}\log(p_{0h}) \\
&= -\frac{1}{2^3}\log_2\left(\frac{1}{2^3}\right) - \frac{1}{2^3}\binom{3}{2}\log_2\left(\frac{1}{2^3}\binom{3}{2}\right) \\
&\quad - \frac{1}{2^3}\binom{3}{1}\log_2\left(\frac{1}{2^3}\binom{3}{2}\right) - \frac{1}{2^3}\log_2\left(\frac{1}{2^3}\right) \\
&= \frac{1}{8}\left(2\log_2\left(2^3\right) + 6\log_2\left(\frac{8}{3}\right)\right) = 1.8113\,\text{bits}.
\end{aligned}$$

Thus tossing the three coins where you care about the order of the results is the most uncertain, followed by rolling the dice. Note that (i) and (ii) have different uncertainties because communicating the outcome of each coin requires passing more information than just reporting the number of heads.

Exercise 9.2 (on page 290): *Use Stirling's approximation, $n! \approx \left(\frac{n}{e}\right)^n \sqrt{2\pi n}$, to show that*

$$\binom{n}{f\,n} \approx \frac{2^{n\,H(f)}}{\sqrt{2\pi\,n\,f(1-f)}}$$

where $H(f) = -f\log_2(f) - (1-f)\log_2(1-f)$ is the entropy of Bernoulli trial with success probability f.

$$\begin{aligned}
\binom{n}{f\,n} &= \frac{n!}{(fn)!\,((1-f)n)!} \\
&\approx \frac{\sqrt{2\pi\,n}}{\sqrt{2\pi\,f\,n}\sqrt{2\pi\,(1-f)\,n}}\left(\frac{n}{e}\right)^n\left(\frac{fn}{e}\right)^{-fn}\left(\frac{(1-f)n}{e}\right)^{-(1-f)n} \\
&= \frac{1}{\sqrt{2\pi\,n\,f\,(1-f)}}\,f^{-fn}\,(1-f)^{-(1-f)n} \\
&= \frac{1}{\sqrt{2\pi\,n\,f\,(1-f)}}\,2^{-f\,n\,\log_2(f)-(1-f)\,n\,\log_2(1-f)}
\end{aligned}$$

which is the result we wish to show. Note that $2^{n\,H(f)} = \mathrm{e}^{n\,H_e(f)}$ where $H_e(f) = -f\log(f) - (1-f)\log(1-f)$, where now we use natural logarithms rather than logs to the base 2. We plot $\log\left(\binom{n}{fn}\right)/n$, versus f, and the logarithm of the approximation for $n = 10$ in Figure A.9. As can be seen the approximation is excellent even for small n

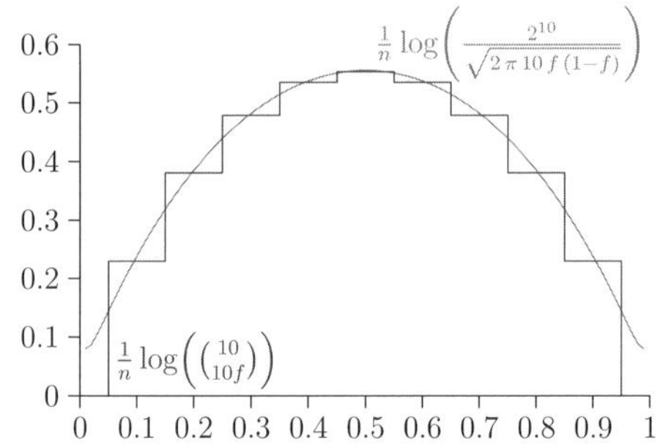

Figure A.9 Plot of $\log\left(\binom{n}{fn}\right)/n$, versus f, and the logarithm of the Stirling approximation for $n = 10$.

Exercise 9.3 (on page 290): *Compute the entropy for a normally distributed random variable $X \sim \mathcal{N}(\mu, \sigma^2)$. Plot the entropy versus σ and show that it can be negative.*

The entropy is given by

$$H_X = -\int_{-\infty}^{\infty} \mathcal{N}(x|\mu,\sigma^2)\, \log\Big(\mathcal{N}(x|\mu,\sigma^2)\Big)\, \mathrm{d}x$$
$$= \int_{-\infty}^{\infty} \mathrm{e}^{-(x-\mu)^2/(2\sigma^2)} \left(\frac{(x-\mu)^2}{2\sigma^2} + \log(\sqrt{2\pi}\,\sigma) \right) \frac{\mathrm{d}x}{\sqrt{2\pi}\,\sigma}$$
$$= \frac{1}{2} + \frac{1}{2}\log(2\pi) + \log(\sigma).$$

We plot this in Figure A.10. We note that when $\sigma < \mathrm{e}^{-1/2}/\sqrt{2\pi}$ the entropy goes negative. As the entropy of a continuous variable doesn't have an information theory interpretation this should not be too worrying.

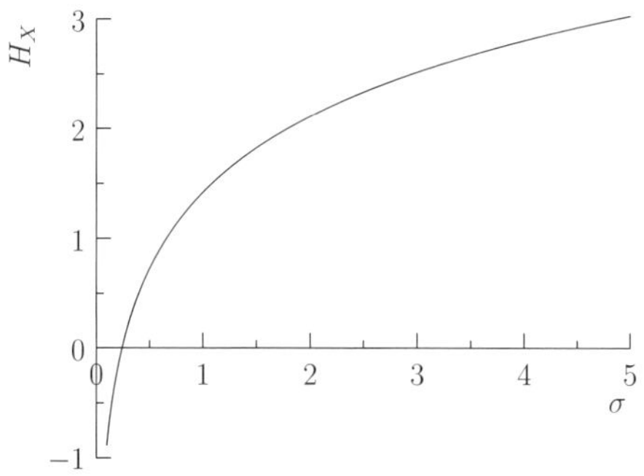

Figure A.10 The entropy of a random variable $X \sim \mathcal{N}(\mu, \sigma^2)$ versus the standard deviation, σ.

Exercise 9.4 (on page 290): *Consider a system whose state, X, we represent by a discrete random variable. We label the possible states by x_i where $i \in \mathcal{I}$ ($\mathcal{I}$ is an index set). The system is in thermal contact with a large system so that it can exchange energy. We observe the average energy of the system is*

$$\mathbb{E}\left[E(X)\right] = \sum_{i\in\mathcal{I}} p_i\, E(x_i) = U$$

where $p_i = \mathbb{P}\left(X = x_i\right)$. Find the maximum entropy probability, p_i, of being in a state x_i. This is the Boltzmann distribution.

We maximise the entropy subject to the constraint on the average energy and that the probabilities sum to one. To do this we consider the Lagrangian

$$\mathcal{L} = -\sum_{i\in\mathcal{I}} p_i \log(p_i) + \lambda_0 \left(\sum_{i\in\mathcal{I}} p_i - 1 \right) + \lambda_0 \left(\sum_{i\in\mathcal{I}} p_i\, E(x_i) - U \right).$$

Setting the first derivative to zero

$$\frac{\partial \mathcal{L}}{\partial p_i} = -1 - \log(p_i) + \lambda_0 + \lambda_1 E(x_i) = 0$$

or

$$p_i = e^{-1+\lambda_0+\lambda_1 E(x_i)}.$$

From the normalisation condition for the probabilities (which gives $\lambda_0 = 1 + \log(Z)$)

$$p_i = \frac{1}{Z} e^{\lambda_1 E(x_i)}, \qquad Z = \sum_{i \in \mathcal{I}} e^{\lambda_1 E(x_i)}.$$

Note that we could have written this down immediately using Equations (9.7) and (9.8). This is the famous *Boltzmann distribution*. The Lagrange multiplier λ_1 is usually written as $-1/(k\,T)$ where the constant T is equal to the temperature measured in Kelvin and k is known as Boltzmann constant. We note that

$$U = \frac{\partial \log(Z)}{\partial \lambda_1} = \sum_{i \in \mathcal{I}} p_i\, E(x_i).$$

It may seem strange that the temperature is a Lagrange multiplier, but it controls the balance between entropy and energy, which is the defining property of temperature.

The Boltzmann distribution is the basis for much of classical (i.e. not quantum) statistical mechanics. It provides an extremely good description of many physical systems (the corrections due to quantum mechanics often only play an important role at very low temperatures).

A.10 Answers to Chapter 10. Collective Behaviour

Exercise 10.1 (on page 306): *Compute a histogram of the time for a random walk, with the walker taking steps of* ± 1 *to return to his/her starting position. If* T *is the return time show that* $\mathbb{P}\left(T = t/2\right) \approx t^{-3/2}/\zeta(3/2)$, *where the normalisation constant* $\zeta(3/2) \approx 2.61$ *is the Riemann zeta function. Use this fact to show that a random walker will return to his/her starting position with probability arbitrarily close to 1, but the expected return time is infinite.*

This is an easy simulation to carry out as we just have to add ± 1 with equal probability and count how many steps until the sum reaches 0. We repeat this many times. Figure A.11 shows a histogram plot on a log-log scale of the frequency of returning to the origin after $t/2$ steps. Note that we can only return to the origin after an even number of steps.

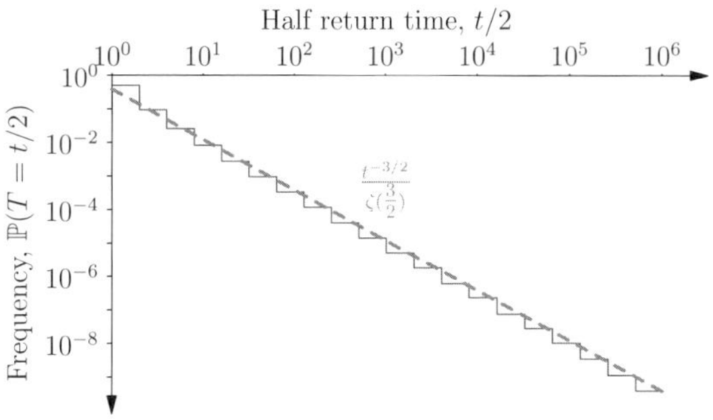

Figure A.11 Histogram plot on a log-log scale of the frequency of returning to the origin after $t/2$ steps of a one-dimensional random walk. Also shown is a histogram of $\mathbb{P}(t) = t^{-3/2}/\zeta(3/2)$.

Assuming $\mathbb{P}\left(T = t/2\right) = t^{-3/2}/\zeta(3/2)$ then the probability of returning in no more than $t/2$ steps is

$$\mathbb{P}\left(T \leq t/2\right) = \sum_{t'=1}^{t} \frac{t'^{-3/2}}{\zeta(3/2)} \approx 2\int_1^t \frac{1}{t^{3/2}}\,\mathrm{d}t = 1 - \frac{1}{\sqrt{t}}.$$

(If we wanted to be more rigorous we could bound the sum by an integral.) In the limit $t \to \infty$ this probability becomes arbitrarily close to 1. The mean return time is

$$\mathbb{E}\left[T\right] = \sum_{t'=1}^{\infty} t' \frac{t'^{-3/2}}{\zeta(3/2)} == \frac{1}{\zeta(3/2)} \sum_{t'=1}^{\infty} \frac{1}{\sqrt{t'}} = \infty$$

since the sum diverges. We thus have the curious situation that the mean time for a random walk to return to the origin is infinite, but it will return with probability arbitrarily close to 1. In fact, it returns to the origin infinitely often. To see this consider a period, t, made up of $t^{1/4}$ subintervals each of length $t^{3/8}$. The probability that a random walk returns to the origin in the subinterval is approximately $1 - c'/t^{3/8}$ for some constant c'. The probability that this happens at least $t^{1/4}$ times is

$$\left(1 - \frac{c'}{t^{3/8}}\right)^{t/4} \approx 1 - \frac{c'}{t^{1/8}}$$

which is arbitrarily close to 1 as $t \to \infty$.

Exercise 10.2 (on page 307): *Simulate the branching process for some distribution and measure the mean number of particles and the variance after five iterations. Compare this with the theoretical results κ_1^5 and $\kappa_2\kappa_1^4(\kappa_1^5 - 1)/(\kappa_1 - 1)$, where κ_1 and κ_2 are the mean and variance of the distribution of children for a single individual.*

This can be tried with any distribution. We use a Poisson distribution, $\mathrm{Poi}(k|\mu)$, which has both mean and variance equal to μ. In Figure A.12 we show the logarithm of the mean and variance of the number of particles averaged 10 000 trials. The solid lines show the results obtained from the generating function analysis.

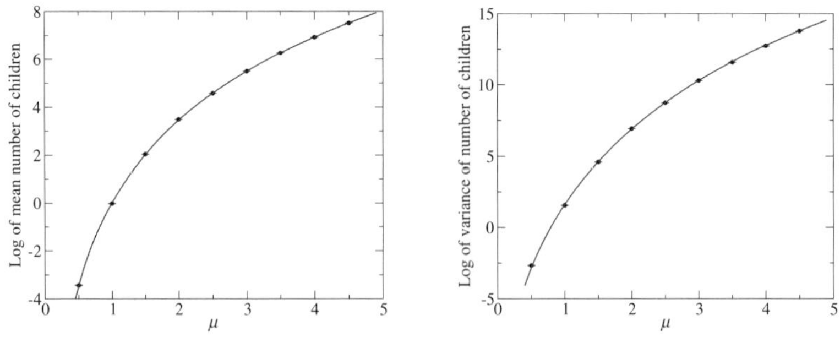

(a) log-mean (b) log-variance

Figure A.12 (a) Shows the logarithm of the mean number of particles generated by a Poisson branching process versus μ. The solid line shows $5\log(\mu)$. (b) Shows the logarithm of the variance in the number of particles generated by the same branching process. The solid line shows $\log(\mu^5(\mu^5-1)/(\mu-1))$.

A.11 Answers to Chapter 11. Markov Chains

Exercise 11.1 (on page 345): *The figure below shows a diagrammatic representation of a Markov model consisting of six states. Write down the transition matrix and compute its eigenvalues and the steady state. What is the characteristic relaxation time to reach the steady state? Perform a Monte Carlo simulation to find the steady-state probabilities empirically and compare this with those obtained from considering the eigenvectors of the transition matrix.*

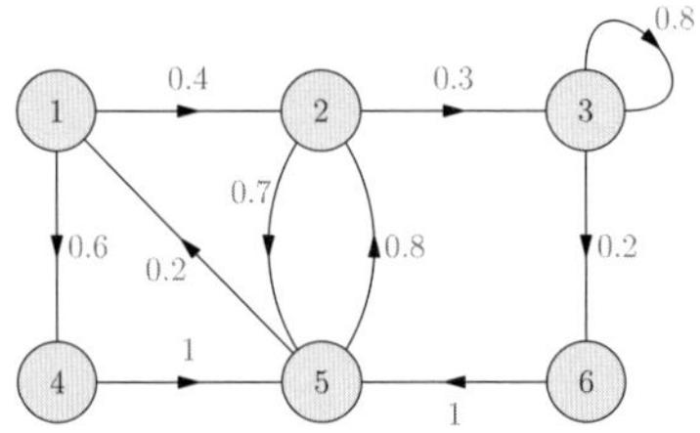

This is easily computed in a language such as MATLAB, Octave, or python where you can quickly write programs to manipulate matrices. The transition matrix can be read off from the figure above. In MATLAB/Octave we can input this as

```
M = [0.0 0.0 0.0 0.0 0.2 0.0;
     0.4 0.0 0.0 0.0 0.8 0.0;
     0.0 0.3 0.8 0.0 0.0 0.0;
     0.6 0.0 0.0 0.0 0.0 0.0;
     0.0 0.7 0.0 1.0 0.0 1.0;
     0.0 0.0 0.2 0.0 0.0 0.0]
```

To compute the eigenvalues and eigenvectors we use

```
[V L] = eig(M);
```

where the eigenvalues are given by `diag(L)`: these are 1, 0.608, -0.534, -0.320, 0.046, and 0. The steady state of the system is given by the eigenvector with eigenvalue 1. It can be found using the command `V(:,1)'/sum(V(:,1))` and is equal to

$$\pi = (0.053,\ 0.233,\ 0.349,\ 0.037,\ 0.264,\ 0.070)^{\mathsf{T}}.$$

The speed of relaxation towards the steady-state distribution is equal to $\lambda_1^t = \mathrm{e}^{t\,\log(\lambda_2)} = \mathrm{e}^{t\,\log(0.608)} = \mathrm{e}^{-0.50\,t}$. Thus this is a rapidly mixing Markov chain where every two steps the components orthogonal to the steady state distribution decay by approximately $\mathrm{e}^{-1} \approx 0.37$.

To perform a Monte Carlo simulation of the Markov model we start in some state, say state 1, and move to a new state according to the transition probabilities. This is then repeated many times. To obtain a sample from the steady-state distribution we need to run long enough to forget the initialisation – because this is rapidly mixing this will happen to high accuracy after 20 steps. Running for 10 million iterations (which took my computer less than a second), I obtain agreement with the result above for the steady state to four decimal places.

Exercise 11.2 (on page 345): *Consider an ergodic Markov chain described by the transition matrix* M *starting from an initial state* $\boldsymbol{p}(1)$*. To compute the expected time to reach a state i we can considering a modified transition matrix* $\hat{\mathsf{M}}$ *where* $\hat{M}_{jk} = M_{jk}$ *if* $k \neq i$ *and* $\hat{M}_{ji} = 0$ *otherwise. That is, in this modified system the dynamics are identical to the original until we reach state i, whereupon the walker disappears. The probability of being in state i at time t is equal to* $\boldsymbol{\delta}_i \hat{\mathsf{M}}^{t-1}\boldsymbol{p}(1)$*, where* $\boldsymbol{\delta}_i$ *is the vector with zero elements everywhere except in the* i^{th} *position where the element is equal to 1. The expected first-passage time is thus given by*

$$T_{fpt} = \sum_{t=1}^{\infty} (t-1)\,\boldsymbol{\delta}_i^{\mathsf{T}} \hat{\mathsf{M}}^{t-1}\boldsymbol{p}(1).$$

This just sums the number of steps taken times the probability of reaching state i in those steps. Show that we can write this as

$$T_{fpt} = \boldsymbol{\delta}_0^{\mathsf{T}}\,(\mathbf{I} - \hat{\mathsf{M}})^{-1}(\mathbf{I} - \hat{\mathsf{M}})^{-1}\,\boldsymbol{p}(1) - 1$$

where $\mathbf{I}$ *is the identity matrix. Further show that this can be simplified to*

$$T_{fpt} = \mathbf{1}^{\mathsf{T}}\,(\mathbf{I} - \hat{\mathsf{M}})^{-1}\,\boldsymbol{p}(1) - 1$$

where $\mathbf{1}$ *is the vector of all 1s. Note that you have to sum a geometric series of matrices and differentiate matrices so this exercise isn't for the faint-hearted.*

We can split the sum as

$$T_{\mathrm{fpt}} = \sum_{t=1}^{\infty}(t-1)\,\boldsymbol{\delta}_i^{\mathsf{T}}\hat{\mathsf{M}}^{t-1}\boldsymbol{p}(1) = \sum_{t=1}^{\infty} t\,\boldsymbol{\delta}_i^{\mathsf{T}}\hat{\mathsf{M}}^{t-1}\boldsymbol{p}(1) - \boldsymbol{\delta}_i^{\mathsf{T}}\sum_{t=1}^{\infty}\hat{\mathsf{M}}^{t-1}\boldsymbol{p}(1).$$

As the original problem was ergodic eventually we reach state i from any starting state so that

$$\boldsymbol{\delta}_i^{\mathsf{T}} \sum_{t=1}^{\infty} \hat{\mathbf{M}}^{t-1} \boldsymbol{p}(1) = 1.$$

Then

$$\begin{aligned} T_{\mathrm{fpt}} &\overset{(1)}{=} \sum_{t=0}^{\infty} t\, \boldsymbol{\delta}_i^{\mathsf{T}} \hat{\mathbf{M}}^{t-1} \boldsymbol{p}(1) - 1 \overset{(2)}{=} \boldsymbol{\delta}_i^{\mathsf{T}} \sum_{t=0}^{\infty} t\, \hat{\mathbf{M}}^{t-1} \boldsymbol{p}(1) - 1 \\ &\overset{(3)}{=} \sum_{t=0}^{\infty} \boldsymbol{\delta}_i^{\mathsf{T}} \frac{\partial \hat{\mathbf{M}}^t}{\partial \hat{\mathbf{M}}} \boldsymbol{p}(1) - 1 \end{aligned}$$

(1) Using the two equations above.
(2) Taking the vector $\boldsymbol{\delta}_i^{\mathsf{T}}$ outside the summation.
(3) Using $\frac{\partial \hat{\mathbf{M}}^t}{\partial \hat{\mathbf{M}}} = t\, \hat{\mathbf{M}}^{t-1}$.

Notice that we are free to extend the sum to 0 as this does not change the sum. (If you don't like to take the derivative with respect to the matrix you can replace $\hat{\mathbf{M}}$ by $a\,\hat{\mathbf{M}}$, take derivatives with respect to a, and at the end set $a = 1$.) Now we can exchange the order of the derivative and the sum, giving

$$T_{\mathrm{fpt}} = \boldsymbol{\delta}_i^{\mathsf{T}} \frac{\partial}{\partial \hat{\mathbf{M}}} \sum_{t=1}^{\infty} \hat{\mathbf{M}}^t \boldsymbol{p}(1) - 1. \tag{A.4}$$

To sum the geometric series we note that

$$\begin{aligned} \left(\sum_{t=0}^{n-1} \hat{\mathbf{M}}^t \right) \left(\mathbf{I} - \hat{\mathbf{M}} \right) &= \left(\mathbf{I} + \hat{\mathbf{M}} + \hat{\mathbf{M}}^2 + \cdots + \hat{\mathbf{M}}^{n-1} \right) \left(\mathbf{I} - \hat{\mathbf{M}} \right) \\ &= \mathbf{I} - \hat{\mathbf{M}}^n \end{aligned}$$

since the intermediate terms $\mathbf{M}$, $\mathbf{M}^2$, …, $\mathbf{M}^{n-1}$ all cancel. As our new transition matrix loses probability we have

$$\lim_{n\to\infty} \hat{\mathbf{M}}^n \boldsymbol{p}(1) = \mathbf{0}$$

so that we can neglect the matrix $\hat{\mathbf{M}}^n$ in the limit when n becomes infinite. Thus,

$$\left(\sum_{t=0}^{\infty} \hat{\mathbf{M}}^t \right) \left(\mathbf{I} - \hat{\mathbf{M}} \right) = \hat{\mathbf{M}}$$

or

$$\sum_{t=0}^{\infty} \hat{\mathbf{M}}^t = \hat{\mathbf{M}} \left(\mathbf{I} - \hat{\mathbf{M}} \right)^{-1}.$$

We can also use the geometric series to rewrite the identity we found above

$$\boldsymbol{\delta}_i^{\mathsf{T}} \sum_{t=1}^{\infty} \hat{\mathbf{M}}^{t-1} \boldsymbol{p}(1) = \boldsymbol{\delta}_i^{\mathsf{T}} (\mathbf{I} - \hat{\mathbf{M}})^{-1} \boldsymbol{p}(1) = 1.$$

Since this holds for any vector $\boldsymbol{p}(1)$ for which $\mathbf{1}^{\mathsf{T}}\boldsymbol{p}(1) = 1$, it must be the case that $\boldsymbol{\delta}_i^{\mathsf{T}}(\mathbf{I} - \hat{\mathbf{M}})^{-1} = \mathbf{1}^{\mathsf{T}}$. Another way to see this is to rewrite it as $\boldsymbol{\delta}_i^{\mathsf{T}} = \mathbf{1}^{\mathsf{T}}(\mathbf{I} - \hat{\mathbf{M}}) = \mathbf{1}^{\mathsf{T}} - \mathbf{1}^{\mathsf{T}}\hat{\mathbf{M}}$. But as the original matrix, $\mathbf{M}$, is a stochastic matrix $\mathbf{1}^{\mathsf{T}}$ is a left-handed eigenvector (all the columns sum to one). In the modified matrix, $\hat{\mathbf{M}}$, all columns sum to one except the ith column where all the elements are set to zero. Thus, $\mathbf{1}^{\mathsf{T}} - \mathbf{1}^{\mathsf{T}}\hat{\mathbf{M}} = \boldsymbol{\delta}_i^{\mathsf{T}}$.

Returning to the mean first-passage time

$$\begin{aligned}
T_{\text{fpt}} &\overset{(1)}{=} \boldsymbol{\delta}_i^{\mathsf{T}} \left(\frac{\partial}{\partial \hat{\mathbf{M}}} (\mathbf{I} - \hat{\mathbf{M}})^{-1} \right) \boldsymbol{p}(1) - 1 \\
&\overset{(2)}{=} \boldsymbol{\delta}_i^{\mathsf{T}} (\mathbf{I} - \hat{\mathbf{M}})^{-1} (\mathbf{I} - \hat{\mathbf{M}})^{-1} \boldsymbol{p}(1) - 1 \\
&\overset{(3)}{=} \mathbf{1}^{\mathsf{T}} (\mathbf{I} - \hat{\mathbf{M}})^{-1} \boldsymbol{p}(1) - 1.
\end{aligned}$$

(1) From Equation (A.4) and replacing the sum of the geometric series by $(\mathbf{I} - \hat{\mathbf{M}})^{-1}$.
(2) Using $\frac{\partial}{\partial \hat{\mathbf{M}}} (\mathbf{I} - \hat{\mathbf{M}})^{-1} = (\mathbf{I} - \hat{\mathbf{M}})^{-2}$.
(3) Using the identity we found that $\boldsymbol{\delta}_i^{\mathsf{T}}(\mathbf{I} - \hat{\mathbf{M}})^{-1} = \mathbf{1}^{\mathsf{T}}$.

Thus we can compute the mean first-passage time by inverting the matrix $\mathbf{I} - \hat{\mathbf{M}}$. Be warned this matrix may be ill-conditioned in which case inverting the matrix can be numerically unstable. This is particularly true for large transition matrices. However, often this gives a quick method for obtaining expected first-passage times. In fact, it does not even require the matrix to be ergodic, it holds provided the system will eventually finish in state i. It can also be extended to multiple end states. We just modify the transition matrix so that the walker 'vanishes' as soon as it reaches any of the end states.

Exercise 11.3 (on page 345): *Compute the expected first-passage time to reach state 6 starting from state 1 for the Markov chain described in Exercise 11.1 using the formula from Exercise 11.2, and compare this with empirical measurements.*

The matrix $\hat{\mathbf{M}}$ is identical to $\mathbf{M}$ except the last column has all components equal to zero. To compute the expected first-passage time we use the command

```
Tfpt = sum((eye(6)-Mhat)\[1 0 0 0 0 0]')-1.
```

For this problem we find $T_{\text{fpt}} = 14.333$. To compute it numerically we start from state 1 and count how many steps it takes to reach state 6. Again repeating this 10 million times I obtained agreement to four significant figures.

Exercise 11.4 (on page 345): *Consider a random variable that can take values in the set $\mathbb{Z}_n = \{0, 1, 2, \ldots, n-1\}$. At each time step the value increases or decreases by 1 modulo n. That is, the system is performing a random walk around the ring. Construct the transition matrix and compute the eigenvalues for $n = 499$. Show that if n is even there are two eigenvalues with modulus 1. Plot a histogram of the eigenvalues and show how the second-largest eigenvalue grows with n.*

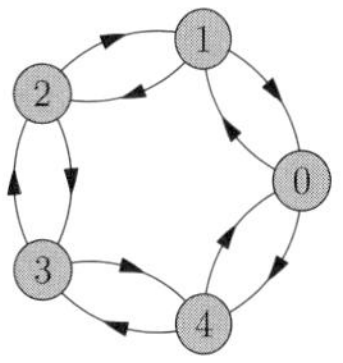

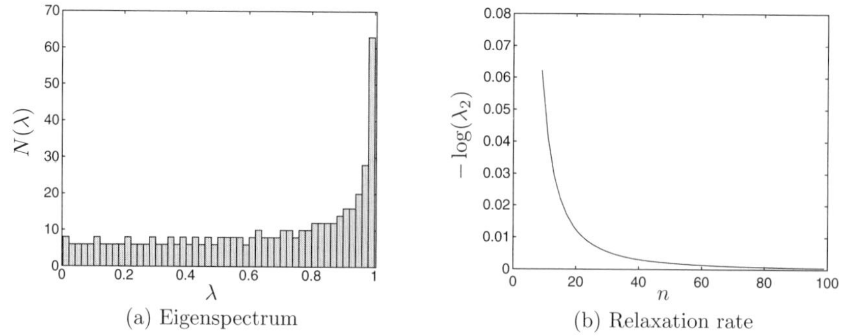

Figure A.13 (a) Spectrum (histogram) of eigenvalues of the transition matrix for $n = 499$. (b) Relaxation time, $-\log(\lambda_2)$, for odd n.

The transition matrix is zero except down the two off-diagonals, $M_{n\,n-1}$ and $M_{n-1\,n}$ which are both equal to 1/2. When n is even then there are eigenvalues ± 1 since the Markov chain is periodic. In Figure A.13(a) we show the spectrum of eigenvalues for this transition matrix with $n = 499$. In Figure A.13(b) we show the logarithm of the second-largest eigenvalue as a function of the ring size. We see for this problem the second-largest eigenvalue becomes very close to 1 as we increase n, indicating that it will take considerable time to reach the equilibrium distribution (which is when all states are visited with equal probability).

Exercise 11.5 (on page 346): *Use Markov Chain Monte Carlo (MCMC) to generate normal deviates where the candidate solution is taken by adding* $U \sim \mathrm{U}(-1/2, 1/2)$. *Compute the time it takes to generate* 10^6 *deviates. Compare this to other methods for generating normal deviates (see Exercise 3.2).*

This took just 32 ms per deviate compared with 51 ms for Box–Muller. However, the deviates are not independent. In fact we have to accept the deviates even if the proposal is rejected otherwise the sample is biased. If we want our deviates to be (approximately) independent we would accept only every fifth deviate or every tenth (depending on how strict we are about our deviates being independent). In this case, this strategy is far slower that Box–Muller.

Exercise 11.6 (on page 346): *Use the clustering algorithm (see Example 11.6) or, if you prefer, the Metropolis algorithm (see Example 11.4) to simulate the two-dimensional Ising model as described in Section 10.4. Compute the mean and variance of the energy and magnetisation per spin. (When using the clustering algorithm you should use the absolute magnetisation below the critical point because the clustering algorithm restores ergodicity which is broken in real magnets and by Metropolis.)*

I used the clustering algorithm on a 64×64 square lattice. In Figure A.14 I show the energy per spin, its variance, and the magnetisation of variance in the magnetisation per spin. Comparing (a), (b), and (c) with the Onsager solution shown in Figure 10.6 we see good agreement. Note that the variance in the magnetisation is equal to the susceptibility, ξ_T times $(kT)^2$, where the

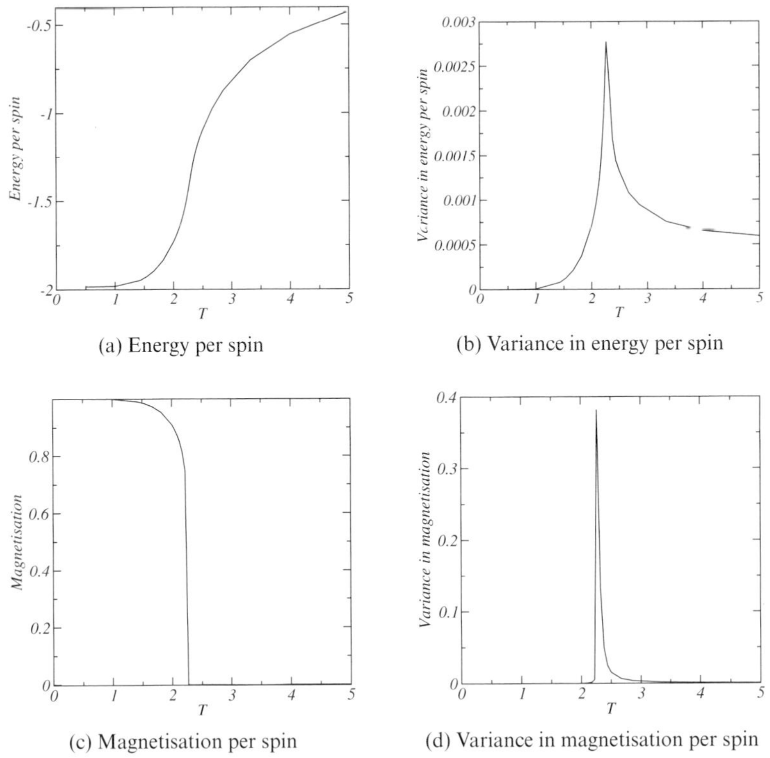

Figure A.14 Results from simulating the two-dimensional Ising model on a 64×64 lattice. Shown is (a) energy, (b) variance in energy, (c) magnetisation, and (d) variance in magnetisation.

susceptibility measures the response of the magnetisation to a change in the magnetic field.

Exercise 11.7 (on page 346): *Use MCMC to obtain an estimate for the posterior given a normal likelihood, $X_i \sim \mathcal{N}(\mu, \sigma)$ using an uninformative prior. (Generate $n = 20$ data points from a normal distribution with $\mu = 1/2$, $\sigma = 2$.) Plot a histogram of the posterior for the mean and show that it is distributed according to a t distribution (see Exercise 8.4 on page 255).*

To perform the MCMC we need to choose a proposal distribution. I choose

$$\mu' \sim \mathcal{N}(\mu, 1) \qquad\qquad \sigma' \sim \mathrm{Exp}\left(\tfrac{1}{\sigma}\right)$$

where

$$\mathrm{Exp}\left(\sigma' \middle| \tfrac{1}{\sigma}\right) = \frac{\mathrm{e}^{-\sigma'/\sigma}}{\sigma}$$

is the exponential distribution (with mean σ). Using this choice of proposal distribution the acceptance ratio is

$$r = \frac{\sigma^2}{\sigma'^2} \mathrm{e}^{-\frac{\sigma}{\sigma'} + \frac{\sigma'}{\sigma}} \prod_{i=1}^{n} \frac{\mathcal{N}(X_i|\mu', \sigma'^2)}{\mathcal{N}(X_i|\mu, \sigma^2)}.$$

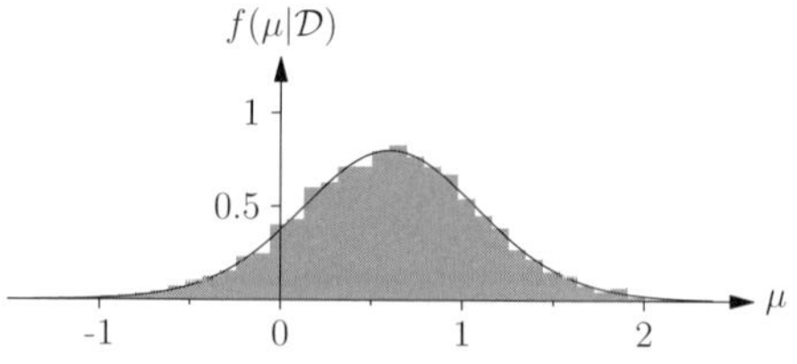

Figure A.15 Histogram of the marginal posterior for the parameter μ. Also shown is the true marginal posterior.

Using a million steps I found for my data $\hat{\mu} = 0.59 \pm 0.51$ and $\hat{\sigma} = 2.30 \pm 0.4$. Note that these are the mean of the posterior plus or minus the variance in the posterior. Figure A.15 shows the histogram of the marginal posterior for μ calculated using MCMC. For this problem there is a conjugate prior so we didn't need to perform MCMC. The marginal distribution of this prior is shown by the solid line in Figure A.15.

Exercise 11.8 (on page 346): *In Example 8.11 on page 228 and Example 8.12 on page 232 we considered the problem of estimating the parameters for a mixture of two Gaussians. Generate data consisting of 500 deviates from $\mathcal{N}(0,1)$ and 200 deviates from $\mathcal{N}(3,1)$. The task is to infer the means and variances of the two distributions assuming these distributions are unknown and we don't know which data point was generated by which distribution. We assume, however, that we know the likelihood of a data point is given by a mixture of Gaussians*

$$f(X_i|\mu_1,\sigma_1,\mu_2,\sigma_2,p) = p\,\mathcal{N}(X_i|\mu_1,\sigma_1^2) + (1-p)\,\mathcal{N}(X_i|\mu_2,\sigma_2^2).$$

Use MCMC to estimate the unknown parameters μ_1, σ_1, μ_2, σ_2, and p, as well as their uncertainty.

This is again a relatively straightforward MCMC problem, although it no longer has a conjugate prior so we cannot obtain a closed-form solution for the posterior. In Example 8.12 we used the expectation maximisation (EM)-algorithm to find the *maximum a posteriori* (MAP) solution. One advantage of MCMC is that it gives us the uncertainty in the parameter estimates. Once again we have to choose an appropriate proposal distribution. In this case we used

$$\mu_i' \sim \mathcal{N}(\mu, 0.01) \qquad \sigma_i' \sim f_0^{\infty}(\mathcal{N}(\sigma_i, 0.01)) \qquad p' \sim f_0^1(\mathcal{N}(p, 0.01))$$

where $f_0^a(x) = x$ if $0 \leq x \leq a$ or $x + \lceil -x \rceil$ if $x < 0$ or $x - \lfloor x \rfloor$ if $x > a$. In other words, we reflect x off the sides of the interval $[0, a]$. This is a symmetric proposal distribution (the probability of the reverse reflection is the same as the reflection itself). In fact, because of the parameters we are using, it is very unlikely that the reflection ever happens. For the data shown in Example 8.11 I found $\hat{\mu}_1 = 0.04 \pm 0.1$, $\hat{\sigma}_1 = 0.98 \pm 0.08$ $\hat{\mu}_2 = 2.8 \pm 0.2$, $\hat{\sigma}_1 = 1.01 \pm 0.08$ and $\hat{p} = 0.69 \pm 0.05$. Note that unlike the MAP solution we can obtain estimates for uncertainty. As the uncertainty due to the limited data is rather large, there is little point in running the MCMC very long to obtain a good estimate of the posterior. A very basic MCMC algorithm is quite sufficient to obtain good estimators for this problem.

Exercise 11.9 (on page 346): *Consider the problem of tracking a financial time series. Assume that a stock has a true value, $X(t)$. Due to market volatility the price of the stock, $Y(t)$, is taken to be equal to the 'true value' plus some random noise*

$$Y(t) = X(t) + \epsilon(t)$$

where $\epsilon(t) \sim \mathcal{N}(0, 1)$. The true value $X(t)$ varies much more slowly and we assume

$$X(t+1) = X(t) + V(t) \qquad\qquad V(t+1) = V(t) + \eta(t)$$

where $\eta(t) \sim \mathcal{N}(0, 10^{-4})$. Generate some time series data from this model. The figure below illustrates some time series data for 100 data points.

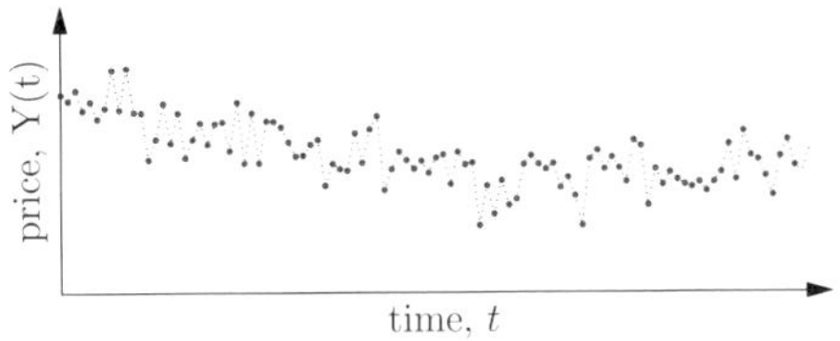

Use a Kalman filter to estimate the true price and its uncertainty. Also use a particle filter to perform the same calculation and compare the two estimates.

Let $\boldsymbol{X}(t) = (X(t), V(t))^{\mathsf{T}}$ describe the state of the system. We model our uncertainty in the state by

$$\boldsymbol{X}(t) \sim \mathcal{N}\left(\begin{pmatrix}\mu_X(t)\\ \mu_V(t)\end{pmatrix}, \begin{pmatrix}\Sigma_{XX}(t) & \Sigma_{XV}(t)\\ \Sigma_{XV}(t) & \Sigma_{VV}(t)\end{pmatrix}\right).$$

We initialise this to

$$\boldsymbol{\mu}(0) = \begin{pmatrix}Y(0)\\ 0\end{pmatrix} \qquad\qquad \boldsymbol{\Sigma}(0) = \begin{pmatrix}1 & 0\\ 0 & 1\end{pmatrix}.$$

The update equations giving the priors at the next time step for the Kalman filter are

$$\tilde{\boldsymbol{\mu}}(t+1) = \mathbf{A}(t)\,\boldsymbol{\mu}(t) \qquad\qquad \tilde{\boldsymbol{\Sigma}}(t+1) = \mathbf{A}(t)\,\boldsymbol{\Sigma}\,\mathbf{A}^{\mathsf{T}}(t) + \mathbf{Q}(t)$$

where

$$\mathbf{A}(t) = \begin{pmatrix}1 & 1\\ 0 & 1\end{pmatrix} \qquad\qquad \mathbf{Q}(t) = \begin{pmatrix}0 & 0\\ 0 & 10^{-4}\end{pmatrix}.$$

The innovation matrix in this case is a 1×1 matrix

$$\mathbf{S}(t+1) = \mathbf{C}(t+1)\,\tilde{\boldsymbol{\Sigma}}(t+1)\,\mathbf{C}^{\mathsf{T}}(t+1) + \mathbf{R}(t+1)$$

where $\mathbf{C}(t+1)$ and $\mathbf{R}(t+1)$ are defined by $Y(t+1) = \mathbf{C}(t+1)\,\boldsymbol{X}(t+1) + \epsilon(t+1)$ with $\mathbb{V}\text{ar}\left[\epsilon(t+1)\right] = \mathbf{R}(t+1)$. In our model

$$\mathbf{C}(t+1) = (1, 0), \qquad\qquad \mathbf{R}(t+1) = (1),$$

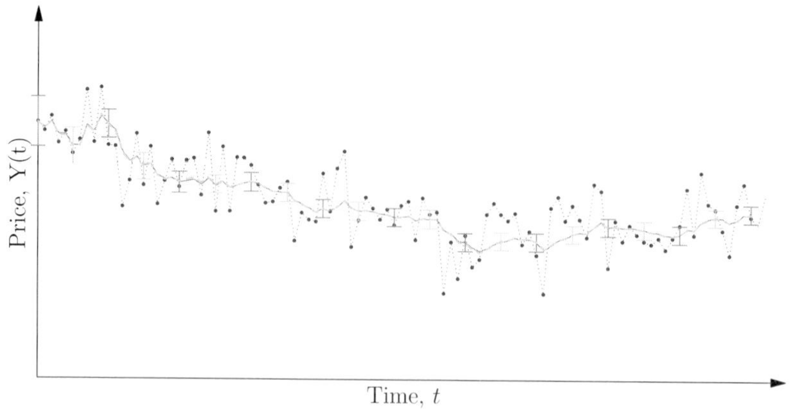

Figure A.16 Example of tracking the 'true' value of a stock using a Kalman filter (solid curve) and particle filter (dashed curve overlapping the Kalman filter). The dots show the current price (observations). The error bars show the errors (they are only shown for every tenth data point).

so that $\mathbf{S}(t+1) = \tilde{\Sigma}_{XX}(t+1) + 1$ (this is a 1×1 matrix). The optimal Kalman gain matrix is $\mathbf{K}(t+1) = \tilde{\mathbf{\Sigma}}(t+1)\,\mathbf{C}^{\mathsf{T}}(t+1)\mathbf{S}^{-1}(t+1)$, which for this system is

$$\mathbf{K}(t+1) = \frac{1}{1+\tilde{\Sigma}_{XX}(t+1)} \begin{pmatrix} \tilde{\Sigma}_{XX}(t+1) \\ \tilde{\Sigma}_{XY}(t+1) \end{pmatrix}.$$

The mean and covariance for the posterior after seeing the observed price is

$$\boldsymbol{\mu}(t+1) = \tilde{\boldsymbol{\mu}}(t+1) + \mathbf{K}(t+1)\left(\boldsymbol{Y}(t+1) - \mathbf{C}(t+1)\,\tilde{\boldsymbol{\mu}}(t+1)\right)$$
$$\mathbf{\Sigma}(t+1) = \left(\mathbf{I} - \mathbf{K}(t+1)\,\mathbf{C}(t+1)\right)\,\tilde{\mathbf{\Sigma}}(t+1)$$

or for our model

$$\boldsymbol{\mu}(t+1) = \begin{pmatrix} \tilde{\mu}_X(t+1) \\ \tilde{\mu}_V(t+1) \end{pmatrix} + \frac{(Y(t+1) - \tilde{\mu}_X(t))}{1+\tilde{\Sigma}_{XX}(t+1)} \begin{pmatrix} \tilde{\Sigma}_{XX}(t+1) \\ \tilde{\Sigma}_{XY}(t+1) \end{pmatrix}$$
$$\mathbf{\Sigma}(t+1) = \tilde{\mathbf{\Sigma}}(t+1) - \frac{1}{1+\tilde{\Sigma}_{XX}(t+1)} \begin{pmatrix} \tilde{\Sigma}_{XX}(t+1) & 0 \\ \tilde{\Sigma}_{XY}(t+1) & 0 \end{pmatrix} \tilde{\mathbf{\Sigma}}(t+1).$$

The particle filter is easy to write. We maintain a population of states $(\boldsymbol{X}_i(t) | i = 1, 2, \ldots, n)$ where $\boldsymbol{X}_i(t) = (X_i(t), V_i(t))$. We update them according to the update rules and weight them according to $w_i(t) = w_i(t-1) \times \mathcal{N}(X_i(t) - Y(t), 1)$ with $w_i(0) = 1$. After some time we resampled from the population of particles. We used a population of 1000 and resampled after five time steps.

In Figure A.16 we show the Kalman filter predictions on the true value of a stock as a solid curve. The error bars show the predicted error in the estimate. We also show the result of the particle filter as a dashed curve. For this problem (since the updates are linear and the noise normally distributed) the particle filter should, in the limit of a large number of particles, be identical to the Kalman filter (which we see it is to a high accuracy).

This model could be used to decide when to buy and sell shares, but a couple of points are worth making. Here we have assumed that we know the model perfectly (i.e. the size of the fluctuations in the 'observation' and the change in the 'velocity' of the true price). In a real problem these would have to be estimated. Our model is highly simplified. For example, we have assumed that the noise in our model is independent, while in real financial data it is likely that the noise has some auto-correlation. Modelling this is beyond the ability of the standard Kalman filter, but it could be modelled by a particle filter. Further, as we will see in Chapter 12, normally distributed noise is a poor model for the fluctuations found in financial data. Finally, it is worth bearing in mind that our estimate of the true value of a stock is based entirely on the valuation of the buyers and sellers and may be completely unrelated to the true value of the company. One hopes there are a few people who know what they are doing and value the stock accords with its true worth, but history tells us this may be wishful thinking.

A.12 Answers to Chapter 12. Stochastic Processes

Exercise 12.1 (on page 389): *We want to show that for a positive-definite matrix* K' *and its inverse* $\mathsf{L}' = \mathsf{K}'^{-1}$ *given by*

$$\mathsf{K}' = \begin{pmatrix} \mathsf{K} & \boldsymbol{k} \\ \boldsymbol{k}^\mathsf{T} & k \end{pmatrix} \qquad \mathsf{L}' = \mathsf{K}'^{-1} = \begin{pmatrix} \boldsymbol{L} & \boldsymbol{l} \\ \boldsymbol{l}^\mathsf{T} & l \end{pmatrix}$$

the Equations (12.1) on page 354, and (12.2) *on page 354 are identical. That is,*

$$\boldsymbol{k}^\mathsf{T}(\mathsf{K} + \sigma^2\mathsf{I})^{-1}\boldsymbol{y} = \frac{-\boldsymbol{l}^\mathsf{T}\left(\mathsf{L} + \sigma^{-2}\mathsf{I}\right)^{-1}\boldsymbol{y}}{l - \boldsymbol{l}^\mathsf{T}\left(\mathsf{L} - \sigma^{-2}\mathsf{I}\right)^{-1}\boldsymbol{l}}$$

$$k - \boldsymbol{k}^\mathsf{T}(\mathsf{K} + \sigma^2\mathsf{I})^{-1}\boldsymbol{k} = \frac{1}{l - \boldsymbol{l}^\mathsf{T}\left(\mathsf{L} - \sigma^{-2}\mathsf{I}\right)^{-1}\boldsymbol{l}}.$$

This is difficult to show using matrix identities. Instead, for a randomly generated positive-definite matrix K' *and for a particular value of* σ *and* $\boldsymbol{y}$ *show, using your favourite matrix manipulation language (e.g. MATLAB, Octave, python), that these two expressions give the same answer (up to machine accuracy).*

We provide a MATLAB/Octave program to show this. We create a positive definite matrix by first generating a random matrix A and we then set $\mathsf{K}' = \mathsf{A}\,\mathsf{A}^\mathsf{T}$ (we call this `Kp`), from this we extract K (`Km`), $\boldsymbol{k}$ (`kv`), and k (`ks`). We compute the variance term and mean term. We then repeat this for the inverse matrix. The two expressions give identical answers up to machine precision.

```
n = 10;
A = rand(n,n);
y = rand(n-1,1);
sigma = 10;

Kp = A*A';
Km = Kp(1:n-1,1:n-1);
kv = Kp(1:n-1,n);
ks = Kp(n,n);
```

```
varK = ks - kv'*inv(Km+sigma*eye(n-1))*kv
meanK = kv'*inv(Km+sigma*eye(n-1))*y

Lp = inv(Kp);
Lm = Lp(1:n-1,1:n-1);
lv = Lp(1:n-1,n);
ls = Lp(n,n);
varL = 1.0/(ls - lv'*inv(Lm+(1.0/sigma)*eye(n-1))*lv)
meanL = -varL*lv'*inv(Lm+(1.0/sigma)*eye(n-1))*y/sigma.
```

It is more computationally efficient using the covariance matrix K' rather then its inverse. By using the Cholesky decomposition of K we can efficiently find the distribution $f(a(\boldsymbol{x}^*) = a|\mathcal{D})$ at different values of $\boldsymbol{x}^*$ (see, for example, Rasmussen and Williams (2006)).

Exercise 12.2 (on page 389): *Simulate a discrete time model for the evolution of stock prices*

$$\Delta S(t) = \mu\, S(t)\, \Delta t + \sigma\, S(t)\, \sqrt{\Delta t}\, \eta$$

where $\eta \sim \mathcal{N}(0,1)$ *with* $\mu = \sigma = \Delta t = 0.1$, *starting from* $S(0) = 1$ *up to* $t = 10$ *and* $t = 20$. *Repeat this multiple times and plot histograms of* $S(10)$, $S(20)$. *Compare this with the theoretical result for the system*

$$dS(t) = \mu\, S(t)\, dt + \sigma\, S(t)\, dW(t)$$

given by

$$\begin{aligned} f_{S(t)}(S(10) = t) &= \text{LogNorm}\left(s \middle| \left(\mu - \tfrac{\sigma^2}{2}\right) t, \sigma^2 t\right) \\ &= \frac{1}{s\,\sigma\,\sqrt{2\,\pi\,t}} \mathrm{e}^{-\left(\log(s) - (\mu - \sigma^2/2)t\right)/(2\,\sigma^2\,t)}. \end{aligned}$$

In Figure A.17 we show histograms of the stock prices $S(10)$ and $S(20)$, obtained from simulating the discrete time model. We also compare these with the continuous time model. Note that the fluctuations in the stock prices grow substantially from $t = 10$ to $t = 20$.

Exercise 12.3 (on page 389): *Consider a particle experiencing normally distributed perturbations in a quartic potential*

$$dX(t) = -X^3(t)\, dt + dW(t).$$

Compute the stationary distribution $f(x) = \lim_{t\to\infty} f_{X^t}(x,t)$ *using Equation (12.12) on page 375, and compare it with a simulation of a discrete time system.*

Using Equation (12.12) we have

$$f(x) = \frac{\mathrm{e}^{-2\int_{-\infty}^{x} y^3\, \mathrm{d}y}}{\int_{-\infty}^{\infty} \mathrm{e}^{-2\int_{-\infty}^{z} y^3\, \mathrm{d}y}\, \mathrm{d}z} = \frac{2^{\frac{3}{4}}}{\Gamma\left(\frac{1}{4}\right)} \mathrm{e}^{-x^4/2}.$$

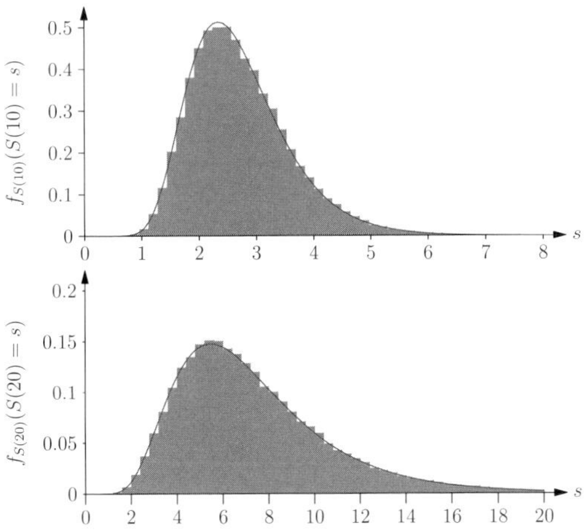

Figure A.17 Histograms of the stock prices $S(10)$ and $S(20)$ obtained from 10^5 runs. Also shown in blue is the theoretical result for the continuous time version.

Figure A.18 shows this distribution and a histogram of values obtained from the simulation of

$$\Delta X(t) = -X^3(t)\,\Delta t + \sqrt{\Delta t}\,\eta$$

with $\eta \sim \mathcal{N}(0, 1)$ and $\Delta t = 0.01$. Note that with a significantly larger Δt it is too easy to get large jumps taking you away from the high probability solutions and we get noticeable deviations from the steady-state distribution of the stochastic differential equation.

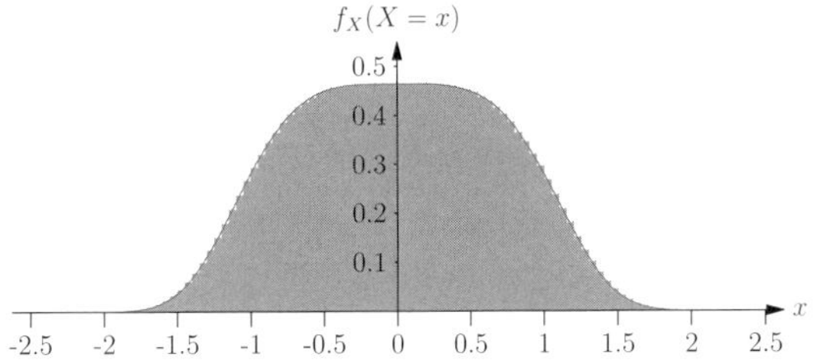

Figure A.18 Histogram of the stationary distribution of $X(t)$ computed from the discrete time simulation and from solving the corresponding Fokker–Planck equation.

Exercise 12.4 (on page 390): *Gillespie simulations are extremely easy to code. Simulate the set of reactions*

$$\emptyset \xrightarrow{k_1=1} X \qquad \emptyset \xrightarrow{k_2=0.5} Y \qquad X + Y \xrightarrow{k_3=0.0001} X \qquad X \xrightarrow{k_4=0.01} \emptyset$$

starting from the initial condition $n_X = n_Y = 0$ (assuming $\Omega = 1$). Write down approximate stochastic differential equations for $dn_X(t)$ and $dn_Y(t)$ assuming that the number of reactions that occur in a short time interval are given by a Poisson distribution.

The simulation takes only a few lines of code to write. Figure A.19 shows a Gillespie simulation of these reactions.

The two species fluctuate around a fixed point. Around the fixed point the number of molecules are quite high so that the propensities vary rather slowly over time. It is therefore a reasonable approximation to assume the propensities

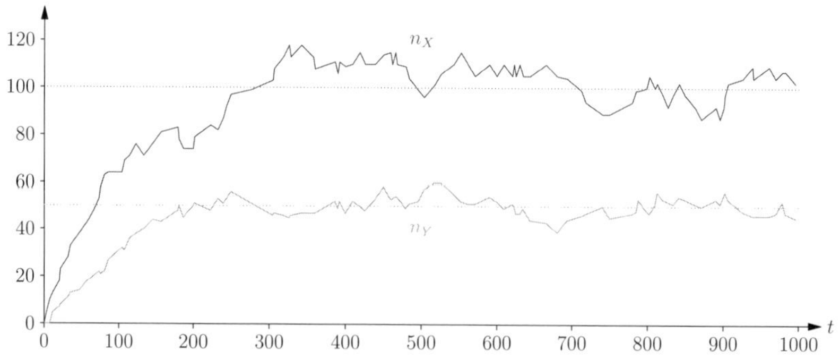

Figure A.19 Gillespie simulation of the reactions $\emptyset \xrightarrow{k_1=1} X$, $\emptyset \xrightarrow{k_2=0.5} Y$, $X + Y \xrightarrow{k_3=0.0001} X$, and $X \xrightarrow{k_4=0.01} \emptyset$, starting from $n_X = n_Y = 0$.

are fixed for a short time period. We consider the expected change in the species number in time Δt

$$\mathbb{E}\left[\Delta n_X(t)\right] \approx (k_1 - k_4\, n_X(t))\,\Delta t \qquad \mathbb{E}\left[\Delta n_Y(t)\right] \approx (k_2 - k_3\, n_X(t)\, n_Y(t))\,\Delta t.$$

(Note that we are ignoring the change in $n_X(t)$ and $n_Y(t)$ caused by the reaction.) As the reactions are governed by a Poisson process, the fluctuations in the number of reactions are equal to the mean number of reactions, thus

$$\mathbb{V}\mathrm{ar}\left[\Delta n_X(t)\right] = (k_1 + k_4\, n_X(t))\,\Delta t \qquad \mathbb{V}\mathrm{ar}\left[\Delta n_Y(t)\right] = (k_2 + k_3\, n_X(t)\, n_Y(t)\,\Delta t.$$

(Notice that we don't care whether the reactions create or destroy a species, the variance is still positive.) Therefore, we can approximate our system by the stochastic differential equations

$$\begin{aligned} \mathrm{d}n_X(t) &= \left(k_1 - k_4\, n_X(t)\right)\mathrm{d}t + \sqrt{k_1 + k_4\, n_X(t)}\,\mathrm{d}W(t) \\ \mathrm{d}n_Y(t) &= \left(k_2 - k_3\, n_X(t)\, n_Y(t)\right)\mathrm{d}t + \sqrt{k_2 + k_3\, n_X(t)\, n_Y(t)}\,\mathrm{d}W(t). \end{aligned}$$

We could write down the corresponding Fokker–Planck equation (although it's not particularly easy to solve). Instead we can seek the fixed point where $\mathbb{E}\left[\mathrm{d}n_X(t)\right] = \mathbb{E}\left[\mathrm{d}n_Y(t)\right] = 0$. Averaging the stochastic differential equations

$$\begin{aligned} \mathbb{E}\left[\mathrm{d}n_X(t)\right] &= \left(k_1 - k_4\, \mathbb{E}\left[n_X(t)\right]\right)\mathrm{d}t = 0 \\ \mathbb{E}\left[\mathrm{d}n_Y(t)\right] &= \left(k_2 - k_3\, \mathbb{E}\left[n_X(t)\, n_Y(t)\right]\right)\mathrm{d}t = 0 \end{aligned}$$

the first equation gives $\mathbb{E}\left[n_X(t)\right] = k_1/k_4 = 100$. The second equation only tells us

$$\mathbb{E}\left[n_X(t)\, n_Y(t)\right] = \mathbb{E}\left[n_X(t)\right]\, \mathbb{E}\left[n_Y(t)\right] + \mathbb{C}\mathrm{ov}\left[n_X(t),\, n_Y(t)\right] = k_2/k_3.$$

If we assume that $\mathbb{C}\mathrm{ov}\left[n_X(t),\, n_Y(t)\right]$ is small then $\mathbb{E}\left[n_Y(t)\right] \approx k_2/(\mathbb{E}\left[n_X(t)\right]\, k_3) = 50$. These numbers are close to what we see in the simulation (after the initial transient phase).

Appendix B

Probability Distributions

Contents

For reference we give tables of probability distributions together with their mean, variancc, and characteristic functions. We split the distributions into three: univariate discrete distribution, univariate continuous distributions, and multivariate discrete and continuous distributions.

Table B.1 Discrete univariate distributions.

Name	Range	Probability Mass Function	Mean	Variance	Characteristic Function, $\phi(\omega)$
Bernoulli	$X \in \{0, 1\}$	$\text{Bern}(x\|p) = p^x(1-p)^{1-x}$	p	$p\,(1-p)$	$1 + p\,(\text{e}^{\text{i}\omega} - 1)$
Binomial	$K \in \{0, 1, \ldots, n\}$	$\text{Bin}(k\|n, p) = \binom{n}{k} p^k\,(1-p)^{n-k}$	$n\,p$	$n\,p\,(1-p)$	$\left(1 + p\,(\text{e}^{\text{i}\omega} - 1)\right)^n$
Geometric	$K \in \{1, 2, \ldots\}$	$\text{Geo}(k\|p) = p\,(1-p)^{k-1}$	$\frac{1}{p}$	$\frac{1-p}{p^2}$	$\frac{p\,\text{e}^{\text{i}\omega}}{1-(1-p)\,\text{e}^{\text{i}\omega}}$
Geometric*	$K \in \{0, 1, 2, \ldots\}$	$\text{Geo}^*(k\|p) = p\,(1-p)^k$	$\frac{1-p}{p}$	$\frac{1-p}{p^2}$	$\frac{p}{1-(1-p)\,\text{e}^{\text{i}\omega}}$
Hypergeometric	$K \in \{0, \ldots, \text{min}(n,m)\}$	$\text{Hyp}(k\|N, m, n) = \frac{\binom{m}{k}\binom{N-m}{n-k}}{\binom{N}{n}}$	$\frac{n\,m}{N}$	$\frac{n\,m\,(N-m)\,(N-n)}{N^2\,(N-1)}$	$\frac{\binom{N-m}{n}}{\binom{N}{n}}\,{}_2F_1(-n,-m;N-m-n+1;\text{e}^{\text{i}\omega})$
Negative Binomial	$K \in \{0, 1, 2, \ldots\}$	$\text{NegBin}(k\|r, p) = \binom{k+r-1}{k} p^k\,(1-p)^r$	$\frac{p\,r}{1-p}$	$\frac{p\,r}{(1-p)^2}$	$\left(\frac{1-p}{1-p\text{e}^{\text{i}\omega}}\right)^r$
Poisson	$K \in \{0, 1, 2, \ldots\}$	$\text{Poi}(k\|\mu) = \frac{\mu^k}{k!}\text{e}^{-k}$	μ	μ	$\exp\left(m(\text{e}^{\text{i}\mu} - 1)\right)$

Table B.2 Continuous univariate distributions.

Name	Range	Probability Density Function	Mean	Variance	Characteristic Function, $\phi(\omega)$
Beta	$0 \leq X \leq 1$	$\text{Bet}(x \mid a, b) = \frac{x^{a-1}(1-x)^{b-1}}{B(a,b)}$	$\frac{a}{a+b}$	$\frac{a\,b}{(a+b)^2(a+b+1)}$	${}_1F_1(\alpha; \alpha+\beta; \mathrm{i}\omega)$
Cauchy	$-\infty < X < \infty$	$\text{Cau}(x) = \frac{1}{\pi(1+x^2)}$	not defined	not defined	$\mathrm{e}^{-\lvert\omega\rvert}$
Chi-squared	$0 \leq X < \infty$	$\chi(x \mid k) = \frac{x^{(k/2)-1}e^{-x/2}}{2^{k/2}\Gamma(\frac{k}{2})}$	k	$2\,k$	$(1 - 2\,\mathrm{i}\,\omega)^{-k/2}$
Exponential	$0 \leq X < \infty$	$\text{Exp}(x \mid b) = \text{Gam}(x \mid 1, b) = b\,\mathrm{e}^{-b\,x}$	$\frac{1}{b}$	$\frac{1}{b^2}$	$\left(1 - \frac{\mathrm{i}\omega}{b}\right)^{-1}$
Gamma	$0 \leq X < \infty$	$\text{Gam}(x \mid a, b) = \frac{b^a\,x^{a-1}\,\mathrm{e}^{-b\,x}}{\Gamma(a)}$	$\frac{a}{b}$	$\frac{a}{b^2}$	$\left(1 - \frac{\mathrm{i}\omega}{b}\right)^{-a}$
Log-normal	$0 < X < \infty$	$\text{LogNorm}(x \mid \mu, \sigma^2) = \frac{\mathrm{e}^{-\frac{(\log(x)-\mu)^2}{2\sigma^2}}}{x\sqrt{2\pi}\,\sigma}$	$\mathrm{e}^{\mu+\sigma^2/2}$	$(\mathrm{e}^{\sigma^2} - 1)\mathrm{e}^{2\mu+\sigma^2}$	No closed-form representation
Normal	$-\infty < X < \infty$	$\mathcal{N}(x \mid \mu, \sigma^2) = \frac{1}{\sqrt{2\pi}\,\sigma}\,\mathrm{e}^{-\frac{(x-\mu)^2}{2\sigma^2}}$	μ	σ^2	$\mathrm{e}^{\mathrm{i}\,\mu\,\omega - \frac{1}{2}\sigma^2\,\omega^2}$
Student's T	$-\infty < T < \infty$	$T(t \mid \nu) = \frac{\Gamma(\frac{\nu+1}{2})}{\sqrt{\pi\,\nu}\,\Gamma(\frac{\nu}{2})}\left(1 + \frac{t^2}{\nu}\right)^{-\frac{\nu+1}{2}}$	0 for $\nu > 1$	$\frac{\nu}{\nu-2}$ for $\nu > 2$	$\frac{K_{\nu/2}(\sqrt{\nu}\lvert\omega\rvert)\cdot(\sqrt{\nu}\lvert\omega\rvert)^{\nu/2}}{\Gamma(\nu/2)2^{\nu/2-1}}$
Uniform	$a \leq X < b$	$\text{U}(x \mid a, b) = \frac{[\![a \leq x < b]\!]}{b-a}$	$\frac{b+a}{2}$	$\frac{(b-a)^2}{12}$	$\frac{\mathrm{e}^{\mathrm{i}\,b\,\omega} - \mathrm{e}^{\mathrm{i}\,a\,\omega}}{\mathrm{i}\,\omega(b-a)}$
Weibull	$0 \leq X < \infty$	$\text{Wei}(x \mid \lambda, k) = \frac{k}{\lambda}\left(\frac{x}{\lambda}\right)^{k-1}\mathrm{e}^{-(x/\lambda)^k}$	$\lambda\,\Gamma\left(1 + \frac{1}{k}\right)$	$\lambda^2\left(\Gamma\left(1 + \frac{2}{k}\right) - \Gamma^2\left(1 + \frac{1}{k}\right)\right)$	$\sum_{n=0}^{\infty}\frac{(\mathrm{i}\,\omega\,\lambda)^n}{n!}\,\Gamma\left(1 + \frac{n}{k}\right)$

Table B.3 Continuous and discrete multivariate distributions. PMF/PDF shows the probability mass function or the probability density function.

Name	Range	PMF/PDF	Mean	Covariance C_{ij}	Characterstic Functions, $\phi(\boldsymbol{\omega})$
Categorical	$\boldsymbol{X} \in \Lambda_1^k$	$\mathrm{Cat}(\boldsymbol{x}\vert\boldsymbol{\mu}) = \prod_{i=1}^k \mu_i^{x_i} [\![\boldsymbol{x} \in \Lambda_1^k]\!]$	$\boldsymbol{\mu}$	$[\![i=j]\!]\,\mu_i - \mu_i\,\mu_j$	$\sum_{i=1}^k \mu_i\,\mathrm{e}^{\mathrm{i}\,\omega_i}$
Multinomial	$\boldsymbol{N} \in \Lambda_n^k$	$\mathrm{Mult}(\boldsymbol{n}\vert n, \boldsymbol{p}) = n! \prod_{i=1}^k \frac{p_i^{n_i}}{n_i!} [\![\boldsymbol{n} \in \Lambda_n^k]\!]$	$n\,\boldsymbol{p}$	$n\,[\![i=j]\!]\,p_i - n\,p_i\,p_j$	$\left(\sum_{i=1}^k p_i\,\mathrm{e}^{\mathrm{i}\,\omega_i}\right)^n$
Dirichlet	$\boldsymbol{P} \in \Lambda^k$	$\mathrm{Dir}(\boldsymbol{x}\vert\boldsymbol{\alpha}) = \Gamma(\alpha_0) \prod_{i=1}^k \frac{x_i^{\alpha_i - 1}}{\Gamma(\alpha_i)}, \quad \alpha_0 = \sum_{i=1}^k \alpha_i$	$\frac{\boldsymbol{\alpha}}{\alpha_0}$	$\frac{\alpha_i(\alpha_0 [\![i=j]\!] - \alpha_j)}{\alpha_0^2(\alpha_0+1)}$	No closed-form solution
Normal	$-\infty < X_i < \infty$	$\mathcal{N}(\boldsymbol{x}\vert\boldsymbol{\mu}, \boldsymbol{\Sigma}) = \frac{1}{\sqrt{\vert 2\pi\boldsymbol{\Sigma}\vert}} \mathrm{e}^{-\frac{1}{2}(\boldsymbol{x}-\boldsymbol{\mu})^\mathsf{T}\boldsymbol{\Sigma}^{-1}(\boldsymbol{x}-\boldsymbol{\mu})}$	$\boldsymbol{\mu}$	Σ_{ij}	$\mathrm{e}^{\mathrm{i}\boldsymbol{\mu}^\mathsf{T}\omega - \omega^\mathsf{T}\boldsymbol{\Sigma}\omega}$
Wishart	$\mathbf{V}$ is positive definite	$\mathcal{W}(\mathbf{V}\vert\mathbf{W}, \nu) = \frac{\vert\mathbf{V}\vert^{(\nu-d)/2-1} \mathrm{e}^{-\mathrm{Tr}(\mathbf{W}^{-1}\mathbf{V})/2}}{\vert\mathbf{W}\vert^{\nu/2} 2^{\nu d/2} \pi^{d(d-1)/4} \prod_{i=1}^d \Gamma\left(\frac{\nu+1-i}{2}\right)}$	$n\,\mathbf{W}$	$\mathbb{V}\mathrm{ar}\left[V_{ij}\right] = n\,(W_{ij} - W_{ii}\,W_{jj})$	$\vert\mathbf{I} - 2\,\mathrm{i}\,\mathbf{W}\,\mathbf{V}\vert^{-n/2}$

Bibliography

M. Abramowitz and I. A. Stegun. *Handbook of Mathematical Functions*, volume 55 of Applied Mathematics Series. Dover, 9th edition, 1964. This is the standard collection of facts about mathematical function. It includes a useful chapter on probabilities.

D. Arclioptas, A. Naor, and Y. Peres. Rigorous location of phase transitions in hard optimization problems. *Nature*, 435(9):759–764, 2005. Nice example of the use of the second-moment method applied to the maximum satisfiability problem.

M. Arjovsky, S. Chintala, and L. Bottou. Wasserstein GAN. *arXiv:1701.07875 [cs, stat]*, January 2017. http://arxiv.org/abs/1701.07875. Paper that introduced Wasserstein generative adversarial networks (GANs).

D. Barber. *Bayesian Reasoning and Machine Learning*. Cambridge University Press, 2011. An accessible machine learning book.

S. Bernstein. On a modification of Chebyshev's inequality and of the error formula of Laplace. Annales scientifiques des Institutions mathématiques savantes de l'Ukraine, 1(4):38–49, 1924. Paper deriving Bernstein's inequality. A very early tail-bound result.

C. Bishop. *Pattern Recognition and Machine Learning*. Springer, 2nd edition, 2007. A very readable description of modern developments in pattern recognition and machine learning.

D. M. Blei, A. Y. Ng, and M. I. Jordan. Latent Dirichlet allocation. *Journal of Machine Learning Research*, 3:993–1022, 2003. Clear article introducing the now famous latent Dirichlet allocation (LDA) method.

S. Boucheron, G. Lugosi, and P. Massart. *Concentration Inequalities*. Oxford University Press, 2013. Comprehensive, state-of-the-art guide to obtaining tail inequalities. Gives good coverage of the basics, but rapidly becomes quite advanced.

S. Boyd and L. Vandenberghe. *Convex Optimization*. Cambridge University Press, 2004. Highly respected text on convexity, although not a quick read.

H. Chernoff. A measure of asymptotic efficiency for tests of a hypothesis based on sum of observations. *Annals of Mathematical Statistics*, 23:493–507, 1952. Paper deriving the famous Chernoff tail bound.

T. A. Cover and J. A. Thomas. *Elements of Information Theory*. Wiley, 2nd edition, 2006. A standard reference on all elements of information theory.

N. Cressie and C. K. Wikle. *Statistics for Spatio-Temporal Data*. Wiley, 2011. Comprehensive although dense review of Bayesian methods for dealing with time series and spatial data.

D. Crisan and B. Rozovskii, editors. *The Oxford Handbook of Nonlinear Filtering*. Oxford University Press, 2011. Collection of many articles and tutorials on using particle filters and other non-linear filtering techniques.

A. P. Dempster, N. M. Laird, and D. B. Rubin. Maximum likelihood from incomplete data via the EM algorithm. *Journal of the Royal Statistical Society, Series B*, 39(1):1–38, 1977. Paper proposing the now classic expectation-maximization (EM) algorithm.

L. Devroye. *Non-Uniform Random Variate Generation*. Springer, 1986. This provides a very extensive discussion of techniques for generating random deviates for different probability distributions. Alas this has been out of print for many years, but it is available online at www.nrbook.com/devroye/.

V. A. Dobrushkin. *Methods in Algorithmic Analysis*. Chapman & Hall/CRC, 2010. Nicely written and comprehensive book on combinatorics and generating functions aimed at studying algorithms.

S. Duane, A. D. Kennedy, B. J. Pendleton, and D. Roweth. Hybrid Monte Carlo. *Physics Letters*, B 195:216–222, 1987. The first paper to introduce hybrid Monte Carlo.

D. P. Dubhashi and A. Panconesi. *Concentration of Measure for Analysis of Randomized Algorithms*. Cambridge University Press, 2009. Fairly comprehensive book covering concentration theorems and the applications to problems in the analysis of algorithms.

R. O. Duda, P. E. Hart, and D. G. Stork. *Pattern Classification*. Wiley, 2001. This is an update of Duda and Hart's original book, written in 1973, which served for many years as the standard reference on anything to do with pattern classification.

R. Durbin, S. Eddy, A. Krogh, and G. Mitchison. *Biological Sequence Analysis*. Cambridge University Press, 1st edition, 1998. This is an extremely readable introduction to hidden Markov models and their extension in the context of biological sequence analysis.

A. Einstein. *Investigations on the Theory of the Brownian Movement*. Dover, 1998. Five key papers by Einstein on Brownian motion in a collection edited by R. Furth. These papers were written around 1905.

W. Feller. *An Introduction to Probability Theory and Its Applications*, volume 1. Wiley, 3rd edition, 1968a. This is the classic on probability theory.

W. Feller. *An Introduction to Probability Theory and Its Applications*, volume 2. Wiley, 2nd edition, 1968b.

A. Field, J. Miles, and Z. Field. *Discovering Statistics Using R*. SAGE, 2012. Very informal but comprehensive guide to using statistics in empirical research. Covers topics such as correlation, regression, analysis of variance (ANOVA), experimental design, and factor analysis. There is a companion book that covers the statistical package SPSS. The book assumes less mathematical sophistication than I've assumed and in consequence it is long.

K. H. Fischer and J. A. Hertz. *Spin Glasses*. Cambridge University Press, 1991. Slightly more tutorial-level introduction to spin glasses and disordered systems.

G. S. Fishman. *Monte Carlo: Concepts, Algorithms and Applications*. Springer Series in Operations Research. Springer, 1996. A very comprehensive guide to practical ways of generating fast random deviates. Possibly showing its age, but nice to dip into.

B. R. Frieden. *Science from Fisher Information*. Cambridge University Press, 1998. An attempt to derive almost all the laws of physics from Fisher Information.

C. Gardiner. *Stochastic Methods*. Springer, 4th edition, 2009. Established classic on stochastic processes, covering a lot of details. It is aimed at the practitioner.

A. Gelman, J. B. Carlin, H. S. Stern, and D. B. Rubin. *Bayesian Data Analysis*. Texts on statistical science. Chapman & Hall, 2003. One of the classics of Bayesian analysis.

S. Geman and D. Geman. Stochastic relaxation, Gibbs distributions, and the Bayesian restoration of images. *IEEE Transactions on Pattern Analysis and Machine Intelligence*, 6:721–741, 1984. A classic paper that introduced many of the ideas in using Markov Chain Monte Carlo (MCMC). It also provided the first proof of the convergence of simulated annealing.

I. J. Goodfellow, J. Pouget-Abadie, M. Mirza, B.G Xu, D. Warde-Farley, S. Ozair, A. Courville, and Y. Bengio. Generative adversarial networks. *arXiv:1406.2661 [cs, stat]*, June 2014. http://arxiv.org/abs/1406.2661. The paper that introduced GANs. In the fast-moving world of deep learning most papers are put onto the open web service arXiv.

R. Graham, D. E. Knuth, and O. Patashnik. *Concrete Mathematics*. Addison-Wesley, 1989. This is Donald Knuth's toolbox and is a tour de force of solving problems in discrete mathematics. A joy to read, although prepare to feel intimidated.

G. Grimmett and D. Stirzaker. *Probability and Random Processes*. Oxford University Press, 3rd edition, 2001a. This is a modern classic, which is packed with a lot of material. It ignores the measure-theoretic foundations. Covers many areas in more detail than this book, although it takes a more mathematical viewpoint, missing some applications altogether.

G. Grimmett and D. Stirzaker. *One Thousand Exercises in Probability*. Oxford University Press, 2nd edition, 2001b. Covers numerous problems. Acts as a companion text to *Probability and Random Processes* by the same authors.

I. Gulrajani, F. Ahmed, M. Arjovsky, V. Dumoulin, and A. Courville. Improved training of Wasserstein GANs. In *Proceedings of the 31st International Conference on Neural Information Processing Systems*, NIPS'17, pages 5769–5779, USA, 2017. Curran Associates Inc.

I. Gulrajani, F. Ahmed, M. Arjovsky, V. Dumoulin, and A. C. Courville. Improved training of Wasserstein GANs. *CoRR*, abs/1704.00028, 2017. http://arxiv.org/abs/1704.00028. Paper explaining how to train Wasserstein GANs properly.

W. Hoeffding. Probability inequalities for sums of bounded random variables. *Journal of the American Statistical Association*, 58(301):13–30, 1963. Classic paper deriving the eponymous inequality, but also obtains some bounds for non-independent random variables.

E. T. Jaynes. The well-posed problem. *Foundations of Physics*, 3(4):477–493, 1973. Beautifully argued article showing how to use invariance principles to obtain an uninformative prior. This article is reproduced in the collection by Rosenkrantz (1982).

E. T. Jaynes. *Probability Theory: The Logic of Science*. Cambridge University Press, 1st edition, 2003. A posthumously published diatribe on Bayesian methods. Ed Jaynes spent most of his career fighting for Bayesian methods to be accepted. He won this battle, but it didn't stop him fighting. This contains many wonderful insights but is not the most balanced account.

J. L. Kelly. A new interpretation of information rate. *Bell System Technical Journal*, 35:917–992, 1956. How to gamble and win, if only the odds are on your side.

D. P. Kingma and M. Welling. Auto-encoding variational Bayes. *arXiv:1312.6114 [cs, stat]*, 2013. http://arxiv.org/abs/1312.6114. This is the classic paper that introduced variational auto-encoders (VAEs).

D. E. Knuth. *The Art of Computer Programming: Fundamental Algorithms*, volume 1. Addison-Wesley/Longman, 3rd edition, 1997a. Knuth's classic text on algorithms.

D. E. Knuth. *The Art of Computer Programming: Seminumerical Algorithms*, volume 2. Addison-Wesley/Longman, 3rd edition, 1997b.

W. T. Ziemba, L. C. MacLean, and E. O. Thorp, editors. *The Kelly Capital Growth Investment Criterion: Theory and Practice*. World Scientific Publishing, 2012. Collection of papers discussing the use of the Kelly criterion in investments, including some key papers.

P. M. Lee. *Bayesian Statistics: An Introduction*. Hodder Arnold, 3rd edition, 2003. One of a host of books on Bayesian statistics. This one is highly recommended by many readers.

M. Li and P. Vitányi. *An Introduction to Kolmogorov Complexity and Its Applications*. Springer, 2nd edition, 1997. A standard text for Kolmogorov complexity. Covers a lot of material.

D. J. C. MacKay. *Information Theory, Inference and Learning Algorithms*. Cambridge University Press, 1st edition, 2003. An eclectic collections of ideas centred around information theory and machine learning. Very readable and full of many original insights.

M. Mézard, G. Parisi, and M. A. Virasoro. *Spin-Glass Theory and Beyond*, volume 9 of Lecture Notes in Physics. World Scientific, 1987. Collection of papers with introductory chapters authored by researchers who drove the current research.

T. P. Minka. Expectation propagation for approximate "Bayesian" inference. In *Uncertainty in Articial Intelligence 17*, pages 362–369, 2001. The paper that introduced expectation propagation as an alternative variational approximation scheme.

C. Moler and C. Van Loan. Nineteen dubious ways to compute the exponential of a matrix, twenty-five years later. *SIAM Review*, 45(1):3–49, 2003. Very readable description of the numerical challenges posed by one important matrix operation.

F. Mosteller. *Fifty Challenging Problems in Probability: With Solutions*. Dover, 1988. What it says on the cover. An entertaining set of puzzles, but tends to stress probability as combinatorics.

R. M. Neal. Probabilistic inference using Markov Chain Monte Carlo methods. Technical Report CRG-TR-93-1, Deptartment of Computer Science, University of Toronto, 1993. A description of MCMC by one of its greatest exponents. Very rich, detailed source.

M. E. J. Newman. *Networks: An Introduction*. Oxford University Press, 2010. Good comprehensive overview of the current work on networks.

M. Opper and D. Saad. *Advanced Mean Field Methods*. Neural Information Processing Series. MIT Press, 2001. As the book title suggests, a series of articles covering advanced mean field methods and variational approximation methods.

J. Pearl. *Probabilistic Reasoning in Intelligent Systems: Networks of Plausible Inference*. Morgan Kaufmann, 1988. The book helped start the Bayesian revolution and introduced graphical models. Still a good read.

W. H. Press, B. P. Flannery, S. A. Teukolsky, and W. T. Vetterling. *Numerical Recipes: The Art of Scientific Computation*. Cambridge University Press, 3rd edition, 2007. The Numerical Recipes series provides both a huge number of practical algorithms as well as enjoyable and very informative descriptions of the algorithms. No one who uses computers for science should be without a copy.

J. G. Propp and D. B. Wilson. Exact sampling with coupled Markov chains and applications to statistical mechanics. *Random Structures and Algorithms*, 9:223–252, 1996. The original paper introducing exact sampling in MCMC.

L. R. Rabiner. A tutorial on hidden Markov models and selected applications in speech recognition. *Proceedings of the IEEE*, 77: 257–286, 1989. Classic tutorial introduction to hidden Markov models.

C. E. Rasmussen and C. K. I. Williams. *Gaussian Processes for Machine Learning*. MIT Press, 2006. Covers the subject mentioned in the title, by some of the initiators.

J. Raymond, A. Manoel, and M. Opper. Expectation propagation. *arXiv:1409.6179*, 2014. These are notes from a lecture by Manfred Opper who, together with Ole Winther, had independently, and around the same time as Thomas Minka, come up with an approximation scheme similar to expectation propagation. This scheme was based on old ideas from developing a mean field theory for spin-glass systems.

J. Rissanen. A universal prior for integers and estimation by minimum description length. *Annals of Statistics*, 11(2):416–431, 1983. One of the first papers on minimum description length by the idea's founding father.

R. D. Rosenkrantz. *E. T. Jaynes: Papers on Probability, Statistics and Statistical Physics*, volume 158 of Synthese Library. D. Reidel, 1982. A collection of articles by Ed Jaynes. All *beautifully* argued. Gives a feel for the fight to make Bayesian inference respectable.

D. Stauffer and A. Aharony. *Introduction to Percolation Theory*. CRC Press, 2nd edition, 1997. A nice introduction to percolation and its many applications.

J. M. Steele. *The Cauchy–Schwarz Master Class: An Introduction to the Art of Mathematical Inequalities*. Mathematical Association of America, 2004. Nice introduction to inequalities, although focused on all inequalities not just those related to probabilities.

R. H. Swendsen and J.-S. Wang. Nonuniversal critical dynamics in Monte Carlo simulations. *Physical Review Letters*, 58:86–88, 1987. First paper describing how to speed up MCMC for spin systems using cluster techniques.

N. G. van Kampen. *Stochastic Processes in Physics and Chemistry*. North Holland, [1981] 2001. Very clearly written book concentrating on the use of stochastic processes in the physical sciences by one of the leading proponents. This was influential amongst scientists and is still a good read.

C. J. van Rijsbergen. *The Geometry of Information Retrieval*. Cambridge University Press, 2004. Short, thought-provoking book advocating the use of probabilistic quantum mechanics as an appropriate model for information retreival.

D. J. Watts. *Six Degrees: The Science of a Connected Age*. Vintage, 2003. Popular account of the structure found in networks.

D. Williams. *Weighing the Odds*. Cambridge University Press, 2001. This book concentrates on developing the reader's intuition of probability. It explores some of the darker recesses of probability where it requires sharp analysis to get consistant and correct results. I have skipped such examples, as my experience has never brought me into such dark corners. However, for the mathematically or philosophically minded reader these examples pose interesting problems to ponder. The book also covers classical statistics (from a mathematical point of view). The author is a strong advocate of using confidence intervals in preference to hypothesis testing. Also (and unusually) the book provides a chapter on probabilistic quantum mechanics. This is a subject that I haven't covered as I consider it to be rather specific to physics, although there has been attempts to use this formalism elsewhere – see, for example, van Rijsbergen (2004).

U. Wolff. Collective Monte Carlo updating for spin systems. *Physical Review Letters*, 62:361–364, 1989. Paper describing a fast-cluster MCMC algorithm for spin systems, developing the work of Swendsen and Wang.

J. M. Ziman. *Models of Disorder: The Theoretical Physics of Homogeneously Disordered Systems*. Cambridge University Press, 1979. Classic book covering a wide variety of models of disordered systems. Missing more recent developments.

Index